PLANKTON OF INLAND WATERS

A DERIVATIVE OF ENCYCLOPEDIA OF INLAND WATERS

PLANKTON OF INLAND WATERS

A DERIVATIVE OF ENCYCLOPEDIA OF INLAND WATERS

EDITOR

PROFESSOR GENE E. LIKENS
Cary Institute of Ecosystem Studies
Millbrook, NY, USA

ELSEVIER

AMSTERDAM • BOSTON • HEIDELBERG • LONDON • NEW YORK • OXFORD
PARIS • SAN DIEGO • SAN FRANCISCO • SINGAPORE • SYDNEY • TOKYO
Academic Press is an imprint of Elsevier

ACADEMIC
PRESS

Academic Press is an imprint of Elsevier
525 B Street, Suite 1900, San Diego, CA 92101-4495, USA
30Corporate Drive, Suite 400, Burlington, MA01803, USA
32 Jamestown Road, London, NW1 7BY, UK
Radarweg 29, PO Box 211, 1000 AE Amsterdam, The Netherlands

British Library Cataloguing in Publication Data
A catalogue record for this book is available from the British Library

Library of Congress Catalog Number: Applied

ISBN: 9780123819949

For information on all Academic Press publications
visit our website at elsevierdirect.com

EDITOR

Professor Gene E. Likens is an ecologist best known for his discovery, with colleagues, of acid rain in North America, for co-founding the internationally renowned Hubbard Brook Ecosystem Study, and for founding the Institute of Ecosystem Studies, a leading international ecological research and education center. Professor Likens is an educator and advisor at state, national, and international levels. He has been an advisor to two governors in New York State and one in New Hampshire, as well as one U.S. President. He holds faculty positions at Yale, Cornell, Rutgers Universities, State University of New York at Albany, and the University of Connecticut, and has been awarded nine Honorary doctoral Degrees. In addition to being elected a member of the prestigious National Academy of Sciences and the American Philosophical Society, Dr. Likens has been elected to membership in the American Academy of Arts and Sciences, the Royal Swedish Academy of Sciences, Royal Danish Academy of Sciences and Letters, Austrian Academy of Sciences, and an Honorary Member of the British Ecological Society. In June 2002, Professor Likens was awarded the 2001 National Medal of Science, presented at The White House by President G. W. Bush; and in 2003 he was awarded the Blue Planet Prize (with F. H. Bormann) from the Asahi Glass Foundation. Among other awards, in 1993 Professor Likens, with F. H. Bormann, was awarded the Tyler Prize, The World Prize for Environmental Achievement, and in 1994, he was the sole recipient of the Australia Prize for Science and Technology. In 2004, Professor Likens was honored to be in Melbourne, Australia with a Miegunyah Fellowship. He was awarded the first G.E. Hutchinson Medal for excellence in research from The American Society of Limnology and Oceanography in 1982, and the Naumann-Thienemann Medal from the Societas Internationalis Limnologiae, and the Ecological Society of America's Eminent Ecologist Award in 1995. Professor Likens recently stepped down as President of the International Association of Theoretical and Applied Limnology, and is also a past president of the American Institute of Biological Sciences, the Ecological Society of America, and the American Society of Limnology and Oceanography. He is the author, co-author or editor of 20 books and more than 500 scientific papers.

Professor Likens is currently in Australia on a Commonwealth Environment Research Facilities (CERF) Fellowship at the Australian National University.

CONTRIBUTORS

M E Azim
University of Toronto, Toronto, ON, Canada

R Barone
Università di Palermo, Palermo, Italy

D Bastviken
Stockholm University, Stockholm, Sweden

C M Borrego
Institut d'Ecologia Aquàtica, Universitat de Girona, Girona, Spain

J M Burkholder
North Carolina State University, Raleigh, NC, USA

A Camacho
University of Valencia, Burjassot, Spain

G M Carr
UNEP GEMS/Water Programme, Gatineau, QC, Canada

J-F Carrias
Université Blaise Pascal, Clermont-Ferrand, France

E O Casamayor
Department of Continental Ecology, Centre d'Estudis Avançats de Blanes-CSIC, Blanes, Spain

P Crill
Stockholm University, Stockholm, Sweden

M T Dokulil
Austrian Academy of Sciences, Mondsee, Austria

A Enrich-Prast
University Federal of Rio de Janeiro, Rio de Janeiro, Brazil

J A Fox
Drew University, Madison, NJ, USA

U Gaedke
University of Potsdam, Potsdam, Germany

Z M Gliwicz
Warsaw University, Warsaw, Poland

V Gulis
Coastal Carolina University, Conway, SC, USA

N G Hairston Jr.
Cornell University, Ithaca, NY, USA

J E Havel
Missouri State University, Springfield, MO, USA

V Istvánovics
Budapest University of Technology and Economics, Budapest, Hungary

C Kaiblinger
Austrian Academy of Sciences, Mondsee, Austria

A D Kent
University of Illinois at Urbana-Champaign, Urbana, IL, USA

L Krienitz
Leibniz-Institute of Freshwater Ecology and Inland Fisheries, Stechlin, Germany

J Kristiansen
University of Copenhagen, Øster Farimagsgade, Copenhagen K, Denmark

K A Kuehn
University of Southern Mississippi, Hattiesburg, MS, USA

C Laforsch
Ludwig-Maximilians-University Munich, Planegg-Martinsried, Germany

B Luef
University of Vienna, Vienna, Austria

S C Maberly
Centre for Ecology & Hydrology, Lancaster, UK

L D Meester
Katholieke Universiteit Leuven, Leuven, Belgium

L Naselli-Flores
Università di Palermo, Palermo, Italy

J Padisák
University of Pannonia, Veszprém, Egyetem, Hungary

P Peduzzi
University of Vienna, Vienna, Austria

J Pernthaler
University of Zurich, Kilchberg, Switzerland

T Posch
University of Zurich, Kilchberg, Switzerland

J A Raven
University of Dundee at SCRI, Dundee, UK

J W Reid
Virginia Museum of Natural History, Martinsville, VA, USA

C S Reynolds
Centre of Ecology and Hydrology and Freshwater Biological Association, Cumbria, UK

R D Robarts
UNEP GEMS/Water Programme, Saskatoon, SK, Canada

L G Rudstam
Cornell University, Ithaca, NY, USA

S Sabater
University of Girona, Girona, Spain

N Salmaso
IASMA Research Center, S. Michele all'Adige, (Trento), Italy

R W Sanders
Temple University, Philadelphia, PA, USA

T Sime-Ngando
Université Blaise Pascal, Clermont-Ferrand, France

H A Smith
Ripon College, Ripon, WI, USA

R W Sterner
University of Minnesota, St. Paul, MN, USA

R J Stevenson
Michigan State University, East Lansing, MI, USA

K Suberkropp
University of Alabama, Tuscaloosa, AL, USA

A Sukenik
The Yigal Allon Kinneret Limnological Laboratory, Migdal, Israel

R Tollrian
Ruhr-University, Bochum, Germany

M Tolotti
IASMA Research Center, S. Michele all'Adige, (Trento), Italy

W F Vincent
Laval University, Quebec City, QC, Canada

R L Wallace
Ripon College, Ripon, WI, USA

C E Williamson
Miami University, Oxford, OH, USA

A C Yannarell
University of Illinois at Urbana-Champaign, Urbana, IL, USA

T Zohary
The Yigal Allon Kinneret Limnological Laboratory, Migdal, Israel

CONTENTS

PROTISTS, BACTERIA AND FUNGI: PLANKTONIC AND ATTACHED

ALGAE (INCLUDING CYANOBACTERIA): PLANKTONIC AND ATTACHED

ZOOPLANKTON

ECOSYSTEM INTERACTIONS

INTRODUCTION TO THE PLANKTON OF INLAND WATERS

The tiny floating, suspended, or weakly swimming biota of inland waters are collectively called plankton. The plankton of aquatic ecosystems are largely dependent on water movements for distribution and are highly diverse in form and function (Wetzel, 2001). The plankton of lakes, reservoirs, and rivers consist of protists, bacteria, fungi, cyanobacteria, algae, and tiny animals. The very small, often microscopic, photosynthetic, or chemosynthetic organisms of the plankton are at the base of aquatic foodwebs, and as such, are exceedingly important to the function of the aquatic ecosystems of inland waters. They are preyed upon by the animal component of the plankton, the zooplankton. Attached or sessile forms of these same groups are also considered in this volume.

Populations comprising the plankton of inland waters are highly diverse taxonomically. But, single-species blooms of cyanobacteria, such as *Microcystis* spp., can produce nuisance or toxic conditions in freshwater and coastal ecosystems in response to excessive enrichment of the key nutrients – nitrogen and phosphorus – (eutrophication) from waste-water runoff and/or atmospheric deposition (e.g., Conley et al., 2009; Allan and Castillo, 2007; Xu et al., 2010).

This volume contains five sections: first, a brief introduction to the plankton of inland waters; second, the protists, bacteria, and fungi, both planktonic and attached; third, the algae (including Cyanobacteria), both planktonic and attached; fourth, the so-called zooplankton, the animal component of the plankton; and fifth, the functional and system interactions of the planktonic and attached forms in aquatic ecosystems.

The articles in this volume are reproduced from the Encyclopedia of Inland Waters (Likens, 2009). I thank the authors of the articles in this volume for their excellent and up-to-date coverage of this important limnological topic.

Gene E. Likens
Cary Institute of Ecosystem Studies
Millbrook, NY
December 2009

References Cited/Further Reading:

Allan JD and Castillo MM (2007) *Stream Ecology: Structure and Function of Running Waters*, 2nd edn. Netherlands: Springer.
Conley DJ, Paerl HW, Howarth RW, *et al.* (2009) Controlling eutrophication: Nitrogen and phosphorus. *Science* 323: 1014–1015.
Likens GE (ed.) (2009) *Encyclopedia of Inland Waters* 3 vols. Oxford, UK: Elsevier/Academic Press.
Wetzel RG (2001) *Limnology: Lake and River Ecosystems*, 3rd edn. San Diego, CA: Academic Press.
Xu H, Paerl HW, Quin B, Zhu G, and Gao G (2010) Nitrogen and phosphorus inputs control phytoplankton growth in eutrophic Lake Taihu, China. *Limnology and Oceanography* 55(1): 420–432.

PROTISTS, BACTERIA AND FUNGI: PLANKTONIC AND ATTACHED

Contents

Archaea

E O Casamayor, Department of Continental Ecology, Centre d'Estudis Avançats de Blanes-CSIC, Blanes, Spain
C M Borrego, Institut d'Ecologia Aquàtica, Universitat de Girona, Girona, Spain

Introduction: Archaea the Unseen Third Domain of Life

Archaea are a relatively newly identified group of prokaryotic microorganisms that constitute the third phylogenetic domain of life together with the more well-known Bacteria and Eukarya. Only a few years ago, archaea were thought to be mostly restricted to extreme and anoxic environments but it has recently been established that archaeal biodiversity, abundance, and metabolic capabilities are substantially larger than the previously assumed. Thus, in a very short time two major changes in our perception of prokaryotic world have occurred. What was the basis for such marked changes now widely accepted? What makes archaea one of the most exciting current topics in microbial aquatic research?

First, over 30 years ago, Carl Woese and colleagues started the revolution by analyzing phylogenetic molecular markers (ribosomal RNA) instead of how organisms look or act. The comparison of 16S rRNA gene sequences showed that a group of prokaryotes had genetic differences as high as those observed between prokaryotes and eukaryotes. These results encouraged more detailed studies, including genome sequencing, leading to the conclusion that life is split in three big Domains instead of the previously recognized two of prokaryotes and eukaryotes.

A second contributor to the revolution in understanding was the recognition that laboratory cultures strongly biased views of the archaeal potentials. The phenotypic range of cultivated archaea indicated that these microorganisms were restricted to habitats with extreme values of temperature, pH, salinity, or anaerobic environments for methanogens. Thus, their metabolic diversity and ecological distribution seemed to be more limited than those of other prokaryotes. After considerable recent effort, a couple of new species of aerobic nonextremophilic archaea have been cultured in the laboratory, enabling detailed study of their metabolism and opening a race to bring into culture some of the most enigmatic microbes in freshwater environments.

Finally, widespread use of environmental ribosomal RNA sequencing has unveiled that unseen archaea were present in freshwater ecosystems and that the vast majority of them (excluding methanogens) were unrelated, or at best distantly related, to counterparts known from culture collections. Therefore, the known metabolic capabilities of the Domain Archaea have increased significantly. Unfortunately, the ecological significance, biochemistry, physiology, and impact on freshwater biogeochemical cycles of archaea still remain largely unknown.

To understand the ecology of archaea, a combination of new cultivation strategies, high-resolution molecular technologies, more detailed geochemical analytical techniques, traditional microbiological methods, and bioinformatics analyses on genomic data will be required. As soon as some of these organisms become cultivated and their metabolic and genetic potentials are studied in detail, a wide range of new physiological and ecological phenotypes will be discovered. In the meantime, scientists are profiting from new molecular genome-based technologies and from some special features of archaea, e.g., the specific archaeal membrane lipids that have ether-linkages instead of ester-linkages typical of bacteria and eukaryotes. Some of these lipids can be used as biomarkers to trace the occurrence of different archaeal communities in ancient sediments or as paleotemperature proxies, useful to extrapolating water temperatures and climatic transitions. Altogether, the study of archaea is certainly a timely and exciting topic with strong evolutionary, ecological, and biogeochemical implications.

Archaeal Habitats in Inland Waters

The Domain Archaea comprises four main groups (Kingdoms): Crenarchaeota and Euryarchaeota are the two main Kingdoms; Korarchaeota (detected only by 16S ribosomal gene sequences obtained from a variety of marine and terrestrial hydrothermal environments, such as the hot spring Obsidian Pool in Yellowstone National Park) and Nanoarchaeota (represented by a nanosized hyperthermophilic symbiont originally found in a submarine hot vent as well as in the Obsidian Pool) are less widespread and diverse.

Cultivated species of crenarchaeota have thermoacidophilic phenotypes and, in theory, occur in peculiar hot freshwater environments with an active sulfur cycle such as sulfureta and thermal springs. Uncultivated mesophilic crenarchaeota are in turn abundant in other natural environments and evidence exists that the largest proportion and greatest diversity of this group is present, surprisingly, in cold environments.

In contrast, euryarchaeota consists of cultured organisms that are more diverse in their physiology, metabolic capabilities, and habitat occurrence. This group includes well-known obligate anaerobic and widespread methanogens, the extreme halophiles from salt lakes, the hyperthermophiles, and the thermoacidophiles (i.e., the Thermoplasmata group lacking cell walls from hot springs and sulfuretic fields). Again, widespread noncultured mesophilic and cold-adapted phylotypes have been described for euryarchaeota in different freshwater ecosystems (**Table 1**).

Springs

Freshwater aerobic archaea have been traditionally associated with fresh waters influenced by hydrothermal activity (hot-water vents and fumaroles). Geothermally heated water percolates through volcanic material, which strongly influences the chemical composition. The emerging water is often enriched in reduced molecules such as sulfide, methane, and H_2, which yield energy for archaeal chemolithotrophic

activity. These high-temperature ecosystems are interesting but unusual freshwater habitats. However, they may be useful model systems for understanding life under extreme environments on earth as well in relation to astrobiological studies. In addition, thermophilic archaea (both euryarchaeota and crenarchaeota) and bacteria (e.g., *Thermus aquaticus*) inhabiting these systems with optimum temperatures around 85 °C, are natural sources of biotechnological products, e.g., DNA polymerases.

Recently, sulfidic streamlets from emerging cold water (around 10 °C) in a nongeothermal environment have been reported to support the growth of a unique microbial community. A string-of-pearls-like, macroscopically visible structure, mainly composed of a nonmethanogenic euryarchaeota, occurs in these streamlets and is viable at temperatures ranging from −2 to 20 °C in close association with a sulfide-oxidizing bacteria. In this case, close links between archaea and the sulfur cycle arise in meso- to psychrophilic environments that compliment the better known associations of thermal environments.

Salt Lakes

Salt lakes, another extreme environment, are significant components of global inland waters. Salt lakes are complex and heterogeneous with distinct variation of salinity, alkalinity, and other physical/chemical as well as biological properties. Athalassohaline lakes are inland saline lakes with ionic proportions different from those lakes with salt composition similar to seawater. The conventional salinity value of $3 \, g \, l^{-1}$ is taken as the dividing line between fresh and saline waters. The range of salinity encountered in inland waters can reach up to $350 \, g \, l^{-1}$ and even beyond in certain lakes. The diversity of aquatic haline environments is enormous around the world, but the prokaryotes thriving in inland saline lakes are poorly known.

Considerable differences are apparent in the structure of archaeal communities along salinity gradients. At the lower end of the range ($\sim$50–70 $g \, l^{-1}$), bacteria are the predominant components of the prokaryotic plankton, and mesohaline uncultured euryarchaeota distantly related to haloarchaea are found. This group is widespread in mesohaline freshwater environments surveyed so far but no representatives are available in culture to allow ecophysiological studies. At the highly saline end ($>$200 $g \, l^{-1}$) the microbial community is dominated by extremely halophilic archaeal cells of square-shaped morphology that account for up to 75% of total prokaryotes beyond $350 \, g \, l^{-1}$. These calcium and magnesium chloride saturated brines are one of the most extreme habitats

Table 1 Distribution of main groups of archaea in inland water ecosystems

Environment	Archaeal group	Comments
Lakes	Nonthermophilic Crenarchaeota (uncultured)	Distribution throughout the whole water column Biogeochemical role unknown
	Euryarchaeota (uncultured, non methanogens)	Distributed mainly in oxic and suboxic zones of the water column Biogeochemical role unknown
	Methanogens	Anoxic hypolimnia and sediments Methanogenesis
	ANME (uncultured anaerobic methane oxidizers)	Mainly in sediments but also in plankton Occurring either as syntrophic consortia with sulphate-reducing bacteria or as archaeal aggregates
Rivers	Eury- and Crenarchaeota	Most clones relate to soil archaea
Estuaries	Eury- and Crenarchaeota	Highly diverse communities due to inputs from different sources (e.g. rivers, coastal waters, marshes, soil).
Marshes	Methanogens Nonthermophilic Crenarchaeota	Mainly associated to rizosphere. Most clones related to soil archaea
Sediments	Methanogens	Highly diverse environments with different physicochemistry and nutrient loads
	ANME Nonthermophilic Crenarchaeota	High archaeal abundance and richness
Sulfureta and hot springs	Thermophilic chemolithotrophic Eury- and Crenarchaeota	Main source for Crenarchaeota cultured strains Extreme thermophiles and acidophiles Important players in the sulphur cycle
Salt and Soda Lakes, Solar Salterns	Halophilic and extreme Halophilic Euryarchaeota	Microbial communities dominated by archaeal representatives at the highest salinities. Important sources for novel genera and species of extreme haloarchaea
Acid Mine Drainage	Acidophilic chemolithotrophic Eury-(mainly Thermoplasmata) and Crenarchaeota	Extreme acidophiles Important players in biogeochemical cycling of sulphur and sulphide metals
Symbionts	Methanogens Crenarchaeota	Anaerobic freshwater protozoa (methanogens) Freshwater sponges?

ANME: Anaerobic methane oxidizers.

in the world, but cell concentrations at the higher salinities are towards the high end of the range found in any natural planktonic system, reaching up to 10^8 haloarchaeal cells ml^{-1} in some cases. High cell densities produce a visible pink-red color, due to their carotenoid pigments. In these environments, haloarchaea are very abundant but grow at very low specific growth rates, similar to a laboratory culture in stationary phase.

Haloarchaea are well known from a wide range of available pure cultures. They use two photosynthetic pigments to successfully develop in haline environments: bacteriorhodopsin (a light-driven proton pump that captures light energy and uses it to move protons across the membrane out of the cell creating a proton gradient that generates chemical energy) and halorhodopsin (which uses light energy to pump chloride through the membranes to maintain osmotic pressure). In addition, haloarchaea contain high concentration of salts internally and exhibit a variety of molecular characteristics, including proteins that resist the denaturing effects of salts, and DNA repair systems that minimize the deleterious effects of desiccation and intense solar radiation. Crenarchaeota lack most of this special enzymatic equipment and are not present in hypersaline environments.

Rivers and Estuaries

Rivers and estuaries transport materials and energy from both terrestrial and aquatic sources to the marine environment. Activity and diversity of microbial communities change strongly along this transit, especially in the mixing zone where fresh and saline waters meet. Rivers are characterized by spatial heterogeneity and variability imposed by differences in water flow and geomorphology. Accordingly, the structure of microbial communities greatly differs among the different river zones and it is also strongly influenced by water velocity. Although there is a wealth of information on the structure of microbial communities in rivers, studies focusing on archaeal occurrence, diversity, and abundance are scarce. However, the small number of studies do provide a valuable comparison between the riverine communities and the associated estuarine/coastal waters (Table 2).

Benthic (sediments and biofilms) microbial riverine communities are complex and active, although archaea usually represent a minor fraction of the prokaryotic assemblages. However, archaeal phylotypes' richness is usually high, mainly in the particle-attached fraction, supporting the idea that rivers act as collectors of allochthonous archaea from catchments and neighboring ecosystems. Studies on archaeal diversity in rivers from diverse geographic locations with very different physichochemical conditions have revealed both euryarchaeota and

Table 2   Overall diversity and distribution of archaeal communities studied in different lotic habitats

Site	Main characteristics	Archaeal diversity[a]	Observations	Year	Source
Columbia (USA)	Temperate river It drains into a estuary	Marine and freshwater Crenarchaeota Euryarchaeota in the estuary	Mainly associated with particulate matter ('particle-attached archaea')	2000	1
Sinnamary (French Guiana)	Tropical river Interrupted by a dam (Petit Saut)	Euryarchaeota (Methanogens and Thermoplasmatales)	Detected in all sampling river stations downstream the dam	2001	2
Duoro (Portugal)	Temperate river Study carried out in estuary sediments	Archaeal community dominated by nonthermophilic Crenarchaeota (marine cluster)	Most of the sequences were obtained from surface sediment layers	2001	3
Rio Tinto (Spain)	Acidic (pH 2.2), high metal content	Euryarchaeota (Thermoplasma and Ferroplasma)	Extreme chemolithotrophic acidophilic archaea	2003	4
Aquifer (Idaho, USA)	Oxic, basalt aquifer	Euryarchaeotal clones related to methanogens and extremophiles. Crenarchaeota closer to freshwater clones	First report on Archaea inhabiting oxic temperate ground water	2003	5
Mackenzie (Canada)	Arctic river (mean temperature 3 °C), particle-rich waters	Mainly Euryarchaeota (methanogens and uncultured)	High diversity compared to other rivers	2006	6
		Marine Group I.1a Crenarchaeota	Clones related to archaea from soil and sediments Possible allochthonous origin		
Yangtze River estuarine region of East China Sea (China)	Temperate estuary	Sequences related to marine clones of both uncultured Euryarchaeota and Crenarchaeota (autotrophic ammonia-oxidizer *Nitrosopumilus maritimus*).	Remarkable spatial differences in archaeal composition	2007	7
	Planktonic samples analyzed		Low abundance but high diverse archaeal communities		

[a]Phylogenetic identity of the main clones of 16S rRNA genes recovered.

Sources

1. Crump BC and Baross JA (2000) Archaeaplankton in the Columbia River, its estuary and the adjacent coastal ocean, USA. *FEMS Microbiology Ecology* 31: 231–239.

2. Dumestre JF, Casamayor EO, Massana R, and Pedros-Alio C (2002) Changes in bacterial and archaeal assemblages in an equatorial river induced by the water eutrophication of Petit Saut dam reservoir (French Guiana). *Aquatic Microbial Ecology* 26: 209–221.

3. Abreu C, Jurgens G, De Marco P, Saano A, and Bordalo AA (2001) Crenarchaeota and Euryarchaeota in temperate estuarine sediments. *Journal of Applied Microbiology* 90: 713–718.

4. Gonzalez-Toril E, Llobet-Brossa E, Casamayor EO, Amann R, and Amils R (2003) Microbial ecology of an extreme acidic environment, the Tinto River. *Applied and Environmental Microbiology* 69: 4853–4865.

5. O'Connell SP, Lehman RM, Snoeyenbos-West O, Winston VD, Cummings DE, Watwood ME, and Colwell FS (2003) Detection of Euryarchaeota and Crenarchaeota in an oxic basalt aquifer. *FEMS Microbiology Ecology* 44: 165–173.

6. Galand PE, Lovejoy C, and Vincent WE (2006) Remarkably diverse and contrasting archaeal communities in a large arctic river and the coastal Arctic Ocean. *Aquatic Microbial Ecology* 44: 115–126.

7. Zeng Y, Li H, and Jiao N (2007) Phylogenetic diversity of planktonic archaea in the estuarine region of East China Sea. *Microbiological Research* 162: 26–36.

crenarchaeota that are highly similar in their 16S rRNA gene sequence with uncultured archaea from soils, rice fields, marshes, and anoxic sediments from lakes (**Table 2**). These comparisons suggest an allochthonous origin for riverine archaea. Methanomicrobiales and uncultured methanogens from soils and anoxic sediments are predominant euryarchaeotal components among the archaea, and high nutrient loading combined with hypoxic conditions in river sediments may favor their growth and activity. In turn, most of the crenarchaeotal sequences obtained from rivers affiliate with either marine planktonic or soil crenarchaeota. Therefore, riverine archaea seems to be more related to both sediment decomposition and passive transport. Exceptions arise in rivers with extreme conditions, such as the Rio Tinto (Spain), an acid river (pH 2.2 along nearly 100 km) where the combination of an active sulfur–iron cycle with high amounts of dissolved metals (Fe, Cu, Zn) favor the presence of the iron oxidizing chemolithoautotroph *Ferroplasma* (Thermoplasmata). Again in this example as for the sulfidic streamlets discussed above, a linkage between archaea and the sulfur–iron metabolism arise in a mesophilic environment.

The structure and dynamics of the microbial communities thriving in estuarine waters are more complex than in rivers due, in part, to the mixing regime of these environments. Estuaries have strong spatial and temporal gradients imposed by the contact between fresh water and marine waters, the geomorphology of the area, the influence zone of the freshwater input, wind mixing, and tidal action. Moreover, the estuary usually receives high inputs of organic matter from the river and from the coastal marine environment. As a consequence, estuarine microorganisms are a mixture of riverine and marine components. Although archaeal phylotypes from plankton and sediments belong mostly to the nonthermophilic marine crenarchaeota, methanogens have also been detected in the sediment (**Table 2**). These results point to a main influence of marine waters on estuarine archaea although remarkable spatial differences are observed among and within systems.

Lakes

Stratified lakes   Stratified lakes with seasonal or permanent oxic–anoxic interfaces have been subject of intense research by microbial ecologists because of the environmental conditions imposed by the vertical physichochemical gradients and pronounced changes in oxygen concentration. Stratification results in well-defined water compartments with different conditions suitable for growth of distinct and highly diverse microbial assemblages that play different roles in biogeochemical cycles. Archaea have been found to change along the vertical profile: archaeal richness, identity, and abundance change between oxic, oxic–anoxic, and anoxic zones for most studied lakes. This is a first indication that uncultured phylotypes are autochthonous and metabolically active *in situ*. Another indication can be found in humic stratified boreal lakes. Humic lakes receive large inputs of allochthonous (terrestrially derived) organic material, and consequently foreign archaeal populations are entering in the lake. However, uncultured planktonic phylotypes are distantly related to their counterparts from boreal forest soils suggesting these populations are lacustrine.

In the upper and well-oxygenated water layers, nonthermophilic freshwater crenarchaeota and several phylotypes of the uncultured euryarchaeota (mainly related to uncultured freshwater and marine clones and also distantly related to the Thermoplasmata and relative groups) have been frequently detected (**Table 3**), although they usually constitute a minor fraction of the picoplankton. At the oxic–anoxic interface, complex microhabitats with sharp gradients that favor the activity of different microbial populations exist. Anoxic layers in the water column as well as anoxic sediments have a combination of anaerobiosis, low redox potentials, and an accumulation of dissolved sulfur- and nitrogen-reduced compounds, as well as methane. Different studies carried out in stratified lakes have shown that archaeal abundance increase with depth. Uncultured euryarchaeota are frequently found at the oxic/anoxic interfaces whereas methanogens and non-thermophilic crenarchaeota are found in the anoxic waters and sediments (**Table 3**).

Cultivation has remained elusive for these new archaeal groups and there is very little understanding of their metabolism and roles within the ecosystem. Concerning nonthermophilic freshwater crenarchaeota, recent findings indicate that although they have been generally found in plankton and sediments from very different lakes, they may be less abundant than their marine and soil counterparts. Also, the richness of nonthermophilic crenarchaeota in suboxic and anoxic water layers is low, yielding very few phylotypes distantly related to the ubiquitous crenarchaeota from marine water and sediments or soils. Phylotypes detected at and below the oxycline affiliate with marine benthic groups. This latter cluster has been properly named as Miscellaneous Crenarchaeota Group (MCG) since it includes a large diversity of sequences retrieved from different environments such as soils, terrestrial environments, deep-paleosoils and forest lakes.

Table 3 Overall diversity and distribution of archaeal communities studied in different lakes

Site	Main characteristics	Archaeal diversity	Observations	Year	Source
Alpine and polar lakes					
Gossenköllesee (Austria); Crater Lake (Oregon, USA); Pyrenean lakes (Spain)	High-altitude, ultraoligotrophic lakes	High archaeal richness among lakes	Archaeal abundance among lakes from 1% to 37% of total prokaryotic counts	1998, 2001, 2007	1, 3, 4, 5
	Completely oxygenated	Spatial segregation between crena- and euryarchaeota	Higher abundance of Archaea in autumn and after ice-cover formation (early winter) in alpine lakes		
	High UV-radiation	The most abundant Crenarchaeota are closely related to nonthermophilic marine planktonic groups	Crenarchaeota more abundant either at the air–water interface and in deep waters (300–500 m depth)		
Fryxell (Antarctica)	Permanently frozen	Mainly Euryarchaeota (methanogens and uncultured)	Coexistence of cold-adapted methanogenic and methanotrophic archaea in the anoxic bottom waters	2006	2
	Active methanogenesis and sulfate reduction in the sediment	Crenarchaeota related to uncultured marine benthic group			
Great and large lakes					
Michigan, Lawrence (WI, USA)	Oligo- to mesotrophic lakes	Euryarchaeota (methanogens)	Crenarchaeotal 16S rRNA up to 10% of total environmental RNA extracted	1997	6, 7
	Sediment samples analyzed	Crenarchaeota related to the marine group			
Laurentian Great Lakes (USA): Erie, Huron, Michigan, Ontario and Superior Onega (Russia); Victoria (Africa)	Temperate to cold waters	All sequences clustered with marine nonthermophilic planktonic Crenarchaeota	Archaeal rRNA accounted for 1 to 10% of total planktonic rRNA	2003	8
	Oligo- to mesotrophic lakes Different climatic and geographic conditions covered Planktonic samples analyzed		Presence of cosmopolitan crenarchaeotal phylotypes		

Location	Habitat	Archaeal groups detected	Findings	Year	Sources
Stratified lakes Sælenvannet (Norway); Vilar (Spain); Pavin (France)	Meromictic sulfide-rich lakes	Euryarchaeota (methanogens, methanogens endosymbionts of anaerobic ciliates, and populations distantly related to Thermoplasmata)	Archaeal abundance increase with depth and maximal abundance below the chemocline	1997	9, 13, 16
	Moderate to cold temperatures	Crenarchaeota (nonthermophilic related to the marine and freshwater groups)	Seasonal dynamics, with higher relative abundance of Crenarchaeota in autumn and winter	2001 2007	
Charca Verde (France)	Freshwater pond	Methanogens, populations distantly related to Thermoplasmata and anaerobic methane-oxidizing archaea	Possible cooccurrence of methanogenic and methanotrophic (ANME-related) archaea in the anoxic water column	2007	18
Valkea Kotinen (Finland)	Sulfide-rich waters Boreal forest lake Anoxic hypolimnion with methane	No Crenarchaecta detected Methanogens and uncultured euryarchaeota distantly related to Thermoplasmata Crenarchaeota of the nonthermophilic freshwater group	Archaea up to 7% of total microscopic counts No significant changes in abundance along season Freshwater crenarchaeota not related to soil crenarchaeota	2000	11
Stratified lakes Solar Lake (Sinai, Egypt)	Hypersaline lake Sulfide-rich hypolimnion Active methanogenesis at the bottom	Methanogens and uncultured populations distantly related to Thermoplasmata No Crenarchaeota detected	Archaeal community dominated by haloarchaea (salinities >10%) Halophilic methanogens present	2000	12
Rotsee (Switzerland); Dagow (Germany); Biwa (Japan); Kinneret (Israel)	Samples from anoxic sediments	Methanogens, methanogenic endosymbionts of anaerobic ciliates and uncultured euryarchaeota distantly related to Thermoplasmata	Archaeal abundance (methanogens) accounted for 1 to 7% of total prokaryotes	1999; 2004; 2007	10, 14, 15, 17
	Mesoeutrophic lakes with anoxic hypolimnion	Crenarchaeota detected only in sulfurous sediments (freshwater nonthermophilic group)	Methanogenic endosymbionts up to 1%		

Sources
1. Pernthaler J, Glockner FO, Unterholzer S, Alfreider A, Psenner R, and Amann R (1998) Seasonal community and population dynamics of pelagic bacteria and archaea in a high mountain lake. *Applied and Environmental Microbiology* 63: 4299–4306.
2. Karr EA, Ng JM, Belchik SM, Sattley WM, Madigan MT, and Achenbach LA (2006) Biodiversity of methanogenic and other *Archaea* in the permanently frozen Lake Fryxell, Antarctica. *Applied and Environmental Microbiology* 72: 1663–1666.

3. Urbach E, Vergin KL, Young L, and Morse A (2001) Unusual bacterioplankton community structure in ultra-oligotrophic Crater Lake. *Limnology and Oceanography* 46: 557–572.

4. Urbach E, Vergin KL, Larson GL and Giovannoni, SJ (2007) Bacterioplankton communities of Crater Lake, OR: dynamic changes with euphotic zone food web structure and stable deep water populations. *Hydrobiologia* 574: 161–177.

5. Auguet JC, and Casamayor EO (2008) A hotspot for cold Crenarchaeota in the neuston of high mountain lakes. *Environmental Microbiology* 10: in press. DOI: 10.1111/j.1462-2920.2007.01498.x

6. MacGregor BJ, Moser DP, Alm EW, Nealson KH, and Stahl DA (1997) *Crenarchaeota* in Lake Michigan sediment. *Applied and Environmental Microbiology* 63: 1178–1181.

7. Schleper C, Holben W, and Klenk HP (1997) Recovery of crenarchaeotal ribosomal DNA sequences from freshwater-lake sediments. *Applied and Environmental Microbiology* 63: 321–323.

8. Keough BP, Schmidt TM, and Hicks RE (2003) Archaeal nucleic acids in picoplankton from great lakes on three continents. *Microbial Ecology* 46: 238–248.

9. Øvreås L, Forney L, Daae FL, and Torsvik V (1997) Distribution of bacterioplankton in meromictic Lake Sælenvannet, as determined by denaturing gradient gel electrophoresis of PCR-amplified gene fragments coding for 16S rRNA. *Applied and Environmental Microbiology* 63: 3367–3373.

10. Falz KZ, Holliger C, Großkopf R, Liesack W, Nozhevnikova AN, Müller B, Wehrli B, and Hahn D (1999) Vertical distribution of methanogens in the anoxic sediment of Rotsee (Switzerland). *Applied and Environmental Microbiology* 65: 2402–2408.

11. Jurgens G, Glöckner F.-O, Amann R, Saano A, Montonen L, Likolammi M, and Münster U (2000) Identification of novel *Archaea* in bacterioplankton of a boreal forest lake by phylogenetic analysis and fluorescent *in situ* hybridization. *FEMS Microbiology Ecology* 34: 45–56.

12. Cytrin E, Minz D, Oremland RS, and Cohen Y (2000) Distribution and diversity of archaea corresponding to the limnological cycle of a hypersaline stratified lake (Solar Lake, Sinai, Egypt). *Applied and Environmental Microbiology* 66: 3269–3276.

13. Casamayor EO, Muyzer G, and Pedrós-Alió C (2001) Composition and temporal dynamics of planktonic archaeal assemblages from anaerobic sulfurous environments studied by 16S rDNA denaturing gradient gel electrophoresis and sequencing. *Aquatic Microbial Ecology* 25: 237–246.

14. Glissmann K, Chin KJ, Casper P, and Conrad R (2004) Methanogenic pathway and archaeal community structure in the sediment of eutrophic Lake Dagow: effect of temperature. *Microbial Ecology* 48: 389–399.

15. Koizumi Y, Takii S, and Fukui M (2004) Depth-related changes in archaeal community structure in a freshwater-lake sediment as determined by denaturing gradient gel electrophoresis of amplified 16S rRNA genes and reversely transcribed rRNA fragments. *FEMS Microbiology Ecology* 48: 285–292.

16. Lehours A.-C, Evans P, Bardot C, Joblin, K, and Fonty G (2007) Phylogenetic diversity of archaea and bacteria in the anoxic zone of a meromictic lake (Lake Pavin). *Applied and Environmental Microbiology* 73: 2016–2019.

17. Schwarz JK, Eckert W, and Conrad R (2007) Community structure of *Archaea* and *Bacteria* in a profundal lake sediment Lake Kinneret (Israel). *Systematic and Applied Microbiology* 30: 239–254.

18. Briée C, Moreira D, and López-García P (2007) Archaeal and bacterial community composition of sediment and plankton from a suboxic freshwater pond. *Research in Microbiology* 158: 213–227.

In contrast with the limited understanding of the MCG group, methanogens are active key players within the carbon cycle in anoxic waters and sediment, and their richness, abundance, and activity in lakes of different trophic status around the world have been extensively documented. Further, methanogenic endosymbionts of anaerobic ciliates have been also observed in oxic–anoxic interfaces and anoxic water zones of stratified lakes. Interest in methanogens is linked to global warming produced by the strong greenhouse effects of methane gas in the atmosphere. Depending on the energy and carbon source used to generate methane, methanogens can be divided in hydrogenotrophic, if they use H_2/CO_2, and acetoclastic, if they use acetate. Interestingly, mixed communities of acetoclastic methanogens (family Methanosaeteceae and relatives) and hydrogenotrophic methanogens (families Methanobacteriaceae, Methanospirillaceae, and Methanomicrobiaceae) alternate in dominance of the community depending mainly on temperature (high temperatures favor hydrogenotrophs) and substrate availability (i.e., H_2, acetate, and simple methylated compounds).

The unexpected discovery of archaea able to anaerobically oxidize methane in deep-sea anoxic sediments is a significant new finding about methane cycling in anoxic environments. The first described anaerobic methane oxidizers (ANME) were found to be in close association with sulfate-reducing bacteria forming a syntrophic consortium whereas the archaeal partner oxidizes methane and further transfers electrons to bacteria, which reduce sulfate to hydrogen sulfide. Anaerobic methane oxidization also occurs in freshwater anoxic hypolimnia and sediments with planktonic single cells, short chains, or small aggregates of ANME archaea reaching up to 1% of total prokaryotes at certain depths. In addition, a new bacteria/archaea consortium that oxidizes methane to carbon dioxide coupled to denitrification has been recently described in a methane-saturated freshwater sediment. Altogether, different consortia of ANME appear to be distributed worldwide. To date, however, no pure cultures of methanotrophic archaea are available.

Great lakes A handful of large lakes have been surveyed for archaeal abundance and 16S rRNA gene diversity. Quantitative rRNA hybridizations revealed that the percentage of archaeal nucleic acid was between 0.7% and 10% of the picoplankton collected from the epilimnion and hypolimnion of the Laurentian Great Lakes in North America, Africa's Lake Victoria, and Lakes Ladoga and Onega in Russia. In addition, analysis of sedimentary archaeal

rRNA and mostly archaeal membrane lipids (crenarchaeol) from the top surface sediments of several large lakes representative of different climatic and physical scenarios (Lake Superior and Lake Michigan in North America, Lake Malawi in the East Africa, and Lake Issyk Kul in Central Asia) suggest ubiquitous and widespread distribution of nonthermophilic planktonic crenarchaeota in large freshwater bodies. The lipid structure was identical to the marine counterparts and would indicate a very close relationship both in phylogeny and metabolism for marine and large lake archaea. The phylogenetic analysis carried out with 16S rRNA genes points in the same direction. Altogether, the limited studies in large lakes have been sufficient to increase the overall absolute abundance of freshwater crenarchaeota several orders of magnitude, and subsequently increase their expected impact in biogeochemical cycling.

High-altitude and polar lakes 16S rRNA gene surveys have indicated that the largest proportion and greatest diversity of archaea exist in permanently cold environments, and the study of cold-adapted archaea is a growing area of research. Little is known about these cold-water archaea in fresh waters except that they are present in considerable numbers up to 10^4–10^5 ml^{-1} (i.e., 5–30% of total prokaryotic plankton). They are found in the Antarctic, the Arctic, and in high mountain lakes but very few data on abundance, composition, and other attributes are available so far. A recent study carried out in Pyrenean lakes points to the hydrophobic surface film at the air–water interphase (neuston) as being a possible hotspot for nonthermophilic crenarchaea (see **Table 3**). In general, these remote systems are difficult to reach and aquatic biota experience extreme physical conditions such as low temperatures and high UV exposure. In addition, many of these lakes are ice covered most of the year and have extreme oligotrophic conditions. Pelagic archaeal population dynamics are related to the dynamics of the lake. Archaea seem to be abundant in the plankton only during autumn thermal mixing and abundances decrease thereafter, as depicted from a detailed monthly study (see **Table 3**). Again, these results indicate that archaea are metabolically active *in situ*. In fact, recent observations indicated that archaeal 16S rRNA genes from high mountain lakes and perennially cold or permanently frozen Antarctic lakes belong to methanogens, uncultured euryarchaea (distantly related to Thermoplasmata), and crenarchaeota. There is a considerable distance between any ribosomal sequence for these aquatic forms relative to terrestrial archaea. These freshwater archaea are likely involved in the major biogeochemical cycles that occur in these ecosystems. Interestingly, freshwater uncultured planktonic crenarchaeota are very abundant at certain depths in several surveyed high mountain lakes and closely similar to the marine crenarchaeal group, which is very abundant in the deep ocean. For example, in Crater Lake (OR, USA), archaea are very abundant in bottom waters reaching up to 20% of total prokaryotes with a single predominant phylotype related to the marine crenarchaeota and in certain Pyrenean lakes (Spain) archaea reached more than 30% of total microscopic counts. As other cases discussed in this article, there is undoubtedly ample room for new and exciting discoveries related to archaea and cold-water habitats in lakes.

Functional Role: the Known and the Knowable

The Known Functional Roles

Much knowledge has been gained on the functional role of archaea *in situ* from the cultured representatives of the three traditional groups available in the laboratory (**Table 4** and **Table 5**). The role that ubiquitous methanogens carry out in the environment is well characterized. Formation of methane in the final degradation step of organic matter under anaerobic conditions is an exclusive attribute of archaea. In addition, methanogens are the only known archaea able to fix N_2. Hyperthermophiles are metabolically quite diverse, ranging from chemoorganotrophs to chemolithoautotrophs. They are aerobes, facultative anaerobes, fermenters, or anaerobes respiring sulfate or nitrate on organic substrates. Several species are at the base of the food web as primary producers of organic matter using carbon dioxide as sole carbon source and obtaining energy by the oxidation of inorganic substances like sulfur and hydrogen. Finally, haloarchaea are mostly aerobic chemoorganotrophs although the ability to use nitrate as a terminal electron acceptor in energy metabolism has been found in several studies.

In relation to the nitrogen cycle, some cultured archaea (both euryarchaeota and crenarchaeota) are capable of denitrification and nitrogen fixation like bacteria. Unlike bacteria, none of the archaea were initally thought to be capable of nitrification. According to this view, the role of archaea in the nitrogen cycle seemed to be important only at a local scale (i.e., extremophilic freshwater environments) although a careful evaluation must be done after microbial ecophysiologists have explored the potential capabilities of uncultured archaea. In fact, recent data indicate the potential for nitrification in both uncultured and recently cultured widespread archaea.

The Knowable Functional Roles

Scientists are just beginning to glimpse the functional diversity of noncultured freshwater archaea, although current data are still scarce and fragmentary (**Table 6**). Although we suspect that most of archaeal cells are active under *in situ* conditions, no clear data is available on the precise physical, chemical, and biological characteristics of their niche. This characterization would be very helpful for developing culture enrichments and further isolation of new archaeal species. Obtaining pure laboratory cultures will facilitate the precise determination of their ecophysiological, biochemical, and genetic properties. At the moment, new, culture-independent approaches are starting to provide a better and more comprehensive understanding of the roles of archaea within the ecosystem and habitats. These studies indicate that the majority of freshwater archaea, especially crenarchaeota, are mesophilic or even psychrophilic in contrast with early views that archaea were primarily themophilic and/or halophilic. In fact, recent evidence suggests that some of the ubiquitous pelagic crenarchaeota play an essential role in biogeochemical cycling in aquatic ecosystems and may act as chemoautotrophs, oxidizing ammonia to nitrate, and fixing inorganic carbon in the dark. Further, the detection of tetraether membrane lipids of freshwater crenarchaeota in both ancient and recent sediments indicates crenarchaeota are potential mediators of CO_2 drawdown from the atmosphere to sediments.

Molecular phylogenetic surveys carried out in natural environments for the last 15 years have provided a better comprehension of archaeal diversity and phylogeny based on the 16S rRNA gene. At present, scientists have a wide array of powerful techniques that will advance future studies. These include (1) metagenomics – the study of large DNA fragments obtained directly from the environment and the use of high-throughput DNA sequencing on microbial enrichments and further genome reconstruction using bioinformatics; (2) comparative genomics – using metagenomes and available genomes in databases to both make inferences about ecology, biology, and evolution, and to search for relevant functional genes linked to archaeal phylogenetic markers in metagenomic libraries; (3) functional genes surveys – monitoring both at the DNA level and at the mRNA level to demonstrate potential and *in situ* expression, respectively; and (4) isotopically labeled substrate tracking combined with fluorescence *in situ* hybridization (FISH) and microautoradiography for detection of nutrient uptake by specific archaeal cells in mixed communities.

Table 4 Summary of the basic metabolic capabilities, required growth conditions and occurrence of the main groups of *Euryarchaeota* in inland waters

Group	Metabolic process	Energy source	Carbon source	e^- donor	e^- acceptor	Conditions	Occurrence	Cultured representatives
Methanogens	Methanogenesis	O.M. (acetate, format, methanol, ...)	O.M.[a]	O.M.	O.M.	Extreme anaerobiosis	Anoxic waters of stratified lakes	*Methanococcus, Methanobacterium, Methanosphaera*
		H_2	CO_2	H_2	CO_2	Low redox potential	Sediments, rice fields; Anoxic microhabitats in solar salterns, soda lakes and salt marshes; Sewage digestors	Most methanogens
Methanotrophs	Anaerobic oxidation of methane	CH_4	O.M.	CH_4	SO_4^{2-}; Nitrite	Extreme anaerobiosis; Low redox potentials; Sulfate-reducing bacteria[b]; Anoxic conditions; Active sulphur cycle	Anoxic sediments	None
Halophiles	Aerobic respiration	O.M. (sugars, aminoacids, glycerol)	O.M.	O.M.	O_2	Moderate to extreme salinities (from 7% to 37% NaCl)	Salt lakes	*Haloarcula, Haloferax, Halorubrum, Natrinema*
	Anaerobic respiration		O.M.		Nitrate, DMSO, organic acids	High Ph	Soda lakes	*Halobacterium, Haloarcula, Haloferax*
	Fermentation	Sugars, aminoacids	O.M.	O.M.	O.M.	Oxic, anoxic	Hypersaline Antarctic lakes	*Halobacterium*
	Nonphotosynthetic photoheterotrophy[c]	Light	O.M.	O.M.	–		Solar salterns	*Halobacterium*

[a]Organic matter.

[b]Anaerobic methane oxidation involves a symbiosis between archaeal cells and sulfate-reducing bacteria forming syntrophic consortia with different structures and cell numbers. Some studies have shown that methanotrophic archaea can oxidize methane without bacterial consortia.

[c]Under suboxic or anoxic conditions, light-induced isomerization of retinal in Bacteriorhodopsins causes a translocation of protons across the membrane that is used to generate ATP.

Table 5   Summary of the basic metabolic capabilities, required growth conditions, and occurrence of the main groups of Crenarchaeota in inland waters

Group	Metabolic process	Energy source	Carbon source	e^- donor	e^- acceptor	Conditions	Occurrence	Cultured representatives
Thermophiles	Aerobic chemolithotrophy	S^0	CO_2	S^0	O_2	High temp. (40–97 °C); Acidic pH (1–5.5); Oxic to anoxic conditions	Oxic–anoxic microhabitats in thermal sulphur springs	*Acidianus, Sulfolobus*
	Aerobic respiration	O.M.[a] (sugars, peptides, aminoacids, alcohols)	O.M.	O.M.	O_2		Oxic zones of thermal sulphur springs	*Thermoproteus, Sulfolobus*
	Anaerobic chemolithotrophy	H_2	CO_2	H_2	S^0, Nitrate, Fe^{3+}		Suboxic–anoxic zones in thermal sulphur springs	*Acidianus, Pyrodictium*
	Anaerobic respiration	O.M. (sugars, peptides, aminoacids, alcohols)	O.M.	O.M.	S^0, Nitrate, Nitrite		Suboxic–anoxic zones in thermal sulphur springs	*Pyrococcus, Thermoproteus. Thermococcus, Pyrobaculum*
	Fermentation	O.M. (sugars, peptides, aminoacids, alcohols)	O.M.[b]	O.M.	S^0		Anoxic zones in thermal sulphur springs	*Pyrococcus, Thermococcus*

[a]Organic matter.
[b]Not completely oxidized to CO_2.

Table 6 Summary of the putative metabolic capabilities, required growth conditions, and occurrence of the main groups of nonthermophilic Crenarchaeota detected so far in inland waters

Group	Putative metabolic process	Energy source	Carbon source	e^- donor	e^- acceptor	Conditions	Occurrence	Cultured representatives
Nonthermophiles	Aerobic chemolithotrophy	NH_4^+	CO_2	NH_4^+	O_2	O_2 and reduced nitrogen compounds (NH_4^+) Oligotrophy	Close relatives detected in great lakes and alpine lakes	Candidatus Nitrosopumilus maritimus[a]
	Aerobic chemoorganotrophy?; Mixotrophy?	O.M.[b] (aminoacids)	O.M.	O.M.	O_2	Oxic conditions Oligo- to mesotrophic		None
	Anaerobic chemolithotrophy?	H_2, H_2S	CO_2	H_2, H_2S	S^0, Nitrate, Fe^{3+}	Suboxic to anoxic conditions	Suboxic–anoxic zones in stratified lakes (meta- and hypolimnia)	None
						Reduced sulphur compounds Meso- to Eutrophic	Sediments	
	Anaerobic chemoorganotrophy?	O.M. (sugars, peptides, aminoacids, alcohols)	O.M.	O.M.	S^0, Sulfate, Nitrite	Suboxic to anoxic conditions	Suboxic–anoxic zones in stratified lakes (meta- and hipolimnia)	None
						Meso- to eutrophic	Sediments	

[a]Isolated from marine environment.
[b]Organic matter.

Microbial ecologists have already started to apply these methodologies on marine and soil uncultured archaea as well as those from inland waters. New and exciting results are continuously arising. In one example, the ANME archaea are offering clues for the complete understanding of the methane cycle, as indicated earlier. In another example, nonthermophilic crenarchaeota from the ocean and soils have been recently proposed to be chemolithoautotrophs, oxidizing ammonia using an archaeal ammonia monooxygenase (AOA), which is phylogenetically distantly related with the same enzyme present in bacterial nitrifiers (AOB). Pure cultures and complex mixtures in laboratory enrichments of nonthermophilic crenarchaeota are starting to become available for ecophysiological experiments. Field data suggest that planktonic communities either contain both autotrophic and heterotrophic archaeal members or are largely composed of cells with a mixotrophic metabolism. Archaea may also be capable of nitrification based on emerging information. The present knowledge on uncultured archaea will benefit very soon from all these methodological improvements leading to the elucidation of the precise role of these microorganisms in the biogeochemical cycling of inland waters.

Concluding Remarks

Overall, archaea are a common component of freshwater plankton and different groups of uncultured archaea tend to occupy different ecological niches, although the metabolism and physiological roles remain challenging questions. Our knowledge has largely increased as the number of studies performed and the diversity of biotopes sampled have increased. In general, archaea seem to be within a range $<10\%$ of total prokaryotic plankton and might be playing important and previously unrecognized roles in inland water ecosystems. From the few studies available, it can be deduced that archaea show spatial and temporal differences in abundance, and more detailed spatiotemporal surveys will unveil the conditions under which archaea reach higher numbers in freshwater habitats. Further studies combining ecology and several of the molecular and genomic tools available, as well as traditional microbiology, are still needed to enlarge our view on the ecology of archaea in lakes, rivers, and streams. As soon as some of these organisms become cultivated and their metabolic and genetic potentials are studied in detail, new physiological and ecological phenotypes will be discovered. It seems clear that the roles that archaea play in the ecosystems have been grossly underestimated and that scientists have begun to realize this fueling the recent burst forward in new knowledge.

Glossary

Acetoclastic – Type of methanogens that produce methane using acetate as energy and carbon source.

Ammonia monooxygenase (AMO) – The key functional enzyme responsible for the conversion of ammonia to hydroxylamine (which is further converted to nitrite by hydroxylamine oxidoreductase).

ANME – ANME stands for anaerobic methane oxidation. A process carried out by methanotrophic archaea in a sort of syntrophic consortium with sulfate-reducing bacteria. These consortia were first described in deep-sea anoxic sediment but recent data have demonstrated their presence in anoxic waters and sediments of some freshwater habitats.

Bacteriorhodopsins – Integral membrane proteins containing retinal present in many halophilic archaea. Bacteriorhodopsins use light energy to generate a transmembrane proton motive force (proton pumps) subsequently converted into ATP. Some halophiles contain *halorhodopsins*, modified bacteriorhodopsins that act as light-driven chloride pumps.

Chemolithoautotroph – An organism that obtains energy from reduced inorganic compounds and carbon from CO_2.

Chemoorganoautotroph – An organism that obtains energy from organic compounds and carbon from CO_2.

Chemoorganoheterotroph – An organism that obtains energy and carbon from organic compounds.

Crenarchaeol – Glycerol dialkyl glycerol tetraether thought to be solely produced by 'cold' crenarchaeota (i.e., nonthermophilic). Excellent biomarker and paleotemperature proxy.

Crenarchaeota – One of the four phyla (kingdoms) of the Domain Archaea. Members of this group are either extremophiles or nonextremophiles inhabiting the most diverse environments.

Euryarchaeota – One of the four phyla (kingdoms) of the Domain Archaea. Methanogens and halophiles are the most known members of this phylum.

Extremophile – An organism that occupy environments judged by human standards as harsh because of both physical and chemical extremes. Literally 'extreme-loving.'

Hydrogenotrophic – Type of methanogens that produce methane using H_2 and CO_2 as energy and carbon sources, respectively.

Korarchaeota – One of the four phyla (kingdoms) of the Domain Archaea. Members of this kingdom have been identified only by environmental DNA sequences in high-temperature environments.

Methylotrophic – Type of methanogens that produce methane using simple C1 compounds as energy and carbon source.

Nanoarchaeota – The most recent phylum (kingdom) discovered within the Domain Archaea. Members of this phylum are very small in size and they are all symbionts of hyperthermophilic archaea.

Phylotype – In the context of microbial ecology refers to a uncultured microorganisms only described by their ribosomal gene sequence (mainly 16S or 18S)

Picoplankton – The portion of the plankton comprised between the 2–0.2 μm size range.

Sulfureta – Natural environments usually found around active volcanoes very rich in sulfur-reduced compounds.

Syntrophy – Cooperation of two or more microorganisms that combine their metabolic capabilities to degrade a substance not capable of being degraded by either one alone.

Further Reading

Brock TD (1997) Prokaryotic diversity: Archaea. In: Madigan MT, Martinko JM, and Parker J (eds.) *Biology of Microorganisms*, 8th edn, pp. 635–740. Englewood Cliffs, NJ: Prentice Hall.

Cavicchioli R (2006) Cold-adapted archaea. *Nature Reviews Microbiology* 4: 331–343.

Chaban B, Ng SYM, and Jarrell KF (2006) Archaeal habitats— From the extreme to the ordinary. *Canadian Journal of Microbiology* 52: 73–116.

DeLong EF (1998) Everything in moderation: archaea as 'nonextremophiles' *Current Opinion in Genetics and Development* 8: 649–654.

Dworkin M, Falkow S, Rosenberg E, Schleifer K-H, and Stackebrandt E (eds.) (2006) *The Prokaryotes—A Handbook on the Biology of Bacteria*, Vol. 3: *Archaea*, 3rd edn. New York: Springer-Verlag.

Garcia J-L, Patel BKC, and Ollivier B (2000) Taxonomic, phylogenetic and ecological diversity of methanogen archaea. *Anaerobe* 6: 205–226.

Hershberger KL, Barns SM, Reysenbach AL, et al. (1996) Wide diversity of Crenarchaeota. *Nature* 384: 420.

Huber R, Huber H, and Stetter KO (2000) Towards the ecology of hyperthermophiles: Biotopes, new isolation strategies and novel metabolic properties. *FEMS Microbiology Reviews* 24: 615–623.

Nicol GW and Schleper C (2006) Ammonia-oxidizing Crenarchaeota: Important players in the nitrogen cycle? *Trends in Microbiology* 14: 207–212.

Oren A (1994) The ecology of the extremely halophilic archaea. *FEMS Microbiology Reviews* 13: 415–440.

Powers LA, Werne JP, Johnson TC, et al. (2004) Crenarchaeotal membrane lipids in lake sediments: A new paleotemperature proxy for continental paleoclimate reconstruction? *Geology* 32: 613–616.

Rudolph C, Wanner G, and Huber R (2001) Natural communities of novel archaea and bacteria growing in cold sulfurous springs with a string-of-pearls-like morphology. *Applied and Environmental Microbiology* 67: 2336–2344.

Schleper C (2005) Genomic studies of uncultivated archaea. *Nature Reviews Microbiology* 3: 479–488.

Woese CR (1987) Bacterial evolution. *Microbiological Reviews* 51: 221–271.

Relevant Websites

http://archaea.ws/index.html – A scholarly journal providing rapid peer review and publication of articles dealing with any aspect of research on the archaea.

http://www.archaea.unsw.edu.au – ArchaeaWeb is an information resource for researchers working with archaea and extremophiles.

http://www.ucmp.berkeley.edu/archaea/archaea.html – An introductory page on the main characteristics of the archaea, maintained by the University of California at Berkeley.

http://tolweb.org/tree?group=Archaea&contgroup=Life_on_Earth – A page dedicated to the archaea in the Tree of Life web project.

Bacteria, Attached to Surfaces

J-F Carrias and T Sime-Ngando, Université Blaise Pascal, Clermont-Ferrand, France

Introduction

Bacteria attached to surfaces can be found in all types of inland water habitats, including fresh water and saline lakes, rivers, streams, ponds, ground waters, springs, cave waters, floodplains, and also extreme environments such as hypersaline lakes and hot springs. Microenvironments for attached bacteria in these ecosystems consist of physical interfaces in both the pelagic and benthic areas. They include submerged plants, rocks, stones, sediments, suspended particles and planktonic cells, and also the air–water interface.

In general, attached bacteria form microbial consortia with a diversity of other micoorganisms such as archaea, cyanobacteria, protists (algae and protozoa), fungi as well as micrometazoa. Biofilms, organic aggregates, and microbial mats represent commons forms of these different associations of attached microbial communities that occur in aquatic ecosystems. These complex assemblages contain numerous species of bacteria and play key roles in driving biogeochemical cycles and controlling the quality of inland waters. Attached bacteria usually predominate among the microorganisms in these consortia. They display a broad taxonomic and metabolic diversity, enabling them to carry out a wide range of biological processes.

Living inside a biofilm, a microbial mat, or an organic aggregate undoubtedly offers bacteria numerous advantages. First, nutrient availability is enhanced inside the consortia and helps sustain high levels of productivity. Second, the structure provides a protected habitat for bacteria against harmful conditions such as desiccation, toxic agents, and UV radiation, as well as against various predators. Third, enhanced interactions and related processes (gene transfer, nutrient exchange, quorum sensing) are promoted with other bacteria, partly owing to codependence on space.

Different Microhabitats for Attached Bacteria in Inland Waters

Biofilms consist of organized, complex communities encased in extracellular polymeric substances (EPSs) that have been synthesized by the associated bacteria. The composition of the polymeric matrix may differ among and within biofilms under different life conditions, but exopolysaccharides are an essential component of the matrix. The bacterial microcolony is the basic living structure unit of a biofilm. In vitro, the initial unit can be produced from a single species or from several species. In natural conditions, microcolonies within biofilms (**Figure 1**) are multispecies communities, and the architecture of the biofilm is the result of complex interactions among bacteria, physicochemical factors, and the activity of protistan and metazoan grazers. Naturally occurring biofilms are diverse and complex three-dimensional structures with biotic and abiotic constituents. Bacteria forming biofilms on rocks or stones (epilithic), on aquatic plants (epiphytic), and in the sediment of streams, rivers, and shallow lakes (epipelic) are known to play an important role not only in the degradation of organic matter, but also in the degradation and transformation of contaminants and metals. Biofilms are also found at the air–water interface. This microlayer is considered an extreme environment and studies on the thriving microbial communities (e.g., the neuston) have revealed that these biofilms present characteristics similar to those found on solid surfaces.

Microbial mats are a particular type of biofilm, also known as photoautotrophic biofilms or cyanobacterial mats. They are present in hot springs, sulfur springs, on the sediments of hypersaline lakes, and in Antarctic lakes. They are made up of several laminated, colored layers of microbes (**Figure 2**) growing at the sediment–water interface of their extreme environments. Light, oxygen, and hydrogen sulfide, the levels of which vary with depth, determine the vertical distribution of the associated communities. Oxygenic cyanobacteria generally dominate the surface layers, whereas anaerobic bacteria, such as anoxygenic photoautotrophs and sulfur-reducing bacteria, are found in the deepest zone of the mat matrix, which can range in thickness from several millimeters to a few centimeters. Prokaryotic communities developing along vertical microgradients are metabolically interdependent. Unlike biofilms, naturally occurring microbial mats generally lack eukaryotic organisms, most likely because of the extreme conditions prevailing in these ecological niches.

Organic aggregates in pelagic environments are usually colonized by various planktonic microbes, especially bacteria. These suspended organic particles and their associated microorganisms generally occur in the size range of $<1\,\mu m$ to several centimeters in plankton. The largest macroscopic aggregates

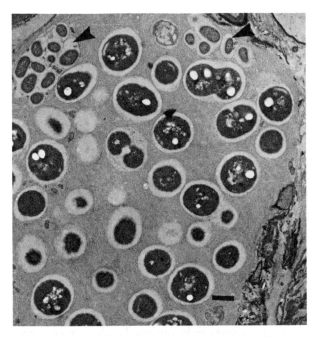

Figure 1 Thin section of a bacterial biofilm showing distinct microcolonies (arrowheads) scattered throughout a dominant family of Gram-negative bacteria. The extracellular polysaccharide that adheres to the cell is also visible as the clear material surrounding bacteria. The inert substratum, to which the biofilm is attached, can be seen at the bottom right-hand corner. Bar = 1 μm. Reproduced from **Figure 1** in Beveridge TJ, Makin SA, Kadurugamuwa JL, and Li Z (1997) Interactions between biofilms and the environment. *FEMS Microbiology Reviews* 20: 291–303. Copyright (1997), with permission from Blackwell publishers.

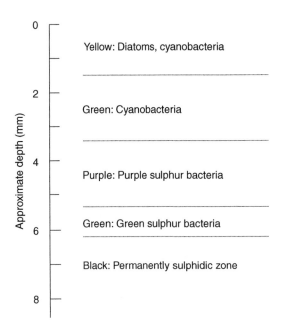

Figure 2 The vertical colour banding of typical cyanobacterial mats. Reproduced from Figure 6.3 in Fenchel T, King GM, and Blackburn TH (2000) *Bacterial Biogeochemistry – The Ecophysiology of Mineral Cycling. Part 6: Microbial Mats and Stratified Water Columns.* London: Academic Press. Copyright (1998), with permission from Elsevier.

(>500 μm in size), known as lake snow or river snow, are generally composed of algae (mainly diatoms), fecal pellets, bacteria, and a variety of organic detritus. Their abundance ranges from <1 to 50–$100\,l^{-1}$ and they form a significant fraction of the sinking particulate organic matter in pelagic environments. The abundance of microscopic aggregates (<500 μm in size), is higher, ranging from 10 to $10^8\,l^{-1}$, depending on the productivity of the ecosystem. Based on the dyes used to stain and visualize these microscopic aggregates, different types of particles have been described: transparent exopolymeric particles (TEP) stainable with an acid solution of Alcian Blue, proteinaceous particles stainable by Coomassie Brillant Blue (Coomassie Stained Particles, CSP), and DAPI Yellow Particles (DYP) stainable with 4′,6-diamidino-2-phenyl-indole (DAPI). TEP, which consist mainly of polysaccharides, are those that have been most closely studied (**Figure 3**). They play an important role in the aggregation of phytoplankton cells, primarily during bloom events, and in the formation of macroscopic aggregates. The bulk of suspended aggregates contain attached bacteria. Their numbers

increase with particle size, and most studies have found that attached bacteria make up about 10% of the total bacterial numbers within the water column. This relative abundance is certainly an underestimate because the bacterial counting method includes a pressure filtration step that can detach some cells from aggregates. Bacteria associated with aggregates are usually larger than free-living cells in the surrounding water and they frequently exhibit higher enzymatic activities per cell. This implies that attached cells are more important in terms of activity than in terms of biomass in the water column, relative to free-living communities.

Epibiotic bacteria are found in association with other aquatic organisms (mainly cyanobacteria) and are generally fixed to the cell surface or live inside the mucilage around the host cell. The phycosphere consists of the microenvironment around the surface of host algae or cyanobacteria. In this zone, organic compounds, mainly polysaccharides, are at elevated concentrations relative to the external medium and provide a resource to the associated bacteria inside the mucilage. Inversely, phytoplanktonic cells benefit from the nutrients released by the activity of associated bacteria. Phycospheres produced by the colonial cyanobacteria *Anabaena* and *Microcystis* have close similarities with organic aggregates. Occurrence of epibiotic bacteria is not a general feature of phytoplankton populations because most phytoplankton

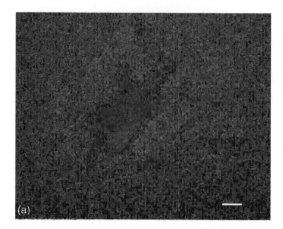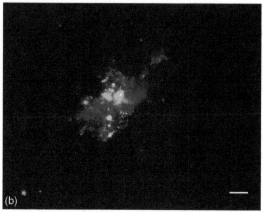

Figure 3   A Transparent Exopolymer Particle (TEP) stained with Alcian blue, a polysaccharide-specific dye, (a) visualized under direct light and (b) DAPI-stained bacteria (small brilliant-blue spots) free-living and associated to the particle under UV light excitation. Bars represent 10 μm. Photographs from JF Carrias.

have developed defense mechanisms to prevent colonization by sessile microorganisms.

Adhesion of Bacteria to Surfaces, Biofilm Formation, and Structure

Because attached bacteria are important in drinking water distribution systems, waste treatment plants, and various industrial settings, and also because they are implicated in numerous infections, the mechanisms of bacterial adhesion to a substratum are well studied. Many of these studies concern biofilms, typically focusing on one or a small number of species grown under laboratory conditions. Consequently, most of the findings cannot be reliably extrapolated to natural environments in which interactions between bacteria and surfaces involve a great diversity of colonization strategies, bacterial surface compositions, and solid surface characteristics. However, recent studies have revealed that distantly related bacterial species show common features during the attachment procedure. Adhesion of a bacterium to surfaces in natural environments is considered to be nonspecific, and the analysis of a large number of both in vitro and natural biofilms, and several studies of microbial mats and organic aggregates, clearly indicate the importance of exopolysaccharides in the formation of the matrix around the bacterial microcolonies. This suggests that common characteristics are probably used by different bacteria to produce the useful elements for an attached lifestyle under different environmental conditions.

Figure 4 sets out current knowledge of the different steps involved in the colonization of surfaces by bacteria. Initially transport of free-living bacteria to the substratum is required before colonization. Three different modes of transport, which are not mutually exclusive, are generally recognized: (1) diffusive transport related to the Brownian motion of the bacteria that can be observed under a microscope, (2) liquid flow by convection, which is a faster mode of transport of cells, and is undoubtedly an important factor enhancing bacterial adhesion in natural waters, and (3) active transport, whereby bacteria are motile by their flagella and encounter surfaces; this active transport may also involve chemotaxis. The next step is the initial attachment of the cells to the substratum (**Figure 4**). Physicochemical processes are involved in the initial adhesion and are related to the surface properties of either the substratum or the cells. This short period has been the subject of different theoretical attachment models, based mainly on colloid and surface chemistry, designed to gain insight into the control mechanisms of bacterial adhesion to surfaces. During the attachment process, bacteria move on the substratum by twitching motility and can detach and be free-living plankton, or synthesize EPSs to form a cell monolayer and develop in a biofilm. At this point, the attachment becomes irreversible. Genetic studies of the adhesion of *Pseudomonas aeruginosa* have revealed that cells change phenotype when they become sessile and that this change is achieved through gene expression.

The development of the biofilm involves the growth and multiplication of attached cells within bacterial colonies and also the colonization of new bacterial species. Structural features on the cell surface (fimbriae, flagella) are involved in the firm attachment of bacteria, while EPSs, which consist of polysaccharides, extracellular DNA, and proteins, play an important role in the coaggregation of cells. During its maturation, the biofilm is greatly influenced by abiotic factors, which partly determine its structure. A mature biofilm consists of a complex

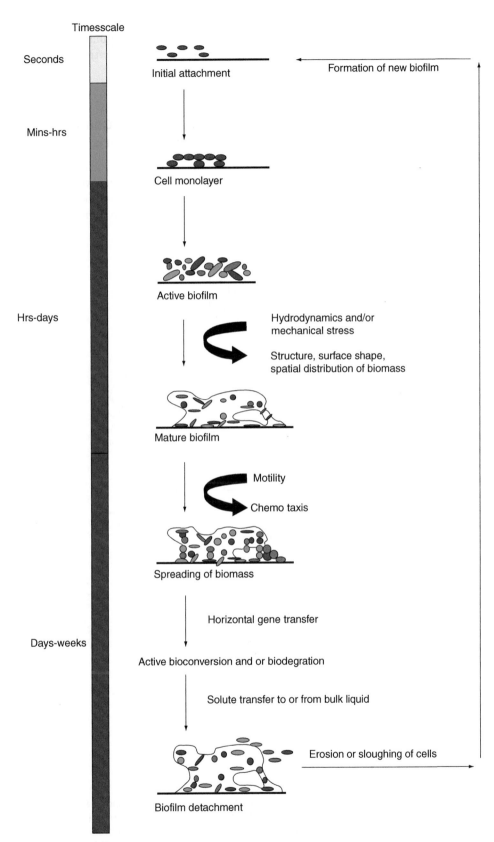

Figure 4 Schematic representation of steps involved in the formation of a biofilm and main factors influencing its evolution. Reproduced from **Figure 1** in Singh R, Paul D, and Jain RK (2006) Biofilms: Implications in bioremediation. *Trends in Microbiology* 14: 389–397. Copyright (2006), with permission from Elsevier.

three-dimensional organization with multispecies microcolonies embedded in a matrix with interconnecting channels and water flows. Bacteria can move, allowing chemotaxis processes considered to be an important feature of bacterial biodegradation within the biofilm. In addition, the proximity of bacterial cells enhances the possibility of gene transfer and of coordinated behavior due to signaling molecules (quorum sensing). All of these biological factors are important in the maturation and maintenance of biofilms. Finally, other microorganisms (algae, fungi, protozoa) colonize the surface to form a mixed mature biofilm as found in many freshwater ecosystems, in which complex interactions take place (**Figure 5**). Some of the bacteria regularly detach from the consortia allowing the colonization of new niches to create another biofilm.

Composition and Community Structure of Attached Bacteria

Early investigations of the composition and diversity of bacterial communities were based on microscopic observations and the isolation and cultivation of cells. Detailed studies of the morphology, physiology, and

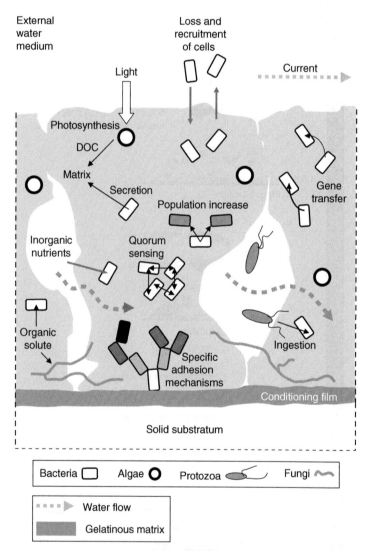

Figure 5 Biological interactions within a mixed biofilm: a small-scale microbial ecosystem found as a surface layer on rocks and stones in rivers and the littoral zone of lakes, the thickness of the biofilm ranges from a few micrometers (µm) to several millimetres (mm). Different microorganisms within the biofilm (not draw to scale) are indicated by the symbols. In a mature biofilm, the gelatinous matrix is typically highly structured, occurring as columns with interstitial spaces through which water can percolate. Water flow also occurs across the surface of the biofilm. Solid arrows simply relate organisms to particular activities. Reproduced from figure 1.5 in Sigee DC (2005) *Freshwater Microbiology – Biodiversity and Dynamic Interactions of Microorganisms in the Aquatic Environments.* Chichester: Wiley. Copyright (2005), with permission from Wiley.

Table 1 Some commonly occurring prokaryotes which are known to attach to surfaces in aquatic environments

Method of analysis	Examples and modifications	Use(s)	Refs
Direct observation	Epifluorecent microscopy, Confocal Scanning Laser Microscopy	Morphological observation, enumeration	1
Molecular methods	*In situ* hybridization, comparative sequence analysis	Community analysis, taxonomy	1, 3
Fluorescent labeling	Fluorescence in situ hybridization in combination with microautoradiography and microsensors	Characterization of biofilms communities	2, 4
Detection of gene expression	Reporter protein assay, in vivo expression technology (IVET), recombination based IVET (RIVET)	Investigation of gene activity	2, 5
PCR	Direct *in situ* PCR	Characterization of genetic and phylogenetic properties	6

Reproduced from Table 1 in Singh R, Paul D, and Jain RK (2006) Biofilms: Implications in bioremediation. *Trends in Microbiology* 14: 389–397. Copyright (2006), with permission from Elsevier.

Sources

1. Wimpenny J, Manz W, and Szewzyk U (2000) Heterogeneity in biofilms. *FEMS Microbiology Reviews* 24: 661–667.
2. Aoi Y (2002) *In situ* identification of microorganisms in biofilm communities. *Journal of Bioscience Bioengineering* 94: 552–556.
3. Woese C and Fox GE (1977) Phylogenic structure of the prokaryotic domain: The primary kingdoms. *Proceedings of the National Academy of Sciences of the United States of America* 74: 5088–5090.
4. Ito T, Nielsen JL, Okabe S, Watanabe Y, and Nielsen PH (2002) Phylogenic identification and substrate uptake patterns of sulphate reducing bacteria inhabiting an oxicanoxic sewer biofilm determined by combining microautoradiography and fluorescent in situ hybridization. *Applied and Environmental Microbiology* 68: 356–364.
5. Chalfie M, Tu Y, Euskirchen G, Ward WW, and Prasher DC (1994) Green fluorescent protein as marker for gene expression. *Science* 263: 802–805.
6. Tani K, Kurokawa K, and Nasu M (1998) Development of a direct *in situ* PCR method for detection of specific bacteria in natural environments. *Applied and Environmental Microbiology* 61: 4074–4082.

genetics of the isolated microorganisms have been carried out in the laboratory from pure cultures. These bacteria often belong to the most commonly occurring prokaryotes in the field, and some of them are known to attach to surfaces or to form microbial mats (**Table 1**). A large number of bacterial species from microbial mats have been characterized, particularly those of hot springs. These species include different anoxygenic phototrophs (purple sulfur bacteria, purple nonsulfur bacteria, green sulfur bacteria, green nonsulfur bacteria), some chemoautotrophs (colorless sulfur bacteria), sulfur and sulfate-reducing bacteria, and a variety of thermophilic and halophilic prokaryotes. Some widespread chemoorganotrophic species with specialized appendages or stalks used for attachment have also been examined in detail by standard methods.

However, most of the microorganisms have not been cultivated. Less than 1% of bacterial cells from natural environments are amenable to standard culture techniques. To open this 'black box,' new methods to access the composition and structure of aquatic bacterial communities have been developed, mostly using molecular tools. Molecular techniques (**Table 2**) such as terminal restriction fragment length polymorphism (T-RFLP) and denaturing gradient gel electrophoresis (DGGE) of PCR amplified 16S rDNA have been applied for prokaryotic diversity and community structure analysis. In addition, fluorescent in situ hybridization (FISH), using 16S rRNA-targeted oligonucleotide probes combined with epifluorescence

microscopy has become a useful tool for analyzing bacterial composition and counting bacterial groups in aquatic environments. These molecular approaches were first applied to whole bacterial communities in aquatic habitats and have revealed an enormous diversity of microbial species in the natural environment. These studies also led to the discovery of novel uncultivated prokaryotes.

More recently, more and more studies have focused on the differences between attached and free-living bacteria and on the community structure of biofilm-associated bacteria in aquatic environments. Specific probes have been applied to the study of the microbial colonization of biofilms and micro- and macroaggregates in both marine and freshwater ecosystems. These studies are few in number, but have provided important information on the diversity and community structure of bacteria attached to surfaces.

The majority of the studies using molecular tools for attached bacteria analysis in freshwater lakes and rivers indicate high abundances of the beta-subgroup of the Proteobacteria on suspended particles and in biofilms. *In situ* examination of the bacterial community on river and lake snow aggregates as well as in river biofilms indicates that beta-proteobacteria often constitute 20–50% of total bacterial abundance. Alpha-proteobacteria and Bacteroidetes (former Cytophaga/Flavobacteria) represent a large proportion of the total bacteria on suspended particles, but generally do not exceed the abundance of beta-proteobacteria. Because members of the Bacteroidetes

Table 2 Techniques commonly used for analysis of attached bacterial communities

Genus	Class	Main characteristics	Habitat/microhabitat
Caulobacter	Alphaproteobacteria	Gram-, rod-shaped, flagella, stalk, aerobic, chimioorganotroph	Widespread in freshwater habitats
Hyphomicrobium	Alphaproteobacteria	Gram-, rod-shaped, hyphae formation, budding reproduction, aerobic, chimioorganotroph	Widespread in freshwater and marine habitats
Sphaerotilus	Betaproteobacteria	Gram-, filamentous, microaerophilic, chimioorganotroph	Flowing freshwater, sewage, wastewater treatment plants
Pseudomonas	Gammaproteobacteria	Gram-, rod-shaped, flagella, aerobic, chimioorganotroph	Widespread: freshwater and saline habitats, soil, plant and animal tissue
Beggiatoa	Gammaproteobacteria	Gram-, filamentous, microaerophilic, chimiolithotroph, gliding motility on surfaces	Microbial mats in sediments of freshwater and saline lakes, sulphur-springs, marine habitats
Thiocapsa	Gammaproteobacteria	Gram-, spherical shape, anaerobic, photolithotrophic, purple sulphur bacteria	Microbial mats in sediments of freshwater and saline lakes, sulphur-springs, marine habitats
Chlorobium	Bacteroidetes	Gram-, chains of spherical cells, anaerobic, photolithotrophic, green sulfur bacteria	Microbial mats in sulphide-rich habitats (hot springs, mud, sediments)
Chloroflexus	Chloroflexi	Gram-, filamentous, anaerobic, photoorganotroph, green nonsulfur bacteria	Microbial mats in sediment of freshwater and saline lakes, hot springs, marine habitats
Halobacterium	Halobacteria	Gram-, Rod-shaped, aerobic, chimioorganotroph, can use photosynthesis to survive in anaerobic conditions	Microbial mats in hypersaline environments
Halococcus	Halobacteria	Gram-, cocci, aerobic, chimioorganotroph,	Microbial mats in hypersaline inland waters

are specialized in the degradation of complex macromolecules, it is generally assumed that a high proportion of this group is related to the presence of refractory materials associated with suspended aggregates or biofilms. The community structure of free-living bacteria in lakes and rivers differs from that of bacteria on suspended particles and on sediments and usually shows a lower proportion of beta-proteobacteria. In contrast to fresh water, no clear dominance of any single bacterial group on particles has been reported in marine systems. Temporal changes in the bacterial composition of freshwater aggregates have also been observed, suggesting a succession of different functional groups that may operate sequentially in the degradation of organic matter. However, the factors influencing the spatiotemporal changes in attached-bacterial community composition remain largely unknown. The use of group-specific probes gives low phylogenetic resolution and provides only a rough description of the bacterial composition. Investigations using a large number of probes at the genus and species-specific levels in combination with other methods (**Table 2**), such as the use of microsensors, confocal scanning laser microscopy (CSLM), and microautoradiography, will make it possible to characterize in detail the composition, physiological activities,

and spatial organization of biofilm and aggregate-associated bacteria. The distribution of uncultured nitrifying bacteria of the genera *Nitrosospira* and *Nitrospira* in biofilms was recently studied under laboratory conditions using FISH in combination with CSLM and microsensor measurements. This approach was also used in freshwater sediments, suggesting that these genera occur frequently in the environment, but that they inhabit microniches different from those occupied by the other well-known genera of nitrifying bacteria; *Nitrosomonas* and *Nitrobacter*.

Control of Attached Bacteria

Attached bacteria in different microhabitats (e.g., biofilms and suspended particles) are assumed to be governed by the same controlling factors as free-living bacteria. The major factors controlling abundance, biomass, and growth of attached bacteria include the availability of organic substrates and nutrients, grazing pressure mainly from protozoa, and lysis from viral communities. However, the relative importance of these factors is not well known for attached bacteria compared with free-living bacteria. In addition, physicochemical factors such as the light environment (UV radiation) and oxygen concentrations can affect bacterial biomass in aquatic systems, but their impact on attached cells remains largely unknown.

Most studies of grazer-down control of attached bacterial communities have focused on predation by protozoa. Some amoeboid, flagellated, and ciliated protozoa are able to graze bacteria on surfaces and it has been shown that they can reduce bacterial density substantially and affect biofilm architecture. Based on studies conducted in marine systems, it is likely that a large fraction of bacteria attached to pelagic aggregates are grazed by protozoa in inland waters. Accordingly, these microbial grazers are potentially important players in the recycling of nutrients and the maintenance of community growth in attached bacterial microhabitats. Their presence can enhance the degradation of organic matter, although the relative contribution of bacteria–grazer interactions in attached communities (compared with free-living ones) remains largely unknown. Some studies have indicated that the aggregation of free-living bacteria and the formation of bacterial colonies increase when protozoan bacterivory is intense. This is considered as one of the known defense mechanisms of aquatic bacteria against predation. Predators might thus be viewed as stimulators of the early formation of aggregates and biofilms, although this hypothesis needs to be specifically addressed and examined in the context of diverse populations of bacteria and protozoa coexisting in aquatic systems.

The impact of viral lysis on bacterial mortality was recently studied in lakes, but differences between attached and free-living bacteria were not investigated. The relative contributions of viral lysis and protozoan grazing to attached-bacterial mortality are still unknown. In freshwater benthic habitats (sediments and biofilms), a few studies have shown that viral infectivity is less pronounced than in the upper water column, suggesting that bacterial cells in colonies are less exposed not only to grazing but also to viral attacks. However, this inference is still essentially speculation and needs to be confirmed by further work.

The positive effect of nutrients on the development of attached bacteria is confirmed by the general increase in the abundance of these populations with the increasing trophic status of aquatic systems. The biomass and activities of bacteria attached to biofilms or aggregates increase with primary production and dissolved organic carbon (DOC). The concentration of DOC in the water column is often considered a limiting factor for bacterial growth in biofilms. However, in a given ecosystem, the productivity of attached bacteria compared with that of free-living cells and their spatial and temporal distributions are largely unknown, although it is widely accepted that in shallow lakes and in running waters, most bacterial productivity is associated with surfaces. The importance of resources, predation, and viral lysis in the regulation of attached bacteria in inland waters thus remain to be evaluated, in the context of within and between-system variability.

Activities of Attached Bacteria and their Potential Contributions to the Degradation of Organic Matter

Bacterial communities of inland waters play key roles in the degradation of organic matter and the cycling of important elements such as carbon, nitrogen, phosphorus, and sulfur. Because of the ubiquity of bacteria in natural ecosystems, it is likely that attached bacteria are involved in the overall bacterial processes. Although the role and metabolic diversity of bacteria associated with microbial mats in hypersaline aquatic environments is reasonably well documented, our knowledge of the functional significance of bacteria associated with biofilms and suspended particles in fresh water remains incomplete. We will now focus on the role of attached bacteria in the degradation of organic matter and the related biogeochemical cycles in fresh waters, and refer readers to the list of further readings for information on the diversity and functional role of attached bacteria in microbial mats, hypersaline lakes, and other extreme environments, such as hot springs.

Bacteria on Freshwater Organic Aggregates

One of the most important roles of bacteria in freshwater environments arises from their ability to respire and recycle organic matter into nutrients and CO_2. In addition, bacteria utilize both dissolved and particulate organic matter (DOM and POM) from external and internal sources. Surfaces are often considered as a major site for the respiration and recycling and hence aerobic heterotrophic processes on freshwater aggregates and biofilms are important. These processes include POM solubilization, DOM hydrolysis and uptake, respiration and biomass production within the organic matrix, and both substrate uptake from and release into the surrounding waters. Attached bacteria are often metabolically more active than free-living bacteria and exhibit higher cell-specific uptake rates of free amino acids and monosaccharides, and higher ectoenzymatic hydrolysis rates of aminopeptidase, phosphatase, and glucosidase. Proteins are generally decomposed more rapidly than polysaccharides by attached bacteria. This differential use has important implications for the decomposition of the organic matter in microhabitats (e.g., macroaggregates, biofilms) dominated by diatoms, because dissolution of diatom silica frustules takes place only when they are colonized by bacteria, which are probably able to hydrolyze the cell-wall proteins inside the frustules.

In freshwater diatom-derived aggregates, bacterial respiration is often positively correlated with POM content, and increases with the age of the aggregates. This contrasts with bacterial growth efficiency, which decreases from fresh to aged aggregates. The turnover of the aggregate-associated POM based on respiration losses is substantially higher than if calculated only on the basis of bacterial biomass production. Particle-associated bacterial biomass production on aggregates is usually <30% of total bacterial production in fresh waters. However, total bacterial production is much higher in samples with suspended aggregates than in samples with no aggregates, and attached bacterial production generally increases with aggregate size. In addition, the rates of POM solubilization and net release of labile substrates into the surrounding waters generally exceed the carbon demand of aggregate-associated bacteria. As a result of high hydrolytic activities of attached bacteria, and high metabolic activities of bacterial grazers, concentrations of organic matter (dissolved free and combined amino acids, DOC, etc.) and of inorganic nutrients (e.g., C, N, P, Fe, and Si) are significantly higher in the pore water of the aggregates than in the surrounding waters. Besides recycling within aggregates, the highly enriched labile DOM and nutrients may also be partly released into the surrounding waters. Hence, the activity of attached bacteria in pelagic systems can have two major consequences: (1) stimulation of the growth rates of free-living microorganisms in the vicinity of aggregates, resulting in the occurrence of hot spot microenvironments with high productive potentials and (2) regulation of the degradation rates of aggregates and of the vertical sinking flux of organic matter from the water column to the sediments (**Figure 6**).

Bacteria in Freshwater Biofilms

Few studies have focused on the measurements of the activities of bacteria associated with biofilms. Most of our knowledge on this topic comes from studies in streams where contributions by biofilm bacteria, due to their high abundance and diverse metabolic capabilities, dominate the assimilation and flux of DOM, the dominant form of organic carbon in aquatic systems. Bacterial activity in biofilms is dependent on the source, quality, type, and quantity of DOM and nutrients. Different types of DOM and inorganic nutrients induce various responses in biofilm bacterial communities, but this is obvious only when different taxa are distinguished and seasonal changes are taken into account. Attached bacterial activity is more directly

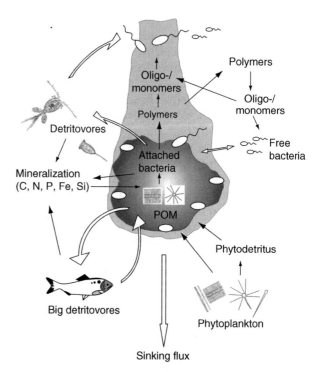

Figure 6 Loss processes and microbial decomposition pathways of macroscopic-organic aggregates. POM: Particulate organic matter. Reproduced from **Figure 5** in Simon M, Grossart H-P, Schweitzer B, and Ploug H (2002) Microbial ecology of organic aggregates in aquatic ecosystems. *Aquatic Microbial Ecology* 28: 175–211. Copyright (2002), with permission from Inter-Research.

dependent on DOM, primarily the low molecular weight (LMW) organic substrates, than to inorganic nutrients, known to influence epilithic bacteria indirectly via the algal compartment. Bacterial vs. algal competition for inorganic nutrients seems to be less marked in biofilms than in plankton. Algal-released DOM, which comprises largely labile, easily utilizable LMW compounds, is thus considered as the main source of carbon for heterotrophic bacteria in biofilms. In the absence of significant amounts of toxins, bacterial productivity, biomass, biovolume, and enzyme production are enhanced by algal exudates. The mutual reliance of bacteria and algae in biofilms is further facilitated by their close spatial proximity. However, in canopy-covered streams, algal-derived resources are often insufficient for the metabolic needs of biofilm bacteria, partly because of reduced light penetration. There is also evidence of photoinhibition of algae in biofilm communities in open streams. In addition to algal exudates, allochthonous DOM such as leachate from litter fall is an important source of organic material in many streams, but not all bacteria species in biofilms can utilize leaf leachate as a carbon source. Phenolic compounds contained within leachates can actually inhibit the growth of some bacteria.

Conclusion and Future Research

Bacterial studies in inland waters have focused mainly on the pelagic compartment, and most often only free-living bacteria are considered. It is only relatively recently that bacteria attached to surfaces have been taken into account, even though the attached lifestyle is considered common in aquatic microbial ecology. However, the mechanisms that control the adhesion and detachment of natural aquatic bacteria and their ecological significance need to be further investigated, including the relative importance of the physical, chemical, and biological factors involved. Compared with the free-living mode, our knowledge of the diversity and the functional roles of attached bacteria in inland waters is scant, partly because of the broad diversity of colonizable habitats for these microorganisms. More sampling efforts are needed, combined with the use of modern molecular and analytic tools, to gain insight into species-specific community composition, seasonal and spatial organization, and the related metabolisms of the microbial associations. Top-down and bottom-up control of attached populations and interactions (competition) within these populations and with other biological compartments are still imperfectly understood. In this regard, the relative importance of grazing and viral lysis should be considered in future studies. Attached bacteria need to be considered as a biological compartment in ecosystem studies, because their potential role in processing organic matter and cycling carbon is probably important, possibly more important than free-living bacteria. This need is especially important for lotic ecosystems and shallow lakes where attached bacteria largely dominate the total bacterial community in terms of both biomass and productivity. In addition, the importance of attached bacteria in degrading pollutants must be addressed by in situ investigations, because most of the related findings derive from experiments conducted in controlled laboratory conditions. Overall, there is shortage of in situ studies on the role of attached bacteria in cycling or sequestering major elements (C, N, P) in inland waters and sediments, and also in the degradation of pollutants. This knowledge is critical for better management of inland water ecosystems, the major resource for life on the earth.

Glossary

Alpha-proteobacteria – A class of proteobacteria, a major group of bacteria.

Autoroph – Organism that produces complex organic compounds from simple inorganic molecules and an external source of energy, such as light (photoautotroph) or chemical reactions of inorganic compounds (chemoautotroph).

Bacteroidetes – A phylum of bacteria that are widely distributed in the environment.

Beta-proteobacteria – A class of proteobacteria, a major group of bacteria.

Bottom-up control – A bottom-up control of a population is established when an increase of food supply increases the abundance of this population.

Chemoorganotroph – Organism that obtains carbon from organic compounds and energy from the oxidation of inorganic compounds.

Chemotaxis – Phenomenon in which organisms direct their movements according to chemical substances in their environment.

Consortium (Plural: consortia) – A microbial consortium is a group of different species of microorganisms with different metabolic activities. The organisms interacts each other and all benefit from the activities of others.

Cytophaga – A genus of rod-shape, gram-negative bacteria.

Epibiotic – Able to grow or to attach on the surface of a living organism.

Epilithic – Able to grow or to attach on the surface of stones and rocks.

Epipelic – Able to grow or to attach on the surface of mud, sediment.

Epiphytic – Able to grow or to attach on the surface of plants or algae.

Fimbriae – Thin projection of the bacterial cell that is used to adhere to surfaces.

Flagella – Projection from the cell body forming a filament surrounded by the plasma membrane and usually used for locomotion.

Flavobacteria – Bacteria of the genus Flavobacterium. They are found in soil and freshwater environments.

Hot-spot – In aquatic microbial ecology, a hot-spot is a microenvironment of high microbial abundance, diversity and activities, usually formed from suspended detritus.

Mucilage – Various gelatinous substances (generally polysaccharides) secreted by plants and microorganisms (especially algae) into their surrounding environments.

Neuston – Organisms living at the air–water interface.

Phycosphere – Zone surrounding an algal cell within which others microorganisms are influenced by algal activity.

Quorum sensing – Ability of bacteria to communicate and coordinate their behavior via signaling molecule.

Raptorial – Organism adapted for seizing prey.

Top-down control – A top-down control of a population is demonstrated by an increase in the abundance of this population when the pressure of its predators is reduced.

See also: Bacteria, Bacterioplankton; Bacteria, Distribution and Community Structure; Cyanobacteria; Microbial Food Webs; Protists; Viruses.

Further Reading

An YH and Friedman R (2000) *Handbook of Bacterial Adhesion – Principles, Methods, and Applications.* Totowa: Humana Press Inc.

Berkeley RCW, Lynch JM, Melling J, Rutter PR, and Vincent B (1980) *Microbial Adhesion to Surfaces.* London: Ellis Horwood limited.

Cohen Y and Rosenberg E (1989) *Microbial Mats: Physiological Ecology of Benthic Microbial Communities.* ASM editions: Washington, DC.

Fletcher M (1996) *Bacterial Adhesion – Molecular and Ecological Diversity.* New York: Wiley-Liss, Inc.

Jefferson KK (2004) What drives bacteria to produce a biofilm? *FEMS Microbiology Letters* 236: 161–173.

Krumbein WE, Paterson DM, and Zavarzin GA (2003) *Fossil and Recent Biofilms: A Natural History of Life on Earth.* Kluwer Academic Publishers: 481p.

Lawrence JR and Neu TR (2003) Microscale analyses of the formation and nature of microbial biofilm communities in river systems. *Reviews in Environmental Science and Bio/Technology* 2: 85–97.

Marshall K (2006) Planktonic versus sessile life of prokaryotes. In: Dworkin M, Falkow S, Rosenberg E, *et al.* (eds.) *The Prokaryotes – A Handbook on the Biology of Bacteria. Volume 2: Ecophysiology and Biochemistry,* 3rd edn., pp. 3–15. New York: Springer.

Ofek I, Sharon N, and Abraham SN (2006) Bacterial adhesion. In: Dworkin M, Falkow S, Rosenberg E, *et al.* (eds.) *The Prokaryotes – A Handbook on the Biology of Bacteria. Volume 2: Ecophysiology and Biochemistry,* 3rd edn., pp. 3–15. New York: Springer.

Rajbir S, Debarati P, and Rakesh KJ (2006) Biofilms: Implications in bioremediation. *Trends in Microbiology* 14: 389–397.

Sigee DC (2005) *Freshwater Microbiology – Biodiversity and Dynamic Interactions of Microorganisms in the Aquatic Environments.* Chichester: Wiley.

Simon M, Grossart HP, Schweitzer B, and Ploug H (2002) Microbial ecology of organic aggregates in aquatic ecosystems. *Aquatic Microbial Ecology* 28: 175–211.

Van Loosdrecht MCM, Lyklema J, Norde W, and Zehnder AJB (1990) Influence of interfaces on microbial activity. *Microbiological Reviews* 54: 75–87.

Ward DM, Ferris MJ, Nold SC, and Bateson M (1998) A natural view of microbial biodiversity within hot spring cyanobacterial mat communities. *Microbiology and Molecular Biology Reviews* 62: 1353–1370.

Wimpenny J, Manz W, and Szewzik U (2000) Heterogeneity in biofilms. *FEMS Microbiology Reviews* 24: 661–671.

Relevant Websites

http://www.erc.montana.edu/.
http://www.biofilmsonline.com.
http://www.microbeworld.org/.

Bacteria, Bacterioplankton

R D Robarts, UNEP GEMS/Water Programme, Saskatoon, SK, Canada
G M Carr, UNEP GEMS/Water Programme, Gatineau, QC, Canada

Introduction

To the old adage of 'out of sight, out of mind' could be added 'and unimportant.' With respect to bacterio-plankton in inland waters this was true for many scientists interested in aquatic ecology, but not all, and is certainly not true today. Bacteria have been known to inhabit natural waters since the seventeenth century although their numbers, biomass, distribution, and roles in these systems were poorly understood until the latter part of the twentieth century. Even in classical textbooks of the 1930s, such as P.S. Welch's 'Limnology,' a whole chapter (albeit short) was devoted to bacteria and other microorganisms. Welch clearly had an intuition about the potential importance of bacteria in lakes and lamented the fact that very few lakes have received any study of their native bacteria. Indeed, E.P. Odum in his book 'Fundamentals of Ecology' in 1963 noted that, "Because of the technical difficulties of study, microbial ecology is, unfortunately, often completely omitted from the general college course in ecology."

The situation with respect to our knowledge and understanding of bacterioplankton in inland waters changed drastically beginning in the 1960s with publications describing and applying radioactively labeled compounds to measure the turnover and uptake of specific dissolved organic compounds in lakes. Thereafter, new protocols were developed to measure in situ the rates of bacterioplankton production in terms of cell numbers, biomass, and carbon. In addition, the introduction of the use of fluorescent dyes in microscopy led to the routine measurement of bacterial numbers and biomass and later to quantitative measurements of different aspects of microbial metabolism such as respiration and enzymatic activities. The methodology to study bacterioplankton has continued to evolve and thus there has been a revolutionary advance in our knowledge and understanding of the distribution and size of bacterioplankton populations in lakes, reservoirs, wetlands, rivers, and ground waters and of their role in biogeochemical and energy cycles in these systems over the past three decades. In addition, since microbes are the first link joining the biotic and abiotic components of aquatic systems, they are excellent and sensitive indicators of human impacts on these systems.

Bacterial Numbers and Biomass

Early methods to estimate the size of bacterioplankton populations were based on plate culturing techniques. But it had been accepted for a long time that these methods greatly underestimated the size of the populations. A number of direct-count methods (counting of stained bacteria with a microscope) were devised in conjunction with the use of centrifuges, chemical flocculation, or filtration in order to concentrate bacterial cells. Limnologists in the 1920s believed that direct counting of bacteria could not be undertaken without concentrating cells because lake populations were too low for accurate enumeration. Such concentrating mechanisms also created problems so there was some resistance to the use of direct counts. In the 1930s a concentration method was developed that involved evaporating water samples under reduced pressure. It was claimed that this method produced bacterial counts 2–4000 times higher than those from culture plates. For example, researchers reported bacterial numbers of $(1–6) \times 10^6 \, ml^{-1}$ in the Russian Lake Glubokoye. As can be imagined, such numbers for an unpolluted lake were viewed as being overly high and would produce a visible turbidity (which it did not), especially when others were reporting numbers of only 740–32 600 cells ml^{-1} in lakes using a direct-count method at around the same time. However, a survey of bacterial numbers in Wisconsin lakes found that they ranged widely from 19×10^3 to $2 \times 10^6 \, ml^{-1}$ using this new method.

Today with the widespread use of epifluorescence microscopy, and more recently with the use of flow cytometers, we know that bacteria occur in many unpolluted inland waters at concentrations of millions per milliliter. Bacteria in most natural aquatic systems range from 10^4 to 10^8 cells ml^{-1} of water, although cell concentrations as high as $10^9 \, ml^{-1}$ have been reported (**Figure 1**). In a 1984 cross-system overview of bacteria in surface waters, cell concentrations of $(0.1–13.4) \times 10^6 \, ml^{-1}$ were reported for a range of lakes. Even a very remote and pristine Arctic lake, Lake Taymyr on the Taymyr Peninsula in Russia, has bacterial numbers that varied between 1.3×10^6 and 4.8×10^6 cells ml^{-1}. Moreover, there have been several studies that found that

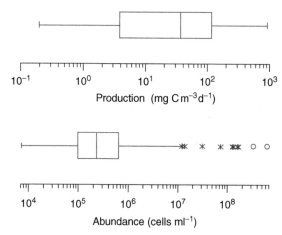

Figure 1   Distribution of bacterioplankton production (top panel) and abundance (bottom panel) measurements from inland waters, as reported in the primary literature and summarized in Carr GM and Morin A (2002) Sampling variability and the design of bacterial abundance and production studies in aquatic environments. *Canadian Journal of Fisheries and Aquatic Sciences* 59: 930–937. Box plots show median (mid-line) and 25th and 75th percentiles. Whiskers extend to data points that fall within 1.5 times the midrange. Asterisks denote data points that extend beyond 1.5 times the midrange, and open circles are data points that extend beyond 3 times the midrange.

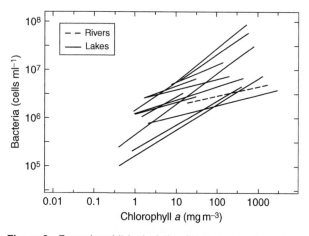

Figure 2   Range in published relationships between chlorophyll *a* and bacterioplankton abundance. Adapted from Gasol, JM and Duarte CM (2000) Comparative analysis in aquatic microbial ecology: how far do they go? *FEMS Microbiology Ecology* 31: 99–106, with permission.

bacterial abundance in the anoxic hypolimnia of stratified lakes can be as much as twice that of the oxic strata.

Several studies have tried to identify limnological parameters that regulate the variations in bacterial numbers between different aquatic systems (**Figure 2**) and seasonally in a specific system. In one such study, bacterial numbers were significantly correlated

(log-transformed data) with chlorophyll *a* concentrations for lakes in North America and Europe. This was not surprising as it had been assumed for some time that phytoplankton would be a major provider of substrates for bacterial production and growth, either at the time of their senescence and death or by the extracellular release of dissolved organic compounds during photosynthesis. However, one of the disconcerting conclusions was that the increase in the number of bacteria did not increase with the increase in chlorophyll *a* concentrations as lake trophy changed, since the slopes of the regressions were less than 1. As a general rule of thumb, oligotrophic waters have bacterial concentrations $<1.7 \times 10^6$ cells ml^{-1}, mesotrophic waters $(1.7–6.5) \times 10^6$ cells ml^{-1}, and eutrophic waters $>6.5 \times 10^6$ cells ml^{-1}.

Results of within-system correlation analyses of bacterial numbers have been variable. In some studies, no correlations have been found whereas in others significant correlations between bacterial numbers and water temperature, primary production, and chlorophyll *a* concentration have been reported. This should not be surprising as total counts of bacterial cell numbers with epifluorescence microscopy techniques do not discriminate between different physiological groups, between live and dead cells, or between metabolically active and nonactive cells (**Figure 3**). Another factor that influences bacterioplankton dynamics in aquatic systems is losses due to processes such as grazing (protistan and invertebrate), parasitism, and sedimentation. However, even today few lakes have been sufficiently studied to provide firm conclusions on the relative importance of environmental factors on population sizes.

Metabolically Active Bacterioplankton

Generally, the number of metabolically active bacteria is more variable between plankton communities than the total number of bacteria (**Figure 4**). As with total bacterial concentrations, the number of active cells in bacterioplankton populations has been correlated to indicators of lake trophy such as total phosphorus, chlorophyll, and dissolved organic carbon (DOC). However, the numbers of these two groups do not increase in the same way over enrichment gradients, resulting in a proportionately higher number of active cells in eutrophic versus oligotrophic waters. The mean of published numbers of active bacteria in lakes is about 22% with a rare study reporting as much as 100% active cells. For aquatic systems, generally, the numbers range from <1% in oligotrophic waters to >90% for highly productive systems. Differences in methodology may account for

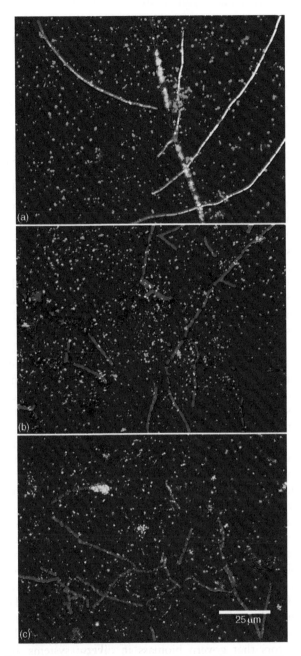

Figure 3 Photomicrographs of bacterioplankton cells from Wascana Creek, Saskatchewan, filtered onto 0.2-μm pore-size cellulose nitrate filters: (a) stained with Syto 9 for total cell counts (Molecular Probes Inc., Eugene, OR, USA; cells appear green), (b) the same preparation stained with the commercial Live Dead stain for determination of ratio of live to dead cells, live cells appear green while putative dead cells red, and (c) filtered cell preparation stained with 5 mmol l^{-1} cyanoditolyltetrazolium chloride (Polysciences Inc, Warrington, PA, USA) incubated for 60 min and counterstained with Syto 9; yellow cells are active while green cells are not metabolically active as determined by this assay. Magenta cells and filaments are cyanobacteria imaged using autofluorescence of their photosynthetic pigments. Imaging was carried out using the 522 + 16, 598 + 16 excitation lines of an MRC 1024 laser microscope (Zeiss, Jena, Germany). (Photos: J.R. Lawrence and G.W. Swerhone, NWRI, Environment Canada, Saskatoon, unpublished.)

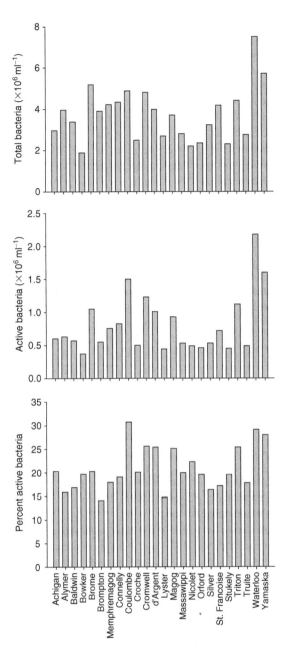

Figure 4 Variation in the total number of bacterial cells, metabolically active cells and the percentage of metabolically active bacteria in 24 lakes of southern Québec, Canada. The average number of active cells in the lakes was 20.9%. Data are from del Giorgio P A and Scarborough G (1995) Increase in the proportion of metabolically active bacteria along gradients of enrichment in freshwater and marine plankton: implications for estimates of bacterial growth and production rates. *Journal of Plankton Research* 17: 1905–1924.

some of this variance, but research, first in marine systems and later in fresh waters, indicated that bacterial grazers may selectively remove larger, actively growing cells from a population leaving it dominated by small, slow-growing cells. Studies of the long-term seasonal variations in metabolically active bacteria in a particular system have shown that their contribution to the total bacterioplankton population can also be large, but such studies are rare. Hopefully, with the recent use of flow cytometry to determine metabolically active bacteria, more data will become available that will provide new insights on metabolically active bacteria in inland waters and the factors that influence them. Such data have significant consequences for the measurement of bacterioplankton growth and production measurements, which will be outlined here.

Bacterial Biomass

Bacterial Biovolumes

Spatial and temporal changes in bacterial cell volume have been reported from a wide range of aquatic habitats, although the number of studies has been limited because of the difficulty in obtaining such data. To model energy and material flows, it is essential to know the factors that influence bacterial cell volume, as conversion of bacterial cell numbers to carbon, nitrogen, or phosphorus units is dependent on the derivation of accurate 'numbers to volume' conversion factors. Therefore, a major challenge for aquatic microbial ecologists is to identify the factors that control cell volume and hence bacterial biomass. Factors that have been identified include carbon supply from algae, phosphorus concentration, predation, and water temperature.

To derive a value for bacterioplankton biomass, data are needed on the number of cells and their biovolume. This is not as straightforward as it might seem. The methods used to obtain biovolume include epifluorescence microscopy, scanning electron microscopy (SEM), scanning confocal laser microscropy (in conjunction with fluorescent dyes; **Figure 3**) and more recently flow cytometry, also using fluorescent dyes to estimate cell dimensions. These methods have problems ranging from halo effects caused by fluorescence to cell distortions caused by sample preparation for SEM, all of which can alter the calculation of biovolumes, especially with the very small cells typical of bacterioplankton. Median bacterial cell volumes usually range between 0.013 and 0.200 μm^3. Although not confirmed yet, several studies that have looked at bacterial cell size have reported that this decreases with increasing trophic status. Furthermore, there have been some studies comparing bacterial cells between the upper oxic and lower anoxic parts of lakes. These have shown that bacteria from the anoxic hypolimnion can be between 2 and 10 times larger than bacteria in the oxic layer. The reasons for this difference and its implications on lake metabolism require further investigation.

Bacterial Biomass

To derive a carbon biomass value for bacterioplankton, most researchers have either used a constant value for a cell, usually between 10 and 20 fg C cell^{-1}, or a volume conversion factor that has up to a fivefold range of a commonly used value of 121 fg C μm^{-3}. Several studies have produced values in the range of 350–720 fg C μm^{-3}, whereas others reported much lower values of between 32 and 160 fg C μm^{-3}. One reason for the higher values may be the underestimation of cell volumes due to shrinkage effects caused by fixatives and air drying, although this is not applicable to all published values. Smaller cells tend to have higher carbon and lower water contents than larger cells. Therefore, what is clear is that the assumption of a constant ratio of carbon:biovolume is not correct. Some studies have shown that the carbon:biovolume ratio of bacterioplankton not only varied with cell size but also had temporal and geographical variations, indicating that factors such as species composition, nutritional state, growth rate, and other factors played a role.

While it is now clear that bacterial biomass contributes a larger portion to the total planktonic biomass in unproductive freshwater systems than in more productive systems, what is not clear is *why*. The most widely accepted explanations include allochthonous carbon inputs being important in oligotrophic systems, decreased bacterivory, and bacterioplankton access to nutrients that are not available to phytoplankton (i.e., very low concentrations and organic forms) or a combination of these. Correct estimates of bacterioplankton biomass distribution are a fundamental requirement in aquatic microbial ecology, and more data that lead to a sound understanding of the factors that govern biomass in diverse systems are needed.

Bacterial Production

The use of ^{14}C-glucose in the early 1960s and subsequent development of methods to determine rates of bacterial uptake, respiration, and turnover of organic compounds at natural substrate concentrations later in the decade provided much data on the decomposition and flow of organic carbon through food webs in a wide range of natural systems. Since bacteria degrade a large number of organic substrates, further developments introduced the use of labeled sugars, amino acids, organic acids, lignocellulose, and other dissolved and particulate substrates. These early studies demonstrated that bacteria were actively metabolizing organic matter but did not provide quantitative estimates of growth rates and production.

While the growth of autotrophic bacteria that fix carbon dioxide for their primary source of carbon can potentially be measured using $^{14}CO_2$, until about 20 years ago there had not been a reliable method to determine the growth of natural assemblages of heterotrophic bacteria that utilize organic substrates for a carbon source. The two most commonly used methods were the incorporation of [^{3}H-*methyl*]thymidine ([^{3}H]TdR) into bacterial DNA, and the incorporation of ^{3}H-leucine into bacterial protein. Labeled leucine is now generally more widely used than labeled thymidine because of its greater sensitivity and the simpler assumptions and calculations required to derive cell and carbon production estimates.

Bacterial production has been measured in a wide range of aquatic systems from the Arctic to Antarctic. Not surprisingly, there is huge variation in the rates reported (**Figure 1**). Some of this variation is due to the factors used to convert the rates of label incorporation into cell and carbon production units, particularly with the more complicated conversion of thymidine incorporation rates. Several studies have concluded that there is no significant difference in production rates determined with labeled leucine and thymidine, whereas others have concluded that there is a difference with thymidine-derived rates being lower. In addition, there are potential problems associated with isotope dilution for both methods, which is usually not measured on the assumption that saturating concentrations of tracer have been added. If saturating concentrations are not used then production will be underestimated.

Volumetric rates of bacterioplankton production have been reported to range from 0.4 to >900 mg C m^{-3} d^{-1} in hypertrophic lakes (**Figure 1**). Cross-system analysis of bacterial production has found significant correlations with phytoplankton production, chlorophyll *a* concentration, bacterial numbers, and total phosphorus concentration. Within-systems correlations have been found with water temperature, and several studies have also demonstrated that bacterioplankton can be limited by nutrients (phosphorus) and/or by the availability of dissolved organic substrates. Bacterial-specific growth rates vary between 0.017 and 8.7 day^{-1}, producing doubling times ranging from hours to weeks. Values less than 0.01 day^{-1} have been reported for cold waters of <6 °C and under winter ice, giving

doubling times of several years. Specific growth rates have been correlated with water temperature, producing Q_{10} values (i.e., the rate of change in growth rate associated with a 10 °C increase in temperature) of 2–4 in various studies. However, specific groups of aquatic bacteria can have much higher Q_{10} values so that even small water temperature changes can promote large increases in some bacterial processes.

To compare bacterial production and other data, e.g., on the uptake of organic substrates, from different systems or different parts of systems, microbial ecologists normalize them by dividing by the total cell abundance. This scaling process is intended to permit a more robust comparison of data generated from waters with markedly different bacterioplankton population sizes. However, this calculation of cell-specific production, uptake, and specific growth rates (to give population doubling times) in most instances does not enhance data analysis but complicates it. This is because, as noted earlier, a varying proportion of a bacterioplankton population is metabolically inactive and may account for a significant amount of the variation in these rates as well as produce overestimates of bacterial turnover times. At least one study has concluded that if production is scaled to the number of active bacteria in a population instead of the total population abundance, then specific production rates are fairly similar between systems and may, in fact, be higher in unproductive systems that have the lowest bacterial densities. As more data become available for a wide range of aquatic systems, the veracity and implications of this tentative conclusion to aquatic system functioning will become clearer.

In some lakes, rivers, and wetlands, daily heterotrophic bacterial production exceeds daily autotrophic primary production and such systems are considered to be net heterotrophic. To calculate the ratio of the daily rates of primary to bacterial production, the assumption is made that bacterial production is constant over a diel cycle, although studies have shown this not to be the case. Since most studies of bacterial production do not include diel studies, hourly production rates are multiplied by 24. It is also necessary to take into account respiration losses as production measured by labeled leucine or thymidine are considered to be net production. This is done using a growth yield factor which, like many other conversion factors in aquatic microbial ecology, has a very wide range of <0.15–0.9. These calculations provide an estimate of the amount of phytoplankton carbon production and therefore an indication of whether a system is net heterotrophic or autotrophic.

The concept of net heterotrophic systems is still a controversial topic amongst aquatic microbial ecologists. However, evidence is mounting that net heterotrophic systems are prevalent in most rivers and in oligotrophic to mesotrophic lakes, and that these systems act as a source of CO_2 to the atmosphere. Moreover, the occurrence of net heterotrophy in so many systems implies even tighter connections between biogeochemical processes in aquatic systems and adjacent watersheds, as organic carbon from a watershed is needed to fuel heterotrophic processes in the receiving waters.

Role of Bacterioplankton in Inland Waters

The traditional role attributed to bacteria in aquatic ecosystems was that of decomposer of organic matter. But bacteria not only remineralize nutrients back to the water column through decomposition processes, they also store organic carbon, are a food source to other microbes, and are important in the cycling of phosphorus and nitrogen. While early scientific papers recognized the importance of the 'microbial ooze' in the trophic dynamics of aquatic ecosystems, it is only recently that we have begun to be able to quantify the influence of microbes on biogeochemical processes. In the last few decades, bacteria have been viewed as important components of the microbial loop, which contains multiple trophic levels and is important in the cycling of matter and in dissipating energy through respiration. The contribution of bacteria to biogeochemical cycles is only beginning to be fully appreciated.

Decomposition

As decomposers, bacterioplankton degrade dissolved organic matter (DOM), assimilate the byproducts into their cell (bacterial production), and remineralize carbon and nutrients back to the water through respiration. Organic carbon that originates from phytoplankton, either through cell death or excreted during photosynthesis, is usually assumed to be the primary source of dissolved organic carbon (DOC) that fuels bacterial metabolism. Organic carbon of algal origin is typically composed of simple sugars and is rapidly remineralized by bacteria. In contrast, organic matter that enters from allochthonous sources is a complex mixture of high molecular weight and lignified organic compounds that are not as readily decomposed.

Extracellular enzymes are required to mediate degradation of polymeric organic macromolecules into smaller compounds that can be assimilated by the cell. The products of the enzymatic reactions are believed to limit rates of assimilation of organic matter by bacteria and, hence, bacterial growth. As such, estimates of extracellular enzyme activity (EEA) may be used to quantify decomposition rates. For example, studies of EEA and bacterial production found that high bacterial productivities were typically associated with the availability of simple sugars such as saccharides and that during the summer bacterial production was fueled by algal lysates and exudates generated through the microbial loop, whereas production in spring and autumn appeared to be fueled by allochthonous carbon. Such studies support the observation that algal products are the 'preferred' organic carbon source to fuel bacterial metabolism, but that external carbon sources may also fuel decomposition under certain conditions.

Bacterial community structure has been shown to affect the rate of processing of dissolved organic matter (DOM) in aquatic systems. There are differences among major phylogenetic groups in terms of utilization of high versus low molecular weight DOM and in terms of enzyme activities, and changes in phytoplankton community structure have also been shown to produce a response in bacterioplankton community composition. Differences in bacterial composition between lakes have also been found to be correlated to variables that reflect the relative loading of allochthonous versus autochthonous carbon, whereas seasonal changes in community composition within lakes correlated to patterns in algal abundance, temperature, and DOC.

Although a common observation is that sediments overlain by anoxic waters are rich in organic matter content, it is also generally thought that anoxia slows decomposition rates in lakes. There are published studies both supporting and refuting this view. Examination of this contradiction across a series of lakes showed that bacterial production was in fact greater in the anoxic hypolimnion with lower water temperatures than in the epilimnion. The mean ratio of anoxic production to aerobic production was 1.6 for lakes that ranged from ultraoligotrophic to eutrophic. The doubling time of bacteria in the colder anoxic waters was lower than in the warmer oxic waters. This was offset by the greater bacterial abundance and biomass in the anoxic waters. The significance of these findings to not only other lakes but also to our general, and still poor, understanding of the role of bacteria in anoxic zones in lake functioning is a research area requiring further investigation.

It is important to bear in mind that bacteria are not the sole decomposers in aquatic systems: fungi are responsible for the decomposition of high molecular weight and/or lignified compounds, whereas bacteria are primarily responsible for the final decomposition of lower molecular weight polymeric compounds and polysaccharides. A recent study experimentally demonstrated an association between bacteria and fungi, where bacterial decomposition of allochthonous organic matter was dependent on intermediate decomposition products that were produced by fungi. Still, in other studies, fungi have been shown to be the dominant organisms in decomposition.

Nutrient Uptake and Cycling

Bacterioplankton are typically assumed to be limited by the availability of organic carbon and, since algal carbon appears to be the preferred substrate for bacterial metabolism, bacterial production should be closely linked to algal production and/or biomass. At coarse, cross-system scales, this relationship does hold and, as noted above, reasonably good correlations have been detected between bacterial abundance and chlorophyll *a* and between bacterial production and net phytoplankton production. However, there have been many studies that have demonstrated an uncoupling between phytoplankton and bacterioplankton abundance and production. The reasons for the uncoupling appear to rest, at least in part, on the availability of inorganic nutrients, and of phosphorus in particular.

Bacteria are efficient competitors for inorganic nutrients such as phosphorus and nitrogen (mostly as ammonium) that limit algal growth, particularly in low nutrient environments. It has been shown through size fractionation studies that bacteria are usually responsible for at least 50% of the uptake of inorganic phosphorus in lakes. The actual proportion of the inorganic phosphorus pool that is taken up by bacteria does vary, probably in part as a function of phytoplankton biomass and availability of organic carbon. The proportion of ammonium that is taken up by bacteria in freshwater systems has not been intensively studied, mostly because freshwater ecosystems are usually assumed to be phosphorus limited, but in marine systems bacteria can be responsible for ~30% of ammonium uptake.

Although bacteria may be able to incorporate phosphorus more rapidly than do algae, algal biomass is higher than bacterial biomass in more enriched systems and algae are better able to store phosphorus in their cells as phospholipids. Thus, over longer time periods (days to weeks versus minutes and hours), algae may be able to incorporate more inorganic phosphorus than do bacteria. If bacterial biomass is high, as in oligotrophic systems, then algal growth may be limited because of bacterial uptake of nutrients. However, grazing on bacteria by protozoa in the microbial loop will remineralize organic matter and release

inorganic P and N, making these elements once again available to phytoplankton. Also, short-term physical-forcing events can temporally reverse this situation if nutrients are brought up into the water column through the entrainment of P-rich bottom sediments and/or interstitial water. Similarly, annual overturn in stratified lakes can introduce nutrients to the upper water column.

In a cross-system comparison of lake bacterioplankton and phytoplankton, bacterial abundance was more strongly correlated to total phosphorus than to algal abundance, suggesting a degree of uncoupling between algal and bacterial growth. At least one study has shown that inorganic phosphorus additions can stimulate bacterial but not algal production, whereas the addition of both nitrogen and phosphorus increased both algal and bacterial production, suggesting that bacteria are better scavengers for nutrients than are algae when one or more nutrients are limiting.

Effect of Environmental Perturbations on Bacterioplankton

Bacterioplankton community composition, production, and abundance vary seasonally in most freshwater systems where temporal patterns have been examined, and these variations appear to track climatic variations that affect numerous biogeochemical processes. Because of their rapid turnover rates and correlation to numerous environmental variables, bacterioplankton can also be expected to be sensitive to environmental change such as climate change and variability, increases in nutrient concentrations or to pollution by a wide range of human-made chemicals.

The impact of human-made chemicals on bacterioplankton metabolism and diversity is not yet well understood. The addition of herbicides to natural ecosystems, in conjunction with inorganic nutrients, has been shown to have a synergistic effect on bacterial production. Antibiotics commonly used to treat bacterial infections and that have increasingly been detected in aquatic ecosystems at low concentrations can have an inhibitory effect on bacterial production, even at very low concentrations. Long-term exposure to toxic metals can alter a bacterial community toward strains that are resistant to heavy metals. Interestingly, a number of bacterial communities that are metal resistant have also been found to be resistant to antibiotics, suggesting an indirect selection for antibiotic resistance in natural ecosystems exposed to heavy metals. This type of indirect selection for resistance is worrisome, given general public health concerns over the evolution of antibiotic resistant bacteria.

See also: Fungi; Microbial Food Webs.

Further Reading

Bird DF and Kalff J (1984) Empirical relationships between bacterial abundance and chlorophyll concentration in fresh and marine waters. *Canadian Journal of Fisheries Aquatic Sciences* 41: 1015–1023.

Cole JJ (1999) Aquatic microbiology for ecosystem scientists: new and recycled paradigms in ecological microbiology. *Ecosystems* 2: 215–225.

Cole JJ and Pace ML (1995) Bacterial secondary production in oxic and anoxic freshwaters. *Limnology and Oceanography* 40: 1019–1027.

Cole JJ, Findlay S, and Pace ML (1988) Bacterial production in fresh and salt water ecosystems: a cross-system overview. *Marine Ecology Progress Series* 43: 1–10.

Cotner JB and Biddanda BA (2002) Small players, large role: microbial influence on biogeochemical processes in pelagic aquatic ecosystems. *Ecosystems* 3: 105–121.

Odum EP (1959) *Fundamentals of ecology.* Philadelphia: W.B. Saunders Company.

Robarts RD and Zohary T (1993) Fact or fiction—Bacterial growth rates and production as determined by [*methyl-*^{3}H]-thymidine? *Advances in Microbial Ecology* 13: 371–425.

Stepanauskas R, Glenn TC, Jagoe CH, *et al.* (2005) Elevated microbial tolerance to metals and antibiotics in metal-contaminated industrial environments. *Environmental Science and Technology* 39: 3671–3678.

Verma B, Robarts RD, and Headly JV (2007) Impacts of tetracycline on planktonic bacterial production in prairie aquatic systems. *Microbial Ecology* 54: 52–55.

Welch PS (1935) *Limnology.* New York: McGraw-Hill Book Company, Inc.

White PA, Kalff J, Rasmussen JB, and Gasol JM (1991) The effect of temperature and algal biomass on bacterial production and specific growth rate in freshwater and marine habitats. *Microbial Ecology* 21: 99–118.

Bacteria, Distribution and Community Structure

A C Yannarell and A D Kent, University of Illinois at Urbana-Champaign, Urbana, IL, USA

Introduction

Microscopic life dominates many ecosystems in terms of biomass. In many aquatic ecosystems, the bulk of the energy production and a significant fraction of the material transformations are carried out by microorganisms. Among the aquatic microbes, the bacteria are key players in the sequestration of inorganic compounds and the remineralization and dissipation of organic material. A large amount of energy and matter in aquatic systems passes through bacterial communities, and the secondary production of bacteria in some systems can rival primary production. This understanding of bacterial ecology at the ecosystem level has generally been attained by regarding bacterial communities from a 'black box' perspective, by assuming a single role or coherent response from the diverse assemblage of bacteria present at any given time and place. However, not all bacteria are the same, and much can be learned by peering into the black box.

This chapter attempts to integrate the current knowledge of lacustrine bacterial community structure and dynamics within the broader context of community ecology. Although they are occasionally mentioned, bacteria from lotic systems are not discussed in detail. We begin with a short summary of the major freshwater bacterial groups and their general habitat distributions. We then review the changes that lake bacterial communities undergo in time and space. Finally, we examine how the bacteria may interact with other organisms in lakes. More information about lake bacteria can be found in article Bacteria, Bacterioplankton of this volume.

Freshwater Bacterial Clades and their Habitats

Not long ago, traditional microbiological cultivation techniques suggested that there was no clear distinction between bacteria found in the soil and those found in aquatic habitats, but we now know that this is not the case. Surveys of 16S ribosomal RNA gene sequences from freshwater habitats provide evidence of a freshwater bacterial assemblage that is distinct from terrestrial communities. Many of the freshwater clades appear to have widespread geographic distributions. This suggests that freshwater habitats, despite their geographic isolation, harbor bacterial species drawn from a remarkably consistent pool of potential colonists, and this may be a general characteristic of freshwater biology.

Many major phyla of bacteria can be found in both marine and freshwater environments, but the taxa that dominate marine and lacustrine communities differ. The transition between freshwater and marine assemblages in an estuary occurs abruptly at the turbidity maximum where fresh and saline water meet, and it appears to be the product of active environmental selection rather than passive water mixing.

What follows is a brief overview of the bacterial taxa that comprise the major freshwater groups. Some of these groups are freshwater-specific taxa, whereas others are adapted to a broad range of habitats. The nomenclature of bacteria is in continual change due to the high discovery rate of new taxa. Because some of these clades are known only through environmental DNA sequences, they have not yet been properly described, and these are designated with alphanumeric strings instead of proper names. Throughout this chapter, the most recent naming conventions are followed, with alternative names for phyla indicated in parentheses (**Table 1**).

Proteobacteria

The purple bacteria, or *Proteobacteria*, are often the dominant prokaryotes in aquatic systems. The *Proteobacteria* lineage contains phototrophs, chemolithotrophs, and chemoorganotrophs, and its members can be found in both oxic and anoxic environments. This phylum consists of several evolutionarily distinct subdivisions. *Alphaproteobacteria* can be found in many environments, including the water columns and sediments of freshwater and marine systems, and they predominate more in marine than freshwater systems. In lakes, they have also been found in particle-associated communities. There are five clusters of *Alphaproteobacteria* with freshwater representatives: LD12 (formerly alpha V; this cluster is closely related to the ubiquitous marine SAR11 cluster), *Brevundimonas intermedia* (formerly alpha II), CR-FL11, *Novosphingobium subarctica* (formerly alpha IV), and GOBB3-C201.

Gammaproteobacteria, like *Alphaproteobacteria*, are also found in a variety of aquatic habitats but are generally more dominant in marine communities. These organisms may be copiotrophs, adapted to life

Table 1 Typical freshwater bacteria

Phyla	Freshwater groups[a]	Alternate name	Freshwater habitats[b]	Other habitats[b]	References
Actinobacteria					
	acI	ACK-M1	Wide geographic distribution in freshwater systems	Estuaries	
	acII	Luna-1, Luna-2	Lakes, rivers	Estuaries	1, 2
	acIII	Actinobacterial cluster 2	Chemocline of lake	Hypersaline lake	
	acIV	CL500-29	Lakes, rivers	Estuaries, marine sediment and soil	
Proteobacteria					
Alphaproteobacteria					
	LD12	alphaV	Lakes, rivers		3–5
	Brevundimonas intermedia	alphaII	Lakes	Marine waters, aquifer	
	CR-FL11		Lakes, rivers		
	Novosphingobium subarctica	alpha IV	Lakes	Soil, sludge	
	GOBB3-C201		Lakes, rivers		
Betaproteobacteria					
	Polynucleobacter necessarius	betaII	Wide geographic distribution in freshwater systems		
	GKS98	betaIII	Lakes, rivers	Saline lakes	
	LD28	betaIV	Lakes, rivers		
	Ralstonia picketii		Lakes, rivers	Estuaries	6, 7
	Rhodoferax sp. BAL47	betaI	Lakes, rivers		
	GKS16	betaI	Lakes		
Gammaproteobacteria					
	Methylobacter psychrophilus	gamma1	Lakes, rivers	Estuaries, soil	8
Deltaproteobacteria					
	Desulfobacteraceae		Sediments, anoxic waters	Salt marsh, marine sediments	9, 10
	Desulfovibrionales		Sediments, anoxic waters	Salt marsh, marine sediments	9–11
	Desulfobulbaceae		Sediments, anoxic waters	Salt marsh, marine sediments	10, 12
	Geobacteraceae		Sediments, anoxic waters	Salt marsh, marine sediments	13
Bacteriodetes					
	FukuN47	cfI	Lakes, rivers		
	LD2	cfIV	Lakes		
	CL500-6		Lakes, rivers		4, 5
	PRD01a001B	cfIII	Lakes, rivers		
	GKS2-216	cfIII	Lakes		

			Sources
Cyanobacteria			
Planktothrix agardhii	Lakes, rivers		
Aphanizomenon flos aquae	Lakes, rivers		
Microcystis	Lakes, rivers		
Synechococcus 6b	Lakes, rivers	Coastal and marine waters	5
Verrucomicrobiales			
FukuN18	Lakes, rivers	Soil	
LD19	Lakes		
Sta2-35	Lakes		
CL120-10	Lakes, rivers		
CL0-14	Lakes, rivers		5

[a]Except for the *Actinobacteria* and *Deltaproteobacteria* groups, the taxonomic groups are named after cultivated organisms or after the name of the longest available 16S rRNA gene sequence from an environmental clone.

[b]Habitats in which these groups have been detected at the time this list was compiled. The common freshwater taxa are not necessarily exclusive to the freshwater environment.

Sources

1. Eiler A and Bertilsson S (2004) Composition of freshwater bacterial communities associated with cyanobacterial blooms in four Swedish lakes. *Environmental Microbiology* 6: 1228–1243.
2. Warnecke F, Amann R, and Pernthaler J (2004) Actinobacterial 16S rRNA genes from freshwater habitats cluster in four distinct lineages. *Environmental Microbiology* 6: 242–253.
3. Glöckner FO, Zaichikov E, Belkova N, *et al.* (2000) Comparative 16S rRNA analysis of lake bacterioplankton reveals globally distributed phylogenetic clusters including an abundant group of *Actinobacteria*. *Applied and Environmental Microbiology* 66: 5053–5065.
4. Van der Gucht K, Vandekerckhove T, Vloemans N, *et al.* (2005) Characterization of bacterial communities in four freshwater lakes differing in nutrient load and food web structure. *FEMS Microbiology Ecology* 53: 205–220.
5. Zwart G, Crump BC, Agterveld MPKV, *et al.* (2002) Typical freshwater bacteria: an analysis of available 16S rRNA gene sequences from plankton of lakes and rivers. *Aquatic Microbial Ecology* 28: 141–155.
6. Hahn MW (2003) Isolation of strains belonging to the cosmopolitan *Polynucleobacter necessarius* cluster from freshwater habitats located in three climatic zones. *Applied and Environmental Microbiology* 69: 5248–5254.
7. Wu QL and Hahn MW (2006) Differences in structure and dynamics of *Polynucleobacter* communities in a temperate and a subtropical lake, revealed at three phylogenetic levels. *FEMS Microbiology Ecology* 57: 67–79.
8. Loy A, Kusel K, Lehner A, *et al.* (2004) Microarray and functional gene analyses of sulfate-reducing prokaryotes in low-sulfate, acidic fens reveal co-occurrence of recognized genera and novel lineages. *Applied and Environmental Microbiology* 70: 6998–7009.
9. Macalady JL, Mack EE, Nelson DC, *et al.* (2000) Sediment microbial community structure and mercury methylation in mercury-polluted Clear Lake, California. *Applied and Environmental Microbiology* 66: 1479–1488.
10. Perez-Jimenez JR and Kerkhof LJ (2005) Phylogeography of sulfate-reducing bacteria among disturbed sediments, disclosed by analysis of the dissimilatory sulfite reductase genes (*dsrAB*). *Applied and Environmental Microbiology* 71: 1004–1011.
11. Li JH, Purdy KJ, Takii S, *et al.* (1999) Seasonal changes in ribosomal RNA of sulfate-reducing bacteria and sulfate reducing activity in a freshwater lake sediment. *FEMS Microbiology Ecology* 28: 31–39.
12. Purdy KJ, Nedwell DB, Embley TM, *et al.* (2001) Use of 16S rRNA-targeted oligonucleotide probes to investigate the distribution of sulphate-reducing bacteria in estuarine sediments. *FEMS Microbiology Ecology* 36: 165–168.
13. Cummings DE, Snoeyenbos-West OL, Newby DT, *et al.* (2003) Diversity of Geobacteraceae species inhabiting metal-polluted freshwater lake sediments ascertained by 16S rDNA analyses. *Microbial Ecology* 46: 257–269.

under high-nutrient conditions, and this could explain their disproportionate representation in cultivation-based studies. There is a single freshwater cluster currently defined for this subdivision, *Methylobacter psychrophilus* (formerly gamma1).

Betaproteobacteria have been found to be the most dominant members of freshwater bacterial communities in a number of studies, and some groups appear to have widespread distributions in lakes from around the world. This prominence in fresh water is in sharp contrast to their near absence in marine waters. Until recently, marine representatives of *Betaproteobacteria* had only been found in coastal waters, but *Betaproteobacteria* sequences have recently been recovered from the Sargasso Sea, where they comprised a minor part of the community. Six freshwater clusters of *Betaproteobacteria* have been identified: *Polynucleobacter necessarius* (formerly betaII), LD28 (formerly betaIV), GKS98 (formerly betaIII), *Ralstonia picketii*, *Rhodoferax* sp. BAL47 (formerly betaI), and GKS16 (formerly betaI). The *P. necessarius* cluster appears to be particularly cosmopolitan in freshwater ecosystems. These *Betaproteobacteria* are among the most frequently detected freshwater taxa, and isolates from this cluster have been recovered from freshwater habitats from a range of climatic zones. In some systems, up to 60% of the bacterioplankton may be affiliated with the *P. necessarius* cluster. This *P. necessarius* cluster appears to be limited to pelagic freshwater environments, and to date, they have not been detected in freshwater sediments or marine or soil systems.

Organisms from the subdivision *Deltaproteobacteria* are generally restricted to anoxic environments and can often be found in the sediments of lakes. Some *Deltaproteobacteria* are sulfate reducers, and they can often be found just below the chemocline where oxic and anoxic waters meet.

Cyanobacteria

The *Cyanobacteria*, sometimes referred to as blue-green algae, are often counted among the phytoplankton, but they more properly belong within the bacterial domain. Like eukaryotic phytoplankton, they perform oxygenic photosynthesis using photosystems I and II, and cyanobacteria are the ancestors of the eukaryotic chloroplasts. *Cyanobacteria* are not the only phototrophic bacteria, but they are generally the dominant bacterial phototrophs in the oxygenated portions of lakes (**Figure 1**). Many *Cyanobacteria* are capable of nitrogen fixation, and thus they can be key players in both the carbon and nitrogen cycles in some systems. Some *Cyanobacteria* are known to produce toxins and form nuisance blooms in eutrophic systems. Freshwater

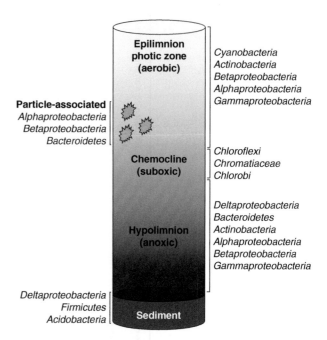

Figure 1 Vertical distribution of dominant freshwater bacterial groups in aquatic environments.

clusters include the *Planktothrix agardhii*, *Aphanizomenon flos aquae*, and *Microcystis* clusters, and the cluster *Synechococcus* 6b contains both marine and freshwater representatives. The *Cyanobacteria* are discussed in more detail in article Cyanobacteria of this volume.

Actinobacteria

The *Actinobacteria*, formerly known as the high G + C gram-positive bacteria, are another group of bacteria that are commonly found in lakes with a wide range of water chemistries. *Actinobacteria* may comprise a large fraction (up to 60%) of the bacterioplankton in some freshwater systems, and their abundance often peaks in late fall and winter. *Actinobacteria* appear to be more tolerant of conditions with low concentrations of organic carbon, and they may be replaced by *Betaproteobacteria* when algal blooms lead to increased carbon levels. Freshwater *Actinobacteria* fall into four distinct phylogenetic clusters: acI, acII, acIII, and acIV. *Actinobacteria* in the acI and acII clades are exclusive to fresh waters. Two acII subclusters, acII-B (formerly Luna-1) and acII-D (formerly Luna-2), are comprised of ultramicrobacterial isolates (cell volume $<0.1 \, \mu m^3$). The acIII cluster (formerly *Actinobacteria* cluster 2) has only been detected in the chemocline of a single European lake and from a hypersaline soda lake in California. AcIII *Actinobacteria* are most closely related to soil organisms. *Actinobacteria* from the acIV cluster may be found in a variety of habitat types: lakes and rivers, as well as estuaries and marine waters and sediments.

Bacteroidetes

This phylum contains members of the genera *Cytophaga*, *Bacteroides*, and *Flavobacterium*, and it has been previously referred to as the *Cytophaga–Flavobacterium–Bacteroides*, or CFB, group. In fresh waters, *Bacteroidetes* comprise a major proportion of particle-associated bacterial communities, and they can also be found at depth in lakes, where they may degrade recalcitrant macromolecules. Major freshwater clusters of *Bacteroidetes* include FukuN47, LD2 (formerly cfIV), CL500-6, PRD01a001B (formerly cfIII), and GKS2-216 (formerly cfIII).

Other Freshwater Phyla

Although less prominent than those phyla already detailed, other bacterial groups have also been found in fresh waters. The *Firmicutes*, or low G + C grampositives, have been found in freshwater sediments, but not in the water column. *Verrucomicrobia* have been recovered from both sediment and water column samples from oligotrophic and eutrophic lakes. *Planktomycetes* are detected at low abundances in lakes, most often attached to particulate matter. Sequences of *Acidobacteria* have been recovered from the sediments of an acidic lake. The phyla *Chloroflexi* (the green nonsulfur bacteria) and *Chlorobi* (the green sulfur bacteria) contain anoxygenic phototrophs that have been shown to have a particular vertical distribution in lakes, with *Chloroflexi* (along with the purple gamma-proteobacterial *Chromatiaceae* populations) typically higher in the water column. The presence of humic substances in the water column appears to select for *Chlorobi* in metalimnetic communities.

Temporal Variation in Bacterial Communities

Lake bacterial communities harbor great genetic diversity, but they generally have low evenness when compared to other communities. That is, at any given time, communities tend to be dominated by only a few different taxa, with the majority of the species present showing very low abundance. Population dynamics of dominant strains show short-lived blooms at different times and different depths, resulting in a 'succession' of dominant community members.

This 'bloom-and-bust' dynamic has led to the suggestion that lake bacterial dynamics are driven by a multitude of rapidly changing niches that are exploited by different species, which are from a large pool of dormant organisms. From a functional perspective, the activity of the dominant microbes is largely responsible for the creation of new ecological niches. These niches are rapidly filled by formerly dormant species, which then create new niches, and thus the functioning, diversity, and dynamics of the system are inseparable phenomena. Of course, new niches may also result from factors that are external to the bacterial communities.

The rapid overturn and exploitation of niches can generate dramatic shifts in bacterial community composition over short time periods, yet bacterial communities are not always changing rapidly. The pacing of change in bacterial communities appears to shift between long periods of stability and periods of rapid turnover in lakes. Algal communities show similar dynamics. Thus, a general property of pelagic communities may be that they undergo a series of successions throughout the year. Available niches are filled, the community adapts to the prevalent environmental conditions, and then shifting conditions prompt rapid species turnover, beginning the process again along a different ecological trajectory.

Data from many bacterial studies suggest that there is a strong seasonal component to the patterns of community change, and thus it is reasonable to look to seasonal events as primary sources of change in bacterial communities. One such source of seasonal change, at least in nontropical climates, is temperature. Bacterial growth rates appear to be temperature dependent in lakes of only up to around 15 °C, but temperature may still be an important influence on bacterial diversity and community composition outside of this range. In some lakes, water temperature is the primary determinant of water density and therefore controls water-column mixing, which has been shown to influence bacterioplankton communities. The physical and chemical changes to the lake environment due to mixing and stratification are likely to be the primary causes of bacterial community turnover in lakes that undergo mixing.

Other seasonal events not related to temperature or mixing dynamics have also been implicated as mechanisms responsible for variation in bacterial communities. Correlations between the seasonal dynamics of grazers and phytoplankton and those of bacteria suggest that biological interactions are important. For lakes with inflowing rivers, seasonal changes in flow regimes can also have an impact on the concentration of nutrients in lakes.

The impact of external seasonal events on bacterial communities is apparent even in communities with differing composition. Bacterial communities in northern temperate lakes displayed similar seasonal dynamics over several years despite the fact that the identity of the important community members in these lakes differed in each year. The influence of these external events is different from the internal dynamics within the community that give rise to succession. The former represents a structuring force that changes over a long time scale relative to the lifespan of the organisms involved; the latter represent interactions and responses to the environment and to the niches available, and these occur at time scales more relevant to individual organisms. The emerging picture of the dynamics of lake bacterial communities is one involving the interplay of fast variables (e.g., disturbance and biological interactions), slow, seasonal variables, and the rapid response of individual populations to the niches that arise from these changes.

Spatial Variation in Bacterial Communities

Spatial heterogeneity is important for the creation and maintenance of biological diversity. Spatial relationships can structure biological interactions and impose constraints upon the flow of matter and energy in ecosystems. For microbes, the spatial relationships are presumed to be more stable in the soil than in pelagic environments, but space may be no less important for the aquatic forms. The small size of bacteria allows them to respond to spatial heterogeneity over a very large range of scales, from micrometers to the global scale.

Algae and particles are hotspots of bacterial activity in freshwater systems, and they can be heavily colonized by bacteria. The availability of exploitable resources may underlie the formation of particle-associated communities in aquatic environments. Bacterial colonization of particulate matter increases following a diatom bloom, with concurrent increases in bacterial production, abundance, and enzyme activity. The secondary production of particle-attached bacteria in lakes can equal or exceed that of the free-living bacteria even when less than 10% of the cells are associated with particles. The composition of particle-associated communities is distinct from free-living bacterial community composition. Thus, organic particles represent distinct niches that are exploited by a unique assemblage of organisms.

Environmental changes occurring with depth are important sources of variation for bacterial communities in lakes. One of the most important factors that change with depth is the presence or absence of available oxygen. The abundance and mean cell size of bacteria are greater in anoxic waters than in oxic waters, and anoxic bacterial communities are more productive overall than oxic communities. Bacterial community composition differs between the epilimnia and hypolimnia of lakes, particularly when the former is oxygenated and the latter is not. Bacterial communities are more similar in the oxic portions of different lakes than within the same lake across the oxycline. These differences have been particularly well studied in meromictic lakes, and community change with depth is more pronounced in these lakes right around the oxycline.

The identity of the major bacterial phototrophs changes with depth, in response to differing light levels and different spectral characteristics of incoming photons. In general, one finds *Cyanobacteria* in the oxic epilimnion, *Chlorobi* and phototrophic *Gammaproteobacteria* such as *Chromatiaceae* near the oxic–anoxic interface, and *Chloroflexi* near the top of the anoxic zone, where they can oxidize H_2S. However, this sequence is often disrupted by seasonal mixing dynamics in lakes that have time-variant stratification regimes. Heterotrophs may also respond indirectly to light; bacterial communities may shift from being P-limited in the epilimnion to being C-limited in low-light waters.

In addition to variation with depth, significant horizontal heterogeneity has been observed in lake bacterial communities due to geographic features such as bays and narrow constrictions between basins, which reduce water flow between the different areas of the lake. Within-lake variability may indicate that different parts of the lake present bacteria with different sets of niches; alternatively, rapid bacterial growth rates may allow communities to exhibit individual dynamics on time scales shorter than the average residence time of the surface waters in these different parts of the lake.

Horizontal variation in bacterial community composition in lakes is generally small compared to the differences seen between lakes. Many studies have reported that different lakes contain different bacterial communities. Variation between lakes has been attributed to differences in water clarity, pH, humic acid content, salinity, productivity and nutrient concentration, differences in water column mixing regimes or water residence time, and the chemistry of toxic metals. Given the wide range of environmental conditions represented by lakes, it is not surprising that community composition exhibits such variation, and lakes may be viewed as individual 'islands' where the communities and environments follow individual trajectories. However, bacterial communities from different lakes are not always distinct, particularly when the lakes represent very similar environments and when there is a great deal of temporal variation in community composition. For example, bacterial communities from lakes with river or stream connections are more similar to each other than to communities from lakes with no connections. In addition, bacterial communities in lakes with short hydraulic retention times resemble those of the inlet and outlet streams, while those in lakes with long hydraulic retention times are distinct from stream communities. This suggests that landscape-level features also influence bacterial communities in lakes.

Interactions with Other Organisms

Patterns in bacterial community composition are tied to the dynamics of other pelagic organisms. Like all members of ecosystems, bacteria interact with many different organisms, and these interactions profoundly affect bacterial community structure and dynamics.

Predation by Protists

Ciliated and flagellated protists, referred to respectively as ciliates and heterotrophic nanoflagellates (HNFs), are involved in a variety of close interactions with bacteria. Bacteria and anaerobic ciliates can have symbiotic relationships, as sometimes seen in lake sediments, where hydrogen exchange between these organisms contributes to the production of methane by bacteria. However, the predator–prey relationship is the most studied bacterial–protist interaction, and protists are generally considered the most effective predators of bacteria in lakes. Ciliates and HNFs can consume a sizable portion of bacterial standing stock each day, and protistan grazing can control the total abundance of bacteria and structure bacterial communities, if the protists themselves are not preyed upon. Small ciliates and HNF impact bacterial populations directly, whereas large ciliates may also consume HNF, making for complex food-web interactions.

Protistan grazing of bacteria displays many features of classical predator–prey interactions. Size-selective grazing by protists can directly impact bacterial community composition (**Figure 2**). Ultramicrobacteria ($<0.1\ \mu m^3$ cell volume) and large bacteria ($>2.44\ \mu m$ in one dimension) can escape HNF grazing, as can certain *Bacteroidetes* and *Alpha* and *Betaproteobacteria*, which can form large filaments in the presence of grazers. Particle-associated bacteria are also protected from grazing by HNF. Protists have been show to obey optimal foraging rules, selectively

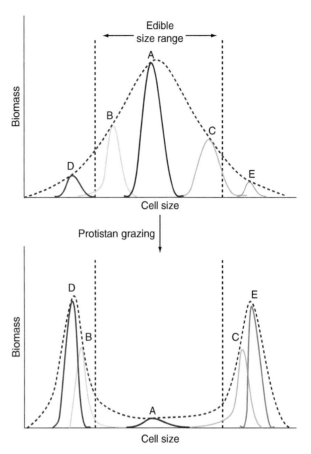

Figure 2 The effects of protistan grazing on bacterial size distributions and bacterial community structure. Different bacterial species are depicted in different colors. Populations in the 'vulnerable' size range (A, B, and C) are disproportionally reduced by grazing (shown in the upper panel), unless they can change their morphology to a less susceptible form (B, C) as shown in the bottom panel. Populations outside the edible size range (D, E) will have a selective advantage under grazing pressure. Reprinted with permission from Macmillan Publishers Ltd: *Nature Reviews Microbiology*, copyright 2005.

consuming the most common and the most metabolically active cells. Thus, protists can serve as keystone predators, affecting bacterial morphology, activity, and community composition. Protistan grazing on lake bacteria can act as an ecological switch, alternatively recycling dissolved nutrients and organic carbon back to the environment or shuttling bacterial biomass to larger organisms via food–web interactions with zooplankton.

Predation by Cladocerans

Cladocerans are considered nonselective filter feeders, and they are capable of consuming much larger prey than protists. Most cladocerans, particularly *Daphnia*, are capable of feeding on bacteria-sized particles and of controlling bacterial population growth. However, cladocerans seem to be less efficient at processing the smallest bacterial cells. Strong grazing pressure by *Daphnia* can shift the bacterial community toward smaller, but more metabolically active cells. In addition, cladocerans can graze upon and suppress populations of nanoflagellates. HNF predation on bacteria is insignificant when large-bodied cladoceran grazers are present. This is not quite a trophic cascade, since the cladocerans themselves can suppress total bacterial abundance. However, certain portions of the bacterial community may be released from selective protistan grazing, leading to shifts within the community. Because cladocerans consume the filamentous bacteria not consumed by HNFs, the taxonomic shifts in bacterial community composition witnessed under HNF grazing should not occur. More importantly, the fast-growing, medium-sized cells targeted by HNFs may be released from pressure by cladoceran grazing, particularly if they are small enough to be inefficiently handled by the cladoceran feeding apparatus.

Microbial food webs are addressed in more detail in article Microbial Food Webs of this volume.

Lysis by Viruses

Viruses may be selective for the cells that they infect or lyse. Viral lysis affects a subset of the bacterial community disproportionately, selecting for populations that are slow growing, generalistic 'K-strategists.' High viral lysis decreases the mean size of cells in bacterial communities, and the frequency of visibly infected cells $<0.3\,\mu m$ in diameter is less than that for the rest of the community. A major effect of viral lysis may be to maintain the clonal diversity of bacterial populations, as viruses preferentially lyse the most abundant, dominant members of the bacterial community. Thus, viruses can structure the community composition of host cells and may help prevent competitive dominance by removing the most rapidly growing members of the bacterial community. The importance of viruses in aquatic systems is addressed in more detail in the article Viruses of this volume.

Interactions with Phytoplankton

In general, bacterial abundance is linked to algal biomass, and bacterial secondary production is linked to net primary production by phytoplankton across aquatic systems; these parameters are also correlated on roughly seasonal scales. Bacterial abundance and productivity sometimes peak during or immediately after phytoplankton blooms.

The make-up of phytoplankton communities can influence bacterial activities and community composition. These changes may be mediated by the concentration and characteristics of the dissolved organic carbon released by different phytoplankton. The diversity and community composition of phytoplankton and bacteria are highly correlated over time. This may represent the response of both bacteria and phytoplankton to the same external drivers, but it may also indicate a direct influence of one community on the other. Tight species–specific interactions between algae and some bacteria can be important factors determining bacterial species presence and absence.

Microbial Interactions Along the Productivity Gradient

Lake productivity has a significant influence on bacterial community composition. The proportion of metabolically active cells increases with increased primary productivity, and primary productivity may determine the number of niches available for different species of bacteria. The diversity of different bacterial clades displays different relationships with primary productivity. *Bacteroidetes* appear to be most diverse at intermediate productivity levels, *Alphaproteobacteria* are most diverse at low and high productivity levels, and the diversity of *Betaproteobacteria* shows no relationship with primary productivity.

Changing community composition over the productivity gradient can also be linked to changing predator–prey relationships. The impact of nanoflagellate grazing changes along the productivity gradient, with grazers able to exert more control over total bacterial growth rates under oligotrophic conditions, and more control over bacterial community composition in eutrophic systems. The shifting importance of predators has important implications for nutrient cycling, energy flows, and system resiliency.

Summary

The dynamics of lake bacterial communities suggest communities in varying states of recovery from disturbance events. The frequent intervention of external events prevents competitive exclusion and maintains community diversity. Species diversity can also be maintained in systems through fine-scale partitioning of niches and the spatial separation of microniches. Groups of bacteria degrading dissolved organic matter can be highly specialized in the roles they play. If community diversity is maintained solely as the result of this kind of niche specialization, then any disturbance that resulted in a loss of diversity would also upset community function, because competitive exclusion within niches would ensure that each role was being played by a single species. If, however, other forces were at work maintaining diversity, then this sort of niche partitioning could represent multiple pathways to the same end process, leading to functional stability.

Nutrient concentration, water color, pH, redox potential, and water temperature all appear to play selective roles to influence bacterial communities within and between lakes, but there is still considerable variation in community composition given environments that share common characteristics. The factors influencing bacterial community composition can be conceptualized as acting in a hierarchical, selective manner, with the physical and chemical properties of lakes operating at high levels, and biotic influences operating at lower levels. The lake bacterial communities ultimately consist of the set of species that successfully navigate this hierarchy at any given time and place.

Units and nomenclature

μm – micrometer
rRNA – ribosomal RNA

Glossary

16S ribosomal RNA sequence – The DNA sequence of the gene encoding the RNA portion of the small ribosomal subunit. The ribosome is the cellular component that translates mRNA sequences into proteins. All living things possess ribosomes, so the genes encoding for ribosomal components can be used to construct phylogenetic trees encompassing all terrestrial forms of life. For practical reasons related to DNA sequencing technologies, the genes encoding the small subunit RNA portions are those most commonly utilized for this purpose.

Anoxygenic phototroph – An organism that can utilize sunlight for energy, but does not produce oxygen as a result. This type of photosynthesis does not make use of the 'photosystem II' reactions, and it is generally performed by anaerobic bacteria such as phototrophic members of the purple bacteria (*Proteobacteria*), and the green sulfur (*Chlorobi*) and green non-sulfur (*Chloroflexi*) bacteria.

Bacterial phylum – One of the major evolutionary lineages of bacteria. These are not Phyla in the classical Linnaean sense of the word, as a bacterial phylum does not refer to a level of organization below the level of Kingdom. In fact, any given bacterial phylum probably encompasses more biological diversity than the combined Kingdoms *Animalia*, *Fungi*, and *Plantae*.

Chemolithotroph – An organism obtaining its energy through the oxidation of inorganic compounds.

Chemoorganotroph – An organism obtaining its energy through the oxidation of organic compounds.

Copiotroph – An organism adapted to a high-nutrient environment. This term is typically used in contrast with the term oligotroph, which is an organism adapted to a low-nutrient environment.

Domain – One of the three primary evolutionary divisions of life. The three domains are *Archaea*, *Bacteria*, and *Eukarya*. Modern molecular evidence suggests that the three-domain model of life should replace the five-Kingdom model associated with the classical Linnaean scheme. Domains *Archaea* and *Bacteria* comprise those organisms formerly assigned to Kingdom *Monera*, while the domain *Eukarya* comprises the Kingdoms *Animalia*, *Fungi*, *Plantae*, and *Protista*.

Environmental DNA sequence – A DNA sequence, often of a gene of interest, obtained from DNA purified from an environmental sample and not from a tissue sample or pure laboratory culture (e.g., of bacteria). Because a pure culture is necessary for the proper description of a bacterial species, those bacteria known only from environmental DNA sequences are considered to be discovered but not described, and they are not assigned a binomial species name.

Nitrogen fixer – An organism that can convert diatomic atmospheric nitrogen (N_2) to ammonia (NH_3).

Phototroph – An organism that can obtain metabolic energy through the capture of photons.

Sulfate reducer – An (anaerobic) organism that can utilize sulfate (SO_4^{2-}) as a terminal electron acceptor, essentially 'breathing' SO_4^{2-} instead of O_2.

Ultramicrobacterium – A very small bacterium, with a cell volume of less than $0.1\ \mu m^3$.

Viral lysis – The rupturing of an infected cell following the production of viruses. This results in cell death and the release of viral particles into the environment.

See also: Bacteria, Bacterioplankton; Cyanobacteria; Microbial Food Webs; Viruses.

Further Reading

Dworkin M, Falkow S, Rosenberg E, Schleifer K-H, and Stackebrandt E (eds.) (1999–2006) The Prokaryotes: An Evolving Electronic Resource for the Microbiological Community. New York: Springer-Verlag. http://141.150.157.117:8080/prok PUB/index.htm.

Glöckner FO, Zaichikov E, and Belkova N (2000) Comparative 16S rRNA analysis of lake bacterioplankton reveals globally distributed phylogenetic clusters including an abundant group of actinobacteria. *Applied and Environmental Microbiology* 66: 5053–5065.

Lindström ES, Kamst-Van Agterveld MP, and Zwart G (2005) Distribution of typical freshwater bacterial groups is associated with pH, temperature, and lake water retention time. *Applied and Environmental Microbiology* 71: 8201–8206.

Martiny JBH, Bohannan BJM, Brown JH, *et al.* (2006) Microbial biogeography: Putting microorganisms on the map. *Nature Reviews Microbiology* 4: 102–112.

Methé BA and Zehr JP (1999) Diversity of bacterial communities in Adirondack lakes: Do species assemblages reflect lake water chemistry? *Hydrobiologia* 401: 77–96.

Nold SC and Zwart G (1998) Patterns and governing forces in aquatic microbial communities. *Aquatic Ecology* 32: 17–35.

Pernthaler J (2005) Predation on prokaryotes in the water column and its ecological implications. *Nature Reviews Microbiology* 3: 537–546.

Pernthaler J and Amann R (2005) Fate of heterotrophic microbes in pelagic habitats: Focus on populations. *Microbiology and Molecular Biology Reviews* 69: 440–461.

Whitaker RJ, Grogan DW, and Taylor JW (2003) Geographic barriers isolate endemic populations of hyperthermophilic Archaea. *Science* 301: 976–978.

Yannarell AC and Triplett EW (2005) Geographic and environmental sources of variation in lake bacterial community composition. *Applied and Environmental Microbiology* 70: 214–223.

Zwart G, Crump BC, Agterveld M, *et al.* (2002) Typical freshwater bacteria: An analysis of available 16S rRNA gene sequences from plankton of lakes and rivers. *Aquatic Microbial Ecology* 28: 141–155.

Zwart G, Hiorns WD, Methé BA, *et al.* (1998) Nearly identical 16S rRNA sequences recovered from lakes in North America and Europe indicate the existence of clades of globally distributed freshwater bacteria. *Systematic and Applied Microbiology* 21: 546–556.

Fungi

V Gulis, Coastal Carolina University, Conway, SC, USA
K A Kuehn, University of Southern Mississippi, Hattiesburg, MS, USA
K Suberkropp, University of Alabama, Tuscaloosa, AL, USA

Introduction

Aquatic fungi are microscopic organisms with mostly mycelial growth and hyphae developing on or within their typically submerged organic substrates of plant or animal origin. Resident aquatic fungi are able to complete their life cycle in fresh waters and often have special adaptations for growth, sporulation, and dispersal in aquatic environments. A number of so-called transient fungi that are blown in from terrestrial ecosystems are regularly reported from fresh waters, but they may or may not be metabolically active under submerged conditions. Emergent macrophytes (e.g., cattail or reed stands) also harbor terrestrial fungi, especially during the aerial standing-dead decomposition stage. These fungal assemblages will also be considered here because of their crucial importance in decomposing plant detritus in wetlands and lake littoral zones. Freshwater fungi are rather difficult to observe and study due to unusual character of the habitat and the intimate association of fungal hyphae with the substrate they colonize. As a result, they were often overlooked by both aquatic ecologists and mycologists alike. Since the 1940s, some interest in the systematics and evolutionary relationships among these fungi has emerged. Furthermore, development and application of quantitative methods within the last two decades have established that fungi play a key role in the decomposition of plant litter in freshwater environments and are important mediators of energy and nutrient transfer to higher trophic levels (e.g., shredding invertebrates). Even though some freshwater fungi or fungus-like organisms are economically or ecologically important parasites of aquatic animals and plants, most freshwater fungi depend on dead coarse particulate organic matter (CPOM), mostly plant litter, as their primary source of energy and nutrients. This dead plant material may be of autochthonous (e.g., submerged or emergent macrophytes) or of allochthonous origin (leaf litter and wood from the riparian zone). This chapter will focus on fungi associated with plant litter in two freshwater ecosystems that have received the most attention: streams and wetlands or marshes. Since fungal activity and fungi-mediated processes differ between the lotic and lentic ecosystems, they will be discussed separately.

Fungal Diversity in Fresh Waters

About 2000 fungal species have been reported to date from submerged decaying substrates, as spores in water, parasites of aquatic plants or animals, or associated with decaying macrophytes in wetlands (**Table 1**). Overall, fungi are thought to be one of the largest groups of eukaryotes on Earth, second only to insects, with only about 5% of species described to date. True fungal diversity in fresh waters, therefore, is likely to be significantly higher than the 2000 species mentioned above. Representatives of all major taxonomic groups of fungi (Chytridiomycota, Zygomycota, Ascomycota, and Basidiomycota) have been observed, isolated, and described from freshwater habitats. Chytrids thrive in aquatic environments and are the only group of fungi producing zoospores and no true mycelium. Even though they are able to degrade recalcitrant compounds, such as chitin or cellulose, their functional role in organic matter decomposition has not been extensively studied. Some chytrid taxa are well-known parasites of algae or invertebrates. One species, *Batrachochytrium dendrobatidis*, has been recently identified as the etiological agent responsible for declines in amphibian populations worldwide. Very little is known about saprotrophic Zygomycota in fresh waters, although they can be detected on submerged substrates using molecular techniques. Representatives of one class within this phylum, the Trichomycetes, are quite common in fresh waters. These fungi are highly specialized obligate symbionts in the digestive tracts of arthropods and are often found in association with nymph or larval stages of aquatic insects. Apparently, aquatic environments have selected against basidiomycetes or their asexual (mitosporic) stages, since they are not commonly observed in fresh waters. The most common and ecologically significant group of fungi in fresh waters is ascomycetes and their anamorphs or mitosporic fungi, i.e., fungi whose sexual or meiosporic stages are unknown. The following discussions of fungal characteristics and processes and their importance in decomposition pertain largely to this group of fungi. Fungal-like organisms, the Oomycetes, are phylogenetically distant from the true fungi, but are traditionally studied by mycologists and are quite common in fresh waters as both saprotrophs on various organic

Table 1 Some estimates of fungal diversity in freshwater habitats

Taxonomic or ecological group	Number of species
Chytridiomycota	576
Trichomycetes	148
Ascomycota (meiosporic)	ca. 500
Basidiomycota	11
Aquatic hyphomycetes (mitosporic)	ca. 300
Aero-aquatic fungi (mitosporic)	90
Miscellaneous mitosporic fungi	405
Fungi on emergent macrophytes	>600[a]
Freshwater or amphibious lichens	>100
Oomycota (Saprolegniales)[b]	138

[a]600 species have been recorded from a single species of emergent macrophyte, *Phragmites australis*.

[b]Oomycota are not true fungi but fungus-like heterotrophs of the Kingdom Stramenopila.

1. Descals E (2005) Diagnostic characters of propagules of Ingoldian fungi. *Mycological Research* 109: 545–555.
2. Gessner MO and van Ryckegem G (2002) Water fungi as decomposers in freshwater ecosystems. In: Bitton G (ed.) *Encyclopedia of Environmental Microbiology*. New York: Wiley (online edition: DOI: 10.1002/0471263397.env314).
3. Shearer CA, Descals E, Kohlmeyer B, *et al.* (2007) Fungal biodiversity in aquatic habitats. *Biodiversity and Conservation* 16: 49–67.
4. Tsui CKM and Hyde KD (eds.) (2003) *Freshwater Mycology*. Hong Kong: Fungal Diversity Press.

substrates as well as parasites of fishes (*Saprolegnia*) and crustaceans (*Aphanomyces*). Their role in decomposition of plant litter is not well understood.

Fungi in Streams and Rivers

Headwater woodland streams receive about 90% of their carbon input from the riparian zone in the form of plant litter (leaves and wood), while in-stream primary producers, such as algae and macrophytes, play a greater role in large rivers or streams without a riparian canopy (e.g., grassland, desert, or tundra streams). Microbial decomposition of CPOM in lotic ecosystems is primarily carried out by fungi, while bacteria assume a greater role in the decomposition of fine particulate organic matter (FPOM, particle size < 1 mm) and dissolved organic matter (DOM). Aquatic hyphomycetes, which are mostly mitosporic ascomycetes, dominate on submerged leaf litter in streams, while both mitosporic fungi and ascomycetes colonize submerged wood. Currently, little is known about the diversity of fungi on submerged macrophytes in lotic waters and their role in litter decomposition.

Aquatic Hyphomycetes

The most well-studied group of fungi in streams are aquatic hyphomycetes or Ingoldian fungi (**Figure 1**).

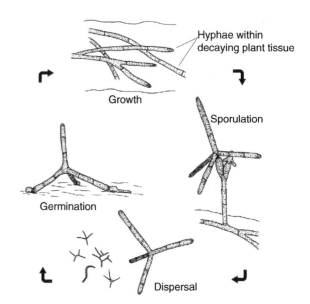

Figure 1 Life cycle of a typical aquatic hyphomycete (*Lemonniera aquatica*). Note that a few aquatic hyphomycetes may have sexual stages (not shown). Conidial shapes of several species of aquatic hyphomycetes are shown under 'dispersal.' Drawing by K. Suberkropp.

They produce large, distinctly shaped spores (conidia) (tetraradiate, sigmoid, or variously branched) that facilitate their dispersal and attachment to submerged substrates in lotic environments (**Figure 2**). Once conidia come in contact with the substrate, mucilage secretion, development of appressoria, and rapid germination (within a few hours) ensure their secure attachment and colonization of a new substrate. Fungal hyphae then grow mostly within the plant litter and finally give rise to conidiophores (conidia-bearing structures) that protrude from the substrate into the water and shed newly formed conidia. Once released, conidia are carried downstream by the current to colonize new substrates. Conidia may also become trapped and concentrated in foam at the air–water interface (neuston) where they can survive for weeks. As autumn-shed leaves or branches fall into the stream, they become colonized by conidia from neuston or by spores drifting in the water column.

Fungal Biomass, Production, and Reproduction

Current estimates of fungal biomass associated with plant litter are based on determining the concentration of ergosterol, a lipid specific to cell membranes of higher fungi. Fungal biomass associated with leaf litter in streams generally increases rapidly following colonization and peaks within 2–8 weeks depending on the type of substrate, temperature, stream water chemistry, and other environmental factors (**Figure 3(a)**). At its maximum, fungal biomass can comprise up to 23%

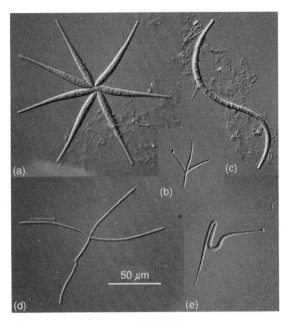

Figure 2 Spores of aquatic hyphomycetes. (a) *Flabellospora* sp., (b) *Alatospora acuminata* s.l., (c) *Anguillospora* sp., (d) unidentified conidium, and (e) *Condylospora* sp. All spores are from a single sample from a stream in Alabama. Differential interference contrast (DIC) micrographs by V. Gulis.

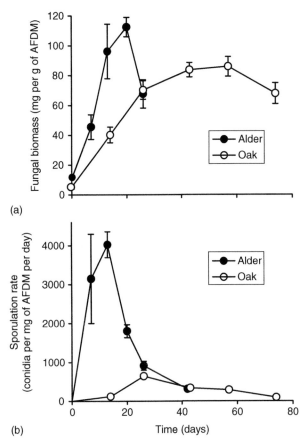

Figure 3 Dynamics of fungal biomass (a) and conidia production of aquatic hyphomycetes (b) associated with two types of decomposing leaf litter in a Portuguese stream. Symbols indicate means $\pm$ 1 SE ($n = 4$–8). On the basis of data from Gulis V, Ferreira V, and Graça MAS (2006) Stimulation of leaf litter decomposition and associated fungi and invertebrates by moderate eutrophication: Implications for stream assessment. *Freshwater Biology* 51: 1655–1669, and unpublished data by the same authors.

of total detrital mass; however, typical values are around 10% (**Table 2**). Fungal biomass usually declines during later stages of decomposition as considerable losses occur due to production of spores, senescence and death of hyphae, grazing by detritivores, and losses of mycelial fragments with or as FPOM. Fungal reproduction follows a similar pattern with maximum conidial production of aquatic hyphomycetes often occurring before the peak in biomass (**Figure 3(b)**). Up to 80% of fungal production may be allocated to conidia. This translates into 2–12% of leaf mass loss converted into fungal spores (or, in other terms, up to 7% of initial litter mass lost as conidia, e.g., **Figure 4**). Since massive numbers of conidia are released into the water column (up to $25\,000\,\mathrm{l}^{-1}$ in some streams), they can be captured (membrane filtration of stream water), counted, and identified based on their unique morphologies. This approach has provided valuable information on diversity and reproductive activity of aquatic hyphomycetes from a variety of streams.

Fungal growth rates (as determined from radiolabelled ^{14}C-acetate incorporation into ergosterol) on submerged leaf litter can be as high as 0.42 per day, but are typically an order of magnitude lower. Fungal production on decomposing leaves in streams has been found to peak at 0.6–16 mg of fungal C per g of detrital C per day, though variation for randomly

collected leaves at various stages of colonization can be very high. Even though fungal biomass on submerged wood may be almost as high as that on leaves, fungal growth rates and, consequently, production are about 5–10 times lower on wood than on leaves (**Table 2**). This can be explained by the poor oxygen supply to the inner layers of wood, the more recalcitrant nature of wood constituents (e.g., higher lignin content), and the lower nutrient (N and P) content of wood than leaf litter.

Fungal biomass and production associated with leaf litter (per m^2 of streambed) and concentration of aquatic hyphomycete conidia in water often show clear seasonal patterns in temperate streams. However, this seasonality is not directly driven by temperature, but rather by seasonal availability of the substrate (i.e., input of autumn-shed leaves). In

Table 2 Comparison of fungal and bacterial biomass and production associated with submerged plant litter in streams and rivers and emergent plant litter in freshwater marshes (standing-dead and submerged)

Parameter	Fungi	Bacteria	Source
Streams and rivers			
Leaf litter			
Biomass (mg of C per g of detrital C)	108–174	2–6	3
Biomass (mg of C per g of detrital C)	70–77	3–4	5
Biomass (mg of C per g of detrital C)	38–140	0.4–0.7	13
Production (mg of C per g of detrital C per day)	1.2–1.4	0.4	1
Production (mg of C per g of detrital C per day)	3.5–9	0.11–0.6	8,9
Production (mg of C per g of detrital C per day)	0.5–7.2	0.04–0.28	13
Wood			
Biomass (mg of C per g of detrital C)[a]	10–119	–	4
Biomass (mg of C per g of detrital C)[b]	2–55	0.4–3.8	11
Production (mg of C per g of detrital C per day)[a]	0.003–0.28	–	4
Marshes			
Emergent macrophytes			
Biomass (mg of C per g of detrital C)	5–34	0.002–0.05	2
Biomass (mg of C per g of detrital C)	8–50	0.1–0.5	6
Biomass (mg of C per g of detrital C)	2–62	0.1–1.7	7
Biomass (mg of C per g of detrital C)	7–63	0.3–0.4	10
Biomass (mg of C per g of detrital C)	20–124	0.02–2.1	12
Production (mg of C per g of detrital C per day)	0.02–1.4	<0.001–0.003	2
Production (mg of C per g of detrital C per day)	0.1–0.3	0.01–0.06	6
Production (mg of C per g of detrital C per day)	0.2–1.3	0.001–0.04	7
Production (mg of C per g of detrital C per day)	0.5–5.0	0.01–0.07	10
Production (mg of C per g of detrital C per day)	0.6–5.0	<0.001–0.005	12

Values are maximum biomass or production estimates from leaf litter decomposition experiments or from wood in streams (see notes) or ranges (min–max) for naturally decomposing emergent macrophytes in marshes.

[a]Data from randomly collected naturally occurring submerged small wood (<40 mm diam) collected seasonally over one year, n = 122.

[b]Data from submerged decaying wood veneers sampled only once in about 3 months after deployment.

Sources

1. Baldy V, Chauvet E, Charcosset JY, and Gessner MO (2002) Microbial dynamics associated with leaves decomposing in the mainstem and a floodplain pond of a large river. *Aquatic Microbial Ecology* 28: 25–36.
2. Findlay SEG, Dye S, and Kuehn KA (2002) Microbial growth and nitrogen retention in litter of *Phragmites australis* compared to *Typha angustifolia*. *Wetlands* 22: 616–625.
3. Gulis V and Suberkropp K (2003) Leaf litter decomposition and microbial activity in nutrient-enriched and unaltered reaches of a headwater stream. *Freshwater Biology* 48: 123–134.
4. Gulis V, Suberkropp K, and Rosemond AD, unpublished data.
5. Hieber M and Gessner MO (2002) Contribution of stream detrivores, fungi, and bacteria to leaf breakdown based on biomass estimates. *Ecology* 83: 1026–1038.
6. Kuehn KA, Lemke MJ, Suberkropp K, and Wetzel RG (2000) Microbial biomass and production associated with decaying leaf litter of the emergent macrophyte *Juncus effusus*. *Limnology and Oceanography* 45: 862–870.
7. Newell SY, Moran MA, Wicks R, and Hodson RE (1995) Productivities of microbial decomposers during early stages of decomposition of leaves of a freshwater sedge. *Freshwater Biology* 34: 135–148.
8. Pascoal C and Cássio F (2004) Contribution of fungi and bacteria to leaf litter decomposition in a polluted river. *Applied and Environmental Microbiology* 70: 5266–5273.
9. Pascoal C, Cássio F, Marcotegui A, Sanz B, and Gomes P (2005) Role of fungi, bacteria, and invertebrates in leaf litter breakdown in a polluted river. *Journal of the North American Benthological Society* 24: 784–797.
10. Sinsabaugh RL and Findlay S (1995) Microbial production, enzyme activity, and carbon turnover in surface sediments of the Hudson river estuary. *Microbial Ecology* 30: 127–141.
11. Stelzer RS, Heffernan J, and Likens GE (2003) The influence of dissolved nutrients and particulate organic matter quality on microbial respiration and biomass in a forest stream. *Freshwater Biology* 48: 1925–1937.
12. Su R, Lohner RN, Kuehn KA, Sinsabaugh R, and Neely RK (2007) Microbial dynamics associated with decomposing *Typha angustifolia* litter in two contrasting Lake Erie coastal wetlands. *Aquatic Microbial Ecology* 46: 295–307.
13. Weyers HS and Suberkropp K (1996) Fungal and bacterial production during the breakdown of yellow poplar leaves in two streams. *Journal of the North American Benthological Society* 15: 408–420.

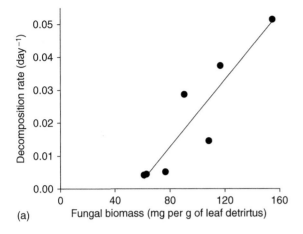

(a)

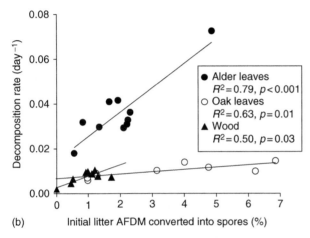

(b)

Figure 4 Relationships between the peak fungal biomass associated with 7 types of leaf litter and their decomposition rates (a) and between the cumulative aquatic hyphomycete spore output (as % of initial litter mass) and decomposition rates of leaf litter and wood (b). Data for panel (a) are from Gessner MO and Chauvet E (1994) Importance of stream microfungi in controlling breakdown rates of leaf litter. *Ecology* 75: 1807–1817. Data for panel (b) are from Ferreira V, Gulis V, and Graça MAS (2006) Whole-stream nitrate addition affects litter decomposition and associated fungi but not invertebrates. *Oecologia* 149: 718–729, and unpublished data by the same authors.

temperate streams, a major peak in fungal activity occurs in late autumn–early winter when the water temperatures are quite low. Interestingly, the life cycles of many detritivorous invertebrates (shredders) are timed to take advantage of these peaks in substrate availability, fungal activity and, hence, enhanced nutritional quality and palatability of litter.

Fungi associated with leaf litter typically account for 95–99% of total microbial biomass, whereas the biomass of bacteria is very low (**Table 2**). Fungi also dominate microbial communities of leaf litter in terms of production, even though bacteria have higher growth and turnover rates than fungi. Fungal production is comparable to or exceeds bacterial production from leaf litter by a factor of 1–627. Fungi also appear to dominate microbial communities colonizing submerged wood in streams (67–97% in terms of biomass). There are some indications that the relative importance of bacteria on submerged macrophytes may be higher than that on leaves and wood. However, more studies of these substrates are needed.

Fungal Role in Decomposition of Submerged Organic Matter and Their Importance at the Ecosystem Scale

Aquatic hyphomycetes have enzymatic capabilities to degrade the major constituents of plant litter. Extracellular enzymes produced by these fungi include enzymes that hydrolyze cellulose (endo- and exoglucanases, exoglucosidase) and hemicelluloses, pectin (polygalacturonase and pectin lyase), proteins, and lipids. Some laboratory studies have shown degradation of lignin or its derivatives by some aquatic fungi. However, ligninolytic activity has not been demonstrated under natural situations. In contrast, activities of multiple enzymes involved in degradation of other plant constituents have been reported from decomposing leaf litter in streams and have been correlated with fungal metabolic activities. Pectin degradation by fungi is probably a key process in the breakdown of leaf litter in streams, since it facilitates leaf maceration and FPOM release.

Fungal activity (e.g., peak biomass, sporulation) associated with leaf litter and wood in streams is positively correlated with final decomposition rates, indicating that fungi play a key role in CPOM dynamics (**Figure 4**). The activity of shredding invertebrates and mechanical fragmentation of leaf litter in streams often depends on the extent of fungal colonization. Increases in N and protein contents of leaf litter and partial digestion of refractory plant polymers due to fungal activity render leaf material a more nutritious and palatable food resource for invertebrate consumers. In feeding experiments, most shredding invertebrates show clear preference for fungal-colonized leaf material versus uncolonized leaves. As softening due to fungal enzymatic activity and shredding by invertebrates continue, leaf litter becomes more susceptible to mechanical fragmentation by water current and abrasion by sediments.

In some studies, fungal diversity has been shown to have a positive effect on fungal biomass and leaf litter decomposition rates; however, the magnitude of these effects is small, suggesting relatively high functional redundancy among species of aquatic hyphomycetes. Interestingly, some shredders show preferences for leaf litter colonized by a particular fungal species.

Fungal production can be significant in stream ecosystems and is of the same order of magnitude as bacterial and secondary invertebrate production. Estimates of fungal production associated with leaf litter in small streams range from 8 to 96 g of C per m^2 of streambed per year and are positively correlated with mean annual standing crop of leaf litter. Consequently, annual fungal production on an areal basis should be lower in larger streams and rivers, which receive lower CPOM inputs and are less retentive. Losses of leaf litter carbon as CO_2 due to microbial respiration (mainly fungal) can also be significant and have been estimated to account for 17–56% of total leaf C loss during decomposition. On the basis of production data and fungal growth efficiencies of 24–60% or direct respiration measurements, fungal assimilation can be calculated. Fungal assimilation (production plus respiration) accounted for 5–40% (in one instance as much as 97%) of the annual leaf litter input in several headwater streams. Indirect fungi-mediated losses of DOM and FPOM during decomposition have not been accounted for in these estimates, and the downstream transport of leaf litter in less retentive streams can be significant. Fungal activity on submerged wood is considerably lower than that on leaves. The available estimates of annual fungal production on small wood (<40 mm diam) from two streams are 4–6 g of C per m^2 of streambed per year. Nevertheless, fungi can assimilate a considerable proportion of annual small wood input to these streams (15–20%).

Factors Affecting Fungal Activity and Microbially Driven Plant Litter Decomposition in Lotic Ecosystems

Fungal activity associated with plant litter is controlled by characteristics of the substrate, environmental variables, and biotic interactions. Fungal reproduction, i.e., sporulation rate of aquatic hyphomycetes, is typically affected to a much greater extent than biomass accrual. The type of leaf litter or wood and, more specifically, the lignin, nitrogen, and phosphorus concentrations can exert strong control over fungal growth and reproduction. Lignin is difficult to degrade enzymatically due to its refractory structure. Hence, plant litter of low carbon quality (high lignin) supports low fungal activity and decomposes slowly. High N and P concentrations in plant litter often stimulate fungal activity. Thus, for a given substrate, fungal activity largely depends on the interplay of these intrinsic factors (i.e., lignin to nutrient ratios). To complicate the matter further, fungal activity

is also affected by dissolved inorganic N and P concentrations in stream water, since fungi are capable of taking up N and P from both the substrate and the overlaying water column. A number of laboratory and field studies have shown that when fungi are limited by N and P supply from the substrate or the water column (e.g., in pristine low-nutrient streams), even small experimental increases in dissolved inorganic nutrients result in considerable increases in fungal biomass, production, and reproduction, and, consequently, accelerated plant litter decomposition. From an environmental perspective, eutrophication of rivers and streams often leads to stimulation of fungal activity, faster plant litter disappearance, and, consequently, reduced resource availability (amount and timing) to higher trophic levels (e.g., shredding invertebrates).

Apart from the dissolved nutrients, other chemical parameters, such as pH, alkalinity, oxygen, pollution, etc., also affect fungal activity in streams. Aquatic hyphomycete diversity is typically higher in soft water streams, while fungal activity and leaf litter decomposition rates are usually greater in hard water streams. Such differences may be due to the greater activity of fungal pectin lyase in hard water streams and, consequently, faster leaf litter softening, maceration, and overall leaf litter mass loss. Organic pollution and sedimentation may also negatively affect fungal diversity and activity through oxygen limitation. Coal mine effluents and elevated heavy metal concentrations in water typically reduce fungal diversity and slow plant litter decomposition. However, some species of aquatic hyphomycetes appear to be resistant to high concentrations of heavy metals in chronically polluted streams or at least are more tolerant than aquatic invertebrates. Decomposition in these streams is generally slow, but this is more likely the result of a decreased presence of shredding invertebrates.

Elevated stream water temperatures positively affect fungal growth and litter degrading activity. Therefore, increasing temperatures as predicted from global climate change scenarios might lead to increased fungal activity and concomitant increases in litter decomposition rates. Such responses may have negative consequences that propagate through higher trophic levels as mentioned earlier for the effects of nutrients.

Diversity of riparian vegetation has a positive but rather small effect on aquatic fungal communities by affecting the diversity of resources available to fungi. Biotic interactions of fungi with stream inhabitants include interactions with bacteria and detritivores.

Fungi obviously compete with bacteria for detrital resources. Some aquatic hyphomycetes produce antibiotics that limit bacterial growth. However, interactions of fungi with bacteria are not well understood. Interactions with shredding invertebrates include direct competition for resources (plant litter), grazing by invertebrates, and direct ingestion of spores, which can pass undamaged through the digestive system facilitating fungal dispersal.

Fungi in Freshwater Marshes and Lake Littoral Zones

Freshwater marshes, including the littoral zone of lakes, are unique transitional habitats that occur at the interface between the terrestrial and aquatic environments. Emergent plants, such as cattails (*Typha*) or common reed (*Phragmites*), are common inhabitants of freshwater marshes. These plants often have very high rates of primary production, making freshwater marshes among the most productive ecosystems on the planet. Most of the living plant biomass is not consumed by herbivores in these ecosystems. Instead, the bulk of the plant material enters the detrital pool following plant senescence and death. Consequently, microbial decomposers (fungi and bacteria) and detritus-feeding consumers (macroinvertebrates) are important groups of organisms involved in the breakdown and mineralization of this plant matter in wetlands.

Characteristics of Emergent Plant Decomposition

Important aspects to consider when examining emergent plant decomposition are the spatial and temporal conditions under which plant litter naturally decomposes. In many emergent plants, abscission and collapse of leaves and culms to the sediments or overlying surface waters does not occur immediately after plant shoot senescence and death. A significant portion of the dead plant mass often remains in an aerial standing-dead position. As a result, the decomposition sequence of emergent plants typically involves two distinct spatial phases separated in time: an initial phase that occurs under aerial standing-dead conditions followed by a second phase that occurs under submerged conditions or on surface sediments. When litter decomposition studies have closely simulated natural decay conditions (i.e., standing-dead initially), fungi have been found to play a dominant role in decomposition in these environments.

Fungi Associated with Emergent Plant Litter

Current knowledge of fungal diversity associated with emergent plant litter largely comes from traditional microscopic studies where fungi associated with litter have been detected and identified by direct observation of reproductive structures (e.g., ascomata) either directly from field collected material or by subsequent culturing techniques in the laboratory. Ascomycetes are the most common fungal taxa encountered, including both hyphomycetous and coelomycetous anamorphs (mitosporic fungi). Basidiomycetes have also been observed on decaying emergent plant litter, but are much less frequent than ascomycetes.

Distinct spatial and temporal changes in fungal taxa have been observed during the decomposition of emergent plant litter. Terrestrial taxa are commonly observed during the initial stages of decomposition (standing-dead). These taxa are frequently replaced by fungi adapted to aquatic environments when shoots collapse to the sediments or surface waters. In addition to temporal shifts, fungi colonizing standing-dead litter also exhibit distinct spatial distribution patterns within the litter canopy. Different fungal taxa may occupy specific plant parts, such as leaves, sheaths, or the nodes and internodes of culms. These colonization patterns may be a result of spatial variation in environmental conditions within the litter canopy (temperature, water availability, etc.) and/or differences in the intrinsic quality of the plant litter substrate, such as the amounts of recalcitrant compounds (lignin) or available nutrients (N and P) within different plant tissues.

Fungal Biomass and Production

As plant litter decomposes, fungal hyphae pervasively invade the substrate, producing a suite of extracellular enzymes that enable fungi to degrade plant polymers, absorb, and assimilate plant litter carbon and nutrients into fungal biomass. In addition, fungi convert plant litter carbon into CO_2 as a result of their respiration. Despite well-documented evidence indicating fungal colonization of emergent plant litter, few studies have quantified the role of fungi in litter decay or their overall importance in wetland carbon and nutrient cycling. The use of the biochemical marker, ergosterol, to estimate litter-associated fungal biomass in freshwater marshes has provided compelling evidence that fungi are quantitatively important in decomposing emergent plant litter and play a pivotal role in marsh carbon and nutrient cycling. Fungal biomass associated with decaying litter increases gradually during shoot senescence and early standing-dead decomposition, with peak fungal biomass often accounting for as much as 10% of the total detrital mass. Differences in fungal biomass have been observed between plant organs (e.g., leaves vs. culms), with generally higher biomass being observed in leaf litter than culms (**Figure 5**). In addition, fungal biomass associated with standing-dead culms has been observed to vary spatially along the culm axis, with higher fungal biomass occurring in the upper portions of culms (**Figure 5**). These differences in fungal biomass between plant organs are likely due to the more recalcitrant nature (lignin concentration) and the lower amounts of nutrients (higher C:N and C:P ratios) in culm tissues compared to leaf tissues.

Significant changes in litter-associated fungal biomass often occur following the collapse of standing-dead litter. Fungal biomass rapidly decreases after submergence. This initial decline is followed by an increase in fungal biomass during later stages of submerged litter decay, possibly reflecting a shift to aquatic or semiaquatic fungal taxa. Despite changes in the environment, fungal decomposers typically account for a major portion of total microbial biomass associated with decaying litter. Simultaneous estimates of litter-associated fungal and bacterial biomass indicate that fungi typically account for >90% of the total microbial biomass on both standing and collapsed litter.

Fungal growth rates associated with both standing and collapsed plant litter are also significant, with peak growth rates as high as 0.10 per day.

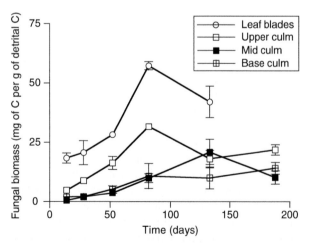

Figure 5 Fungal biomass associated with standing-dead leaf and culm litter of the freshwater emergent macrophyte *Erianthus giganteus*. Symbols indicate means ± 1 SE (*n* = 3). Data from Kuehn KA, Gessner MO, Wetzel RG, and Suberkropp K (1999) Standing litter decomposition of the emergent macrophyte *Erianthus giganteus*. *Microbial Ecology* 38: 50–57.

Corresponding fungal production has been found to peak at 1.0–3.0 mg of fungal C per g of detrital C per day. As observed with fungal biomass (above), fungal production also differs between plant organs, with higher production being observed in leaf than culm litter. Thus, like fungal biomass, fungal production may also vary depending on plant litter quality (lignin and nutrient concentrations). In addition, fungal production associated with emergent plant litter greatly exceeds bacterial production, accounting for >93% of the total microbial production when both microbial groups have been quantified simultaneously.

Factors Affecting Fungal Activity and Microbially Driven Plant Litter Decomposition in Marsh Ecosystems

As in streams, fungal activity associated with emergent plant litter in marsh ecosystems is controlled by a variety of physical and chemical conditions. These conditions are markedly different for fungi inhabiting standing-dead and submerged/sediment litter, as changes in the litter decay environment are accompanied by major shifts in both physical and chemical conditions. As indicated earlier, differences in litter-associated fungal biomass and production are often observed among plant litter organs (leaves vs. culms). Leaf litter typically contains lower concentrations of recalcitrant compounds (lignin) and higher concentrations of both nitrogen and phosphorus than corresponding culm litter. Thus, for a given litter substrate, fungal activity may depend on a combination of these internal factors (i.e., lignin to nutrient ratios, C:N:P ratios, etc.). To date, no studies have examined the potential influence of exogenous nutrient inputs (i.e., eutrophication) on fungal diversity or activity in freshwater marsh systems.

A number of laboratory and field studies have established that water availability is a key factor controlling the metabolic activities of microbial (fungal and bacterial) assemblages in standing-dead litter. Microbial respiration (CO_2 evolution) associated with standing-dead litter fluctuates rapidly upon exposure to wetting or drying conditions. Under laboratory conditions, rates of microbial respiration increase rapidly following addition of water to plant litter (from 10 to >200 μg of CO_2–C per g of AFDM per h within 5 min after being wetted). Respiration continues at high rates until plant litter is exposed to drying conditions. Under natural field conditions, rates of microbial respiration from standing-dead litter exhibit a cyclical diel periodicity (**Figure 6**), with the highest rates occurring at night when water becomes available (i.e., high relative humidity, dew formation) to litter inhabiting microorganisms. In contrast, microbial respiration virtually ceases during the day as a result of increased desiccation stress. These diel patterns suggest that temperature-driven increases in relative humidity and subsequent dew formation are important controlling mechanisms underlying the nighttime increases in water availability and associated microbial respiration in standing litter.

Distinct differences in patterns and rates of microbial respiration have been observed among plant organs (leaves vs. sheaths vs. culms). Maximum respiration rates observed from standing leaf litter are considerably higher (>24%) than rates from corresponding sheath litter experiencing identical environmental conditions (**Figure 6(a)**). Furthermore, maximum respiration rates from standing culm litter are often an order of magnitude lower than rates from both leaf and sheath litter. Variation in rates of microbial respiration from litter substrates can be attributed to differences in litter water absorption patterns, litter quality characteristics, and the extent of fungal colonization (biomass). Maximum rates of respiration among litter fractions are positively correlated with litter-associated fungal biomass (e.g., $r > 0.65$), suggesting that fungi are likely responsible for a major portion of the respiratory carbon release from standing-dead litter.

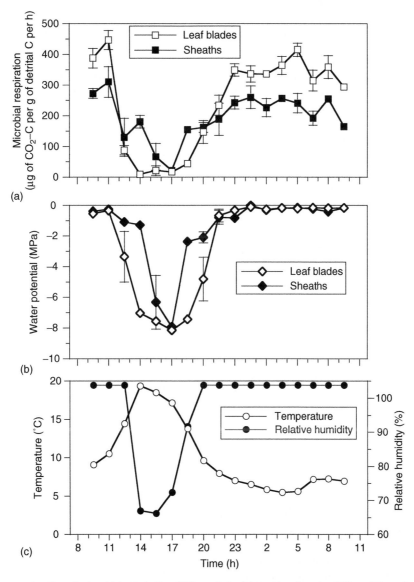

Figure 6 (a) Diel changes in rates of microbial respiration (CO_2 evolution) from standing-dead litter of *P. australis* within a temperate freshwater lake littoral marsh. Corresponding diel changes in plant-litter water potential (b) and temperature and relative humidity (c) are also illustrated. Symbols indicate means $\pm$ 1 SE ($n = 3$). Data from Kuehn KA, Steiner D, and Gessner MO (2004) Diel mineralization patterns of standing-dead plant litter: Implications for CO_2 flux from temperate wetlands. *Ecology* 89: 2504–2518.

Fungal Role in Decomposition of Emergent Plant Litter and their Importance at the Ecosystem Scale

When seasonal estimates of fungal biomass and production per gram of detritus are accompanied by areal (m^{-2}) estimates of emergent plant litter standing crop, the importance of fungi at the ecosystem scale can be estimated. Very few studies have attempted to quantify the impact of fungi at this scale. However, initial data suggest that fungal biomass and annual fungal production associated with wetland emergent plant litter per m^2 can be sizable when compared to other consumers. Because of considerable litter accumulation in freshwater marshes, annual standing stock of fungal biomass can average as much as 18 g of C per m^2. Substantial fungal production on areal basis have also been observed. For example, annual fungal production estimates associated with standing-dead *Typha angustifolia* leaf and stem litter totaled 70 and 45 g of C·per m^2 per year, respectively. When combined, these annual production estimates indicated that roughly 10% of the annual aboveground *Typha* production was transformed and assimilated into fungal biomass. The loss of detrital carbon due to microbial (fungal) respiration (CO_2 evolution) associated with emergent standing litter is also a significant pathway of carbon flow in freshwater marshes. When integrated on an areal basis, estimated daily flux rates of between 1.4 and 3.3 g of C per m^2 per day have been reported for microbial assemblages inhabiting standing-dead *Juncus effusus* litter in a subtropical wetland. These flux rates were similar to or greater than CO_2 flux rates from the wetland sediments. Similarly, daily CO_2 flux rates reported from standing-dead *Phragmites australis* litter in a north temperate freshwater marsh were lower (51–570 mg of C per m^2 per day), but within the range of CO_2 flux estimates reported from wetland sediments in this type of climates. Although few in number, these studies provide some evidence that fungi likely play a key role in wetland carbon and nutrient cycles.

Conclusions

Fungal diversity in fresh waters can be high and includes representatives from all major fungal phyla, with the Ascomycota exhibiting the highest diversity and the Basidiomycota being underrepresented in comparison to terrestrial environments. The microscopic size, intimate association of fungi with their substrates, and scarcity of aquatic mycologists make detection and reliable identification of aquatic fungi difficult. However, the development and application of biochemical and molecular techniques in recent decades has greatly improved our ability to detect and quantify fungi associated with different substrates. These methods should greatly increase our understanding of the occurrence and importance of fungi in freshwater ecosystems in the future. Even though the key role of fungi in decomposition of CPOM and, hence, carbon and nutrient cycling in streams and wetlands is now widely recognized, these aquatic organisms are clearly understudied in comparison to other groups, such as algae or invertebrates.

Major gaps in our knowledge about fungi in freshwaters include

1. the virtual absence of data on freshwater fungi from polar and some tropical regions,
2. a lack of understanding of the relative contribution of chytrids to the decomposition of plant litter in freshwater environments,
3. a lack of data on the relative importance of fungi and bacteria in decomposition of submerged macrophytes and wood,
4. fungi associated with and decomposition of submerged and floating-leaf macrophytes are poorly understood, and
5. more quantitative and modeling studies are needed to understand the effects of global change (temperature, elevated nutrient loading, precipitation) on fungi and fungi-driven processes in fresh waters.

See also: Bacteria, Attached to Surfaces; Microbial Food Webs; Protists.

Further Reading

Bärlocher F (ed.) (1992) *The Ecology of Aquatic Hyphomycetes.* Berlin: Springer-Verlag.

Gessner MO, Gulis V, Kuehn KA, Chauvet E, and Suberkropp K (1997) Fungal decomposers of plant litter in aquatic ecosystems. In: Kubicek CP and Druzhinina IS (eds.) *The Mycota: Vol. IV, Environmental and Microbial Relationships,* 2nd edn., pp. 301–324. Berlin: Springer-Verlag.

Gessner MO and van Ryckegem G (2003) Water fungi as decomposers in freshwater ecosystems. In: Bitton G (ed.) *Encyclopedia of Environmental Microbiology.* New York: Wiley. (online edition: DOI: 10.1002/0471263397.env314).

Graça MAS, Bärlocher F, and Gessner MO (eds.) (2005) *Methods to Study Litter Decomposition: A Practical Guide.* Dordrecht: Springer.

Gulis V, Kuehn KA, and Suberkropp K (2006) The role of fungi in carbon and nitrogen cycles in freshwater ecosystems. In: Gadd GM (ed.) *Fungi in Biogeochemical Cycles,* pp. 404–435. Cambridge: Cambridge University Press.

Shearer CA (1993) The Freshwater Ascomycetes. *Nova Hedwigia* 56: 1–33.

Shearer CA, Descals E, Kohlmeyer B, *et al.* (2007) Fungal biodiversity in aquatic habitats. *Biodiversity and Conservation* 16: 49–67.

Shearer CA, Langsam DM, and Longcore JE (2004) Fungi in freshwater habitats. In: Mueller GM, Bills GF, and Foster MS (eds.) *Measuring and Monitoring Biological Diversity: Standard Methods for Fungi*, pp. 513–531. Washington, DC: Smithsonian Institution Press.

Tsui CKM and Hyde KD (eds.) (2003) *Freshwater Mycology.* Hong Kong: Fungal Diversity Press.

Wong MKM, Goh TK, Hodgkiss IJ, *et al.* (1998) Role of fungi in freshwater ecosystems. *Biodiversity and Conservation* 7: 1187–1206.

Relevant Websites

http://aftol.org – Assembling the Fungal Tree of Life.

http://mycology.cornell.edu – Extensive directory of mycology internet resources.

http://www.life.uiuc.edu/fungi – Freshwater Ascomycete Database (also includes some data on mitosporic fungi).

http://www.indexfungorum.org – Index Fungorum (check current fungal names, search bibliography of systematic mycology).

http://www.msafungi.org – Mycological Society of America.

http://www.nhm.ku.edu~fungi – Trichomycetes, including online monograph.

http://www.botany.uga.edu/zoosporicfungi – Zoosporic Fungi Online.

Protists

R W Sanders, Temple University, Philadelphia, PA, USA

What are Protists?

The protists are an evolutionarily diverse group that includes most of what have been historically called protozoa, including ciliates, amoeba, heterotrophic flagellates, oomycetes, and slime molds plus phytoplankton, periphyton, and algae (red, green, and brown). Protists are primarily unicellular eukaryotes with heterotrophic and/or photosynthetic nutrition that are not plants, animals, or fungi. Like other eukaryotes, protist cells possess at least one membrane-bound nucleus plus endoplasmic reticulum and other organelles that can include mitochondria, chloroplasts, and flagella or cilia. Some protists are colonial or related to multicellular groups, but none develop tissue. The different ecological roles filled by protists within clearly recognized phyla led to some subgroups being historically described by both zoological and botanical nomenclature. For example, many dinoflagellate genera contain both phototrophic and heterotrophic species, and in addition to the occurrence of both photosynthetic and heterotrophic euglenids, some of photosynthetic species can be induced to lose their chloroplasts and survive using dissolved organic matter for energy and carbon.

In earlier taxonomic schemes 'Protista' was considered a Kingdom, but advances in systematics have shown that there are groupings within the protists that are not taxa of single origin (monophyletic) or that do not contain all the descendents of a common ancestor (holophyletic). Hence, protist is not a phylogentically useful term. Rather, it is a term of convenience that describes an assemblage of often distantly related organisms lumped together as a matter of ecological utility. The protists do have lineages that are properly organized to represent evolutionary relationships. Some clades, which are groups of ancestral species and all their known descendents, include only protists. Alveolata (ciliates, dinoflagellates, and apicomplexa), Cercozoa (amoeba and amoebaflagellates with fine pseudopods or filipodia), and Euglenozoa (euglenids, kinetoplastids, and diplonemids) typify purely protistan clades. Other clades, however, indicate a shared ancestry between some protists and plants or animals that is closer than their evolutionary relationship to other protists. For example, the choanoflagellates group with true fungi, chytrids, and animals as Opisthokonts, and the green algae, including common freshwater species like *Chlamydomonas* and *Volvox*, group with the Plantae.

As one might expect from such an evolutionarily and ecologically diverse group, making generalizations about the protists must be a cautionary exercise. Yet, for the practicing field biologist or ecologist, the small size and functional similarities of many of the protists makes grouping them together a rational practice. Biologists often still group the protists by their primary means of locomotion into flagellates, ciliates, and amoeba. These subdivisions are usually easily discernable with light microscopy and are retained here, not because they reflect current understanding of phylogeny, but rather because they often reflect niches and form the basis of much of the study of community ecology in freshwater protists to date.

Morphological Groups

Flagellates

Flagellates have one-to-many flagella that function in motility, attachment to a substrate, and also for feeding. They are phylogenetically diverse (**Table 1**), and are the most numerous of the freshwater protists reaching abundances in eutrophic systems that can approach 10^5 cells ml^{-1} and 10^6 cells cm^{-3} in the plankton and sediments, respectively. Purely heterotrophic free-living flagellates include choanoflagellates and bicosoecids. Choanoflagellates, easily recognized by their distinctive 'collar' of microvilli and a single flagellum, can have a lorica that is sometimes difficult to see with light microscopy. They are most often attached and/or colonial in fresh waters. Bicosoecids are heteroconts – possessing two different types of flagella – and are often attached by one of the flagella to a lorica or to the substrate. Cercazoans have two flagella, but are amoeboflagellates that use pseudopodia for feeding. Some species traditionally considered as amoeba are now also classified as Cercazoa (explained later).

Among the several taxa of photosynthetic flagellates are a number of groups with members that are wholly heterotrophic or that combine photosynthetic and heterotrophic nutrition (mixotrophy). Dinoflagellates, chrysomonads, euglenids, and cryptomonads have numerous phagotrophic or osmotrophic species that lack chloroplasts. Phagotrophy of bacteria or other protists by chloroplast-containing species has also

Table 1 Some common genera of heterotrophic flagellates

Group	Genera	Comments
Opisthokonta		
Choanomonada		Collar flagellates, with a single flagellum surrounded by cytoplasmic 'fingers' used
Monosigidae	*Monosiga*	in feeding on bacteria; mostly sessile/attached; *Salpingoeca* with a cellulose theca
Salpingoecidae	*Salpingoeca*	
Rhizaria		
Cercazoa	*Cercomonas*	Bacterivorous using pseudopods; amoebo-flagellates
Chromalveolata		
Haptophyta	*Chrysochromulina*	Some haptophytes in marine systems produce ichthyotoxin, although this has not been demonstrated in fresh waters
Straminopiles		
Bicosoecida	*Bicosoeca*	Immotile in a lorica, filter feeder on bacteria
Chrysophyceae	*Actinomonas, Ciliophrys*	Pedinellids, single flagellum, usually stalked, filter feeds on bacteria
	Paraphysomonas	One short and one long flagella; silica spines on cell body used to identify species requires electron microscopy
	Spumella (Monas)	With light microscopy nearly identical to *Paraphysomonas*.
	Ochromonas	Typically with chloroplast, which may be reduced in mixotrophic populations using osmotrophy or phagotrophy
Cryptophycea	*Goniomonas, Chilomonas*	Bacterivorous; *Chilomonas* species were recently reassigned, with freshwater members assigned as heterotrophic *Cryptomonas*
	Cryptomonas	Primarily photosynthetic species, some of these with low rates of phagotrophy reported (mixotrophy)
Alveolata		
Dinoflagellata	*Gymnodinium*	Two flagella, one horizontal around cell that gives slow spinning motion when swimming; photosynthetic, predatory and mixotrophic species
Excavata		
Euglenozoa		
Euglenida	*Peranema, Entosiphon, Astasia*	Most euglenids have two flagella emerging from an anterior pocket, with one trailing and usually in contact with the substrate. Colorless species ingest detritus, bacteria and/or are osmotrophic
Kinetoplastea	*Bodo, Rhyncomonas*	Only two common genera of free-living kinetoplastids; generally glide on substrates like euglenids and ingest attached bacteria

The higher ranking follows the 'nameless-rank' system of Adl *et al.* (2005), which considers the taxonomic endings to be accidents of history without the traditional hierarchical meaning of past classification codes. Some rankings are tentative, including the union of Alveolata with Haptophycea, Stromenopiles and Cryptophysea into Chromalveolata. Most of the higher groups include photosynthetic flagellates, some of which combine heterotrophic and phototrophic nutrition.

been reported for dinoflagellates, chrysomonads, and cryptomonads, and is discussed further in this article and elsewhere in this volume.

Many recent ecological field studies separate flagellates into different groups based primarily on size. This is most common in examinations of the microbial food web, where heterotrophic nanoflagellates (HNAN) are enumerated collectively due to their high abundance, importance as bacterivores, and the difficulty of separating taxa in preserved samples. This size-based assemblage originally described plankton that was ≤20 μm in diameter, but flagellates ≤5 μm often dominate this size category. It is difficult to separate individuals this small into different taxonomic groups by light microscopy; even division into functional groups of photosynthetic versus heterotrophic flagellates is problematic without proper fixation and slide preparation methods. Consequently, care must be taken to determine if reported results combined all small protistan cells into one trophic category or functionally divide groups into autotrophic and heterotrophic forms (usually based on the presence or absence of chloroplasts). Heterotrophic flagellates larger than 20 μm are typically much less abundant than nanoflagellates, and include many of the dinoflagellates and euglenids (**Figure 1**).

Amoeboid Protists

Amoebae are plastic cells that lack a fixed external morphology (**Figure 1**). They move primarily using pseudopodia with shapes that vary in different taxa from finely pointed and branching filose extensions to blunt lobopodia. The pseudopodia are used both for movement and for engulfing food particles. Several amoeboid groups have flagella during temporary swimming stages for dispersal, and some of these, including the stalked heliozoans, have affiliation with Cercozoa or Stamenopiles (**Table 2**).

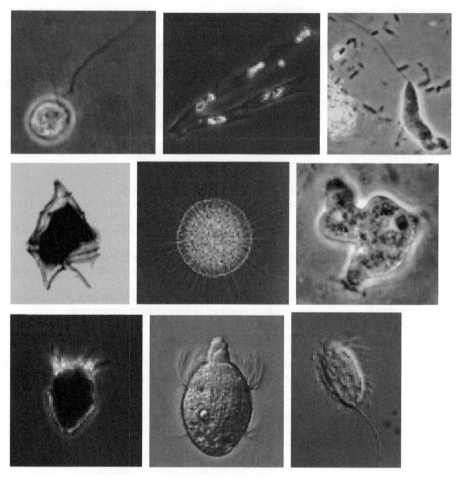

Figure 1 Photomicrographs of representative protists with the major locomotove modes (left-to-right beginning at top left). Flagellates: *Paraphysomonas vestita* (6 μm diameter), *Dinobryon* sp. (30 μm, a colonial mixotrophic alga). *Peronema* sp. (30 μm), (a colorless euglenid) *Peridinium limbatum* (70 μm, an armored dinoflagellate). Amoeboid protists: *Actinosphaerium* sp. (heliozoan, 200 μm), unidentified amoeba (25 μm). Ciliates: *Strombidium* sp. (50 μm), *Didinium* sp. (130 μm), *Cyclidium* sp. (20 μm). Cell size given as diameter or length, excluding flagella or cilia.

Amoebae are common on surfaces and in the sediments of fresh waters (mostly in the top cm), reaching abundances greater than $10^3 \, \text{cm}^{-3}$ in eutrophic lakes. Amoebae are usually less numerous in the plankton, but may also be concentrated in a microlayer associated with the air–water interface. Testate amoebae have vase-shaped coverings from which the pseudopods extend, and some of these species form oil droplets or gas vacuoles that reduce their density and allow them to rise from the sediment into the plankton. Nontestate planktonic amoebae may take on stellate floating forms, a flagellated morphology, or associate with floating detrital aggregates, none of which are taxa specific. Heliozoans are a polyphyletic group that has some amoeboid characteristics in that they form specialized pseudopods strengthened with a microtubular array (axopods). Heliozoans are not unusual in the plankton, but they are primarily benthic predators that use the axopods to

capture mobile prey such as ciliates from the water. Some species are attached to the substrate by stalks and others tumble across the sediment on their axopods.

Most amoebae use pseudopods to capture and/or engulf small prey including bacteria, algae and detritus. Some amoebae, including *Nuclearia delicatula* and some *Mayorella* and *Naeglaria* species are specialized predators of larger particles such as filamentous cyanobacteria, and can rapidly reduce prey abundances. Photosynthetic endosymbionts are associated with several amoebae, including a heliozoan (*Acanthocystis turfacea*). The relationship is even reflected in the species names of some naked (*Chaos zoochlorellae, Mayorella viridis*) and testate (*Paulinella chromatophora*) amoebae. Despite the widespread occurrence of some of these freshwater amoebae–algal symbioses, the degree to which they contribute to the nutrition of amoebae has not been

Table 2 Some common genera of amoeboid protists and their food

Group	Genera	Comments	Food
Amoebozoa			
Tubulinea			
Tubulinida	Amoeba, Chaos[a], Hartmannella		Bacteria and/or small protists
Testacealobosia	Arcella, Difflugia[a]	Testate/shelled, benthic and planktonic stages	Omnivores, algae, detritus
Flabellinea			
Dactylopodida	Mayorella[a]		Predators of protists, detritus
Vannellida	Vannella, Platyamoeba	Benthic and planktonic	Primarily bacteria
Acanthamoebidae	Acanthamoeba	Some species are facultative human pathogens	Primarily bacteria
Excavata			
Heterolobosea			
Vahlkampfilidae	Naegleria, Tetramitus, Vahlkampfia		Bacteria, cyanobacteria
Opisthokonta			
Mesomycetozoa			
Nucleariida	Nuclearia		Bacteria, cyanobacteria
Heliozoans[b]			
Cercozoa			
Nucleohelea			
Clathrulinidae	Clathrulina, Hedriocystis	Benthic, sessile, often stalked, sometimes colonial	Fine particulate matter, including bacteria
Stramenopiles			
Actinophryidae	Actinosphaerium, Actinophrys	Planktonic or mobile benthic	Other protists and rotifers
Centrohelida			
Acanthocystidae	Acanthocystis	Planktonic	Other protists, phytoplankton
Heterophridae	Heterophrys, Chlamydaster	Planktonic	

The higher ranking follows Adl et al. (2005), which considers the taxonomic endings to be accidents of history and without the traditional hierarchical meaning of past codes of classification. Many amoebae have flagellated stages, and flagellated forms dominate the life stages of some groups (e.g., Stramenopiles, Cercazoa).

[a]Some spp. with zoochlorellae or photosynthetic symbionts.

[b]Heliozoans have some general similarities, which include radiating stiffened pseudopodia, but the group is not monophyletic. Some heliozoan species have flagellated and/or free amoeboid stages and the group falls across taxonomic lines that include flagellates or are of uncertain affinities.

well studied. Some amoebae, including *Naegleria fowleri* and one or more *Acanthomoeba* are facultative pathogens of humans – the causative agents of amoeboid meningitis.

Ciliates

Ciliophora is one protistan lineage that was grouped together historically and is still recognized as monophyletic. Cilia are present during some stage of the life cycle in all ciliates, and are arranged in lines (kinities). The arrangement of kinities varies from a sparse distribution to a dense covering of the whole cell, and is an important morphological feature for ciliate identification. Cilia are sometimes arranged compactly to form compound organelles, such as cirri that are used for movement on surfaces, or as membranelles that direct currents toward the cytostome ('mouth'). Similar in structure to eukaryotic flagella, the cilia move water across the cell and contribute to both locomotion and food capture. Another morphologically distinguishing character of ciliates is nuclear dimorphism, in which

each cell has a least one macronucleus and one micronucleus. The macronucleus has multiple copies of a ciliates genome and controls RNA synthesis for cellular regulation. The diploid micronuclei undergo meiosis and are exchanged during conjugation with another cell when ciliates reproduce sexually.

Most ciliates are motile and have adaptations that allow them to move into the plankton (**Figure 1**), but many species are more commonly associated with the benthos. There also are species that use stalks, loricae, or a gelatinous matrix to attach to substrates. Some attached species are found frequently in the water column attached to detritus or as epibionts on planktonic organisms, including fish. Many ciliates are omnivorous, ingesting detritus, bacteria, and other small protists or even metazoans. Others are specialized bacterivores, algivores, or carnivores. Bacterivores and algivores typically generate water currents with cilia that bring prey toward the cytosome; others use cilia to sweep attached or settled particles from detritus or the substrate. Predatory ciliates often use toxicysts or extrusomes, which immobilize active prey. Ciliate species with endosymbiotic algae are fairly common and occasionally may reach biomass levels equal to that of phytoplankton, thus contributing substantially as primary producers. In addition, there are species that retain functional chloroplasts from ingested algal prey (kleptoplasty).

Although identification at the level of 'ciliate' seems straightforward, ciliature, size, and shape vary considerably. One group of sessile and predatory ciliates, the suctoria, has cilia only during its dispersal stages (**Table 3**). Additionally, ciliates overlap in size with several metazoan groups, such as rotifers, gastrotrichs, and turbellarians, which also are ciliated and can be mistaken for ciliates by beginning microscopists.

Reproduction and Population Growth

Protists reproduce asexually by binary fission, which can lead to rapid population growth. For heterotrophic protists, population growth rates are generally positively related to food concentration and temperature up to some critical level, and negatively related to species cell size. But generation times are highly variable and maximum growth rates of some species with relatively large size can approach those of much smaller protists. Populations of many protists can double several times per day in laboratory situations, although division rates of one to several days are probably more realistic for natural populations where food and predators may limit growth. Protistan productivity in natural systems, particularly in the benthos, is not well documented and is likely higher than is generally recognized. The high potential growth rates enable heterotrophic protists to persist during periods of high predation, and to respond rapidly to increases in fast-growing prey such as bacteria and photosynthetic protists.

Sexual reproduction is known for many protists and it takes varied forms, but the essential features are formation of gametic cells or nuclei, fusing of the nuclei and halving of the chromosome number

Table 3 Some common genera of ciliates and their freshwater habitats

Group	Genera	Habitat	Feeding mode/food
Karyorelictea	*Loxodes*	Planktonic and benthic	Omnivores, other protists, bacteria
Heterotrichea	*Blepharisma, Stentor[a]*	Mostly motile, planktonic, but also attached to substrate	Mostly filter feeders on bacteria, small protists
Spirotrichia			
Hypotrichia	*Aspidisca, Euplotes*	Mostly motile, benthic	Other protists and bacteria
Oligotrichia	*Strombidium[a], Halteria*	Mostly planktonic	
Choreotrichea	*Stobilidium*	Planktonic	
Stichotrichia	*Stylonychia*	Mostly planktonic	
Armophorea	*Metopus, Caenomorpha*	Planktonic, anaerobic	Mostly filter feeders on bacteria, typically dependent on methanogenic endosymbionts
Litostomatea Haptoria	*Didinium, Dileptus, Lacrymaria*	Motile; planktonic and benthic	Predators on motile protists
Phyllopharyngea Suctoria	*Tokophyra, Discophyra*	Sessile; cilia only during dispersal stage	Predators on motile protists
Nassophorea	*Nassula, Pseudomicrothorax*	Motile, benthic and planktonic	Algae, cyanobacteria
Colpodea	*Colpoda, Bursaria*	Motile, benthic or planktonic	Mostly filter feeders on bacteria, small protists
Protostomatea	*Coleps[a], Prorodon[a], Urotricha*	Mostly benthic, some planktonic	Omnivores; detritus, protists, decomposing animals
Oligohymenophorea			
Hymenostomata	*Colpidium, Glaucoma, Tetrahymena*	Mostly benthic	Mostly filter feeders on bacteria, small protists
Peniculia	*Frontonia[a], Paramecium[a], Stokesia[a]*	Benthic and planktonic	
Scuticociliata	*Cyclidium, Pleuronema, Lembadion*	Mostly planktonic	
Peritrichia	*Vorticella[a], Ophridium[a], Epistylis*	Sessile, attached by stalk, some epiplanktonic	

The higher ranking follows Adl *et al.* (2005). Most of the groups include species from several habitats, with the most common habitat listed first.
[a]Genera have at least one species with photosynthetic symbionts or sequestered chloroplasts.

at some point in the life cycle. Sex in protists is rarely associated with increases in the number of individuals; it interrupts asexual reproduction and the rapid population growth associated with binary fission. Sex appears to be essential for preventing senescence and extinction of clones of several protists, including *Paramecium*, but is unknown in some others. Autogamy, in which the meiotic products fuse in the same cell, only stalls senescence in *Paramecium*, but is the normal pattern in the heliozoan *Actinophrys*. When sexual reproduction does occur, it is usually linked to environmental change, and in some cases to the formation of cysts.

Cyst formation is not always associated with sex in protists. Some protists undergo binary division in cysts, while others do not increase in number during encystment. Whether or not encystment is associated with reproduction, it is frequently a response to environmental stress. The ciliate *Colpoda* and the amoeba *Acanthamoeba*, common in temporary bodies of water and soil, will encyst due to a variety of signals that are linked to drying. Food shortage, increased salt concentrations, high temperature, and low oxygen all can initiate formation of desiccation-resistant cysts. For species associated with temporary aquatic habitats, this can be an important means of survival as well as dispersal.

Distribution of Protists

In the Pelagic Zone

Protists are able to position themselves in the water column through several mechanisms, including production of gas or lipid vacuoles noted previously for the testate amoebae *Difflugia*. However, swimming with flagella or cilia is the most common way that motile algae, ciliates, and heterotrophic flagellates position themselves in response to a variety of gradients in a stratified water column. Light requirements obligate phototrophs to maintain themselves in the euphotic zone, but many phototrophic and some pigmented heterotrophic organisms, such as ciliates in the genus *Blepharisma*, will swim to avoid bright light. Motile photosynthetic organisms, such as dinoflagellates and ciliates with algal symbionts, are known to migrate toward the surface in the morning and into deeper, potentially nutrient-richer layers later in the day. Most heterotrophic protists tend to respond to other environmental parameters such as oxygen or food abundance. Some heterotrophs, including the ciliates *Metopus* and *Caenomorpha* (**Table 3**), are obligate anaerobes and are found only in anoxic areas of sediments and hypolimnia of

eutrophic waters. The ciliate *Loxodes* prefers low oxygen environments and will migrate out of anoxic sediments and congregate within a specific low oxygen layer in the overlying water. *Loxodes* uses geotaxis linked to oxygen concentration to maintain its position in the oxycline. Algivorous protists will tend to accumulate in layers where light and nutrient levels are best for their prey, just as the greatest abundance of bacterivorous protists, especially heterotrophic flagellates, are associated with areas of higher bacterial production.

Aggregates of detritus occur in the water column (as well as the benthos) and these will often have higher abundances of bacteria. Increases of bacterial prey lead to higher abundances of certain heterotrophic protists on these microhabitats relative to the surrounding water – especially in oligotrophic systems. Occurrence of 'benthic' protists, such as the flagellate *Bodo* and the ciliate *Colpoda*, in the water column may be due to association with these aggregates. Another distinct habitat, the air–water interface, also can be enriched with protists that use the surface as a substrate. Both naked and testate amoebae, such as *Arcella*, various ciliates and flagellates (e.g., *Codonosiga*) can hang below or move atop the water's surface feeding on bacteria or phototrophs growing there.

There tend to be strong seasonal changes in the abundance and species composition of aquatic protists. Phototrophic species often decline in winter and increase again with the longer days and greater light intensity in the spring. Bacterial activity increases with autotroph activity, since the photosynthetic organisms release dissolved organic matter and eventually die, and both of these processes supply substrate for bacteria. Heterotrophic protists will increase in response to abundance of algal and bacterial prey, and can equal or exceed zooplankton biomass during periods when their prey are abundant. Protistan population growth rate also tends to be positively related to temperature so the low photosynthesis and low growth rates of winter lead to minimum abundances, while protistan productivity tends to increase in warmer seasons.

In the Benthos

The benthos in both lotic and lentic systems tends to accumulate organic matter and detritus, which offers both a food source and a habitat for benthic protists. Attached and motile protists can occur in high numbers on and in the sediments. Attached heterotrophic organisms typically create a current to bring suspended food from the water (e.g., peritrich ciliates), or allow material to fall or swim into specialized feeding appendages (suctoria and heliozoans). Phototrophic species can form mats or biofilms that often have high abundances of cyanobacteria with or without protistan phototrophs. Motile protozoa move on and through these biofilms, and also at the surface and within the sediment. In sediments with relatively high organic matter content, gradients of oxygen, redox potential, pH, and other chemicals set up with sediment depth. As noted for the plankton, species that prefer oxic, microaerophyllic, or anoxic environments will move to appropriate depths in the sediment or will even leave the sediment and move into the water column. The sediments are a source of the anaerobic protozoa that appear in the water column when anoxic hypolimnia develop, typically during summer, in stratified lakes.

Ecological Roles and the Dynamics of Protists in Freshwater Systems

Primary Production

Although the influx of terrestrial carbon and its utilization in aquatic systems suggest that many lentic and lotic systems may tend toward net heterotrophy (i.e., ecosystem respiration exceeds primary production), photosynthesis still contributes significantly to the food webs of freshwater systems. Protists, including periphyton and phytoplankton, are major contributors to total aquatic primary production and to eutrophication. Photosynthetic species dominate the protistan biomass in the euphotic zones of lakes, but size, depth and nutrient concentration of a lake determine whether these phytoplankton are the major primary producers. The deeper the lake and the less area in the littoral zone, the more pronounced the phytoplankton contribution to primary production because the bulk of the benthic area is not within the euphotic zone. If dissolved nutrient levels are sufficient, phytoplankton also can reduce the light reaching sediments, and consequently shade the communities of periphytic algae (including photosynthetic protists) and submerged macrophytes. In shallow lakes, eutrophication may thus lead to a switch from a benthic to a pelagic dominance of primary production. In oligotrophic systems, benthic protists can contribute more to whole-lake primary production; light attenuation does not so strongly limit photosynthesis in deeper water and nutrients are available from the sediment. In streams and rivers, water depth and sediment load similarly affect the relative importance of the periphyton and phytoplankton, with periphyton contributing relatively more in low- to middle-order streams where more light reaches the benthos.

Food Webs and Protists

Advances in the last 30 years that produced data indicating high abundances and productivity of bacteria in aquatic systems also triggered hypotheses about the influence of protists on the population dynamics of bacteria. The microbial loop concept, as developed for planktonic marine systems, recognized that bacterioplankton incorporate dissolved organic matter into biomass and that bacteria are subsequently ingested by heterotrophic protists. Thus, fixed organic carbon that was once considered unavailable to most pelagic organisms could be recovered and funneled into the rest of the pelagic food web via a protistan link. Conversely, if there is efficient remineralization by bacteria and protists, little of the organic matter will immediately transfer to higher trophic levels, though this does not mean that it is necessarily lost from the system.

Individual feeding rates of protists tend to be inversely correlated with size; nanoflagellates ingest a few to 100 bacteria per flagellate per hour, while some ciliates ingest >1000 bacteria per hour. In freshwater systems, heterotrophic protists tend to be the major consumers of bacteria, and heterotrophic flagellates alone can sometimes remove more than 100% of the daily planktonic bacterial production. The usual dominance of bacterivory by heterotrophic flagellates reflects their high abundances, but ciliates and mixotrophic protists also can strongly reduce the abundance and alter the species composition of bacterioplankton – as can rotifers and cladocerans. These multiple trophic pathways moving microbial biomass into the higher trophic levels suggest that the 'microbial loop' should be considered a highly dynamic and persistent component of aquatic food webs rather than an independent entity that is distinct from the primary production–herbivore trophic pathway.

The impact of heterotrophic protists feeding on phototrophic protists in the plankton has received less attention than the trophic interactions of protists and bacteria. Many species of ciliates, amoebae, and dinoflagellates ingest phytoplankton, but rotifers and crustacean zooplankton tend to cause more phytoplankton grazing mortality. However, because of their efficiency in capturing small particles, protistan herbivores are relatively more important if pico-autotrophs dominate primary production. But protists also ingest larger cells, and both amoebae and ciliates can reduce populations of larger phototrophs. For example, the amoeba *Nuclearia* has been investigated as a potential biocontrol agent because of its high ingestion rates when feeding on filamentous algae and cyanobacteria. In sediments, protistan herbivory accounts for <5% of periphyton production. Likewise, studies to date suggest that protists ingest only about 5% of the daily bacterial production in sediments despite having feeding rates similar to those measured in the plankton. Under long-term anaerobic conditions bacterivory by benthic protists may be relatively more important as metazoan diversity decreases, but comparative studies are lacking.

Higher (metazoan) trophic levels utilize autotrophic and heterotrophic protists in planktonic and benthic environments. As noted earlier, autotrophic protists have long been recognized as the base of most aquatic food webs. Zooplankton and filter-feeding benthic organisms capture phytoplankton, and browsers/scrapers such as snails and various aquatic insects ingest periphyton. Heterotrophic protists, which can be as abundant as autotrophs and overlap in size, are ingested incidentally by these 'herbivorous' metazoans that feed by either size- or nonselective methods. Some copepods feed selectively on ciliates and it is likely that there are other such discriminating predators. In laboratory experiments, ciliates or HNAN alone are usually a sufficient and complete diet for metazoan survival. However, ciliates and HNAN as a sole food source sometimes were insufficient for robust reproduction suggesting nutritional deficiency. In natural systems, a mixed diet may overcome these limitations on reproduction observed with food consisting of one or a few species.

Predation by metazoans and other protists impacts populations of heterotrophic protists. Ciliates can be important predators of HNAN and other ciliates in the plankton and benthos. Most suctoria and heliozoans also feed on other protists. Predator removal (or addition) experiments in lakes demonstrated that moderate grazing by crustacean zooplankton can balance or exceed protistan population growth, and that selective feeding can alter protistan community structure. And though summer 'clearwater' phases observed in many lakes generally refer to a reduction in phytoplankton, predation by zooplankton also substantially reduces heterotrophic protist populations during these phases as well. In aquatic systems, it is likely that heterotrophic protists contribute to the diet of at least some life stage of most zooplankton, many fish, molluscs, and aquatic insects that also feed on photosynthetic protists.

Nutrient Cycling

The transfer of nutrients through different chemical states and ecological compartments (biotic and abiotic) is tightly linked to microorganisms – especially in aquatic systems where protists dominate primary production. Availability of dissolved phosphorus and/ or nitrogen limits primary production, and recycling is most important where allochthonous input of N and P is low. Heterotrophic protists can be efficient nutrient remineralizers, depending on the nutrient content of their prey, and contribute to phytoplankton nutrient demand in the epilimnion of stratified lakes when measurable dissolved nutrients are low. Recycling often leads to high nutrient retention in benthic systems within the photic zone of lakes and streams. In the benthos, phototrophic and heterotrophic microorganisms, both prokaryotic and eukaryotic, tend to live in close proximity and often within boundary layers that can lead to localized nutrient limitation and a subsequent importance of recycling.

Summary

Protists represent a ubiquitous, though taxonomically ill-defined, group of generally microscopic eukaryotes that include amoeboid, flagellated, and ciliated taxa. They contribute substantially to primary production, food web interactions, and nutrient recycling in aquatic ecosystems. All trophic levels, including primary producers, have protistan members. Heterotrophs feed as bacterivores, herbivores, carnivores, histophages, and parasites, or as omnivores. Several protistan groups have members that combine photosynthesis and heterotrophy (mixotrophy). Protists are the major predators of bacteria and are prey of benthic and planktonic metazoans. Consequently, dissolved organic matter recovered by bacteria is coupled to metazoans by heterotrophic and mixotrophic protists, while photosynthetic protists (phytoplankton and periphyton) are the base of aquatic metazoan food webs.

Glossary

Kleptoplasty – Retention of functional chloroplasts by a heterotrophic organism after ingestion of an autotroph. Unlike symbiotic algae, these chloroplasts lose functionality over time and must be replaced.

See also: Algae; Chrysophytes – Golden Algae; Diatoms; Green Algae; Harmful Algal Blooms; Microbial Food Webs; Other Phytoflagellates and Groups of Lesser Importance; Photosynthetic Periphyton and Surfaces; Phytoplankton Nutrition and Related Mixotrophy; Phytoplankton Population Dynamics: Concepts and Performance Measurement; Phytoplankton Productivity; Trophic Dynamics in Aquatic Ecosystems.

Further Reading

Adl SM, Simpson AGB, Farmer MA, *et al.* (2005) The new higher level classification of eukaryotes with emphasis on the taxonomy of protists. *Journal of Eukaryotic Microbiology* 52: 399–541.

Foissner W, Berger H, and Schaumburg J (1999) *Identification and Ecology of Limnetic Plankton Ciliates.* Munich: Bavarian State Office for Water Management.

Lee JJ, Leedale GF, and Bradbury P (eds.) (2000) *Illustrated Guide to the Protozoa* 2nd edn., 2 vols. Lawrence: Society of Protozoologists.

Patterson DJ (2000) *Free-living Freshwater Protozoa.* Boca Raton, FL: CRC Press.

Taylor WD and Sanders RW (2001) Protozoa. In: Thorp JH and Covich AP (eds.) *Ecology and Classification of North American Freshwater Invertebrates,* 2nd edn., pp. 43–96. San Diego: Academic Press.

Wehr JD and Sheath RG (eds.) (2003) *Freshwater Algae of North America.* San Diego: Academic Press.

Weisse T (2002) The significance of inter- and intraspecific variation in bacterivorous and herbivorous protists. *Antonie van Leeuwenhoek* 81: 327–341.

Relevant Websites

http://megasun.bch.umontreal.ca/protists/protists.html   –   Protist Image Data.

http://www.microbeworld.org/microbes/protista – Microbe World – Protista.

http://www.uga.edu/~protozoa – International Society of Protistologists.

Sulfur Bacteria

A Camacho, University of Valencia, Burjassot, Spain

Introduction

What are Sulfur Bacteria?

Sulfur is an essential component of organic matter. It is included in some protein-forming amino acids that have an important role in configuration of proteins. Sulfur is also present as other biological molecules in reduced forms. Consequently, all living organisms need sulfur as an elemental component, and uptake of sulfur for use in anabolic processes is therefore inherent to all life. Sulfur is acquired either in form of sulfate, with further reduction to sulfide to form the sulfhydryl group, or, in most heterotrophs, directly from the sulfur contained in the organic matter that they consume.

The average sulfur content of organisms is only 0.2% of dry weight. However, some organisms also use sulfur compounds, in much higher amounts than for anabolic processes, as a source of energy, reducing power (electron donor), or as electron acceptor. Since sulfur can be oxidized or reduced, redox reactions involving exchanges of energy and electrons can occur in nature. The release of energy and the capacity of accepting or transferring electrons make sulfur an element suitable for energetic metabolism. This type of metabolism, where most of the sulfur is not used to built up biomass, but to obtain energy or to act as mediators of redox reactions, implies the use of a much larger amount of the element, sulfur, than needed simply for assimilation.

Sulfur bacteria are defined based on the way that they use sulfur compounds in their energy conservation metabolic pathways, and there are three main options in their usage of sulfur. First, reduced sulfur compounds can provide the main electron donor for photoautotrophic growth. Second, sulfur-oxidation can give growth energy. Third, oxidized sulfur compounds can be used as electron acceptors for anaerobic respiration of organic matter. These three means of sulfur use are additional to uptake for anabolic processes, and consequently sulfur bacteria are considered as those prokaryotes using the inorganic sulfur compounds for energetic processes in chemical reactions that change the oxidation status of sulfur.

Nowadays, the classification of the prokaryotes is mainly based on the analysis of the 16S rRNA gene sequence. Increasing chemotaxonomic and phylogenetic data has led to a redefinition of bacterial taxonomy. Grouping within the higher taxa, however, should remain more stable than in recent years because classification based on phylogenetic traits is currently available for most groups of known prokaryotes. With this point of view, the taxonomic assignments given in this chapter should be considered as the current state of knowledge, with likely future changes provided in forthcoming versions of Bergey's Manual of Systematic Bacteriology. In addition to Bacteria, some Archaea, a phylogenetic domain distinct from Bacteria, can also dissimilatorily use inorganic sulfur compounds. These archaea will consequently be considered, although briefly, in this chapter.

Dissimilatory Metabolism in Sulfur Bacteria

Sulfur compounds are used in different ways by sulfur bacteria. Because of a variety of metabolisms noted in these bacteria, concepts regarding these metabolisms should be clarified. Definition of metabolism is based on four aspects: (1) the source of energy, (2) the source of reducing power, (3) the source of carbon, and (4) the electron acceptor used for respiration. Phototrophic organisms gain energy from light, whereas in chemotrophic (i.e., chemosynthetic) organisms energy is obtained from chemical substances. The latter can be differentiated as chemolithotrophic when the source is an inorganic substance and as chemoorganotrophic when the energy sources are organic compounds. The flow of reducing equivalents (electrons) is commonly associated with the energy conservation process. Reducing power and energy (ATP) generation can be jointly considered when referring to the terminology for the use of chemical electron donors and energy sources. For instance, the aerobic oxidation of elemental sulfur (S^0) to sulfate yields energy, part of which can be used as energy source by chemolithotrophic aerobic sulfur-oxidizing bacteria. Regarding reducing power, inorganic chemical compounds are the source of electrons (commonly coupled to energy generation) driving metabolic activities in lithotrophic organisms, whereas organic compounds act as electron donors in organotrophic organisms. With respect to the carbon source, autotrophs use CO_2 as source of carbon, while heterotrophs obtain carbon from organic matter. Finally, aerobic organisms use oxygen as a terminal electron acceptor for respiration, whereas anaerobic organisms use compounds other than molecular oxygen as terminal acceptors, such as sulfate for sulfate-reducing bacteria. Additionally, some compounds can be fermented, meaning that they are degraded with an endogenous organic electron acceptor without using an external electron acceptor. Fermentation yields less energy in comparison with that yielded by a compound oxidized using an external terminal acceptor. The metabolism of an organism can be consequently defined regarding these four physiological aspects, namely the sources of reducing power, of energy, of carbon, and the terminal acceptor for respiration. For example, a plant would then be regarded as an aerobic (oxygen used for respiration) photo- (energy from light) litho- (reducing power from water) autotroph (carbon from CO_2). A plant is then an aerobic photolithoautotrophic organism. Most types of combinations are found among bacteria, but additionally, many bacteria, including some sulfur bacteria, are metabolically versatile, allowing them to profit from different environmental conditions. In this chapter, when possible, the different taxa will be defined by metabolic type recognizing these assignments should be considered with precaution, because some taxa could facultatively follow other metabolic pathways than those reported here under certain conditions.

In aquatic inland environments, sulfide can come from abiotic sources, such as sulfide-rich geological emanations, or from biotic origins, usually from the reduction of sulfate or from the decomposition of proteins containing sulfur-rich amino acids. Among the biogenic sources, the amount of sulfide produced by anaerobic oxidation of organic matter with sulfate as electron acceptor is much larger ($\sim$567 g H_2S per kg of organic matter), compared with the much smaller amount of sulfur released by decomposition using other electron acceptors, where sulfide originate just from the sulfhydryl groups of the proteins ($\sim$10 g H_2S per kg of organic matter). Because sulfate is the most abundant form of sulfur in the biosphere, microbial sulfate-reduction is a key process that drives the biogeochemical sulfur cycle.

Sulfide has a considerable impact on the water and sediment chemistry as well as on the biota of inland aquatic ecosystems. Although used by several sulfur bacteria, sulfide is toxic for many organisms, because it readily reacts with cytochromes, haemoproteins, and other iron-containing compounds. Thus, the presence of sulfide restricts the composition of biological communities. Beyond certain concentrations, this compound is toxic even for sulfur bacteria. Relative tolerance varies among different sulfur-bacterial groups; of these groups, higher tolerance is exhibited by sulfate-reducers. These microorganisms can typically tolerate sulfide up to 10 mM, which is much higher than that commonly experienced in nature. From the chemical point of view, sulfur interacts strongly with the cycles of iron and phosphorus. Sulfur can determine the presence of different iron forms and, partly, phosphorus availability because the formation or dissolution of iron phosphates depends on the presence of sulfide, thereby influencing eutrophication.

Dissimilatory use of sulfur compounds by prokaryotes follow a diverse set of well known metabolic pathways associated with higher taxonomic units. A diversity of dissimilatory sulfur metabolism is evident even within taxa, as some bacteria have metabolic flexibility to perform different types of dissimilatory sulfur metabolism depending on environmental conditions. For example, some photolithotrophs can function in darkness as sulfur-oxidizing chemolithotrophs under aerobic conditions. It is therefore inadequate to establish exclusive definitions of physiological types of sulfur bacteria. However, to offer an overview, organisms are grouped by the ways through which sulfur compounds are used to obtain reducing power and energy, or as a respiratory electron acceptor (**Figure 1**).

Anoxygenic photosynthesis   Anoxygenic photosynthetic sulfur bacteria use light as energy source for fixing inorganic carbon into organic matter, mainly with reduced sulfur compounds and especially H_2S as electron donor. Because water, from which oxygen is produced in the oxygenic photosynthesis, is not used in this case as an electron donor, this photosynthetic process is anoxygenic. Instead, sulfide or other reduced sulfur compounds act as electron donor for photosynthetic carbon fixation and oxidized sulfur compounds such as elemental sulfur or sulfate are produced, as follows:

$$2H_2S + CO_2 \rightarrow CH_2O + H_2O + 2S^0$$

$$H_2S + 2CO_2 + 2H_2O \rightarrow 2CH_2O + H_2SO_4$$

Sulfide-dependent anoxygenic photosynthesis is performed by two main groups of sulfur bacteria – the purple and the green sulfur bacteria. These bacteria, which have photosynthetic pigments allowing light harvesting, thrive in anaerobic environments where sulfide is available. Additionally, anoxygenic photosynthesis with sulfide as electron donor has also been demonstrated for some cyanobacterial strains, although most cyanobacteria commonly perform plant-type oxygenic photosynthesis.

Chemotrophic sulfur oxidation   Chemolithotrophic sulfur prokaryotes use inorganic reduced sulfur compounds as electron donors and energy sources for their metabolism. Although some photosynthetic sulfur bacteria can also show facultative chemolithotrophic growth, the so-called colorless sulfur bacteria are those sulfur-oxidizers not performing anoxygenic photosynthesis, but instead obtaining energy from inorganic sulfur compounds (chemotrophic). The most

| Sulfur-oxidizing prokaryotes | | | Sulfate and sulfur reducing prokaryotes | |

Anoxygenic phototrophic sulfur bacteria

Chlorobiaceae	Chromatiaceae	Ectothiorhodospiraceae
Ancalochloris	Chromatium	Ectothiorhodospira
Chlorobaculum	Allochromatium	Halorhodospira
Chlorobium	Halochromatium	Thioalkalivibrio
Chloroherpeton	Lamprobacter	Thiorhodospira
Clathrochloris	Lamprocystis	
Prosthecochloris	Marichromatium	
	Pfennigia	
	Rhabdochromatium	
	Thermochromatium	
	Thioalkalicoccus	
	Thiocapsa	
	Thiococcus	
	Thiocystis	
	Thiodictyon	
	Thioflavicoccus	
	Thiohalocapsa	
	Thiolamprovum	
	Thiopedia	
	Thiorhodococcus	
	Thiorhodovibirio	
	Thiospirillum	

Colorless sulfur-oxidizing prokaryotes

Bacteria	Archaea
Thiobacillus	Sulfolobus
Thiosphaera	Acidianus
Thermothrix	
Beggiatoa	
Thiothrix	
Thioploca	
Thiodendron	
Thiobacterium	
Macromonas	
Achromatium	
Thiospira	
Thioalkalimicrobium	
Thioalkalispira	

Sulfate reducing prokaryotes

Bacteria	Archaea
Desulfovibrio	Archaeoglobus
Desulfoarculus	
Desulfobacca	
Desulfobacter	
Desulfobacterium	
Desulfobotulus	
Desulfobulbus	
Desulfocapsa	
Desulfocella	
Desulfococcus	
Desulfocella	
Desulfofustis	
Desulforhabdus	
Desulfohalobium	
Desulfomicrobium	
Desulfomonile	
Desulfonatronovibrio	
Desulfonema	
Desulforhopalus	
Desulfosarcina	
Desulfosporosinus	
Desulfotomaculum	
Thermodesulfovibrio	

Sulfur reducing prokaryotes

Bacteria	Archaea
Desulfuromonas	Sulfolobus
Desulfurella	Thermoproteus
Desulfomicrobium	Acidianus
Dethiosulfovibrio	
Desulfitobacterium	
Sulfospirillum	
Geobacter	
Pelobacter	
Desulfurobacterium	
Aquifex	

Figure 1 Selected genera of sulfur prokaryotes (Bacteria and Archaea) thriving in inland waters environments. Current generic names according to Bergey's Manual of Systematic Bacteriology and to The Prokaryotes (see 'Further Reading'), except for green sulfur bacteria, which is based on Imhof J (2003) Phylogenetic taxonomy of the family Chlorobiaceae on the basis of 16S rRNA and fmo (Fenna–Matthews–Olson protein) gene sequences. *International Journal of Systematic and Evolutionary Microbiology* 53: 941–951.

characteristic reaction performed by chemolithotrophic sulfur-oxidizing bacteria is the aerobic oxidation of hydrogen sulfide with oxygen as electron acceptor,

$$H_2S + \frac{1}{2}O_2 \rightarrow S^0 + H_2O,$$

although the use of electron acceptors other than oxygen (e.g., nitrate) has also been reported. Since under aerobic conditions sulfide is chemically oxidized by oxygen, the coexistence of both oxygen and sulfide in nature is restricted to environments where there is a continuous supply of both substances, either by in situ production or by external renewal. The elemental sulfur produced by reducing sulfide can be accumulated inside the bacterial cell or released from the cell. Elemental sulfur can be further oxidized to sulfate when sulfide is not available

$$S^0 + 1\frac{1}{2}O_2 + H_2O \rightarrow H_2SO_4$$

Sulfate and sulfur reduction Oxidized sulfur compounds can be used as terminal electron acceptor for the anaerobic respiration of organic matter by sulfate- and sulfur-reducing bacteria producing hydrogen sulfide (H_2S). These bacteria are commonly heterotrophic, obtaining organic matter from external sources, although some strains can be facultative autotrophs or mixotrophs (use of both organic and inorganic compounds as carbon source). To be used as an electron acceptor, sulfur must have a certain level of oxidation. Among the sulfur compounds, sulfate is not the only one used as electron acceptors, but others such as sulfite, thiosulfate, hyposulfite and elemental sulfur may also be used. Sulfate and sulfur-reducing bacteria obtain energy by coupling the oxidation of organic compounds (or H_2) to the reduction of sulfate or other sulfur compounds to sulfide, which is released into the environment:

$$H_2SO_4 + 2CH_2O \rightarrow 2CO_2 + 2H_2O + H_2S$$

The process requires a nonlimiting electron donor and sufficient sulfate, usually in the range of several millimoles per liter. This dissimilatory sulfate reduction, which is characteristic of sulfate-reducing prokaryotes, differs from the assimilatory sulfate reduction, a widespread capacity in prokaryotes and plants, which generates reduced sulfur for biosynthesis without sulfide being released into the environment.

Inland Waters Environments Where Sulfur Bacteria Thrive

The need for sulfur compounds in a certain redox status determines the type of environment in which various types of sulfur bacteria can thrive. Some of them have requirements of resources that commonly have opposite gradients in nature. For example, oxygen and sulfide react chemically. Sulfide and light sources have opposite gradients in aquatic habitats, as light comes from the surface and sulfide from the anaerobic decomposition of organic matter mainly occurring in bottom layers and the sediments. When resources are supplied from different directions, microorganisms tend to accumulate at the interface often showing very high metabolic rates enhancing the chemical gradient. This is the reason why many sulfur bacteria are typically located in zones where physical and chemical gradients are steep. Contrastingly, other sulfur bacteria such as sulfate-reducers mostly require stable anaerobic conditions.

Some signs of the suitability of an environment for sulfur bacteria can be easily recognized with the naked eye. The characteristic smell of sulfide indicates that this substance is available for sulfur-oxidizers and suggests active sulfate-reduction. The black color in the sediments is due to metal sulfides formed from the biologically released sulfide and available metals such as iron. White patches of elemental sulfur resulting from the oxidation of sulfide in contact with the air also reflect the suitability of the habitat, as well as the characteristics of the substrate when it is rich in gypsum (calcium sulfate). These suitable environments can be recognized in spite of the absence of signals of the presence of sulfur bacteria. Sometimes even the presence of sulfur bacteria can be easily recognized, such as the case of red waters or sediment coloring by purple sulfur bacteria, or the white hair-shape structures formed by filamentous colorless sulfur bacteria. In the following sections, the main types of environments where sulfur bacteria can be found in inland water habitats will be examined, with a brief description of the features of these environments affecting or being affected by sulfur bacteria.

Chemocline and hypolimnia of stratified lakes   The chemocline of stratified lakes   is an environment typically occupied by massive developments of photosynthetic sulfur bacteria. The chemical gradient (so-called chemocline), representing the transition from the well-oxygenated surface waters to the deep anoxic layers, is usually located at the bottom of the metalimnia or the uppermost hypolimnia of thermally stratified lakes (**Figure 2**), and/or in the top of the monimolimnia of meromictic lakes. At the chemocline, strong physical and chemical gradients are established during the stratification period, since the density gradient sustained by differences in water temperature or salinity impedes turbulent diffusion of chemical compounds. This density difference promotes the separation of the top well-oxidized photic epilimnion from the bottom part of the metalimnion and the hypolimnion, characterized by relatively low light availability and microaerobic or anoxic conditions. Anoxygenic photosynthetic bacteria are consequently restricted to a zone in the vertical profile with a continuous supply of hydrogen sulfide. The relative stability of the thermal stratification in lakes, lasting usually for several months, allows these bacteria to develop dense populations just below the depth of oxygen exhaustion, where they can form abundance maxima (plates). Water in these layers can be purple or green in color because of the bacterial photosynthetic pigments (**Figure 2**). Because of differences in their main ecological requirements, light and sulfide, vertically stratified populations of purple sulfur bacteria, are usually located in shallower depths relative to green sulfur bacteria.

In lakes, the depth of the oxic–anoxic interface largely depends on the input of organic matter to bottom layers and the sediments. The higher the input the greater the oxygen demand. Dead organic matter sinks to deep layers and to the sediments, where it is decomposed and mineralized to inorganic compounds. Sulfide can be derived from the degradation of sulfur contained in this organic matter. Since the relative contribution of sulfur to organic matter is low, the amount of sulfide provided in this way depends on the amount of organic matter decomposed. The higher the trophic status (i.e., productivity) of the lake, the more organic matter sinks to bottom layers and the more sulfide is produced. However, when sulfate is available in sufficient quantity it can be used as an electron acceptor under anaerobic conditions, and a much higher amount (around 60 times more) can be produced from the same amount of organic matter. In this case most of the sulfide is obtained from the reduction of sulfate instead of only from desulfuration of organic matter. This is the case of sulfate-rich lakes, such as brackish coastal lakes, lakes located on gypsum

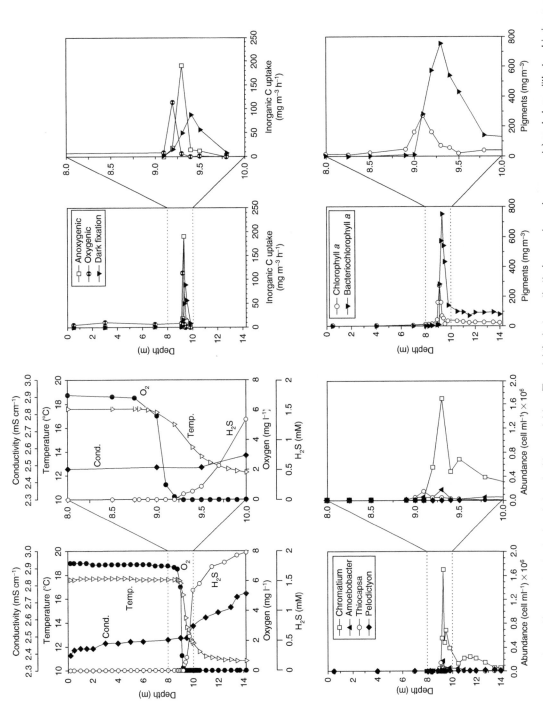

Figure 2 Vertical profiles of (Top left) the main physical and chemical key variables; (Top right) photosynthetic (oxygenic and anoxygenic) and chemolithotrophic inorganic carbon assimilation; (Bottom left) abundance of photosynthetic sulfur bacteria (names based on phenotypic features); and (Bottom right) concentrations of main chloro-pigments, in the sulfate-rich Lake Arcas (Cuenca, Spain) near the end of the stratification period. For each paired graph, the right graph represents a zoom of the left graph centered two meters around the oxic–anoxic interface.

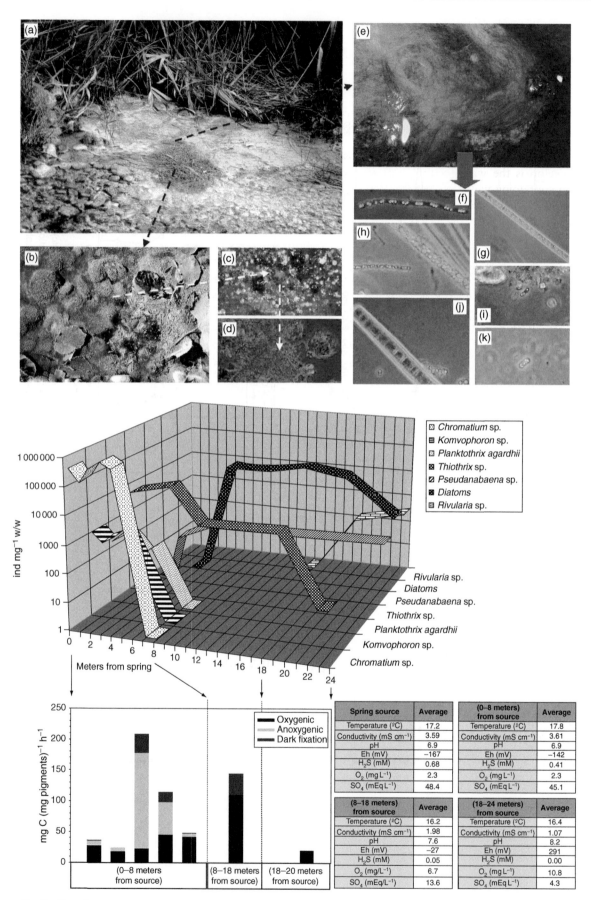

The legend for the 3D plot lists:
- *Chromatium* sp.
- *Komvophoron* sp.
- *Planktothrix agardhii*
- *Thiothrix* sp.
- *Pseudanabaena* sp.
- *Diatoms*
- *Rivularia* sp.

Spring source	Average
Temperature (ºC)	17.2
Conductivity (mS cm⁻¹)	3.59
pH	6.9
Eh (mV)	−167
H₂S (mM)	0.68
O₂ (mg L⁻¹)	2.3
SO₄ (mEq L⁻¹)	48.4

(0–8 meters) from source	Average
Temperature (ºC)	17.8
Conductivity (mS cm⁻¹)	3.61
pH	6.9
Eh (mV)	−142
H₂S (mM)	0.41
O₂ (mg L⁻¹)	2.3
SO₄ (mEq L⁻¹)	45.1

(8–18 meters) from source	Average
Temperature (ºC)	16.2
Conductivity (mS cm⁻¹)	1.98
pH	7.6
Eh (mV)	−27
H₂S (mM)	0.05
O₂ (mg/L⁻¹)	6.7
SO₄ (mEq/L⁻¹)	13.6

(18–24 meters) from source	Average
Temperature (ºC)	16.4
Conductivity (mS cm⁻¹)	1.07
pH	8.2
Eh (mV)	291
H₂S (mM)	0.00
O₂ (mg L⁻¹)	10.8
SO₄ (mEq L⁻¹)	4.3

Figure 3 Continued

karst, or saline endorheic lakes from sulfate rich lands, where sulfate concentrations are high enough to ensure unlimited availability of this electron acceptor for the anaerobic degradation of organic matter. Sulfate reduction in lakes mainly occurs within or in close proximity to sediments.

The water–sediment interface Another type of gradient environment where some sulfur bacteria can develop is the water–sediment interface. Here, oxygen-containing water contacts anaerobic sulfide-releasing sediments. In these environments, colorless sulfide-oxidizing bacteria can find both substances needed for chemolithotrophic growth.

Sediments The sediments of aquatic habitats are a sink for organic matter. Because oxygen diffusion through the sediment is very low, oxygen is primarily consumed by aerobic organisms, thus turning most of the sediments anaerobic. Depending on the supply of organic matter and oxygen exchange with the overlying waters, the anoxic part could be located up to a few centimeters below the sediment surface. Most commonly, oxygen penetrates only the top few millimeters of the sediments. In anaerobic sediments electron acceptors other than oxygen can be used for the anaerobic remineralization of the organic debris as for example in sulfate reduction.

Microbial mats are multilayered microbial communities that can include layers of sulfur bacteria. In inland waters these communities are frequent in shallow sediments of saline lakes or brackish coastal lagoons, and can harbor several types of photosynthetic organisms, such as diatoms, cyanobacteria, and/or photosynthetic sulfur bacteria, as well as other sulfur bacteria such as colorless sulfur-oxidizers and sulfate-reducers. These organisms vertically stratify according to the physical and chemical gradients that are created by their own metabolic activities. Remarkably, the concentrations of oxygen and sulfide change through the vertical profile during the day, depending on the balance between productive and consumptive activities by microorganisms, some of which may track these changes with diel vertical migrations.

The sulfuretum A sulfuretum is a peculiar type of environment usually associated with a spring source of sulfide-rich waters (**Figure 3**). These waters are commonly anoxic and contain sulfide, but they are released into an oxygen-rich environment because of the contact with the atmosphere. Under these conditions oxygen and sulfide coexist, and chemical and biological sulfide oxidation reactions can occur, either by aerobic sulfur-oxidizing bacteria when both substances coexist or by photosynthetic sulfur bacteria where light and sulfide are available. Commonly, the relative amount of sulfide and oxygen changes with the distance to the source, with sulfide-rich anoxic waters at the source point which are progressively enriched in oxygen and impoverished in sulfide away from the source (**Figure 3**). Some of these environments have high temperatures (hot springs) under which characteristic thermophilic species of sulfur bacteria and archaea can develop.

Saline and coastal lakes Waters of athalassohaline (inland saline) lakes and coastal lagoons are usually rich in sulfate as a result of salt accumulation in endorheic basins or by the influence of sea water, respectively. Active sulfate reduction can occur in these lakes with sulfide release from sediments to the water column, which in turn (together with the surface sediments) can be colonized by sulfur-oxidizing photosynthetic or chemolithotrophic bacteria. To balance the outside osmotic pressure in saline environments bacteria accumulate osmotically-active molecules that are compatible with molecular cell structures and metabolic processes. Microbial mats, as described earlier, are usually also found in these environments, especially in the saline inland lakes and

Figure 3 Ecological scheme of a sulfuretum (Fuente Podrida, Valencia, Spain) based on unpublished data. (Top) Macro- and microphotographs, track arrows: (a) source and the surroundings, (b) red patches corresponding to purple sulfur bacteria, (c) detail of a red patch, (d) microphotograph of the patch showing the accumulation of *Chromatium* sp. cells, (e) Detail of the cyanobacterial and colorless filamentous sulfur-oxidizing bacterial mats, (f–h) microphotographs of colorless filamentous gliding sulfur bacteria with sulfur granules, (j) filamentous cyanobacterium, (i,k) *Chromatium* spp. with sulfur granules. (Center) Abundance (cell or filaments per gram of wet weight) and distribution of the dominant microorganisms, *Chromatium* sp. is a purple sulfur bacterium restricted to the anoxic zones of the sulfide-rich waters located close to the spring, *Thiothrix* sp. is a colorless sulfur-oxidizing bacterium distributed through the zones of coexistence of oxygen and sulfide, *Komvophoron*, *Planktothrix*, and *Pseudanabaena* are sulfide-resistant cyanobacteria, *Rivularia* is a cyanobacterium whose oxygenic photosynthesis is inhibited by sulfide, diatoms are sulfide-inhibited eukaryotic algae. (Bottom left) Photosynthetic (oxygenic and anoxygenic) and chemolithotrophic (dark fixation) inorganic carbon assimilation rates in different zones of the sulfuretum, depending from the distance to the sulfide-rich spring. (Bottom right) Average values of selected physical and chemical variables for spring water and waters in several zones located far from the source, where inorganic carbon assimilation was measured. Further information can be found in Camacho A, Rochera C, Silvestre JJ, Vicente E, and Hahn M (2005) Spatial dominance and inorganic carbon assimilation by conspicuous autotrophic biofilms in a chemical gradient of a cold sulfurous spring: The role of differential ecological strategies. *Microbial Ecology* 50: 172–184.

alkaline soda lakes located in areas with dry climatic conditions that facilitate gradual salt accumulation in depressions. Saline inland lakes are often hypersaline (i.e., above the salinity of seawater). Soda lakes are a specific type of salt lake with sodium bicarbonate as dominant salt causing high pH and, among sulfur bacteria, can host mainly alkaliphilic species of purple sulfur bacteria and sulfate-reducers.

Small ponds rich in organic matter   In small ponds accumulating high amounts of organic matter, either from natural or anthropogenic origin, oxygen can deplete in quiet waters or in parts of the water body where oxygenation is diminished. In these cases, when enough sulfate is available, sulfate-reducers oxidize organic matter, producing hydrogen sulfide, which in turn is used by photolithotrophic or chemolitho-trophic sulfur-oxidizing bacteria. Typically, these bacteria form patches in the microenvironments where their requirements are met, such as the lower face of leafs deposited on the bottom of shallow ponds. These kinds of environments include natural habitats, such as wind-sheltered ponds in deciduous forest areas, as well as human-made environments such as waste-water stabilization ponds.

Acidic aquatic environments   Some species of color-less sulfur-oxidizing prokaryotes can live under extremely acidic conditions as low as a pH of 1. In inland waters, these environments are found, for example, in acid mine drainage waters and in streams coming from pyrite-rich areas where sulfuric acid is derived from the oxidation of sulfidic minerals.

Ecological Characterization of Inland Water Sulfur-Using Prokaryotes

Sulfur bacteria are not a phylogenetically-related cluster, with taxa spread among most phylogenetic groups. They also include a wide metabolic diversity. Some taxonomic groups, especially those sharing the same type of basic metabolism, are often closely related. However, even within metabolic types there are polyphyletic origins. For example, anoxygenic photosynthetic bacteria correspond to two different phylogenetic clusters, with purple bacteria (Chroma-tiaceae) corresponding to the γ-proteobacteria and the green sulfur bacteria forming a separate phylum. Still, the treatment of such metabolic groups is the most appropriate approach to understanding the eco-logical significance of sulfur bacteria, as well as their physiological properties that explain their distribu-tion and role in inland aquatic environments. Conse-quently, a functional approach to illustrate the main types of sulfur bacteria will be used here.

Photosynthetic sulfur bacteria   Photosynthetic sulfur bacteria are characterized by a metabolism based on the photosynthetic bioconversion of inorganic carbon into organic matter using reduced sulfur com-pounds, commonly sulfide, as electron donor, and light as energy source. Light availability, both quantity and quality, and sulfide concentrations are the main ecological factors determining the distribution and growth of photosynthetic sulfur bacteria.

Photosynthetic sulfur bacteria can reach very high abundances in aquatic environments where the anoxic sulfide-containing waters overlap the photic zone. The contribution of photosynthetic sulfur bac-teria to primary production in lakes is related to light availability, as well as to other resources such as sulfide or inorganic nutrients. Inorganic carbon fixation rates (per volume of water) of photosynthetic sulfur bacteria are often much higher than those from the overlying phytoplankton (**Table 1**), but usually only the uppermost part of the bacterial population is photosynthetically active (**Figure 2**). Because of the narrow depth range where these high photosynthetic rates occur, the contribution of the phototrophic bac-teria to lake primary production is usually modest, although in some lakes, at least during certain peri-ods, they can account for a majority of the planktonic primary production.

The fate of most carbon fixation by photosynthetic sulfur bacteria is sediment deposition, since predation in the anoxic layers where these bacteria develop is generally quite low. However, diel fluctuations in the oxygen-sulfide interface allow some metazooplank-ton that occupy the microaerobic layers, such as some rotifers or microcrustaceans, to occasionally feed on purple sulfur bacteria. Additionally, anaerobic ciliates in anoxic layers are also potential grazers on phototrophic bacteria, but the studies reported so far on the impact of metazooplankton and protistan grazing on phototrophic bacterial populations have demonstrated that loss rates to grazers are usually low. Other organisms, such as predatory bacteria or viruses, have also been reported to impact on photo-trophic sulfur bacteria.

Phototrophic sulfur bacteria are divided into two main groups, the purple sulfur bacteria and the green sulfur bacteria, although some purple and green nonsulfur photosynthetic bacteria, which will not be considered here, can also use H_2S as an elec-tron donor as an alternative to organic donors.

Purple sulfur bacteria (PSB)   Purple sulfur bacteria (the Chromatiales) are anoxygenic phototrophs that mainly grow photolithoautotrophically in the light using sulfide or elemental sulfur (zero-valent sulfur), among other reduced sulfur compounds, as an electron

Table 1 Inorganic carbon photoassimilation rates of photosynthetic sulfur bacteria in several lakes

Lake	% of inorganic C-photoassimilation	Main anoxygenic phototroph(s)	Inorganic carbon photoassimilation by photosynthetic sulfur bacteria				Source
			$mg\,C\,m^{-3}\,h^{-1}$	$mg\,C\,m^{-3}\,day^{-1}$	$mg\,C\,m^{-2}\,day^{-1}$	$g\,C\,m^{-2}\,year^{-1}$	
Arcas	12 (d)	*Chromatium weissei*	197				3
Banyoles	14 (d)	*Chlorobium phaeobacteroides, Chromatium minus*				18	18
Belovod	9 (d)	*Chromatium okenii*		50–210	55		23
Big Soda	10 (a)	*Ectothiorhodospira vacuolata*	11.4		110–210	50	5
Buchensee	4 (a)	*Amoebobacter purpureus, Pelodictyon phaeoclathratiforme, Chloronema spp.*	1–7				19
Cadagno	25 (d)	*Amoebobacter purpureus*	51		300		4
Cisó	25 (a)	*Thiocystis (Chromatium) minus, Amoebobacter sp.*			147	55.8	11, 12
Cisó	92 (a)	*Chromatium minus*				250	18
Chernyi Kichier		*Pelodictyon luteolum*		60			14
Dagow		*Pelochromatium roseum*	3.7				13
Deadmoose	17.1 (a)	*Lamprocystis roseopersicina*	6.8 – 63.4	60	75	14	16
Faro	45 (d)	*Chlorobium phaeovibrioides*		60	75		24
Fayetteville Green	83 (a)	*Chlorobium phaeobacteroides*		1628	2500	239	7
Fish	1.1 (d)	Photosynthetic sulfur bacteria			13.4		21
Haruna	4.5 (a)–20 (d)	*Chromatium sp.*	184	50	60	3.6	26
Holmsjön		*Chromatium sp.*					17
Kinneret	1–8 (d)	*Chlorobium phaeobacteroides*					2
Kisaratsu	47–85 (d)	*Chromatium sp., Chlorobium sp.*	154	1000	600		26
Knaack	4.7 (a)	*Pelodictyon sp., Clathrochloris sp.*		35		16.7	22
Konon'er		*Pelochromatium roseum, Amoebobacter roseus, Thiocapsa sp.*					14
Kuznechikha	38 (d)	*Chlorochromatium aggregatum, Chloronema sp.*		100			14
Mahoney	40 (d)	*Amoebobacter purpureus*	14–168			33.5	15, 20
Maral-Gëi		*Thiocapsa sp., Chlorobium phaeobacteroides*		120			14
Mary	0.26 (d)	*Pelodictyon sp., Clathrochloris sp. Chlorobium sp.*			1.98		21
Medicine	43 (a)–55 (d)	*Lamprocystis roseopersicina*		2000	190	110	cited in 1
Mirror	3.8 (d)	*Lamprocystis sp., Chromatium sp., Pelodictyon sp., Chlorobium sp.*			59.8		21
Mogil'noe		*Chlorobuim phaeovibrioides, Pelodictyon phaeum*		380			14
Muliczne	9–34 (d)	*Thiopedia rosea*		90	28–157		9
Paul	5.7 (d)	*Prostecochloris sp., Pelodictyon sp.*			16.8		21
Peter	2.9 (d)	Photosynthetic sulfur bacteria			10.5		21
Pluβsee	25 (d)	*Ancalochloris sp., Pelodictyon sp.*		35			cited in 1
Popówka Maia		*Chlorobium limicola*		177			8
Pomyaretskoe	45 (d)	*Chlorobium sp.*		90			14
Repnoe	15 (d)	*Chlorobium phaeovibrioides*		160			14
Rose	6.3 (d)	*Pelodictyon sp., Clathrochloris sp., Chlorobium sp.*			32.9		21

Continued

Table 1 Continued

Lake	% of inorganic C-photoassimilation	Main anoxygenic phototroph(s)	Inorganic carbon photoassimilation by photosynthetic sulfur bacteria				Source
			$mg\,C\,m^{-3}\,h^{-1}$	$mg\,C\,m^{-3}\,day^{-1}$	$mg\,C\,m^{-2}\,day^{-1}$	$g\,C\,m^{-2}\,year^{-1}$	
Smith Hole (annual average)	32 (a)	Chromatium sp.			91		27
(maximum)	92 (d)						
Solar	91 (d)	Chromatium violascens, Prostecochloris aestuarii, (Oscillatoria limnetica)		4960	5960 8015		6
Suigetsu	20 (d)	Chromatium sp., Chlorobium sp.		65	50		26
Valle de San Juan	3 (a)	Amoebobacter roseus, Pelodictyon phaeum		190	38	14	cited in 1
Vechten	3.9–17.5 (d), 3.6 (a)	Chloronema sp., Thiopedia sp., Chromatium sp.	4.1			6	25
Veisovoe	20 (d)	Pelodictyon phaeum		350			14
Waldsea	46 (a)	Chlorobium sp.		1320		32	16
Wadolek	15–62 (d)	Chlorobium limicola		20			8
Zaca	8 (a)	Thiopedia rosea	7.2		19–55		10

Percentage of contribution of inorganic C-photoassimilation by photosynthetic sulfur bacteria to overall lake primary production calculated on (a) annual or (d) daily basis. Taxonomic assignments and units for photosynthetic rates are as given in the data sources.

Sources

1. Biebl H and Pfennig N (1979) Anaerobic CO_2 uptake by phototrophic bacteria. A review. *Archiv für Hydrobiologie Beiheft. Ergebnisse der Limnologie* 12: 48–58.
2. Butow B and Bergstein-Ben Dan T (1992) Occurrence of *Rhodopseudomonas palustris* and *Chlorobium phaeobacteroides* blomms in lake Kinneret. *Hydrobiologia.* 232: 193–200.
3. Camacho A and Vicente E (1998) Carbon photoassimilation by sharply stratified phototrophic communities at the chemocline of Lake Arcas (Spain). *FEMS Microbiology Ecology* 25: 11–22.
4. Camacho A, Erez J, Chicote A, Florin M, Squires MM, Lehmann C, and Bachofen R (2001) Microbial microstratification, inorganic carbon photoassimilation and dark carbon fixation at the chemocline of the meromictic Lake Cadagno (Switzerland) and its relevance to the food web. *Aquatic Sciences* 63: 91–106.

5. Cloern JE, Cole BE, and Oremland RS (1983) Autotrophic processes in meromictic Big Soda Lake, Nevada. *Limnology and Oceanography* 28: 1049–1061.

6. Cohen Y, Krumbein WE, and Shilo M (1977) Solar Lake (Sinai). 2. Distribution of photosynthetic microorganisms and primary production. *Limnology and Oceanography* 22: 609–620.

7. Culver DA and Brunskill GJ (1969) Fayetteville Green Lake, New York. V. Studies of primary production and zooplankton in a meromictic marl lake. *Limnology and Oceanography* 14: 862–873.

8. Czeczuga B (1968) An attempt to determine the primary production of the green sulphur bacteria, *Chlorobium limicola* Nads, (Chlorobacteriaceae). *Hydrobiologia* 31: 317–333.

9. Czeczuga B (1968) Primary production of the purple sulphuric bacteria *Thiopedia rosea* Winogr. (Thiorhodaceae). *Photosynthetica* 2: 161–166.

10. Folt CL, Wevers MJ, Yoder-Williams MP, and Howmiller RP (1989) Field study comparing growth and viability of a population of phototrophic bacteria. *Applied and Environmental Microbiology* 55: 78–85.

11. García-Cantizano J, Casamayor EO, Gasol JM, Guerrero R, and Pedrós-Alió, C (2005) Partitioning of CO_2 incorporation among planktonic microbial guilds and estimation of in situ specific growth rates. *Microbial Ecology* 50: 230–241.

12. Gasol JM, Mas J, Pedrós-Alió C, and Guerrero R (1990) Ecología Microbiana y Limnología en la Laguna Cisó: 1976–1989. *Scientia Gerundensis* 16: 155–178.

13. Glaeser J and Overmann J (2003) Characterization and in situ carbon metabolism of phototrophic consortia. *Applied and Environmental Microbiology* 69: 3739–3750.

14. Gorlenko VM, Dubinina GA, and Kuznetsov SI (1983) The ecology of aquatic microorganisms. Stuttgart: E. Schweizerbart'sche Verlagsbuchhandlung (Nägele U. Obermiller).

15. Hall KJ and Northcote TG (1990) Production and decomposition processes in a saline meromictic lake. *Hydrobiologia* 197: 115–128.

16. Lawrence JR, Haynes RC, and Hammer UT (1978) Contribution of photosynthetic green sulphur bacteria to total primary production in a meromictic saline lake. *Verhandlungen Internationale Vereinung für Theoretishe und Angewandte Limnologie* 20: 201–207.

17. Lindholm T and Weppling K (1987) Blooms of phototrophic bacteria and phytoplankton in a small brackish lake on Aland, SW Finland. *Acta Academiae Aboensis* 47: 45–53.

18. Montesinos E and van Gemerden H (1986) The distribution and metabolism of planktonic phototrophic bacteria. In: Megusar F and Gantar M (eds.) *Perspectives in microbial ecology*, pp. 349–359. Ljubljana: Slovene Society for Microbiology.

19. Overmann J and Tilzer MM (1989) Control of primary productivity and the significance of photosynthetic bacteria in a meromictic lake, Mittlerer Buchensee, West-Germany. *Aquatic Sciences* 51: 261–278.

20. Overmann J, Beatty JT, Krouse HR, and Hall KJ (1996) The sulphur cycle of a meromictic salt lake: The effect of light intensity. *Limnology and Oceanography* 41: 147–156.

21. Parkin TB and Brock TD (1980) Photosynthetic bacterial production in lakes: The effect of light intensity. *Limnology and Oceanography* 25: 711–718.

22. Parkin TB and Brock TD (1981) Photosynthetic bacterial production and carbon mineralization in a meromictic lake. *Archiv für Hydrobiologie* 91: 366–382.

23. Sorokin YI (1970) Interactions between sulphur and carbon turnover in meromictic lakes. *Archiv für Hydrobiologie* 66: 391–446.

24. Sorokin YI and Donato N (1975) On the carbon and sulphur metabolism in the meromictic Lake Faro (Sicily). *Hydrobiologia* 47: 241–252.

25. Steenbergen CLM and Korthals HJ (1982) Distribution of phototrophic microorganisms in the anaerobic and microaerophilic strata of Lake Vechthen (The Netherlands). Pigment analysis and role in primary production. *Limnology and Oceanography* 27: 883–895.

26. Takahashi M and Ichimura KS (1968) Vertical distribution and organic matter production of photosynthetic sulfur bacteria in Japanese lakes. *Limnology and Oceanography* 13: 644–655.

27. Wetzel RG (1973) Productivity investigations of interconnected marl lakes (I). The eight lakes of Oliver and Walter chains, northeastern Indiana. *Hydrobiological Studies* 3: 91–143.

donor for photosynthetic carbon fixation through the Calvin reductive pentose phosphate cycle. Many species are strictly anaerobic and obligate phototrophs, whereas others also grow chemolithoautotrophically or chemoorganoheterotrophically. Purple sulfur bacteria (PSB) include two families of γ-Proteobacteria – the Chromatiaceae and the Ectothiorhodospiraceae. Apart from phylogenetical and other chemotaxonomic differences, the main feature differentiating these families is that elemental sulfur coming from sulfide oxidation accumulates inside the cells of Chromatiaceae and outside the cells of Ectothiorhodospiraceae. PSB have bacteriochlorophyll *a* (many also bacteriochlorophyll *b*) as the main photopigment, and these molecules have strong absorption in the near infrared (**Figure 4**). However these wavelengths, as well as others where bacteriochlorophyll *a* could harvest photons, are strongly absorbed by water or by phytoplankton situated in the overlying waters (**Figure 5**). Consequently, light-harvesting strategies based on bacteriochlorophyll are only successful in shallow water bodies mainly allowing growth at the water–sediment interface. In microbial mats located in shallow environments the relative availability of infrared wavelengths is much higher than in the anoxic layers of lakes. In sediments, however, light attenuation limits the habitat of photosynthetic bacteria to the upper few millimeters of the anoxic zone. In contrast, most planktonic sulfur bacteria colonizing deep lake layers must use other

light-harvesting molecules. The carotenoids with absorption maxima at 480–550 nm are more efficient for light-harvesting at the wavelengths dominating at these depths (**Figure 5**). Among these carotenoids, okenone is the most efficient and is present in meta-hypolimnetic species, although other carotenoids such as spirilloxanthin, lycopene, rhodopinal, or related molecules are produced by various purple sulfur bacteria.

The family Chromatiaceae include species from freshwater environments (although most tolerate moderate salinities), as well as salt-requiring species distributed in marine or saline inland waters environments. In inland waters, these bacteria thrive in anoxic stagnant water bodies and/or sediments, where enough light arrives to allow phototrophic growth. Some Chromatiaceae are adapted to high temperature environments and cold habitats. The Chromatiaceae commonly present peculiar cell inclusions or structures, such as sulfur globules, gas vesicles, and storage polymers of polysaccharides, volutine (polyphosphate), and poly-β-hydroxybutyrate, which can also influence cell density. Sulfur accumulation by Chromatiaceae represents a competitive advantage of these bacteria over other anoxygenic photosynthetic bacteria that deposit elemental sulfur outside the cells. The intracellular sulfur granules serve not only as electron donors for photosynthesis in the absence of dissolved sulfide, but are also used as electron acceptors for endogenous fermentation of stored carbohydrates

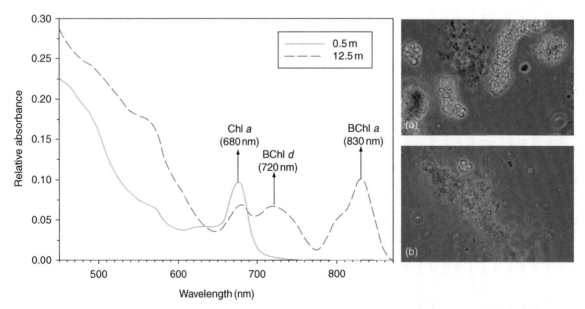

Figure 4 In vivo absorption spectra of selected samples from the meromictic Lake La Cruz (Cuenca, Spain). Note the strong absorption by chlorophyll *a* at 680 nm in the epilimnetic sample (0.5 m) where only algae are present, compared with the hypolimnetic sample from 12.5 m showing the near infrared absorption maxima of bacteriochlorophylls from purple sulfur bacteria (photograph a: *Lamprocystis purpurea*, formerly *Amoebobacter purpureus*) at 830 nm corresponding to bacteriochlorophyll *a*, and green sulfur bacteria (photograph b: *Chlorobium clathratiforme*, formerly *Pelodictyon clathratiforme*) at 720 nm corresponding to bacteriochlorophyll *d*.

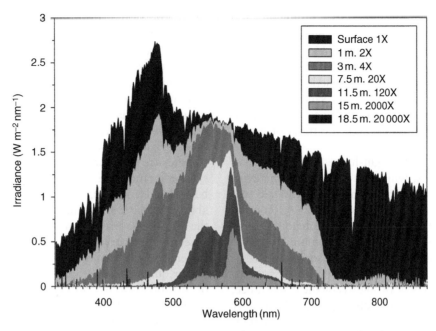

Figure 5   Spectral distribution of light availability at different depths of the meromictic Lake La Cruz (Cuenca, Spain) measured with a spectroradiometer. Note the magnification level for the scale at each depth.

under dark anoxic conditions. Although some species are obligate phototrophs using sulfide or elemental sulfur as the only electron donor, others have the capacity for complementary growth or maintenance strategies, which provides metabolic flexibility. These alternative metabolic strategies can serve to cope with the changing conditions at the oxic–anoxic interfaces of the environments where these bacteria develop. Some Chromatiaceae, mainly freshwater species, move by polar flagella, whereas other planktonic species can change their buoyancy by means of cell inclusions or structures. Motility or buoyancy can help change their location for finding suitable conditions.

The second phylogenetic group of PSB is the family Ectothiorhodospiraceae. This family includes usually halophilic and/or alkaliphilic purple sulfur bacteria that also grow under anaerobic conditions in the light with reduced sulfur compounds as photosynthetic electron donors. Although its main metabolic way of life is photoautotrophic with deposition of elemental sulfur globules outside the cell, some species can also grow photoheterotrophically, or under microaerobic or aerobic conditions in the dark. Ectothiorhodospiraceae move usually with polar flagella, but only one species, *Ectothiorhodospira vacuolata*, is known to produce gas vesicles. Some species of Ectothiorhodospiraceae require very saline and/or alkaline growth conditions and thrive at high salt concentrations; this is the case of *Halorhodospira halophila*, which is the most halophilic eubacterium known, and can grow in salt-saturated solutions.

Among inland waters, the main habitats of Ectothiorhodospiraceae are anoxic waters with sulfide and light and surface layers of anoxic sediments from saline and hypersaline environments with alkaline pH, such as salt and soda lakes. Soda lakes, from where the alkaliphilic members of the genus *Thioalkalivibrio* have been isolated, often show pH values of 10–11. In saline lakes these bacteria accumulate compatible solutes, such as glycine betaine, ectoine, trehalose, and/or sucrose, to cope with the high osmotic pressure.

Green sulfur bacteria (GSB)   Green sulfur bacteria (the family Chlorobiaceae) are anoxygenic phototrophic bacteria that grow only under strictly anoxic conditions. They form a phylogenetically isolated group within the domain Bacteria (Phylum Chlorobi), with freshwater strains clustering separately from marine strains. They are obligate anaerobic photolithotrophs, that use sulfide or elemental sulfur and, in some species, thiosufate, molecular hydrogen or reduced iron, as electron donor for anoxygenic photosynthesis. Some species can also photoassimilate a few organic substances. The electrons from the reduced sulfur compound are used for the assimilatory reduction of CO_2 via the reductive tricarboxylic acid cycle. Oxidation of sulfide results in the formation of sulfur globules, which are deposited outside the cell.

Typically, taxa of GSB have been distinguished according to biochemical or morphological features, such as pigment composition, cell morphology, and the capacity to form gas vesicles. However, the results of recent studies based on the 16S rRNA gene sequence analysis indicate that most of these phenotypic traits are of little phylogenetic significance, although other features such as the salt requirements for growth, type of cell fission, filamentous morphology, and gliding motility might be traits of higher phylogenetic relevance.

Chlorobiaceae use light-harvesting complexes called chlorosomes, which are attached to the cytoplasmic membrane carrying the photosynthetic pigments. Green-colored species (**Figure 4**) contain bacteriochlorophyll *c* or *d* and the carotenoid chlorobactene. Brown-colored strains contain bacteriochlorophyll *e* and the carotenoids isorenieratene or β-isorenieratene. The type and intracellular concentration of pigments is of high ecological relevance, since Chlorobiaceae are distinguished from any other phototrophic microorganism by strong adaptation to low light intensities, allowing them to colonize anoxic deep waters or sediments with very low light availability. For instance, the amount of carotenoids

in brown-colored strains of green sulfur bacteria can be several times higher than those of their green-colored species. This is an important feature regarding the competition within Chlorobiaceae in natural environments, because of the selective absorption of certain wavelengths in surface waters, light availability in deeper layers may correspond to a spectral range that is more efficiently harvested by carotenoids (**Figures 4** and **5**). Other traits of ecological significance related to the light-harvesting and photosynthetic efficiencies are the degree of alkylation of bacteriochlorophylls, which increases as light availability diminishes, leading to a shift of the position of the absorption maximum (**Figure 6**). The size of the antenna light harvesting complexes can be 10 times larger for GSB than for PSB. This, together with the higher quantum yield (mol of C assimilated per mol of quanta absorbed) for CO_2 fixation that is almost double that of PSB, and the lower ATP requirements per molecule of CO_2 fixed by the tricarboxilic acid cycle (used by green sulfur bacteria) compared with the Calvin cycle (used by purple sulfur bacteria), explains the dominance of GSB over PSB in low

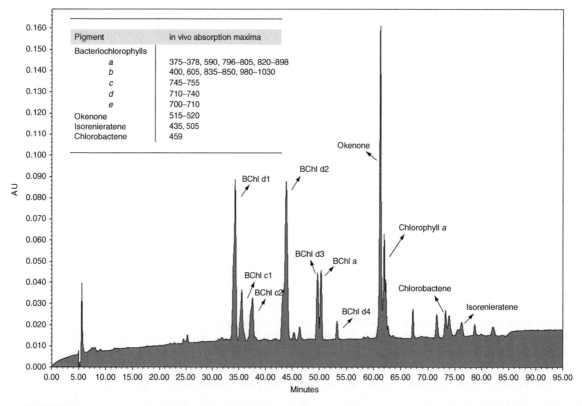

Figure 6 HPLC chromatogram of an acetone extract of photosynthetic pigments from an hypolimnetic sample taken at 12.5 m depth of the meromictic Lake La Cruz (Cuenca, Spain), the same sample as for the in vivo spectrum shown in Figure 4. Note the different pigments corresponding to purple sulfur bacteria (bacteriochlorophyll *a* and okenone) and to green sulfur bacteria (bacteriochlorophyll *d* homologues with different degree of alkylation and carotenoids chlorobactene and isorenieratene). Table on the top of the figure shows the in vivo absorption maxima of these pigments.

light environments. In green sulfur bacteria, the bacteriochlorophyll synthesis is strongly regulated by light intensity, and both the concentrations of pigments and chlorosomes can be multiplied under light-limiting conditions.

GSB are found in various inland waters habitats such as the anoxic hypolimnia of lakes, the bottom layers of microbial mats, or in sulfur springs. Because of their limited physiological capacities compared with purple bacteria, since growth of most green sulfur bacteria depends solely on anoxygenic photosynthesis, these bacteria develop in the narrow zone of overlap between the opposing gradients of light and sulfide. Planktonic lacustrine habitats are commonly colonized by species capable of forming gas vesicles, which aid buoyancy regulation. GSB lack flagellar motility, although gliding motility has been observed for some benthic species. Sulfur springs, including hot springs, are another habitat where GSB occur.

In addition to better performance under low light compared with PSB, the affinity of green sulfur bacteria for sulfide is much higher. Also, sulfide inhibition of growth also occurs at much higher concentrations than for purple bacteria. These abilities explain either colonization of habitats with low but stable sulfide concentration or alternatively those with very high sulfide concentrations by different GSB species. Sulfide consumption by GSB thriving underneath populations of PSB may also limit sulfide diffusion from the sulfide-forming deep layers to the photic zone. This can act as a detoxification mechanism, but also could promote sulfide limitation to the purple sulfur bacteria located in the uppermost layers of the hypolimnia. Other interactions among PSB and GSB are determined by the selective light absorption by the populations situated in shallower waters, commonly purple bacteria, since light penetrating the layers of purple sulfur bacteria is predominantly violet-blue light coinciding with the short wavelength absorption maximum of bacteriochlorophylls *c* and *d* and of chlorobactene. As a consequence, green-colored forms of Chlorobiaceae, which hold this pigment assemblage, are often found underneath Chromatiaceae in stratified lakes, whereas brown-colored species, with a light-harvesting strategy mainly based on carotenoids, thrive in lakes or sediment layers where the selective absorption in shallower layers by algae favors penetration of wavelengths that can be absorbed by isorenieratene.

Colorless sulfur-oxidizing bacteria Since the first microbiologists, such as Winogradsky and Beijerinck, the name 'the colorless (pigment lacking) sulfur bacteria' has been used to designate prokaryotes that are able to use reduced sulfur compounds as sources of

energy and reducing power for growth (thus chemolithotrophic sulfur bacteria). Because of their large size, gliding filamentous colorless sulfur bacteria, such as *Beggiatoa* and *Thiothrix*, characterized by the intracellular deposits of elemental sulfur, were first described in the nineteenth century. Colorless sulfur bacteria comprise a heterogeneous diversity of prokaryotes with few phylogenetic relationships among them, indicating an evolutionary convergence (instead of a divergence from a common ancestor) for the possession of the relevant metabolic pathways supporting growth on reduced sulfur compounds.

The capacity to use reduced sulfur compounds as sources of energy and/or reducing power is widely spread among the different phylogenetic groups of prokaryotes, including not only taxa from the Domain Bacteria but also the Domain Archaea. Although most colorless sulfur bacteria are Proteobacteria (alpha, beta, gamma and epsilon subdivisions); some are included in other bacterial higher taxa, and some, such as *Sulfolobus*, belong to the Archaea (phylum Crenarchaeota). To illustrate the heterogeneous phylogenetic distribution of the colorless sulfur bacteria, according to their 16sRNA gene sequence, members classically belonging to the genus *Thiobacillus* are now recognized to be widespread among proteobacteria; for instance, the former *T. novellus* belonging to the alpha group (α-proteobacteria), the genus-type species, *T. thioparus*, is a member of β-proteobacteria, whereas others, such as *T. thiooxidans* are members of the γ-proteobacteria. Additionally to the commonly recognized as colorless sulfur-oxidizers, there are several taxa that also have the capacity for obtaining energy from the oxidation of sulfur compounds as secondary metabolism (e.g., species of *Pseudomonas*, *Alcaligenes*, etc.).

A wide variety of metabolic possibilities is found among the colorless sulfur bacteria, including chemolithoautotrophy, chemolithoheterotrophy, and chemoorganoheterotrophy. Most colorless sulfur bacteria use oxygen as a terminal electron acceptor for respiration, and they are consequently aerobic, although some are also capable of growing or surviving anaerobically, for instance with nitrate as an electron acceptor (e.g., *Thiobacillus denitrificans*) or by fermentation of intracellular stored carbohydrates. Aerobic sulfur-oxidizing bacteria thrive in gradient environments, where both sulfide and oxygen are available, such as sulfur springs and sulfureta, which are probably the most characteristic environments for colorless sulfur-oxidizing bacteria such as the filamentous gliding genera *Beggiatoa* and *Thiothrix* (**Figure 3**). Although most colorless sulfur bacteria are mesophiles, some are thermophiles thriving in geothermal environments (hot springs) with sulfide emanations. Among them, temperature tolerance is generally higher for the Archaeal taxa compared to Bacteria. Concerning pH, although most are neutrophiles, some are acidophiles, especially some Thiobacilli, and the archaea *Sulfolobus* and *Acidianus*. Interestingly, *Thiobacillus ferrooxidans* can gain energy from the oxidation of iron as well as from sulfur oxidation. Other sulfur-oxidizing bacteria are alkaliphiles, such as the γ-proteobacteria *Thioalkalimicrobium*, thriving in soda lakes.

Sulfate- and sulfur-reducers   As for most of the other functional groups of sulfur prokaryotes, sulfate and sulfur-reducing microorganisms can be grouped as functional groups which are not necessarily phylogenetically consistent. Sulfate-reducers use sulfate as electron acceptor for anaerobic respiration of a wide variety of organic compounds, mostly low molecular weight compounds or H_2. Sulfur-reducers use reduced sulfur forms (or other low oxidation sulfur compounds) as the main electron acceptor for the respiration (oxidation) of organic compounds or H_2. Sulfide is the final product of sulfate or sulfur-reducing respiratory processes. Organic compounds used as electron donors by sulfate-and/or sulfur-reducers include acetate, formate, propionate, butyrate and other fatty acids, acetaldehyde, ethanol and other alcohols, malate, fumarate, succinate, aminoacids, methylated N- and S-compounds, and polar aromatic substances. These organic compounds, many originating from fermentation performed by other bacteria, serve as electron donors for dissimilatory sulfate-reduction, but also as carbon sources for growth. In the latter case, sulfate-reducing bacteria act as terminal degraders of organic matter in anoxic environments, completing the decomposition process started by other organisms. Additional electron acceptors that can also be used are, for example, sulfite and thiosulfate.

Among the domain Bacteria, most sulfate-reducers belong to the δ-proteobacteria, for example the type genus *Desulfovibrio*. Other taxa belong to other groups, such as the spore-forming *Desulfotomaculum*, branched with the Gram-positive bacteria, or even other thermophilic genera recognized as deeply branching lines within the domain Bacteria. The domain Archaea also includes thermophilic sulfate-reducers (e.g., *Archaeoglobus* spp., phylum Euryarchaeota) found in hot springs. Most sulfur-reducers also belong to the δ-proteobacteria branching with sulfate-reducers (e.g., *Desulfuromonas*) although some sulfur-reducers belong to the ε-proteobacteria or to Archaea (*Desulfurolobus* and *Acidianus*). Sulfate and sulfur reducers group of prokaryotes includes mostly mesophilic species but also archaean thermophilic taxa and thermophilic bacteria, such as

the sulfate-reducer *Thermodesulfobacterium* and the sulfur-reducer *Desulfurella*. Some psychrophilic strains have also been described.

It is commonly accepted that most sulfate-reducers are strict anaerobes. Some species, however, can thrive under oxic conditions for a certain time, whereas sulfur-reducing bacteria include both strict and facultative anaerobes. When considering exposure to oxygen, the existence of anoxic microniches within the oxic environments harboring strict anaerobes has been argued to explain active sulfate-reduction in oxic environments. Alternatively, an intermediate metabolism of sulfur compounds avoiding the harmful effects of oxygen has also been described for sulfate-reducers found in oxic environments. However, there is also evidence that some sulfate-reducers are able to use oxygen as electron acceptors for the oxidation of hydrogen under microaerobic conditions.

In inland waters environments, the main habitats for sulfate and sulfur-reducing bacteria are the sediments of sulfate-rich water bodies. These sediments are commonly anaerobic just below a few millimeters under the water–sediment interface. In the absence of oxygen and nitrate, sulfate yields energy as an electron acceptor, and organic matter can be remineralized mainly via sulfate-reduction, allowing for the release of sulfide that diffuses to the water column. From the point of view of free energy changes involved in the respiratory energy generation processes, the energetic yield of anaerobic respiration of organic compounds with sulfate as an electron acceptor is much lower (around 5–10%) than that of the aerobic respiration (oxygen as an electron acceptor). Around 90% of the organic substrate is devoted to energy generation in sulfate reduction and only the remaining being used for biomass generation, thus yielding low growth rates. However, energy and growth yields for sulfate-reduction are higher than for other types of anaerobic respiration of organic matter, such as the methanogenesis. This explains why, under anaerobic conditions in sulfate-rich aquatic environments, sulfate-reduction is a main degradative process for organic compounds.

In sedimentary microbial communities such as the microbial mats, sulfate-reducers are abundant especially in deep layers below the phototrophic microorganisms, providing sulfide to the overlying populations of sulfur-oxidizing bacteria (phototrophic and/or chemotrophic). Interestingly, sulfide diffuses upwards, and consumption by sulfur-oxidizers determines diel changes in the concentration of sulfide to which the overlying microorganisms are exposed. Sulfate and sulfur-reducing bacteria can also be commonly found in other inland waters environments, such as the anoxic hypolimnia of stratified lakes as well as flooded soil such as rice paddies.

Mutualistic Interactions among Sulfur Bacteria

Since sulfur bacteria, depending on the type, use oxidized or reduced sulfur compounds, it is obvious that they are indirectly related in nature by the biogeochemical transformation of these compounds. However, direct interactions including physical contact or close trophic relationships can also occur among sulfur bacteria. An example are photosynthetic consortia, which are stable structural associations of green sulfur bacteria surrounding motile chemotrophic bacteria (thought to be sulfate or sulfur-reducers) situated in the central part of the consortium, which are supposed to be capable of syntrophic growth based on the exchange of inorganic sulfur and organic compounds. This association represents the most evolved symbiosis described so far between prokaryotes. The bacterial partners differ greatly from their free-living counterparts. The green bacteria are phylogenetically different from all known green sulfur bacteria whereas the central bacteria in the consortia studied so far does not belong to the δ-proteobacteria as do the classical sulfate-reducers, but to β-proteobacteria. These consortia are commonly found in the planktonic anaerobic environments of freshwater habitats. Flagellar motility provided by the central cell may also be of ecological relevance allowing the phototrophic non-motile partners to reach well illuminated water layers.

Applied Issues

Some sulfur bacteria are of applied interest. Purple sulfur bacteria and colorless sulfur-oxidizers commonly appear in mixed microbial communities in sewage treatment processes, and could also be used for sulfide removal. Sulfur bacteria can also be used for production of biopolymers such as poly-β-hydroxybutyrate, and of molecular hydrogen. Because of their high tolerance, Ectothiorhodospiraceae can be used under alkaline and saline conditions for some of these purposes. In some cases, such as in the bulking processes of activated sludge in wastewater treatments plant, bacteria such as *Thiothrix* cause problems in the industrial process. In mining activities acidophilic sulfur bacteria, such as *Thiobacillus ferrooxidans*, *T. thiooxidans*, and *T. acidophilus*, are used in the recovery of metals by leaching from ores that are too poor for conventional metallurgical extraction. By this process, recoveries of up to 70% of copper from low-grade ores are

possible. Some sulfate-reducers, such as *Desulfovibrio desulfuricans*, are capable of reducing uranium and can be used for the concentration and removal of radioactive uranium.

See also: Algae; Archaea; Bacteria, Bacterioplankton; Bacteria, Distribution and Community Structure; Chemosynthesis; Comparative Primary Production; Competition and Predation; Cyanobacteria; Diel Vertical Migration; Microbial Food Webs; Phytoplankton Population Dynamics: Concepts and Performance Measurement; Phytoplankton Productivity; Protists; Role of Zooplankton in Aquatic Ecosystems; Trophic Dynamics in Aquatic Ecosystems.

Further Reading

Abelson JN, Simon MI, Peck HD, and LeGall J (eds.) (1994) *Methods in Enzymology*, vol. 243: *Inorganic Microbial Sulfur Metabolism*. New York: Academic Press.

Barton LL (ed.) (1995) *Sulfate-Reducing Bacteria*. New York: Springer.

Blankenship RE, Madigan MT, and Bauer CE (eds.) (1995) *Anoxygenic Photosynthetic Bacteria*. Dordrecht, The Netherlands: Kluwer.

Boone DR and Castenholz RW (eds.) (2001) The Archaea and the deeply branching and phototrophic bacteria. In: Garrity GM (editor-in chief) *Bergey's Manual of Systematic Bacteriology*, 2nd edn., vol. 1. New York: Springer-Verlag.

Brenner DJ, Krieg NR, and Staley JT (eds.) (2005) The Proteobacteria. In: Garrity GM (editor-in chief), *Bergey's Manual of Systematic Bacteriology*, 2nd edn., vol. 2. New York: Springer-Verlag.

Drews G and Imhoff JF (1991) Phototrophic purple bacteria. In: Shively JM and Barton LL (eds.) *Variations in Autotrophic Life*, pp. 51–97. London: Academic Press.

Dworkin M, Falkow S, Rosenberg E, Schleifer KH, and Stackebrandt E (eds.) (2006) *The Prokaryotes: A Handbook on the biology of Bacteria*, 3rd edn., 7 vols. New York: Springer-Verlag.

Glaeser J and Overmann J (2004) Biogeography, evolution, and diversity of epibionts in phototrophic consortia. *Applied and Environmental Microbiology* 70: 4821–4830.

Holmer M and Storkholm P (2001) Sulphate reduction and sulphur cycling in lake sediments: A review. *Freshwater Biology* 46: 431–451.

Madigan M and Martinko J (2005) *Brock: Biology of Microorganisms*. Upper Saddle River, NJ: Pearson Education.

Overmann J (1997) Mahoney Lake: A case study of the ecological significance of phototrophic sulfur bacteria. In: Gwynfryn Jones J (ed.) *Advances in Microbial Ecology*, vol. 15, pp. 251–288. New York: Plenum Press.

Overman J and van Gemerden H (2000) Microbial interactions involving sulfur bacteria: implications for the ecology and evolution of bacterial communities. *FEMS Microbiology Reviews* 24: 591–599.

Peschek GA, Löffelhardt W, and Schmetterer G (1999) *The Phototrophic Prokaryotes*. New York: Kluwer/Plenum.

Stal LJ and Caumette P (1994) *Microbial Mats: Structure, Development and Environmental Significance*. NATO ASI Series, vol. 35. Berlin: Springer.

van Gemerden H and Mas J (1995) Ecology of phototrophic sulfur bacteria. In: Blankenship RE, Madigan MT, and Bauer CE (eds.) *Anoxygenic Photosynthetic Bacteria*, pp. 49–85. Dordrecht, The Netherlands: Kluwer.

Relevant Websites

http://microbes.arc.nasa.gov – NASA microbiology website.

http://commtechlab.msu.edu/sites/dlc-me/ – Digital Learning Center for Microbial Ecology (DLC-ME), Michigan State University.

http://microbewiki.kenyon.edu/index.php/MicrobeWiki – Wiki resource on microbes and microbiology at Kenyon College.

http://www.bacterio.cict.fr/foreword.html – List of Prokaryotic names with standing in nomenclature.

http://bip.cnrs-mrs.fr/bip09/index.html – Evolution of prokaryotic electron transport chains.

http://www.photosynthesisresearch.org/ – International Society of Photosynthesis Research.

http://serc.carleton.edu/microbelife/index.html – Microbial Life – Educational Resources. Science Education Resource Center Carleton College, Northfield, MN.

http://www.pol-us.net/ – Photobiology online. Sponsored by European and American Societies for Photobiology.

http://www.sciencedirect.com/ – Science Direct, Elsevier website on research resources.

http://www.microbionet.com.au/ – Microbio Net. Sciencenet Multimedia Publishing House Pty Limited.

http://www.dsmz.de/ – German Collection of Microorganisms and Cell Cultures.

http://www.cells.de/cellsger/1medienarchiv/Zellfunktionen/Memb_Vorg/Photosynthese/Anoxigene_PST/index.jsp – Videos and other educational resources on microbiology hosted by the German Federal Ministry of Education and Research.

Viruses

P Peduzzi and B Luef, University of Vienna, Vienna, Austria

Introduction

Viruses are typically viewed as microorganisms responsible for diseases in other, larger organisms. During the last one-and-a-half decades, it has become increasingly clear that they are (together with bacteria) the most abundant microorganisms and life forms, and that only a relatively small number are pathogenic to humans; the vast majority play a critical role in aquatic ecosystems. Ample research, plenty of it being conducted in marine systems, has revealed that viruses are involved in the cycling of nutrients and carbon; however, the impact on their hosts' distribution and on the genetic information harbored in aquatic organisms is still at the dawn of comprehensive scientific understanding. Therefore, this field has a tremendous potential for further development and is already widely accepted as an integral and promising discipline in the inland water sciences as well. This chapter will not cover the survival and distribution of human, animal, and plant pathogenic viruses, since topics related to such disease-causing viruses have been reviewed extensively elsewhere. Rather, this chapter will provide a basic insight into aquatic virus ecology, presenting a short overview on current knowledge. It also provides suggestions about important and excellent sources for further reading.

Virus Abundance and Production

Enumeration of Viruses in Aquatic Samples

The great mass of virus particles in aquatic systems is composed of bacteriophages, although viruses of eukaryotes are also significant and contribute to total viral biomass. The size of their capsids (protein shell, see **Figure 5a**) ranges from <20 up to 400 nm, the typical diameter being 30–70 nm. The preponderance of either smaller or larger viruses in any environment may reflect differences in the host community structure (different prokaryotes, algal hosts, etc.). Two differing approaches are usually applied to enumerate viruses. Indirect titer determination by plaque assay, combined with the most-probable-number method ('viable' counts), is routinely used in water analysis but rarely in virus ecology. Such culture-based methods are inefficient for enumerating natural virus assemblages and strongly underestimate virioplankton abundance, with the exception of studies on cyanophage ecology. The second approach is based on direct total counting of viruses in the respective water sample. The discovery of great virus abundance in natural waters is linked to the development of these direct counting techniques, which are also used to accurately enumerate bacterial assemblages. For many years, transmission electron microscopy (TEM) has been used to visualize and characterize phages. Therefore, TEM provided early evidence of high virus numbers in natural aquatic samples. In the late 1970s, high virus abundance ($>10^4$ ml^{-1}) was reported in the waters of Yaquina Bay (USA). Nonetheless, the true numbers were apparently underestimated because of methodological constraints. At the same time 10^5–10^7 bacteriophages ml^{-1} were reported from sewage water, using improved techniques. In these early studies, the findings indicated that viruses were not likely to be an important factor in microbial food webs and aquatic processes. A major breakthrough in virus ecology resulted from new methods and investigations conducted almost exclusively in marine waters, that began in the late 1980s. An improved TEM- and centrifugation-based method of particle collection from liquid suspension was adopted (**Table 1**), which provided accurate counts and much higher estimates of viral abundance in aquatic environments.

At present, viral direct counts in water samples can be conducted using three methods: TEM after counterstaining with uranylactetate (**Figure 1**), epifluorescence microscopy (EFM; **Figure 2**), and flow cytometry (FCM). The latter two methods require staining with nucleic-acid-specific fluorochromes. For TEM and EFM, samples have to be concentrated either by ultracentrifugation onto grids or by filtration techniques. In general, TEM counts proved to be lower than EFM counts because of technical reasons, and the latter method has gained greater acceptance for estimating total viral abundance. FCM requires expensive equipment and expertise in handling the instrument, although a large number of samples can be processed without any concentration procedure. Virus particles are also found to be associated with sediment and suspended material. Such particle- or aggregate-associated viruses can be counted via two different approaches: extraction from the particulate matter (e.g., via sonication), combined with EFM or FCM, or direct inspection of the particles (only feasible with suspended material) with EFM or confocal laser scanning microscopy (**Figure 3**). A list of the available techniques is provided in **Table 1**.

Table 1   Techniques used to enumerate viruses in aquatic samples and associated with particulate material

Method	Strength	Source
A. Planktonic viruses		
Plaque forming unit assay ('viable counts'; PFU)	Potential for isolation and characterization of viruses	18
Transmission electron microscopy (TEM)	Direct counts with potential for sizing and characterization	1, 8, 14, 19
Epifluorescence microscopy (EFM)	Direct counts, easy applicability for reasonably large sample numbers	8, 10, 16, 20
Flow cytometry (FCM)	Rapid total counts, large sample numbers and volumes	3, 4, 13
B. Particle-associated viruses.		
Epifluorescence microscopy after sonication	Direct counts, easy applicability for reasonably large sample numbers	5, 7, 9
Epifluorescence microscopy without sonication	Direct inspection of particles and associated viruses	11, 15
Confocal laser scanning microscopy (CLSM) microscopy (CLSM)	Direct inspection of particles in three dimensions (optical sectioning), low background fluorescence	12
Flow cytometry (FCM)	Rapid total counts, large sample numbers and volumes	6
C. Comparison of methods		2, 8, 17, 19, 20

1. Bergh O, Borsheim KY, Bratbak G, and Heldal, M (1989) High abundance of viruses found in aquatic environments. *Nature* 340: 467–468.
2. Bettarel Y, Sime-Ngando T, Amblard C, and Laveran H (2000) A comparison of methods for counting viruses in aquatic systems. *Applied and Environmental Microbiology* 66: 2283–2289.
3. Brussard C, Marie D, and Bratbak G (2000) Flow cytometric detection of viruses. *Journal of Virological Methods* 85: 175–182.
4. Chen F, Lu JR, Binder BJ, Liu YC, and Hodson RE (2001) Application of digital image analysis and flow cytometry to enumerate marine viruses stained with SYBR Gold. *Applied and Environmental Microbiology* 67: 539–545.
5. Danovaro R, Dell'anno A, Trucco A, Serresi M, and Vanucci S (2001) Determination of virus abundance in marine sediments. *Applied and Environmental Microbiology* 67: 1384–1387.
6. Duhamel S and Jacquet S (2006) Flow cytometric analysis of bacteria- and virus-like particles in lake sediments. *Journal of Microbiological Methods* 64: 316–332.
7. Fischer UR, Kirschner AKT, and Velimirov B (2005) Optimization of extraction and estimation of viruses in silty freshwater sediments. *Aquatic Microbial Ecology* 40: 207–216.
8. Hara S, Terauchi K, and Koike I (1991) Abundance of viruses in marine waters: Assessment by epifluorescence and transmission electron microscopy. *Applied and Environmental Microbiology* 57: 2731–2734.
9. Helton RR, Liu L, and Wommack KE (2006) Assessment of factors influencing direct enumeration of viruses within estuarine sediments. *Applied and Environmental Microbiology* 72: 4767–4774.
10. Hennes KP and Suttle CA (1995) Direct counts of viruses in natural waters and laboratory cultures by epifluorescence microscopy. *Limnology and Oceanography* 40: 1050–1055.
11. Luef B, Aspetsberger F, Hein T, Huber F, and Peduzzi P (2007) Impact of hydrology on free living and particle-associated microorganisms in a river floodplain system (Danube, Austria) *Freshwater Biology* 52: 1043–1057.
12. Luef B, Neu TR, and Peduzzi P (2005) Examination of glycoconjugate, bacterial and viral distribution in riverine aggregates by Confocal Laser Scanning Microscopy. *Geophysical Research Abstracts* 7: 6080.
13. Marie D, Brussard CPD, Thyrhaug R, Bratbak G, and Vaulot D (1999) Enumeration or marine viruses in culture and natural samples by flow cytometry. *Applied and Environmental Microbiology* 65: 45–52.
14. Mathews J and Buthala DA (1970) Centrifugal sedimentation of virus particles for electron microscopic counting. *Journal of Virology* 5: 598–603.
15. Peduzzi P and Luef B (2008) Viruses, bacteria and suspended particles in a backwater and main channel site of the Danube (Austria). *Aquatic Sciences* 70: 186–194.
16. Noble RT and Fuhrman JA (1998) Use of SYBR Green I for rapid epifluorescence counts of marine viruses and bacteria. *Aquatic Microbial Ecology* 14: 113–118.
17. Shibata A, Goto Y, Saito H, Kikuchi T, Toda T, and Taguchi S (2006) Comparison of SYBR Green I and SYBR Gold stains for enumerating bacteria and viruses by epifluorescence microscopy. *Aquatic Microbial Ecology* 43: 223–231.
18. Suttle CA (1993) Enumeration and isolation of viruses. In Kemp PF, Sherr B, Sherr E, and Cole JJ (eds.) *Handbook of Methods in Aquatic Microbial Ecology*, pp. 121–134. Boca Raton: Lewis Publishers.
19. Weinbauer MG and Suttle CA (1997) Comparison of epifluorescence and transmission electron microscopy for counting viruses in natural marine waters. *Aquatic Microbial Ecology* 13: 225–232.
20. Wen K, Ortmann AC, and Suttle CA (2004) Accurate estimation of viral abundance by epifluorescence microscopy. *Applied and Environmental Microbiology* 70: 3862–3867.

Virioplankton Abundance in Inland Waters

Abundance of planktonic viruses in aquatic systems typically ranges between $<10^4$ and $>10^8\,ml^{-1}$ with numbers usually being higher in fresh water than in marine systems (**Table 2**). Bacterioplankton abundance in natural waters is typically less variable than the corresponding viral numbers, and virus abundance apparently increases with the productivity of the system. Viral abundance in the water column of estuaries or very productive lakes was found to be as high as $10^8\,ml^{-1}$, and can be strongly enriched in the surface microlayer or in/on suspended particles.

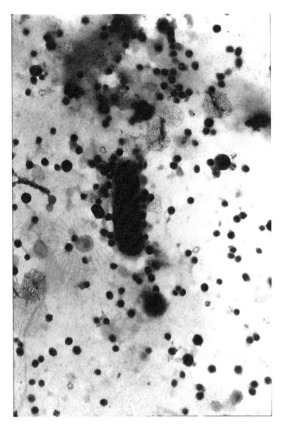

Figure 1 Transmission electron micrograph of viruses and a bacterium collected by ultracentrifugation onto a formvar-coated copper grid.

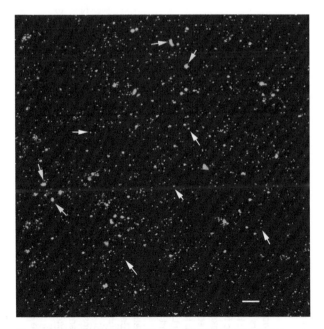

Figure 2 Confocal laser scanning microscopy maximum intensity projection showing viruses and bacteria on a filter. White arrows point to viruses, yellow ones to bacteria. Length of the calibration bar is 5 μm.

Figure 3 Three-dimensional volume reconstruction of a floating aggregate from the Danube River (Austria). Confocal laser scanning microscopy was performed using a Leica TCS-SP. Dual channel presentation of glycoconjugates and nucleic-acid-stained bacteria and virus-like particles (VLPs). Color allocation: nucleic acid – green, glycoconjugates – red, yellow – bacteria and VLPs in or in contact with lectin-specific extracellular polymeric substances. Arrows point to VLPs. Length of the calibration grid is 5 μm. The picture was taken by B. Luef at the Helmholtz Centre for Environmental Research – UFZ Magdeburg, research group of T. Neu.

Exceptionally high virioplankton abundance (up to $2 \times 10^9 \, ml^{-1}$) has recently been observed in the alkaline, hypersaline Mono Lake, California (**Table 2**). Very low virus numbers in natural inland waters are reported from polar freshwater lakes (down to $1.6 \times 10^5 \, ml^{-1}$); almost undetectable (TEM direct counts) values ($<2 \times 10^4 \, ml^{-1}$) were observed in a high-mountain lake associated with intense solar radiation (**Table 2**).

Inland waters provide a high degree of spatial variability that apparently influences virus distribution strongly. Depth gradients are commonly observed in many water bodies; hence, viral distribution follows vertical patterns similar to those of other planktonic organisms. Various kinds of discontinuities in the water column (e.g., thermoclines, pycnoclines, chemoclines, fronts) often lead to changes in viral abundance. Lateral discontinuities such as interlinked water bodies of a river-floodplain can also harbor viruses in varying numbers.

As in marine systems, temporal variability of total viral abundance in inland waters is significant at scales ranging from minutes to months. This variability provides evidence for the dynamic nature of viruses in aquatic systems. Such dynamics are often linked to the variability in potential host abundance (bacterioplankton, phytoplankton).

Table 2 Viral abundance, virus to bacterium ratio (VRB), burst size, frequency of visible infected cells (FVIC), frequency of infected cells (FIC), frequency of mortality due to viral lysis (FMVL), viral production and lysogenic bacteria in various inland waters

Location	Viral abundance (×10⁷ ml⁻¹)	VBR	Burst size	FVIC (%)	FIC (%)	FMVL (% of total bacterial population)	Viral production (×10⁶ VLPs l⁻¹ h⁻¹)	Lysogenic bacteria (% of total bacterial population)	Source
Pelagic									
Running waters									
Ria de Aveiro (Portugal)	2.4–25 (TEM)	4.7–55.6 (18)				49–74			1
Brisbane River/ Moreton Bay, Noosa River (Australia)	0.5–30 (EFM)	3–37					<1–2300		19
Charente River (France)	3.9–9.8^a (EFM)	11.3–14.3^a							3
River Danube (Austria)	2.1 (EFM)								28
ditch	14.6 (TEM)								18
Djeuss Stream (Senegal)	1.1 (EFM), 2.6 (TEM)	4.1	34	0.3	2.1^d	2.5	410	7.1	8
Mahoning River (USA)	0.2–2.0 (EFM)	0.4–12.2							3, 26
Yangtze river estuary (China)	0.1– 1.7 (FCM)	1.5–72.0 (8.7)							23
Lakes, ponds, reservoirs									
Antarctic lakes	0.4–3.4 (EFM)	0.4–141.1							24
Antarctic freshwater lakes (Lake Druzhby, Crooked Lake)	0.02–0.2 (EFM)	1.2–8.4	2–15	10.5–66.7		38–251		ND – 73^e ND – 11.7^f	36, 37
Antarctic lakes (Highway Lake, Pendant Lake, Ace Lake)	0.9–12.0 (EFM)	18.6–126.7							29
Lake Austin (Texas)	14.2 (EFM), 6.1 (TEM)								18
Lake Aydat (France)	26 (FCM)		30.2	1.1			423.7		7
Lake Bourget (France)			8–35		3.4–5.7	3.7–6.3			21
Lake Constance (Germany)	1–4 (TEM)		21–121	0.1–1.7 (0.6)	<1–17	< 34	4–104		17
Cryoconite holes (Arctic)	0.2–0.3 (0.26) (EFM)	8.4–23.2 (13.6)	2–6 (3)	5.1–20.0 (11.3)					37
Lake Erie (Canada, USA)	3.7–37.9 (EFM)	11.5–129.5	8–22 (11.6)	1.63		12.1–23.4 (17.7)			46

Location									Ref.
Fisklösen (Sweden)	0.9–5.4 (EFM)	5.2–24.4	6–21		10–43				43
Lake of Ganzirri (Italy)	0.005–75.4 (13.8) (EFM)	0.4–117 (14)							42
Gäddtjärn, (Sweden)	0.8–4.7 (EFM)	2.9–12.1	6–18		9–41				43
Gossenköllesee (Austria)	<0.002 × 0.5 (TEM)	0.1–10.8	3–45	0.9–2.3					20
Lake Hallwill (Switzerland)	1.5–2.9 [c] (EFM)	1.2–2.3 [c]	7–200 (38)		8–32	Up to 66	0.08–13		12
shallow inland waters (Senegal)	1.5–3.9 (EFM), 1.0–2.8 (TEM)	3.5–13.6	10.3–36.8	0.3–1.1	2.1–7.8 [d]	2.5–9.1	50–1270	ND–2.5	8
Lake Kalandsvannet (Norway)	1.9–20.2 (TEM)			2–16					16
Lakes of the McMurdo Dry Valleys (Antarctica)	0.02–5.6 (EFM)	0.6–53						2.0–62.5	27
Mono Lake – hypersaline (California)	14–190 (EFM)	5.8–47 (19)	10–560 (100)	0.8	2.1–18.3				9, 22
Lake Pavin (France)	0.3–5.0 (EFM)	5.6–14.8	13.2–27.7 [a]	0.5–3.5	5–79	11.8–82.5 [a,b] / 3.5–33.7 [b]	79.7	0.1–16	6, 7, 10
Lake Plußsee (Germany)	1.1–8.8 (EFM), 25.4 (TEM)								5, 44, 45
Priest Pot (UK)	0.7–20.3 (FCM)		5–500	0.5–6.4		ND–107	41.6–1037.5		15
Reservoirs (Sri Lanka)	3.1–7.7 (EFM)	9.7–23.8		1.6–4.4	10.8–26.9	13.2–46.1			34
Rímov Reservoir (Czech Republic)	1.6 (EFM)		19						39
Lake Retba – hypersaline (Senegal)	72 (EFM), 32 (TEM)	5.4	60.2	0.5	3.6 [d]	3.5	1110	0.6	8
Lake Saelenvannet (Norway)	2–30 (TEM)	50							41
Sep Reservoir (France)	0.6–13 (2.3) (EFM)	2.1–51.7 (5.5)		0.5–3.7 (1.6)	1.0–31.5 (12.3)	1–60 (20.1)			35, 38
Sproat Lake (Canada)	ca. 1 (TEM)	3							25
Lake Superior (USA)	0.02–0.9 (TEM)		0.1–2.7					0.1–7.4	40
22 lakes (Quebec, Canada)	4.1–25 (11) (TEM)	4.9–77.5 (20–25)							30
24 lakes (Sweden)	1.2–2.4 (1.7) (EFM)	13–73 (34.8)						ND–31.8	2
Floodplains, wetlands									
Alte Donau (Austria)	1.7–11.7 (5.0) (TEM)	4–39 (19)	18–48	2.8–9 (5)	10–63 (28)	26–125			13
Floodplain segment of the River Danube (Austria)	0.3–10.2 (2.6) (EFM)	2.9–33.9 (13.3)							28

Continued

Table 2 Continued

Location	Viral abundance ($\times 10^7$ ml^{-1})	VBR	Burst size	FVIC (%)	FIC (%)	FMVL (% of total bacterial population)	Viral production ($\times 10^6$ VLPs l^{-1} h^{-1})	Lysogenic bacteria (% of total bacterial population)	Source
Kühwörther Wasser, River Danube backwater (Austria)	1.2–6.1 (TEM)	2.0–17.0	15.5–38.0		5.4–21.6	10.8–43.2			32
Talladega Wetland (USA)	0.009–0.12 (EFM)	0.02–2.5							11
Others									
Barton Spring (Texas)	0.5 (EFM), 0.4 (TEM)								18
Hot Springs (California)	0.007–0.7 (EFM)						40–60		9
marsh	57.4–70.0 (EFM), 22.7 (TEM)								18
Benthic									
Running waters									
Brisbane River/ Moreton Bay, Noosa River (Australia)	viral abundance: $0.2\text{–}4.8 \times 10^9$ cm^{-3} sediment (EFM)	2–65							19
River Danube (Austria)	0.2×10^7 ml^{-1} (suspended solids) (EFM)								28
Djeuss Stream (Senegal)	15.5×10^7 ml^{-1} (EFM), 8.5×10^7 ml^{-1} (TEM) (sediment)	1.3		<0.1				1.9	8
Esino River (Italy)	9.8×10^8 ml^{-1} (sediments) (EFM)	0.8	16.6			18.4	77000	1.5	33
Mahoning River (USA)	$1.7\text{–}6.7 \times 10^8$ AFDM in particulate samples (EFM)	sediment: 0.7–1.2 leaf: 0.2–0.3							26
Mahoning River (USA)	$4.71\text{–}8.91 \times 10^6$ g^{-1} sediment $4.81\text{–}21.8 \times 10^8$ g^{-1} leaf (EFM)	sediment: 0.1–0.5 leaf: 0.4–2.6							4
Lakes, ponds, reservoirs									
Lac Gilbert (Québec)	$6.5 \times 10^8\text{–}1.8 \times 10^{10}$ ml^{-1} sediment (TEM)	0.5–32.9 [c]							31
Lake Hallwil (Switzerland)	$0.2\text{–}0.9 \times 10^8$ cm^{-2} biofilm [c] $3.5\text{–}10.6 \times 10^7$ mg^{-1} C$_{org}$ plant litter [c] $1.9\text{–}5.3 \times 10^9$ cm^{-3} sediment [c] (EFM)	biofilm: 0.2–0.5 [c] plant litter: 0.8–1.8 [c] sediment: 0.8–1.9 [c]		<0.1	≤0.4	<1			12

	Viral abundance			FMVL	VBR	Source
Shallow inland waters (Senegal)	20.7–74 × 10⁷ ml⁻¹ (sediment) (EFM)	0.6–6.3	<0.1		0.3–1.0	8
Lake Retba – hypersaline (Senegal)	56.2 × 10⁷ ml⁻¹ (EFM), 15.3 × 10⁷ ml⁻¹ (TEM), (sediment)	9.1	<0.1		4.2	8
Floodplains, wetlands						
Floodplain segment of the River Danube	0.02–0.9 × 10⁷ ml⁻¹ (0.24 × 10⁷ ml⁻¹) (suspended solids) (EFM)	0.2–12.9 (3.5)				28
Kühnwörther Wasser, River Danube backwater (Austria)	4.3 × 10⁹–7.2 × 10⁹ particles ml of wet sediment⁻¹ (5.8 × 10⁹) (EFM)	0.9–3.2 (1.9)		0.0–39.4 (13.9)[b]		14
Talladega Wetland (USA)	8.6 × 10⁴–5.1 × 10⁶ (1.3 × 10⁶) cm⁻² *J. effuses* 2.8 × 10⁴–4.5 × 10⁷ (1.1 × 10⁷) cm⁻² sediment 3.9 × 10⁵–2.6 × 10⁷ (1.1 × 10⁷) cm⁻² wood (EFM)	*J. effusus:* 0.06–0.9 sediment: 0.03–0.75 wood: 0.01–0.9				11

Data are given as range (average) VBR: virus to bacterium ratio.

FIC, frequency of infected cells; FVIC, frequency of visible infected cells; FMVL, frequency of mortality due to viral lysis; EFM, epifluorescence microscopy; TEM, transmission electron microscopy; FCM, flow cytometry; AFDM, ash-free dry mass; ND, not detectable.

[a] Range of means.
[b] Virus-mediated bacterial mortality (% of bacterial production).
[c] The values were read off a graph.
[d] Calculated, using the formula: FIC = 7.11 × FVIC.
[e] Calculated with a burst size of 4.
[f] Calculated with a burst size of 26.

Sources

1. Almeida MA, Cunha MA, and Alcântara F (2001) Loss of estuarine bacteria by viral infection and predation in microcosm conditions. *Microbial Ecology* 42: 562–571.
2. Anesio AM, Hollas C, Granéli W, and Laybourn-Parry J (2004) Influence of humic substances on bacterial and viral dynamics in freshwaters. *Applied and Environmental Microbiology* 70: 4848–4854.
3. Auguet JC, Montanié H, Delmas D, Hartmann HJ, and Huet V (2005) Dynamic of virioplankton abundance and its environmental control in the Charente Estuary (France). *Microbial Ecology* 50: 337–349.
4. Baker PW and Leff LG (2004) Seasonal patterns of abundance of viruses and bacteria in a Northeast Ohio (USA) stream. *Archiv für Hydrobiologie* 161: 225–233.
5. Bergh Ø, Børsheim KY, Bratbak G, and Heldal M (1989) High abundance of viruses found in aquatic environments. *Nature* 340: 467–468.
6. Bettarel Y, Amblard C, Sime-Ngando T, Carrias, J.-F, Sargos D, Garabétian F, and Lavandier P (2003) Viral lysis, flagellate grazing potential and bacterial production in Lake Pavin. *Microbial Ecology* 45: 119–127.
7. Bettarel Y, Sime-Ngando T, Amblard C, and Dolan J (2004) Viral activity in two contrasting lake ecosystems. *Applied and Environmental Microbiology* 70: 2941–2951.
8. Bettarel Y, Bouvy M, Dumont C, and Sime-Ngando T (2006) Virus-bacterium interactions in water and sediment of West African inland aquatic systems. *Applied and Environmental Microbiology* 72: 5274–5282.
9. Breitbart M, Wegley L, Leeds S, Schoenfeld T, and Rohwer F (2004) Phage community dynamics in hot springs. *Applied and Environmental Microbiology* 70: 1633–1640.
10. Colombet J, Sime-Ngando T, Cauchie HM, Fonty G, Hoffmann L, and Demeure G (2006) Depth-related gradients of viral activity in Lake Pavin. *Applied and Environmental Microbiology* 72: 4440–4445.
11. Farnell-Jackson EA, and Ward AK (2003) Seasonal patterns of viruses, bacteria and dissolved organic carbon in a riverine wetland. *Freshwater Biology* 48: 841–851.
12. Filippini M, Buesing N, Bettarel Y, Sime-Ngando T, and Gessner MO (2006) Infection paradox: high abundance but low impact of freshwater benthic viruses. *Applied and Environmental Microbiology* 72: 4893–4898.
13. Fischer UR and Velimirov B (2002) High control of bacterial production by viruses in a eutrophic oxbow lake. *Aquatic Microbial Ecology* 27: 1–12.

14. Fischer U, Wieltschnig C, Kirschner AKT, and Velimirov B (2003) Does virus-induced lysis contribute significantly to bacterial mortality in the oxygenated sediment layer of shallow oxbow lakes? *Applied and Environmental Microbiology* 69: 5281–5289.

15. Goddard VJ, Baker AC, Davy JE, Adams DG, Ville MMD, Thackeray SJ, Maberly SC, and Wilson WH (2005) Temporal distribution of viruses, bacteria and phytoplankton throughout the water column in a freshwater hypereutrophic lake. *Aquatic Microbial Ecology* 39: 211–223.

16. Heldal M and Bratbak G (1991) Production and decay of viruses in aquatic environments. *Marine Ecology Progress Series* 72: 205–212.

17. Hennes KP and Simon M (1995) Significance of bacteriophages for controlling bacterioplankton growth in a mesotrophic lake. *Applied and Environmental Microbiology* 61: 333–340.

18. Hennes KP and Suttle CA (1995) Direct counts of viruses in natural waters and laboratory cultures by epifluorescence microscopy. *Limnology and Oceanography* 40: 1050–1055.

19. Hewson I, O'Neil JM, Fuhrman JA, and Dennison WC (2001) Virus-like particle distribution and abundance in sediments and overlying waters along eutrophication gradients in two subtropical estuaries. *Limnology and Oceanography* 46: 1734–1746.

20. Hofer JS and Sommaruga R (2001) Seasonal dynamics of viruses in an alpine lake: importance of filamentous forms. *Aquatic Microbial Ecology* 26: 1–11.

21. Jaquet S, Domaizon I, Personnic, S Ram, ASP, Hedal M, Duhamel S, and Sime-Ngando T (2005) Estimates of protozoan- and viral-mediated mortality of bacterioplankton in Lake Bourget (France). *Freshwater Biology* 50: 627–645.

22. Jiang S, Steward G, Jellison R Chu W, and Choi S (2003) Abundance, distribution and diversity of viruses in alkaline, hypersaline Mono Lake, California. *Microbial Ecology* 47: 9–17.

23. Jiao N, Zhao, Y Luo T, and Wang X (2006) Natural and anthropogenic forcing on the dynamics of virioplankton in the Yangtze river estuary. *Journal of the Marine Biological Association of the United Kingdom* 86: 543–550.

24. Kepner RL, Wharton RA, and Suttle CA (1998) Viruses in Antarctic lakes. *Limnology and Oceanography* 43: 1754–1761.

25. Klut ME and Stockner JG (1990) Virus-like particles in an ultra-oligotrophic lake on Vancouver Island, British Columbia. *Canadian Journal of Fisheries and Aquatic Sciences* 47: 725–730.

26. Lemke M, Wickstrom C, and Leff L (1997) A preliminary study on the distribution of viruses and bacteria in lotic environments. *Archiv für Hydrobiologie* 141: 67–74.

27. Lisle JT and Priscu JC (2004) The occurrence of lysogenic bacteria and microbial aggregates in the lakes of the McMurdo Dry Valleys, Antarctica. *Microbial Ecology* 47: 427–439.

28. Luef B, Aspetsberger F, Hein T, Huber F, and Peduzzi P (2007) Impact of hydrology on free-living and particle-associated microorganisms in a river floodplain system (Danube, Austria). *Freshwater Biology* 52: 1043–1057.

29. Madan NJ, Marshall WA, and Laybourn-Parry J (2005) Virus and microbial loop dynamics over an annual cycle in three contrasting Antarctic lakes. *Freshwater Biology* 50: 1291–1300.

30. Maranger R and Bird DF (1995) Viral abundance in aquatic systems: a comparison between marine and fresh waters. *Marine Ecology Progress Series* 121: 217–226.

31. Maranger R and Bird DF (1996) High concentrations of viruses in the sediments of Lac Gilbert, Québec. *Microbial Ecology* 31: 141–151.

32. Mathias CB, Kirschner AKT, and Velimirov B (1995) Seasonal variations of virus abundance and viral control of bacterial production in a backwater system of the Danube River. *Applied and Environmental Microbiology* 61: 3734–3740.

33. Mei ML and Danovaro R (2004) Virus production and life strategies in aquatic sediments. *Limnology and Oceanography* 49: 459–470.

34. Peduzzi P and Schiemer F (2004) Bacteria and viruses in the water column of tropical freshwater reservoirs. *Environmental Microbiology* 6: 707–715.

35. Ram ASP, Boucher D, Sime-Ngando T, Debroas D, and Romagoux JC (2005) Phage bacteriolysis, protistan bacterivory potential, and bacterial production in a freshwater reservoir: coupling with temperature. *Microbial Ecology* 50: 64–72.

36. Säwström C, Anesio MA, Granéli W, and Laybourn-Parry J (2007) Seasonal viral loop dynamics in two large ultraoligotrophic antarctic freshwater lakes. *Microbial Ecology* 53: 1–11.

37. Säwström CW Granéli, Laybourn-Parry J, and Anesio AM (2007) High viral infection rates in Antarctic and Arctic bacterioplankton. *Environmental Microbiology* 9: 250–255.

38. Sime-Ngando1 T and Ram ASP (2005) Grazer effects on prokaryotes and viruses in a freshwater microcosm experiment. *Aquatic Microbial Ecology* 41: 115–124.

39. Simek K, Pernthaler J, Weinbauer MG, Hornak K, Dolan JR, Nedoma J, Masin M, and Amann R (2001) Changes in bacterial community composition and dynamics and viral mortality rates associated with enhanced flagellate grazing in a mesoeutrophic reservoir. *Applied and Environmental Microbiology* 67: 2723–2733.

40. Tapper MA and Hicks RE (1998) Temperate viruses and lysogeny in Lake Superior bacterioplankton. *Limnology and Oceanography* 43: 95–103.

41. Tuomi P, Torsvik T, Heldal M, and Bratbak G (1997) Bacterial population dynamics in a meromictic lake. *Applied and Environmental Microbiology* 63: 2181–2188.

42. Vanucci S, Bruni V, and Pulicanò G (2005) Spatial and temporal distribution of virioplankton and bacterioplankton in a brackish environment (Lake of Ganzirri, Italy). *Hydrobiologia* 539: 83–92.

43. Vrede K, Stensdotter U, and Lindström ES (2003) Viral and bacterioplankton dynamics in two lakes with different humic contents. *Microbial Ecology* 46: 406–415.

44. Weinbauer MG and Höfle MG (1998) Significance of viral lysis and flagellate grazing as factors controlling bacterioplankton production in a eutrophic lake. *Applied and Environmental Microbiology* 64: 431–438.

45. Weinbauer MG and Höfle MG (1998) Size-specific mortality of lake bacterioplankton by natural virus communities. *Aquatic Microbial Ecology* 13: 103–113.

46. Wilhelm SW and Smith REH (2000) Bacterial carbon production in Lake Erie is influenced by viruses and solar radiation. *Canadian Journal of Fisheries and Aquatic Sciences* 57: 317–326.

Virus Abundance in Sediments and on Particulate Matter

Data on the abundance of viruses in limnetic sediments are scarce, although occurrence is no doubt ubiquitous. One cubic centimeter of freshwater sediment can harbor as many as 10^9 virus particles. Abundances decrease with sediment depth. A high age is estimated for some of the phages in marine sediments, which thus serve as a potential reservoir for viruses. A similar situation is expected also for inland water sediments. Viruses show some dependency on the type of particulate matter (e.g., particle size and quality). In a riverine wetland, surface type and potential host bacterial cells influenced the proliferation and distribution of viruses (references 11 and 26 in **Table 2**). There is very little data, particularly for inland waters, on the abundance of viruses associated with suspended particulate material. Interactions between suspended matter and the natural virus assemblage are also poorly documented. In riverine system, for example the Danube River, where suspended matter is often a prominent factor, a variable and occasionally significant proportion of viruses – between 0.4% and 35% – was found to be associated with particulate material (reference 28 in **Table 2**).

Virus-to-Bacterium Ratio

When bacterial numbers are determined concomitantly, a virus-to-bacterium ratio (VBR) is often used to describe the relationship between viruses and their prokaryote hosts. This VBR indicates a typical predominance of virus numbers over bacteria, with values generally falling between 3 and 100; a few exceptions are listed in **Table 2**. The VBR is apparently higher in limnetic than in marine systems, with an average value of around 20. Values for freshwater sediments are typically lower (0.03 to >10; sometimes up to 65); ratios of up to 13 have been determined on floating particles in lakes and river systems (reference 28 in **Table 2**).

Lytic and Lysogenic Viruses

Viruses display different types of life cycles, the most common ones being lytic and lysogenic infections (**Figure 5b**). During the lytic cycle, the virus lets the host metabolism more or less immediately (after a latent period) produce new virus progeny, which are released during host cell lysis. In the lysogenic cycle, the genome of the temperate virus, which is replicated along with the host, remains within the host cell in a dormant stage (prophage) until a lytic cycle is induced. Measuring the proportion of lysogenic host cells in the community may be important to understand whether certain environments or environmental situations favor one or the other of these two viral life cycles. The relative importance of these two pathways is crucial for the affected host community and probably also for related processes such as carbon and nutrient release into the aquatic environment (i.e., for the microbial loop). Methods for assessing the proportion of these two viral life cycles can be obtained from references in **Table 3** (17, 22). Values for lysogenic bacteria (as percent of the total bacterial population) in inland waters range from undetectable to 73%.

Burst Size

The number of virus particles released during lysis of host cells is termed burst size. The size of this viral 'offspring' is a critical parameter in regulating population dynamics of both the cellular host and the viruses themselves. Estimates of burst size are used to relate in situ viral production to lysis rates of host cells and virus-mediated mortality of prokaryotes. Burst size estimates can be obtained by one-step growth curve experiments or, better, by in situ observation (using TEM technology) of virus particles within intact (**Figure 4**) or thin-sectioned cells. Another approach is to use batch cultures and calculate the burst size necessary to balance viral production with measurements of viral decay, or from the observed net production of viruses over the incubation period. In many studies burst size has been reported as an average of inspected cells or as a range of individual observations. A wide range of burst size values can be found in the literature (up to 560!), probably also reflecting differences in the methodologies applied. Phage and host species as well as environmental parameters such as temperature and trophic state of the aquatic system seem to be important. Further, burst size may also be positively correlated to the size of host cells and, moreover, tends to be higher in fresh water than in marine environments. Examples from inland water environments are provided in **Table 2**.

Viral Production, Host Infection, and Mortality

Since the early 1990s, different methods have been developed (largely in marine waters) to assess viral proliferation and virus-induced mortality of microbial host communities. Such methods are crucial for evaluating the role of naturally occurring aquatic viruses in microbial food webs and biogeochemical cycles. A detailed comparison, including the advantages and disadvantages of the various techniques, is available elsewhere, but **Table 3**

Table 3 Techniques used to estimate viral production (VP) in aquatic samples

Method	Strength/Drawback	Source
Net changes of viral abundance over time	No manipulation/only net changes, difficult to interpret	1
Viral decay	No sample filtration, easy to perform/assesses only abiotic decay sources	3, 5, 6
Contact rates	Rapid, easy calculation of VP/conversion factor needed, no direct observation of virus abundance changes	9, 15, 16
Radioisotope incorporation	Measures directly VP rates, reveals flux of bacterial to viral biomass/conversion factor needed, requires working with radioactive material	4, 8, 11, 13, 14
Frequency of visibly infected cells	No incubation required/conversion factor needed, no direct observation of virus abundance changes	2, 4, 11, 12, 18
Fluorescently labeled viruses as tracer	Easy to perform, direct observation of virus abundance changes/lengthy tracer preparation, extensive sample enumerations	7, 10, 11
Virus dilution approach	Easy to perform, direct observation of virus abundance changes/extensive sample manipulation, portion of ambient bacterial community likely lost during filtration	7, 18, 19, 20, 21
Comparison of Methods		17, 20, 22

Sources

1. Bratbak G, Heldal M, Norland S, and Thingstad TF (1990) Viruses as partners in spring bloom microbial trophodynamics. *Applied and Environmental Microbiology* 56: 1400–1405.
2. Binder B (1999) Reconsidering the relationship between virally induced bacterial mortality and frequency of infected cells. *Aquatic Microbial Ecology* 18: 207–215.
3. Fischer UR, Weisz W, Wieltschnig C, Kirschner KT, and Velimirov B (2004) Benthic and pelagic viral decay experiments: a model-based analysis and its applicability. *Applied and Environmental Microbiology* 70: 6706–6713.
4. Fuhrman JA and Noble RT (1995) Viruses and protists cause similar bacterial mortality in coastal seawater. *Limnology and Oceanography* 40: 1236–1242.
5. Guixa-Boixereu N, Lysnes K, and Pedros-Alio C (1999) Viral lysis and bacterivory during a phytoplankton bloom in a coastal water microcosm. *Applied and Environmental Microbiology* 65: 1949–1958.
6. Heldal M and Bratbak G (1991) Production and decay of viruses in aquatic environments. *Marine Ecology Progress Series* 72: 205–212.
7. Helton RR, Cottrell MT, Kirchman DL, and Wommack KE (2005) Evaluation of incubation-based methods for estimating virioplankton production in estuaries. *Aquatic Microbial Ecology* 41: 209–219.
8. Kepner RL, Wharton RA, and Suttle CA (1998) Viruses in Antarctic lakes. *Limnology and Oceanography* 43: 1754–1761.
9. Murray AG and Jackson GA (1992) Viral dynamics: a model of the effects of size, shape, motion and abundance of single-celled planktonic organisms and other particles. *Marine Ecology Progress Series* 89: 103–116.
10. Noble RT and Fuhrman JA (2000) Rapid virus production and removal as measured with fluorescently labeled viruses as tracers. *Applied and Environmental Microbiology* 66: 3790–3797.
11. Noble R and Steward GF (2001) Estimating viral proliferation in aquatic samples. In Paul JH (ed.) *Methods in Microbiology* 30: pp 67–84. San Diego: Academic Press.
12. Proctor LM, Okubo A, and Fuhrman JA (1993) Calibrating estimates of phage-induced mortality in marine bacteria: ultrastructural studies of marine bacteriophage development from one-step growth experiments. *Microbial Ecology* 25: 161–182.
13. Steward GF, Wikner J, Smith DC, Cochlan WP, and Azam F (1992) Estimation of virus production in the sea: I. method development. *Marine Microbial Food Webs* 6: 57–78.
14. Steward GF, Smith DC, and Azam F (1996) Abundance and production of bacteria and viruses in the Bering and Chukchi Sea. *Marine Ecology Progress Series* 131: 287–300.
15. Suttle CA and Chan AM (1994) Dynamics and distribution of cyanophages and their effect on marine *Synechococcus* spp. *Applied and Environmental Microbiology* 60: 3167–3174.
16. Waterbury JB and Valois FW (1993) Resistance to co-occurring phages enables marine *Synechococcus* communities to coexist with cyanophages abundant in seawater. *Applied and Environmental Microbiology* 59: 3393–3399.
17. Weinbauer MG (2004) Ecology of prokaryotic viruses. *FEMS Microbiology Ecology* 28: 127–182.
18. Weinbauer MG, Winter C, and Höfle MG (2002) Reconsidering transmission electron microscopy based estimates of viral infection of bacterioplankton using conversion factors derived from natural communities. *Aquatatic Microbial Ecology* 27: 103–110.
19. Wilhelm SW, Brigden SM, and Suttle CA (2002) A dilution technique for the direct measurement of viral production: a comparison in stratified and tidally mixed coastal waters. *Microbial Ecology* 43: 168–173.
20. Winget DM, Williamson KE, Helton RR, and Wommack KE (2005) Tangential flow diafiltration: an improved technique for estimation of virioplankton production. *Aquatic Microbial Ecology* 41: 221–232.
21. Winter C, Herndl GJ, and Weinbauer MG (2004) Diel cycles in viral infection of bacterioplankton in the North Sea. *Aquatic Microbial Ecology* 35: 207–216.
22. Wommack KE and Colwell RR (2000) Virioplankon: viruses in aquatic ecosystems. *Microbiology and Molecular Biology Reviews* 64: 69–114.

provides a brief compilation of the currently available approaches.

Infection of prokaryotes, for example, can be assessed using electron microscopic observations of the frequency of visibly infected cells (**Figure 4**). Because mature phages are visible inside host cells only in the final stage of lytic viral infection, this has to be converted to a frequency of infected cells (FIC)

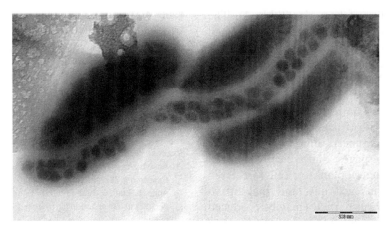

Figure 4 Transmission electron micrograph of two infected bacteria collected in a floodplain environment of the Danube River (Austria). Virus capsids are visible inside the cells. Picture was taken at a Zeiss EM 902 TEM (80 kV), × 30 000.

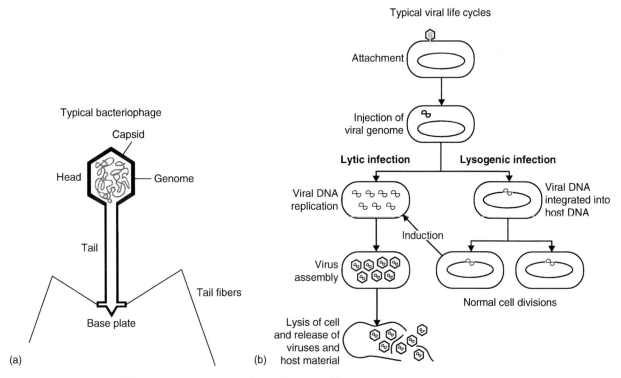

Figure 5 Viral structure (a) and main components of virus life cycles (b).

on the basis of conversion factors or nonlinear relationships (reference 2 in **Table 3**). The general assumption is that only about 0.5–11% of the lytic infected prokaryote cells are visibly infected. Some values from inland waters are even lower, whereas Arctic and Antarctic lakes display exceptionally high values (up to 67; **Table 2**). Virus-induced mortality of prokaryotes can be assessed based on FIC data and models. An early model assumed that under steady-state conditions after cell division one of the two daughter cells will die sooner or later from viruses. If viruses kill 50% of the cells, then they are responsible for 100% of the mortality. In this simple model FIC multiplied by two will yield mortality; however, refined models later replaced this simple approach. Another method is to quantify the virus-related loss of labeled DNA (tritiated thymidine) taken up in the bacterial size fraction in incubations of natural water samples. Experimental manipulation of viral and bacterial abundance was also used to assess the effect of variable encounter rates between bacteria and viruses on viral production and on their hosts. Only very few studies in inland waters have tried to compare methods for estimating virus-induced mortality of bacteria (**Table 3**). The impact of phages on prokaryotes can be expressed as a percent of removal of host standing stock or production. The calculated frequency of mortality due to viral lysis (FMVL) in inland waters ranges from undetectable to values greater than 100% (**Table 2**); this indicates both dynamic and variable interaction between lytic viruses and their hosts, as well as methodological uncertainties. Despite these uncertainties, the data suggest that typically between 10% and 20% of bacterioplankton cells are lysed through viral infection each day, and that up to 2.3×10^{9} planktonic virus particles may be produced per liter per hour (for sediments only a single, even higher value is available; see **Table 2**).

Viral Diversity

Despite the wide distribution of viruses, very little is known about their diversity. Assessing this diversity in the various environments is still a challenge for several reasons: (i) conceptual difficulties in defining a viral species, (ii) unculturable bacteria of natural systems because hosts have to be cultured before phages can be isolated, and (iii) no common molecule suitable to apply molecular phylogeny for viruses.

Several methods are available to assess viral diversity in natural systems (**Table 4**). For example, historically bacteriophage taxonomy was largely based on their morphology. Head size distribution and tail (presence or absence, length) are the commonly used parameters for characterizing viruses. This method has been widely used for decades in medicine, in the domain of public health and to characterize water quality.

In an ecological study conducted in Mono Lake, plaque assays were done to show that cultured isolates represented the minority of species, whereas uncultured viruses represented the dominant fraction of the community. To reveal the diversity of uncultured bacteriophages, viral DNA was used to construct clone libraries.

A community approach has been developed based on the separation by size of viral genomes using pulsed field gel electrophoresis (PFGE). A diverse viral population, on the basis of a broad range of viral genome sizes, was found in river and lake environments, with genome sizes ranging from ~9 to >400 kb with most of the DNA in the <23–60 kb size range. A combination of PFGE and TEM revealed that small viruses (capsid and genome size) probably dominate freshwater viral communities.

Cyanophages are apparently more closely related to each other and can be studied with cyanophage-specific primers targeted against a region of the

Table 4 Techniques used to assess viral diversity in aquatic samples

Method	Based on	Source
Bacteriophage taxonomy	Morphotypes	1, 6, 8, 9, 13
Culture approach	Plaque assays	10
Assessment of community complexity	PFGE	2, 3, 7, 12
Gene analysis	*gp20, mcp* gene	4, 5, 11, 14

Sources

1. Ackermann HW (2003) Bacteriophage observations and evolution. *Research in Microbiology* 154: 245.
2. Agis M, Luef B, and Peduzzi P (2004) Variability of virioplankton diversity in a river floodplain system. *Geophysical Research Abstracts* 6: 4985.
3. Auguet JC, Montanié H, and Lebaron P (2006) Structure of virioplankton in the Charente estuary (France): Transmission electron microscopy *versus* pulsed field gel electrophoresis. *Microbial Ecology* 51: 197–208.
4. Baker AC, Goddard VJ, Davy J, Schroeder DC, Adams DG, and Wilson WH (2006) Identification of a diagnostic marker to detect freshwater cyanophages of filamentous cyanobacteria. *Applied and Environmental Microbiology* 72: 5713–5719.
5. Dorigo U, Jacquet S, and Humbert J-F (2004) Cyanophage diversity, inferred from g20 gene analyses, in the largest natural lake in France, Lake Bourget. *Applied and Environmental Microbiology* 70: 1017–1022.
6. Fischer UR and Velimirov B (2002) High control of bacterial production by viruses in a eutrophic oxbow lake. *Aquatic Microbial Ecology* 27: 1–12.
7. Jiang S, Steward G, Jellison R, Chu W, and Choi S (2004) Abundance, distribution and diversity of viruses in alkaline, hypersaline Mono Lake, California. *Microbial Ecology* 47: 9–17.
8. Lu J, Chen F, and Hodson RE (2001) Distribution, isolation, host specificity, and diversity of cyanophages infecting marine *Synechococcus* spp. in river estuaries. *Applied and Environmental Microbiology* 67: 3285–3290.
9. Pina S, Creus A, González N, Gironés R, Felip M, and Sommaruga R (1998) Abundance, morphology and distribution of planktonic virus-like particles in two high-mountin lakes. *Journal of Plankton Research* 20: 2413–2421.
10. Sabet S, Chu W, and Jiang SC (2006) Isolation and genetic analysis of haloalkaliphilic bacteriophages in a North American soda lake. *Microbial Ecology* 51: 543–554.
11. Short CM and Suttle CA (2005) Nearly identical bacteriophage structural gene sequences are widely distributed in both marine and freshwater environments. *Applied and Environmental Microbiology* 71: 480–486.
12. Stewart GF (2001) Fingerprinting viral assemblages by pulsed field gel electrophoresis (PFGE) In: Paul JH (ed.) *Methods in Microbiology, Volume 30 Marine Microbiology*, 85–103. Academic Press Inc., U.S.
13. Weinbauer MG and Höfle MG (1998) Size-specific mortality of lake bacterioplankton by natural virus communities. *Aquatic Microbial Ecology* 13: 103–113.
14. Wilhelm SW, Carberry MJ, Eldridge ML, Poorvin L, Saxton MA, and Doblin MA (2006) Marine and freshwater cyanophages in a Laurentian Great Lake: evidence from infectivity assays and molecular analyses of g20 genes. *Applied and Environmental Microbiology* 72: 4957–4963.

capsid assembly protein gene (*gp20*) and major capsid protein probes. Techniques such as denaturing gradient gel electrophoresis and clone libraries were used to assess genotype richness of cyanophages.

Recent advances are primarily due to the availability of molecular markers (the g20 gene fragment) that can be used to detect cyanophages in natural systems by polymerase chain reaction.

Since genomics, functional genomics, and proteomics are on the upsurge, new techniques arising from these areas are likely to shed more light on the ecology of viruses. In the near future, new methodologies will no doubt allow deciphering of genomes and proteomes from natural communities.

Impact on the Environment; Conclusions

In the last decade, concepts related to aquatic food webs and related processes of carbon cycling have undergone substantial changes by the inclusion of microorganisms into models. With developing knowledge in viral ecology, processes associated with viruses are increasingly incorporated into aquatic food web models. For example, viral lysis of heterotrophic and autotrophic microbes can play a substantial role in the cycling of dissolved organic matter (DOM). The importance of viruses in this and other processes has given a richer, more realistic, but also much more complex view of aquatic food webs. Thus, viruses are now viewed as ubiquitous components also of inland aquatic environments, typically reaching abundances greater than 10^9 particles per liter. At this concentration, viruses commonly represent around 5% of the carbon fixed in bacteria, and viral biomass is some 25-fold larger than the carbon pool of protists.

In studies on the role of viruses in food web processes, the abundance of the natural community was often experimentally manipulated by enriching (with viral concentrates) or removing the virus-size fraction in order to assess the impact of viruses on phytoplankton and prokaryotic abundance and production. These manipulations caused changes in most of the investigated host populations; for example, increased viral numbers typically led to decreased bacterial abundance. Such results suggest that viral lysis can induce changes in the relative importance of different functional groups in the food web. At least in marine environments viral infection is, on average, as significant as grazing for bacterial mortality. For freshwater systems this may not be valid in such a generalized way, and the relative importance of viral mortality for bacteria appears to vary with space and time and with environmental settings. For example, virus-induced mortalities of bacteria from anoxic lake environments, in a solar saltern, and in polar lakes are among the highest reported so far (**Table 2**). In some inland waters the effect of viral lysis on food web processes and biogeochemical

cycles apparently varies in an unpredictable way. In general, the importance of virus-induced mortality appears to increase with increasing prokaryotic abundance and productivity.

The conversion of biomass to DOM due to viral lysis probably leads to the retention of the elements such as C, N, and P in the epilimnion (upper mixed layer) for a prolonged period. By this conversion, biomass and production is lost to grazers who might otherwise consume bacteria, cyanobacteria, and other microbes lysed by viruses. The DOM produced by lysis, however, remains available to noninfected prokaryotes and thus fuels the microbial loop. As a consequence, prokaryotic respiration and production as well as transformation of organic matter and regeneration of nutrients should be enhanced. For the marine environment it has been estimated that 6% to 26% of the organic carbon fixed by photosynthesis ends up in the dissolved organic carbon pool and is thus diverted from the grazing food chain (via a 'viral shunt'). A similar process can be expected for inland waters, although comprehensive data are still lacking. Through this enhanced bacterial carbon respiration and release of recycled inorganic nutrients, viral lysis might be a critically significant pathway for phosphate recycling in P-limited freshwater environments. Viral lysis apparently can also affect the rate of carbon fixation, both by prokaryotic and eukaryotic primary producers. Phytoplankton production either can be lost to grazers because of viral lysis, or might even be stimulated because of DOM consumption of lysis products by bacteria and remineralization of inorganic nutrients. Thus viral activity has, in certain situations, the potential to maintain higher overall levels of biomass and productivity in the system than would occur in the absence of viruses. The effect of lysogeny on biogeochemical processes and on the ecological performance and composition of bacterioplankton communities is less well understood. For example, mass lysis of hosts due to prophage induction may locally increase utilizable DOM sources and drastically reduce a particular host species.

Viral infection has the potential to influence the diversity of their pro- or eukaryotic hosts. This is expected because (i) viruses are generally specific to certain hosts, and (ii) infection is density-dependent: infection is more likely at high host densities because of increased contact rates. One of the underlying concepts is that competitive dominants are controlled by viruses once they become abundant. Thus, species diversity may be sustained by 'killing the winner,' which allows the poorer competitors to coexist (see 'Further Reading'). Phages and cyanophages can control heterotrophic bacteria and cyanobacterial blooms in freshwater environments as well. Accordingly, such mass lysis events may also cause a change in the composition of the bacterial community. Some effects of viruses on host biomass, production, and diversity are compiled in **Table 5**.

The potential for horizontal gene transfer mediated by viral activity is poorly investigated. Transduction rates (i.e., the virus-mediated gene transfer between a donor and a recipient host cell followed by phenotypic expression) have been reported from lake water with *Pseudomonas aeruginosa* as host cells. Applying a numerical model for the Tampa Bay Estuary, up to 1.3×10^{14} transduction events per year were calculated. Overall, viral and host DNA released during viral lysis is thought to contribute to the pool of extracellular DNA as a reservoir of genetically encoded information. This is no doubt an important factor for genetic variety and prokaryote evolution.

Less progress has been made in determining the mechanisms controlling viral diversity. An idealized food web model was used to evaluate the impact of viral loss by protozoan grazing on diversity regulation and the biogeochemical function: The 'kill the killer of the winner' (KKW) hypothesis considers viral loss by protozoans, interactions among uninfected and infected cells, free-living viruses, and nutrients in the model (see 'Further Reading'). The KKW process has a negative effect on bacterial species richness because of the VBR reduction. Furthermore, the latent period of viruses influences the intensity of the KKW process. A longer latent period reduces bacterial species richness by increasing the risk that the viruses in the host cells will be killed by grazing.

The previously underappreciated role of viruses in aquatic systems has changed towards a view that viruses have a critical influence on the performance of microbial food webs, on the diversity of their hosts, and on biogeochemical cycles. Since inland waters are among those environments that are most severely impacted by human activities, understanding of microbially mediated processes, including viruses that are active at the smallest scales of biology, is essential to predict the responses of ecosystems to environmental change.

Table 5 Viral impact on bacterial biomass, production and diversity

Location	Major results	Source
Lake Baroon (Australia)	Cyanophage Ma-LBP was a major natural control mechanism of *Microcystis aeruginosa* abundance	5
Lake Erie (Canada, USA)	Marine viruses (or at least their genetic material) can survive movement into fresh water; potential for changes in microbial community structure	8
Lake Pavin (France)	Viral lysis is a significant factor for producing shifts in the composition of heterotrophic eubacteria and cyanobacteria but not for their abundance (viruses infected some of the most abundant members of the prokaryotic community)	2
Lake Plußsee (Germany)	Absence or presence of viruses results in changes in abundance and growth rates of bacterial populations as well as in changes of the community composition	6
Rimov reservoir (Czech Republic)	Viral impact appeared to vary among different bacterial groups: filamental bacteria appeared resistant to viral infection; population-specific growth and removal rates among bacteria	3
	Effects on bacterioplankton structure were negligible	1
	Viruses suppressed several bacterial groups; viral lysis accelerated *Flectobacillus* growth; specific effects of flagellates versus viruses on bacterial community composition	4, 7

Sources

1. Hornak K, Masin M, Jezbera J, Bettarel Y, Nedoma J, Sime-Ngando T, and Simek K (2005) Effects of decreased resource availability, protozoan grazing and viral impact on a structure of bacterioplankton assemblage in a canyon-shaped reservoir. *FEMS Microbiology Ecology* 52: 315–327.

2. Jardillier L, Bettarel Y, Richardot M, Bardot C, Amblard C, Sime-Ngando T, and Debroas D (2005) Effects of viruses and predators on prokaryotic community composition. *Microbial Ecology* 50: 557–569.

3. Simek K, Pernthaler J, Weinbauer MG, Hornak K, Dolan JR, Nedoma J, Masin M, and Amann R (2001) Changes in bacterial community composition and dynamics and viral mortality rates associated with enhanced flagellate grazing in a mesoeutrophic reservoir. *Applied and Environmental Microbiology* 67: 2723–2733.

4. Simek K, Weinbauer MG, Hornak K, Jezbera J, Nedoma J, and Dolan JR (2007) Grazer and virus-induced mortality of bacterioplankton accelerates development of *Flectobacillus* populations in a freshwater community. *Environmental Microbiology* 9: 789–800.

5. Tucker S and Pollard P (2005) Identification of cyanophage Ma-LBP and infection of the cyanobacterium *Microcystis aeruginosa* from an Australian subtropical lake by the virus. *Applied and Environmental Microbiology* 71: 629–635.

6. Weinbauer MG and Höfle MG (1998) Distribution and life strategies of two bacterial populations in a eutrophic lake. *Applied and Environmental Microbiology* 64: 3776–3783.

7. Weinbauer MG, Hornák K, Jezbera J, Nedoma J, Dolan JR, and Šimek K (2007) Synergistic and antagonistic effects of viral lysis and protistan grazing on bacterial biomass, production and diversity. *Environmental Microbiology* 9: 777–788.

8. Wilhelm SW, Carberry MJ, Eldridge ML, Poorvin L, Saxton MA, and Doblin MA (2006) Marine and freshwater cyanophages in a Laurentian Great Lake: evidence from infectivity assays and molecular analyses of g20 genes. *Applied and Environmental Microbiology* 72: 4957–4963.

Knowledge Needs

Since virus ecology is still in its infancy, a listing of research needs may be overly subjective and raises more questions than the available space this section allows. Nevertheless, we try to summarize at least a few key issues, acknowledging that this largely represents the authors' point of view. Some of the limitations to current knowledge are valid for both marine and inland water systems; some are specific for inland aquatic environments.

a. The role of viruses as controlling agents of their host population compared to grazing still needs to be more widely studied, particularly in highly dynamic inland water environments.

b. The potential significance of viral life strategies such as lysogeny (as opposed to lytic) in biogeochemical cycles remains to be studied extensively.

c. The methods for estimating virus-induced mortality of host cells need to be refined.

d. More ecosystems (particularly different types of inland waters such as running waters) need to be studied thoroughly.

e. Since viruses are highly dynamic, more studies are needed on shorter time scales, e.g., to understand how processes operate and vary over diurnal cycles.

f. The chemical nature, molecular size, and bioreactivity of viral lysis products of bacteria are largely unknown; this is of importance, e.g., to understand the relevance of such compounds as sources of dissolved organic carbon to prokaryotes.

g. The potential role of liberated viral lysozymes in the transformation of DOM remains unknown; further, the hypothesis that lysozymes are a weapon against competing viruses by sweeping receptors on host cells needs to be investigated.

h. There are no good data on the release of dissolved DNA due to host lysis and on the significance of this release for genetic exchange.

i. Viral diversity has to be elucidated more comprehensively. This is a major issue not only for aquatic virus ecology. Information on the virus genome and proteome should provide invaluable information to develop theories on viral ecology.

j. Methodological progress is desired to study viruses attached to surfaces or within organic matrices (such as suspended particles, floating aggregates, biofilms) in the aquatic environment. Whether attached viruses lose infectivity or are protected from decay, and whether the presence and nature of particulate matter affects the frequency of lytic vs. lysogenic cycles (virus reproduction strategy) need to be investigated. Differences in the community structure of attached vs. ambient water viruses and hosts need to be documented. The potential viral impact on particle dissolution and aggregate formation should be assessed.

k. The virus ecology of sediments and at the aquatic–terrestrial interface needs to be developed; additional data on viral abundance and activity from different types of benthic environments are needed.

l. The information gathered from fine- or medium-scale studies has to be scaled up so that large-scale questions can be addressed, such as whether viruses are stabilizing or destabilizing agents in ecosystems and for biogeochemical cycles.

Focusing on these topics will provide further important insights into the structure and function of aquatic microbial food webs and into carbon cycling. This knowledge is important for understanding inland waters as important active components of the global hydrological cycle.

Abbreviations

AFDM	ash-free dry mass
CLSM	confocal laser scanning microscopy
DGGE	denaturing gradient gel electrophoresis
DOC	dissolved organic carbon
DOM	dissolved organic matter
EFM	epifluorescence microscopy
FCM	flow cytometry
FIC	frequency of infected cells
FMVL	frequency of mortality due to viral lysis
FVIC	frequency of visible infected cells
gp20	capsid assembly protein gene
mcp	major capsid protein
PCR	polymerase chain reaction
PFGE	pulsed field gel electrophoresis
PFU	plaque forming unit
TEM	transmission electron microscopy
VBR	virus to bacterium ratio
VDC	viral direct counts
VLPs	virus-like particles
VP	viral production

See also: Bacteria, Attached to Surfaces; Bacteria, Bacterioplankton; Bacteria, Distribution and Community Structure; Cyanobacteria; Microbial Food Webs; Protists.

Further Reading

Ackermann H-W and DuBow MS (eds.) (1987) *Viruses of Prokaryotes Vol. 1 General Properties of Bacteriophages*. Boca Raton: CRC Press.

Bratbak G and Heldal M (1995) Viruses – The new players in the game; their ecological role and could they mediate genetic exchange by transduction? *Molecular Ecology of Aquatic Microbes* 249–264.

Frank H and Moebus K (1987) An electron microscopic study of bacteriophages from marine waters. *Helgoländer Meeresuntersuchungen* 41: 385–414.

Goyal SM, Gerba CP, and Bitton G (eds.) (1987) *Phage Ecology.* New York: Wiley.

Hurst C (ed.) (2000) *Viral Ecology.* San Diego: Academic Press.

Paul J (ed.) (2001) *Methods in Microbiology.* San Diego: Academic Press.

Reisser W (1993) Viruses and virus-like particles of freshwater and marine eukaryotic algae – A review. *Archiv für Protistenkunde* 143: 257–265.

Suttle CA (2000) Cyanophages and their role in the ecology of cyanobacteria. In: Whitton BA and Potts M (eds.) *The Ecology of Cyanobacteria. Their Diversity in Time and Space*, pp. 563–589. The Netherlands: Kluwer Academic.

Suttle CA, Chan AM, and Cottrell MT (1990) Infection of phytoplankton by viruses and reduction of primary productivity. *Nature* 347: 467–469.

Thingstad TF and Lignell R (1997) Theoretical models for the control of bacterial growth rate, abundance, diversity and carbon demand. *Aquatic Microbial Ecology* 13: 19–27.

Weinbauer MG (2004) Ecology of prokaryotic viruses. *FEMS Microbiology Reviews* 28: 127–181.

Weinbauer MG and Rassoulzadegan F (2003) Are viruses driving microbial diversification and diversity? *Environmental Microbiology* 6: 1–11.

Wilhelm SW and Suttle CA (1999) Viruses and nutrient cycles in the sea. *BioScience* 49: 781–788.

Wommack KE and Colwell RR (2000) Virioplankon: Viruses in aquatic ecosystems. *Microbiology and Molecular Biology Reviews* 64: 69–114.

Relevant Websites

http://www.tulane.edu/~dmsander/garryfavweb.html – All the virology on the WWW.

http://www.nlv.ch/Virologytutorials/definition.htm – Background on viruses.

http://www.mbio.ncsu.edu/ESM/Phage/phage.html – Phage Page at North Carolina State University.

http://web.bio.utk.edu/wilhelm/SCOR.html – Scientific Committee on Oceanic Research, WG 126 on Role of Viruses in Marine Ecosystems.

http://www.mansfield.ohio-state.edu/~sabedon/ – The Bacteriophage Ecology Group.

http://www.virology.net/Big_Virology/BVHomePage.html – The Big Picture Book of Viruses.

http://plantpath.unl.edu/facilities/virology/ – World of Chlorella Viruses, University of Nebraska.

ALGAE (INCLUDING CYANOBACTERIA): PLANKTONIC AND ATTACHED

Contents

Algae

L Krienitz, Leibniz-Institute of Freshwater Ecology and Inland Fisheries, D-16775 Stechlin, Germany

Definition of the Term 'Algae'

Algae represent a highly diverse consortium of ancient plants comprising different evolutionary lineages of mostly photoautotrophic organisms. The different groups of algae are of polyphyletic origin and epitomize the majority of existing divisions of plants. Algae are thallophytic; their vegetative body is not organized in roots and leafy stems like that of the kormophytes. Many algae are living in solitary cells, colonies, filaments, or primitive vegetation bodies and do not have a vascular system. In contrast to the phanerogams (plants producing seeds), the algae are cryptogams that propagate by 'concealed' or 'hidden' reproductive strategies. Following a conception of subdivision of living organisms into five kingdoms (Monera, Protoctista, Fungi, Animalia, and Plantae), the prokaryotic algae (blue-green algae, Cyanobacteria, Cyanoprokaryota) are placed in the Monera (Eubacteria) and the eukaryotic algae in the Protoctista. Hence, the algae do not belong to the kingdom of Plantae. Nevertheless, it is widely accepted (because of the photosynthesis as the mutual characteristic) to interpret algae as 'lower plants' in distinction to the vascular 'higher plants.'

The eukaryotic algae possess membrane-bound organelles such as nuclei, mitochondria, and plastids. The prokaryotic Cyanobacteria do not exhibit such organelles; their DNA and photosynthetic thylakoids lie free in the cytoplasm. The combination of both eukaryotic algal groups and prokaryotic Cyanobacteria to the 'algae' demonstrates the heterogeneous and artificial character of this biological term. On the other side, the inclusion of Cyanobacteria in the group of algae is justified by several ecophysiological and evolutionary biological arguments. Like the eukaryotic algae, Cyanobacteria are able to photosynthetic oxygen production, and the theory of endosymbiosis has given evidence for evolution of chloroplasts by incorporation of Cyanobacteria into eukaryotic host cells. Endosymbiosis is discovered as a major force in the evolution of diverse algal lineages. Algae exhibit a fascinating diversity of life-forms and strategies in response to the environment. Most algae possess chlorophyll *a*. As primary producers, they use the sunlight energy to convert inorganic substances into simple organic compounds, and, provide the principal basis of food webs on the Earth. Furthermore, they produce oxygen that is essential for heterotrophic organisms. Algae populate a wide range of habitats from soil to water, whether it is cold or warm, alkaline or acidic, hyper- or hyposaline and we could cite a range of extremes. In this chapter, we focus on the algae of all types of inland waters.

Recent Status in the Systematics of Algae

The systematics of algae has been going through a dramatic phase of change for centuries. The recent situation is marked by the quest of a compromise between the conventional (artificial) system and the phylogenetic system, at least partly based on molecular genetics. During the last two or three centuries, scientists selected criteria ('folders') mainly on basis of morphotypes. After the introduction of DNA sequencing and phylogenetic analyses in the systematics, deficiencies of the conventional system became evident. Molecular phylogenetic studies revealed that the classical approaches using morphological characters for the circumscription of taxa do not adequately reflect the phylogenetic relationships. It is a challenge to cover step by step the conventional by modern phylogenetical approaches in the frame of interpretation of the organismic base of ecosystem processes. Molecular phylogenetic analyses have shown that one and the same morphotype (for example the spherical 'green ball') evolved independently in different lineages of algae. Such a simple organism as the UGSO (unidentified green spherical object) may be similarly grazeable for zooplankton, and, on the other hand can comprise dozens of organisms differing in their physiological demands and life strategies. On the other hand, highly diverse morphotypes can belong to one and the same phylogenetic lineage, especially those limnologists who exclusively depend on microscopic methods and who are bound to 'clearly' designated morpho-taxa are unsatisfied about the discrepancies between classical and modern systematics. Future efforts have to enable limnologists to apply more intensively the molecular-phylogenetic tools under field conditions.

To increase the knowledge in algal systematics, a polyphasic approach is widely suggested: (1) conventional morphological and ecophysiological study of algae under field conditions, (2) isolation of unialgal cultures for morphological, ontogenetic, biochemical, and physiological studies under laboratory conditions, (3) molecular phylogenetic analyses, and (4) establishment of detection methods of taxa directly in the field samples. First, promising results on environmental samples were provided by methods that are based on denaturating gradient gel electrophoresis (DGGE) and clone libraries initially established for bacteria/Cyanobacteria, but now these are applicable for eukaryotic algae too.

Generally, the sequences of the nuclear encoded small subunit (SSU) of ribosomal RNA (rRNA) (in prokaryotic cells the 16S rRNA [around 1540 nucleotides long] and in eukaryotic cells the 18S rRNA [around 1750 nucleotides long]) are the commonly used sections of the genomes to reveal the phylogenetic relationships of organisms and to construct the phylogenetic tree. Additionally, for deeper insights, it could be essential to analyze further parts of the genomes such as the highly variable internally transcribed spacer (ITS) regions of the ribosomal DNA, and the plastid encoded ribulose-1,5-bisphosphate carboxylase 3-phospho-D-glycerate carboxy-lyase (rbcL) gene.

Increasingly, analyses of SSU rRNA sequences are used in combination with other gene segments. For systematics in cyanobacteria it can be interesting, additionally to the 16S rRNA, to analyze the intergenic spacer region of the phycocyanin gene (PC-IGS). The sequencing techniques and phylogenetic analyses are in a process of permanent refining, and more and more parts at the genomes are used to benefit the systematics. A broad collection of data on DNA sequences of algae are available at databases such as the GenBank at the National Center for Biotechnology Information (NCBI). The molecular phylogenetic findings considerably support the critical re-evaluation of the results of conventional structural analyses. For current limnology and systematics, it is important to build more bridges between classical and modern as well as between field and laboratory methods.

Characterization of the Main Algal Groups that Occur in Inland Water Habitats

Prokaryota

Division Cyanophyta (Cyanobacteria, Cyanoprokaryota), Prochlorophyta The cells of Cyanobacteria are prokaryotic: they do not possess membrane-covered organelles such as mitochondria or nuclei and, they do not have an endoplasmatic reticulum. The photosynthetically active pigment, chlorophyll *a* (chlorophyll *b* and *c* are absent), is located in unstacked thylakoids, which lie free in the cytoplasm. The accessory pigments, phycocyanin and allophycocyanin (blue) and the phycoerythrin (red), are organized in phycobilisomes on the outer surface of the thylakoids. The most important reserve material is cyanophycean starch. The cell wall of Cyanobacteria consists of peptidoglycan and is often covered by mucilage sheaths of hydrated polysaccharides. Several members of Cyanobacteria possess aerotopes consisting of gas-vesicles (old and incorrect term: gas vacuoles) and they can regulate buoyancy in the water body in combination with the physiological conditions of the cytoplasm. Cyanobacteria reproduce asexually. Sexual reproduction has not been observed; however, parasexual processes of transformation and conjugation allow genetic recombination. Flagellated cell stages have never been found at any phase of ontogeny.

Cyanobacteria represent the most ancient photoautotrophic organisms on Earth. Phylogenetic analyses indicate an early evolutionary explosive radiation with a long history of separate lineages. Unicellular and simple filamentous forms evolved polyphyletically and scatter throughout several clades, whereas the heterocytic taxa are of monophyletic origin. Consequently, a modern 'suprageneric' system of cyanobacteria suggests several subclasses each involving both coccoid and trichal taxa (Synechococcophycidae nom. prov., Oscillatoriophycidae nom. prov.) and an other, homogeneous, subclass combining the Nostocales and Scytonematales (Nostocophycideae nom. prov.). The latter subclass reflects the ability to form heterocytes as a fundamental phylogenetic marker.

A number of prokaryotic oxyphototrophs (formerly designated as division Prochlorophyta) are characterized by possession of chlorophylls *a* and *b*, which are not masked by accessory pigments. Phylogenetic analyses revealed that these organisms do not represent a separated monophyletic lineage but they belong to different clades of coccoid and trichal Cyanobacteria.

The nomenclature of Cyanobacteria is ruled both by the International Code of Botanical Nomenclature (ICBN) and the International Code of Nomenclature of Prokaryotes (ICNP). The inconsistent rules of the different codes have caused problems in the naming of Cyanobacteria. Therefore, a unified nomenclatural approach for these organisms, acceptable for botanists and bacteriologists, was initiated at the 16th Symposium of the International Association for Cyanophyte Research (IAC) in Luxembourg in 2004.

Most of the known species of Cyanobacteria (around 2000) are living in fresh water. However, numerous taxa, as a relict of their precambrien history are adapted to extreme environments such as hot springs (**Figure 1**), saline waters (**Figure 2**), icefields, and edaphic habitats.

Eukaryota

Division Rhodophyta The chloroplast of red algae is surrounded only by one double-membrane and contains chlorophyll *a* in unstacked thylakoids. In phycobilisomes on the thylakoid-surface the accessory pigments, phycoerythrin and phycocyanin, are located. Floridean starch is the dominating storage product. Red algae's asexual and sexual reproduction includes unique features; flagellated stages are always absent.

Red algae are predominantly living in marine environments. In inland habitats only few taxa (around 200 species worldwide) are found and mostly distributed attached to rocks in clean brooks and rivers as well as in lakes.

Division Heterokontophyta The principal characteristic of Heterokontophyta is represented in the heterokont flagellation of cells which bear a long pleuronematic and a shorter acronematic flagellum. The pleuronematic flagellum acts for locomotion and

Figure 1 Cyanobacteria in extreme inland water environments: Orange-colored mats of the filamentous cyanobacterium *Phormidium* in a hot spring at the shoreline of Lake Bogoria (Kenya). Photo: L. Krienitz.

is directed forwards during swimming. This flagellum is equipped with two rows of stiff hairs (mastigonemes). The acronematic flagellum points backwards along the cell and is for navigation. A basal swelling of this flagellum together with an eyespot in the chloroplasts near the anterior part of the cell forms the photoreceptor apparatus. The chloroplasts contain chlorophylls *a* and *c*; no chlorophyll *b* is present. Dominating accessory pigments are fucoxanthin (in the Chrysophyceae, Bacillariophyceae, Phaeophyceae, and some Raphidophyceae) or vaucheriaxanthin (in the Xanthophyceae, Eustigmatophyceae, and some Raphidophyceae). In the chloroplasts the thylakoids are grouped in stacks of three. The chloroplast envelope consists of several layers: the own double membrane, a fold of the endoplasmatic reticulum as an outer layer, and a girdle lamella as an inner layer. The main reserve polysaccharide is chrysolaminarin. Members of this division have a multitude of vegetative and sexual reproduction types. The Heterokontophyta represents a natural group of algae characterized by several ultrastructural criteria such as a distinct transitional helix between the flagellar shaft and the basal body of the flagellum. Some differences in fine structures and in the content of accessory pigments serve for differentiation of the classes of Heterokontophyta.

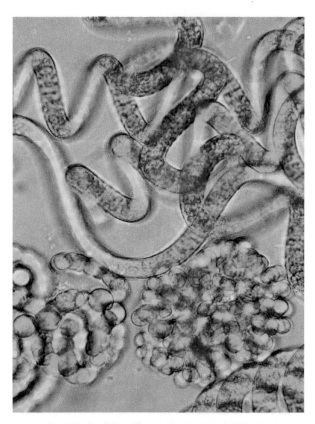

Figure 2 Cyanobacteria in extreme inland water environments: Phytoplankton of the alkaline Rift Valley Lake Nakuru (Kenya), the food source of Lesser Flamingo populations, dominated by the oscillatorian *Arthrospira* and nostocalean *Anabaenopsis* species. Photo: L. Krienitz.

Class Chrysophyceae (golden algae) Members of this class are characterized by its golden color because of the accessory pigments masking the chlorophyll. The cells are mostly monadoid; however, also amoeboid, palmelloid, and coccoid life forms are known, which live solitary or in colonies, and have filaments or primitive thalli. Chrysophytes produce different types of cell wall covers: numerous taxa are characterized by cellulosic envelopes (loricae) or by silica scales, whereas others miss the outer cell wall. Several golden algae are able to develop endogenous silicified resting stages.

More than 1000 chrysophycean species (stomatocysts) have been described, which are restricted mostly to freshwater habitats.

Combined molecular and conventional approaches resulted in splitting different classes off from the Chrysophyceae; especially, the separation of silica-scaled Synurophyceae which is widely accepted.

Class Xanthophyceae (yellow green algae) Xanthophycean chloroplasts are yellow-green because of the dominance of vaucheriaxanthin and diatoxanthin; the brown accessory pigments are absent. Most of the 500 taxa are unicellular or colonial algae (Mischococcales), some are filamentous (Tribonematales) and others are multinucleate siphonal (Vaucheriales). The cell walls often are divided in two overlapping parts. Recent investigations showed that from the coccoid freshwater species (e.g., members of the genera *Goniochloris* and *Tetraedriella*) a considerable number of species are to be transferred to the class of Eustigmatophyceae.

Class Eustigmatophyceae The major ultrastructural characteristics of the Eustigmatophyceae are a uniquely constructed eyespot outside the chloroplast and the absence of a girdle lamella in chloroplasts. Violaxanthin is the dominant 'light harvesting' pigment. The propagation proceeds exclusively asexually. The number of species is about a hundred, but it is pending due to ongoing inclusion of several xanthophycean taxa. Members of the picoplanktonic genus *Nannochloropsis* are well known from marine habitats. Members of this genus have recently come into focus of limnologists because in lakes of different trophic state this polyunsaturated fatty acid rich algae can establish mass developments.

Class Bacillariophyceae All members of this algal class develop a siliceous cell wall (frustule), which is constructed by two overlapping halves (hypotheca and epitheca). The frustule of the Pennales is elongated, bilaterally symmetrical, and the frustule of the Centrales has radial symmetry. During vegetative propagation, the two halves are forced apart, and each of the daughter cells synthesizes a new hypotheca, resulting in a successive decrease of cell size down to about 50% in a given population. After reaching a minimal cell size, sexual reproduction starts and results in the development of expanding auxospores, which results in a substantial enlargement of the cells. During the sexual phase, the only flagellated stage (male gametes) can be observed. The spermatozoids carry only one flagellum (the pleuronematic). Investigations combining ecological and genetic methods have given evidence that in some diatoms a dimorphism can be observed.

Bacillariophyceae are extremely widespread and species rich in marine and freshwater habitats (around 100 000 species).

Class Raphidophyceae This class comprises large unicellular flagellates, which lack cell walls, photoreceptors, and a flagellar transitional helix. In fresh waters, mainly *Goniostomum* spp. are of ecological interest, because they form massive blooms in oligotrophic bog lakes and have recently invaded nutrient-rich waters, especially in Sweden and Finland.

Class Phaeophyceae (brown algae) More than 99% of brown algal species were found in marine habitats. Only about seven species are living in fresh waters and appear as thalli of microscopic filaments, tufts, or crusts. Most of them are rare and endangered.

Division Haptophyta Almost all of Haptophyta are unicellular flagellates. Additionally to the two apical or lateral flagella, a filamentous appendage (haptonema) with a special substructure has been developed. The cell wall is covered by scales. Most of the haptophytes are marine, only few freshwater species are known belonging especially to the genus *Chrysochromulina*.

Division Cryptophyta The majority of Cryptophyta are dorsiventrally shaped unicellular flagellates which bear two flagella of different length. The longer flagellum is equipped with two rows of mastigonemes, whereas the shorter flagellum bears only a single row of mastigonemes. The flagellae emerge at an anterior invagination (gullet) at the ventral side of the cell. The cells are covered by a stiff proteinaceous periplast composed of polygonal plates. Under the plasmalemma, numerous trichocysts are placed which eject mucilage upon stirring the cell. The chloroplast does not contain a girdle lamella; the thylakoids have a relatively thick lumen filled with the accessory pigments phycocyanin and phycoerythrin, which mask the chlorophylls *a* and *c*. Storage material is starch. A special structure, the so-called nucleomorph, support the hypothesis of origin of the chloroplasts by secondary endosymbiosis. The propagation of cryptomonads proceeds mostly by longitudinal cell division; sexual reproduction is only rarely observed. Only about 100 species of freshwater Cryptophyta are known. Combined ontogenetical and phylogenetic studies provided evidence for a life history-dependent dimorphism in *Cryptomonas* which includes *Campylomonas* morphotypes.

Division Dinophyta Most of Dinophyta are unicellular flagellates armoured with a robust envelope (theca) of polysaccharides. The theca is divided into an apical (epicone) and an antapical half (hypocone) and consists of polygonal plates of different species-specific shape and size. The theca contains a transverse furrow (girdle, cingulum) and a backwards pointed longitudinal groove (sulcus) which host the flagella. The two different flagella emerge at the intersection of cingulum and sulcus. The axoneme of the transverse flagellum is helically formed. This flagellum contributes to rotation of the cell. The longitudinal flagellum acts as a rudder navigating the cell and pushing it in forward movement. The chloroplasts are equipped with a girdle lamella and contain chlorophylls *a* and *c*, which are masked by brown accessory pigments (β-carotene, peridinin). Storage material is starch. Dinophyta exhibit a haplontic life cycle, and only the zygotes have a diploid nucleus. The big interphase nucleus (dinokaryon) has a unique structure characterized by highly condensed chromosomes of helicoidal, garland-like shapes which are even visible under the light microscope. Only about 10% of Dinophyta are living in fresh waters where nearly 200 species are known.

Division Euglenophyta Nearly all euglenophytes are unicellular flagellates with two flagella arising from the bottom of a bottle-shaped invagination (ampulla). In most species, one of the flagellae is very short and does not protrude the ampulla. The long flagellum, emergent at the anterior cell pole, drives the cell in a spirally undulating forward movement. The flagellum surface is covered by delicate hairs of three different lengths. The cells of euglenophytes are surrounded by a striped pellicle (periplast), which lies within the cytoplasm. Numerous members of *Euglena* are able to a metabolic change of their cell-shape whereas members of the genus *Trachelomonas* are equipped with a rigid lorica. The chloroplasts contain chlorophylls *a* and *b* and accessory pigments (β-carotene, neoxanthin, diadinoxanthin) which do not mask the chlorophylls. The storage material is plate-, rod- or ring-shaped paramylon (β-glucan).

Euglenophyta are photoautotrophic, phagotrophic, or saprotrophic. Their propagation occurs exclusively by longitudinal cell divison; sexual reproduction has not been observed. They occur predominantly in organically rich fresh waters. More than 800 species are known. The Euglenophyta represent a very old algal lineage, which is closely related to the Kinetoplasta, heterotrophic protozoans, which include the pathogenic *Trypanosoma*.

Division Chlorophyta The Chlorophyta are characterized by isokont (two, four, or more) flagella, which lack mastigonemes. However, some green algal taxa have flagellae with delicate hairs or scales. The flagellar basal bodies are associated with four cruciately arranged microtubular roots. The cruciate pattern of the roots and its position in relation to the basal bodies differ from group to group and are categorized in three main configurations: (1) basal bodies directly opposed (DO), (2) basal bodies in a counterclockwise

(CCW) position; or (3) in a clockwise (CW) position. The chloroplasts are surrounded only by an own double membrane. The thylakoids are stacked (two to six or more thylakoids per stack). The chloroplasts are green and contain chlorophylls *a* and *b* as well as accessory pigments such as xanthophylls, which do not mask the chlorophyll. The most important reserve polysaccharide is starch. This division comprises of about 4000–5000 freshwater species with numerous unicellular or colonial planktonic or epi- and periphytic algae, and many filamentous or thallic benthic taxa.

Class Chlorophyceae Traditionally, the class Chlorophyceae included nearly all green algae. After adjustment of structural versus molecular criteria, the system of green algae was subjected to a substantial reorganization. Therefore, we have to consider the Chlorophyceae in a much narrower sense.

The order Chlamydomonadales comprises mostly unicellular or colonial flagellates, with two or four flagellae with a CW basal body orientation. According to modern considerations, this order comprises taxa formerly belonging to Dunaliellales, Volvocales, Chlorococcales *sensu stricto*, i.e., the relationship of *Chlorococcum*, Tetrasporales, and Chlorosarcinales.

The order Sphaeropleales comprises vegetatively nonmotile unicellular or colonial taxa with biflagellate zoospores having DO basal body orientation. After molecular phylogenetic studies, the scope of this order has been remarkably extended after inclusion of numerous autospore-forming taxa of the classical order Chlorococcales *sensu lato*. For example, the species-rich coccoid families of Hydrodictyaceae, Scenedesmaceae, and Selenastraceae now belong to the Sphaeropleales.

All taxa of the order Oedogoniales grow in filaments on submerged surfaces in fresh waters. The oedogonialean zoids (either asexual zoospores or male gametes) exhibit a unique stephanokont flagellation developing an anterior ring of flagella.

The order Chaetophorales contains filamentous taxa producing quadriflagellate zoids with upper and lower pairs of CW orientated basal bodies.

Class Ulvophyceae The status of this class has been ambiguous and comprises predominantly marine filamentous and thallous green algae with bi- or quadriflagellated zoids of CCW basal body orientation. In fresh water, members of the orders Ulotrichales (trichal) and Cladophorales (siphonocladal) can form massive benthic algal communities.

Class Trebouxiophyceae Most members of Trebouxiophyceae are coccoid unicellular or colonial algae propagating asexually by autospores or zoospores. The flagellated stages investigated showed a CCW basal body orientation. However, numerous autospore-forming taxa are known, where zoids have never been observed during reproduction and can only be classified in trebouxiophytes by means of molecular data.

Most members of the order Trebouxiales are edaphic or live in symbiosis with fungi in lichen thalli. This is also valid for some taxa of the order Chlorellales; however, predominantly these coccoid algae exhibit a multitude of planktonic morphotypes. The families Oocystaceae and Chlorellaceae are to be extended to include different taxa that were, because of their morphology, formerly classified in other families.

Class Prasinophyceae This line comprises of solitary green flagellates and a few coccoids from marine and freshwater habitats. The surface of the cells and flagellae are mostly covered by organic scales. Prasinophytes are widely heterogeneic in their morphology, ultrastructure of flagellar root systems, and accessory pigments. Molecular phylogenetic studies provided evidence for the paraphyletic character of the Prasinophyceae. In fresh waters, especially members of the genera *Tetraselmis* and *Pyramimonas*, are widely distributed.

Division Charophyta This division includes both charophyte green algae and embryophytes (land plants) and has been often termed Streptophyta. When motile cells are developed, they are biflagellated with asymmetrically inserted flagella and two dissimilar flagellar roots. The Charophyta possess several enzyme systems (such as cellulose synthetase) that are not found in other green algae.

Class Zygnemophyceae (conjugates) This class developed only two different live forms, the coccoid and the filamentous organization. Flagellated stages are lacking. Sexual reproduction proceeds by conjugation involving the fusion of two amoeboid gametes. The class is rich in species (more than 5000) which are restricted only to freshwater habitats.

Members of the order Zygnematales are unicellular or mostly filamentous and were represented by the large genera *Mougeotia*, *Spirogyra*, and *Zygnema*.

The order Desmidiales comprises solitary or in amorphous colonies living taxa. The desmid cell is constructed by two symmetrical halves (semicells) which overlap in a central, often constricted isthmus.

Class Charophyceae (stoneworts) The siphonocladal thalli of Charophyceae are macroscopic and differentiated into nodes and whorls of branches bearing internodes. Sexual reproduction is by oogamy; the antheridia and oogonia are surrounded by a layer of sterile cells. The biflagellated male gametes are covered with tiny organic scales. Members of Charales are restricted to freshwater and brackish habitats.

Three Case Studies in Combining Algal Systematics and Ecology

Dimorphism in Centric Diatoms

Traditional morphological studies on the diatom *Cyclotella comensis* suggested different taxa in this form group. The resolution of this taxonomical question was only possible after combination of the conventional LM and SEM methods with comparison of field and cultured material and with the assistance of sequencing methods. It was proved that the supposed different taxa belong to one and the same species. *C. comensis* undergoes a dimorphic population cycle in fresh waters (**Figure 3**). Under inclusion of phases of sexual reproduction in the course of the annual diatom successions of several European lakes two different morphotypes were observed: the morphotype *comensis* (cells 4–15 μm, with undulated central area of the valves) and the morphotype *minima* (cells 2.5–7 μm, with plain valves). On clonal cultures these morphological differences were confirmed (**Figures 4** and **5**). However, the sequences of the ITS-2 spacer region of the SSU rRNA were

identical between the clones and proved the dimorphism in this species. This combined approach is a useful tool to resolve questions in the highly complicated systematics of diatoms and gives new perspectives for interpretation of succession patterns in lakes.

Polyphyletic Origin of Eukaryotic Picoplankton and Consequences for the Food Web

Because of a very effective volume to surface ratio, autotrophic picoplankton is a highly productive member of food webs in marine and freshwater ecosystems. The eukaryotic algal picoplankton (smaller than 3 μm in diameter), commonly designated as 'green spheres,' are equipped with a minimal set of organelles such as nucleus, chloroplast, mitochondrium, and several vacuoles. These organisms are very similar to each other. The differentiation of these picoplankton algae by light microscopy is very difficult, sometimes impossible. Even under the transmission electron microscope the differentiation is based only on very few characteristics such as lamellate vesicles within the vacuoles (**Figures 6** and **7**). However, molecular phylogenetic research has proved that the 'green spheres' have been evolved in different lineages. In marine environments, members of the prasinophyceaen genera *Bathycoccus* and *Pycnococcus* were detected. In hypersaline inland waters *Picocystis*, a very unique picoplankton (probably belonging to a new class of Chlorophyta near the prasinophyte lines) was discovered. In fresh waters, members of the green algal genera *Choricystis*, *Nannochloris*

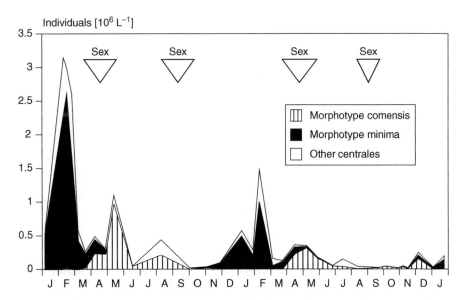

Figure 3 Dimorphism in the centric diatom *Cyclotella comensis*: Succession of the two different morphotypes in Lake Wummsee (Germany). Adapted from Scheffler *et al.* (2003) *Arch. Hydrobiol. Spec. Issue Advanc Limnol.* 58: 157–173.

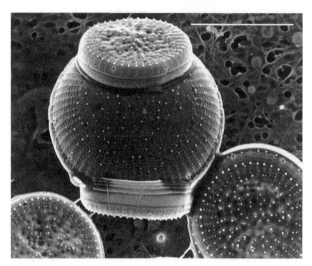

Figure 4 Scanning electron micrographs of *Cyclotella comensis* (scale bar = 5 μm). *C. comensis* initial cell. The frustules of the former mother cell remained attached at the big initial cell. Adapted from Scheffler *et al.* (2005) *Diatom Res.* 20: 171–200.

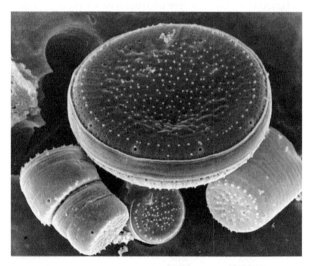

Figure 5 Scanning electron micrographs of *Cyclotella comensis* (scale bar = 5 μm). *C. comensis* clonal culture with two morphotypes; the four small cells represent the morphotype *minima*, and the single large cell belongs to the morphotype *comensis*. Adapted from Scheffler *et al.* (2005) *Diatom Res.* 20: 171–200.

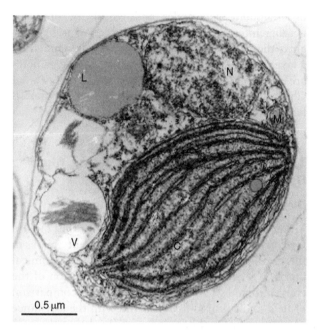

Figure 6 Transmission electron micrographs of two members of the eukaryotic picoplankton from inland waters: *Nannochloropsis limnetica* (Eustigmatophyceae). Adapted from Krienitz *et al.* (2000) *Phycologia* 39: 219–227. *Abbreviations*: L, lipid droplet; V, vacuole with lamellate vesicles.

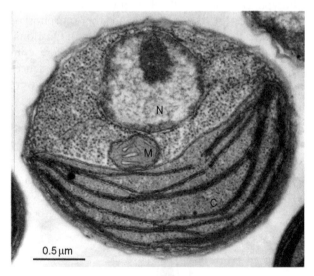

Figure 7 Transmission electron micrographs of two members of the eukaryotic picoplankton from inland waters: *Pseudodictyosphaerium jurisii* (Trebouxiophyceae). Photo: L. Krienitz. *Abbreviations*: N, nucleus; C, chloroplast; M, mitochondrium.

(Trebouxiophyceae), *Mychonastes*, and *Pseudodictyosphaerium* (Chlorophyceae) are very common. Biochemical studies showed substantial differences between the picoplanktonic taxa, especially in the content of accessory pigments and fatty acids. This was documented on the content of polyunsaturated fatty acids in members of the genus *Nannochloropsis* (Eustigmatophyceae), predominantly found in marine habitats but now discovered also in fresh waters of different trophic states (in the oligotrophic Lake Baikal

as well as in hypertrophic village ponds). *Nannochloropsis* has a 50–100-fold higher content in eicosapentaenoic acid than other freshwater picoplankton such as *Choricystis* and *Pseudodictyosphaerium*. This essential fatty acid is a precursor of prostaglandine synthesis, and prostaglandins are themselves precursors for a number of compounds known as tissue hormones. From these results a high impact of the picoplanktonic food algae for consumer populations in water food webs can be concluded. However, to allow a reasonable evaluation of food quality of picoplankton populations, a meticulous determination of these primary producers is essential and this can only be done adequately by modern methods.

Phenotype versus Genotype in Chlorellaceae in Reflection of Ecosystem Conditions

Chlorella is one of the archetypical forms of nanno-planktonic coccoid green algae. As model organisms, culture strains of this genus have conquered laboratories of plant physiology, algal test systems, and mass cultivation. *Chlorella vulgaris*, the type species, is distributed in small polytrophic inland water bodies. More than 100 *Chlorella* species have been described from fresh water, marine and soil habitats, but most of them need to be revised and placed to other algal groups. According to recent investigations, only three 'true' spherical *Chlorella* species are included in this genus. However, after sequence analyses, a growing number of morphologically different algae are found to cluster within the *Chlorella* clade. Therefore, several ellipsoidal, needle-shaped, coenobial, spined or mucilage-covered taxa formerly grouped in different

algal families are accepted as closest relatives to the spherical chlorellae (**Figure 8**). These findings raise the question to which amount these morphological forms represent 'only' phenotypical adaptations to ecosystem conditions such as grazing pressure and buoyant life strategy in phytoplankton? Ecophysiological experiments with *Micractinium* have exhibited a wide range of morphological flexibility. In dense cultures this algae produces solitary 'green spheres,' which exactly fit to the *Chlorella* phenotype. However, under grazing pressure and transferred medium from *Brachionus* cultures, *Micractinium* produced strong bristles. This approach of combination of morphological, onthogenetical, ecophysiological with phylogenetic considerations provides a wide and interesting scope of limnological and phycological activity to elucidate the interaction of structure and function in freshwater ecosystems.

On the example of algal systematics and ecology, a vital picture of actual essentials in life sciences can be drawn. At present, limnologists are passing a challenging period of upheaval from conventional methods to modern molecular phylogenetic tools. They have to hold the balance between both types of work, because the most creative results require respectful interdisciplinary cooperation between 'young' and 'old' approachers. The limnologists have the privilege to work in ecosystems, and ecosystems are the environments where the evolution is taking place.

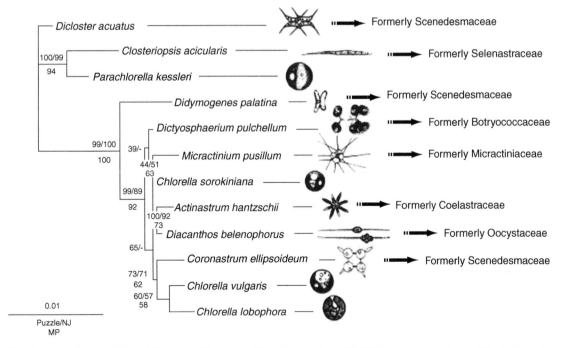

Figure 8 Phylogenetic tree of the relationship of the genus *Chlorella* based on 18S rRNA sequence analyses. Adapted from Luo *et al.* (2005) *Verh. Internat. Verein. Limnol.* 29: 170–173.

Nomenclature

ITS	internally transcribed spacer regions of the ribosomal DNA
PC-IGS	intergenic spacer region of the phycocyanin gene
rbcL	ribulose-1,5-bisphosphate carboxylase 3-phospho-D-glycerate carboxy-lyase gene
rRNA	ribosomal RNA
SSU	nuclear encoded small subunit of ribosomal RNA
NCBI	National Center for Biotechnology Information
DGGE	denaturating gradient gel electrophoresis
SEM	scanning electron microscope
TEM	transmission electron microscope

Basal Body Orientation of the Flagellae
DO	basal bodies directly opposed,
CCW	basal bodies in a counterclockwise position,
CW	clockwise position of basal bodies.

Glossary

Prokaryota

Eukaryota

Division Cyanophyta (Cyanobacteria, Cyanoprokaryota), Prochlorophyta

Division Rhodophyta

Division Heterokontophyta

Class Chrysophyceae (golden algae)

Class Xanthophyceae (yellow-green algae)

Class Eustigmatophyceae

Class Bacillariophyceae

Class Raphidophyceae

Class Phaeophyceae (brown algae)

Division Haptophyta

Division Cryptophyta

Division Dinophyta

Division Euglenophyta

Division Chlorophyta

Class Chlorophyceae

Order Chlamydomonadales

Order Sphaeropleales

Order Oedogoniales

Order Chaetophorales

Class Ulvophyceae

Class Trebouxiophyceae

Class Prasinophyceae

Division Charophyta

Class Zygnemophyceae

Class Charophyceae

See also: Chrysophytes – Golden Algae; Cyanobacteria; Cyanoprokaryota and Other Prokaryotic Algae; Diatoms; Green Algae; Harmful Algal Blooms; Other Phytoflagellates and Groups of Lesser Importance.

Further Reading

Canter-Lund H and Lund JWG (1995) *Freshwater Algae. Their Microscopic World Explored.* Bristol: Biopress Ltd.

Christensen T (1980–1994) *Algae. A Taxonomic Survey.* Odense: AiO Print Ltd.

Graham LE and Wilcox LW (2000) *Algae.* Upper Saddle River: Prentice-Hall.

Hoffmann L, Komárek J, and Kastovsky J (2005) System of cyanoprokaryotes (cyanobacteria) – state in 2004. *Algological Studies (Cyanobacterial Research 6)* 117: 95–115.

Huber-Pestalozzi G (ed.) (1938–1983) *Das Phytoplankton des Süßwassers* (9 vols.). Stuttgart: E. Schweizerbart'sche Verlagsbuchhandlung (Nägele u. Obermiller).

John DM, Whitton BA, and Brook AJ (eds.) (2002) *The Freshwater Algal Flora of the British Isles.* Cambridge University Press.

Komárek J and Anagnostidis K (1998) *Cyanoprokaryota. Chroococcales.* In: Ettl H, Gärtner G, Heynig H, and Mollenhauer D (eds.) *Süßwasserflora von Mitteleuropa, band 19/1.* Jena: Gustav Fischer.

Komárek J and Anagnostidis K (2005) *Cyanoprokaryota. Oscillatoriales.* In: Büdel B, Gärtner G, Krienitz L, and Schagerl M (eds.) *Süßwasserflora von Mitteleuropa, band 19/2.* München: Elsevier GmbH.

Kristiansen J (2005) *Golden Algae.* Rugell: A.R.G. Gantner Verlag K.G.

Lee RE (1999) *Phycology.* Cambridge University Press.

Lewis LA and McCourt RM (2004) Green algae and the origin of land plants. *American Journal of Botany* 91: 1535–1556.

Margulis L, Corliss JO, Melkonian M, and Chapman DJ (eds.) (1990) *Handbook of Protoctista.* Boston: Jones and Bartlett Publishers.

Round FE, Crawford RM, and Mann DG (1990) *The Diatoms. Biology and Morphology of the Genera.* Cambridge University Press.

Van den Hoek C, Mann DG, and Jahns HM (1995) *Algae. An Introduction to Phycology.* Cambridge University Press.

Wehr JD and Sheath RG (eds.) (2003) *Freshwater Algae of North America.* San Diego/London/Massachusetts: Academic Press.

Relevant Websites

http://www.algaebase.org/ – AlgaeBASE.

http://www.algaterra.net – AlgaTerra Information System on terrestrial and limnic Micro Algae.

http://www.calacademy.org/research/diatoms/ – California Academy of Sciences Diatom Collection.

http://www.cyanodb.cz/ – CyanoDB: The on-line database of cyanobacterial genera.

http://www-cyanosite.bio.purdue.edu/index.html – Cyanosite (a web site for cyanobacterial research).

http://www.desmids.nl/ – Desmids (Desmidiaceae) in the Netherlands.

http://www.ncbi.nlm.nih.gov/ – GenBank of the National Center for Biotechnology Information (NCBI).

http://ucjeps.berkeley.edu/INA.html – Index Nominum Algarum (INA).

http://www.fritschalgae.info/ – The Fritsch Collection of illustrations of freshwater algae.

http://www.phycology.net/ – The Phycology.Net.

The Phycogeography of Freshwater Algae

J Padisák, University of Pannonia, Veszprém, Hungary

Species Concepts in Phytoplankton

The species concept for most phycologists is based on the morphological characters and hence the term 'species' means morphospecies. On the other hand, for evolutionary biologists the term means biological species that can be defined as a reproductive community of populations (reproductively isolated from others) that occupy a specific niche in Nature. If we accept the above definition of species, any talk about biological species in groups (Cyanobacteria, Euglenophyta) where sexual reproduction has not been observed yet is meaningless. Nevertheless, recent concepts argue that it would be unproductive and inconvenient to restrict the term 'species' exclusively to one or the other.

Most recognized freshwater algal species appear to be cosmopolitan: they may be found all over the world if studied extensively by specialists. In terms of evolutionary biology, it means that either they have highly efficient means of dispersal or their morphological characters are very static through long evolutionary times. Some species are indeed very static in their morphological characters ('good species') others apparently vary within a wide range. Former taxonomic concepts weighted minor morphological differences by giving them a taxonomic, usually intraspecific, rank. This practice, in lack of a sufficient species concept, must lead to infinite separation, finally to individuals. Recent investigators tend to view species formerly considered separate as being synonymous. Without a deeper discussion of species concepts applicable for algae, it is necessary to realize that any phycogeographical discussion is ultimately dependent on the species concept applied.

Since most freshwater habitats are ephemeral, each algal species should have a genetic ability to switch from the vegetative reproduction to a sexual cycle or from vegetative to dormant stage when (or even before) the habitat becomes too unfavorable. Populations of ephemeral habitats are exposed to extremes of environmental variables. Variation of temperature, either on daily or on seasonal scale, is not significantly more moderate than that in the surrounding terrestrial habitats. They have to cope with high irradiation, including the UV wavelengths, pH and conductivity also often vary in large range. These environmental exposures (especially tolerance of high temperature variation, desiccation, and high irradiation) certainly ease transport of propagules. Algae of permanent waters (large, exorheic lakes), especially phytoplankton, do not have to develop such superior tolerance properties, which, on the other hand, may limit their resistance to transport conditions. There are evidences that desiccation and subsequent rewetting in cyanobacterial trichomes leads to lysis of cells in aquatic species while aerophytic ones are able to tolerate this treatment in combination with low or high temperature treatment.

Dispersal of Algal Species

The occurrence of so many common freshwater algae throughout the world is certainly a reflection of ease of transport; yet for the majority, there is little information on transport mechanisms. Among factors shaping geographic distribution of phytoplankton species, temperature is usually considered to be of prime importance. If the potential area of certain species would be paramountly determined by temperature, we would have only a limited number of distribution types like pantropical, temperate, and circumpolar. But it is not so. We therefore have to suppose that other factors are also important and the role of migratory barriers (seas, arid areas, mountain chains) also have to be considered. It has to be stated that because of incomplete taxonomy and a general lack of understanding of the autoecology, distribution and speciation of freshwater phytoplankton, there have been serious obstacles for detailed phytogeographical analyses: our knowledge has been incomplete and fragmentary.

Active mechanisms (e.g., the species spreads through the efforts of individuals) in phytoplankton dispersal can be ruled out; nevertheless, it is interesting to note that epilithic algae might be able to move from one habitat to another along wet rock surfaces. Passive dispersal agents involve water, air, different animals, and man.

Water, River Courses

Travel along river courses has been the most evident way of dispersal. The river flora, in general, does not contain any species that would not be found in lakes and it is similarly true that phytoplankton species are differently adapted to survive lotic conditions. In the northern hemisphere, unicellular centrics, chlorococcalean green algae, are the best fitted to live in the main courses; in South America, chain-forming centrics and desmids are often found in rivers; and blue-green algae are common in Australian rivers. Riverine dead zones offer a chance for less well-adapted species to survive. Shorelines of most large rivers, especially in the industrialized world, are regulated, as a consequence of which they are reduced to a simple flow channel. In the past, river channels used to be wild offering a high proportion and diversity of dead zones and therefore a wider habitat diversity for many species to survive. The recent and very successful dispersion of the cyanoprokaryote, *Cylindrospermopsis raciborskii* (Woloszynska) Seenayya and Subba Raju, may be largely attributed to its ability to tolerate long travel along river courses. Inocula to colonize newly constructed reservoirs derive primarily from the catchment area. Species not occurring in the catchment appear after a longer time.

According to estimates, more than 50 000 reservoirs with dams higher that 15 m exist at present in all continents. Many were constructed in arid regions that used to represent migratory barriers for aquatic biota. The worldwide impact of algal invasions has not been documented (with the exception of *C. raciborskii*). It was clearly demonstrated for zooplankton that reservoirs (especially in the arid regions) serve as step-stones, this way easing dispersal.

Animals

Gut contents of aquatic animals, ranging between filter feeders to fish, often contain dozens of planktonic algae and they may pass undamaged through the intestines. As lakes can be viewed as islands on the terrain, dispersal from one lake to another needs overland transport, during which desiccation is the most immediate danger. Water beetles and water living mammals can carry algae from one pond to a nearby one; dragonflies may be more effective transporters and waterfowl are of prior importance. In judging the efficiency of transport on feet or feathers of birds, the ability of algae to survive desiccations has to be compared with the travel distance achievable within this time span. In some experiments, most externally transported algae survived for 4 h, there were some left for 8 h, but after that there were hardly any left. Transport via the digestive parts of birds may be more efficient because desiccation is ruled out. Specialized life cycle forms (cysts, akinetes) as well as species with thick, resistant cell walls will better survive longer transport, either externally or internally, than do vegetative forms (known cases include desmids where a correlation between widespread distribution and more resistant resting stages was shown).

In Antarctica, most dispersal must have been carried out by birds since the air is almost sterile.

Airborne Algae

Airborne dispersal of algae has for long been a focus of interest. Ehrenberg found 18 species of diatoms in dust from air collected by Charles Darwin on the board of the HMS Beagle, 300 km from the nearest coast, but he did not test viability. During his transatlantic flight in 1933, already Lindberg collected samples from the air. Later (in 1937), Overeem filtered air from a maximal height of 2000 m downwards and found a maximum concentration of algae at 500 m. The samples contained aerophilic green algae like *Chlorococcum*, *Stichococcus*, *Pleurococcus*, *Hormidium*, and some Cyanoprokaryotes. Later investigations have demonstrated the presence of planktonic algae (*Chlorella*, *Chalmydomonas*, *Nostoc*, *Anabaena*, *Planctonema*) in aerosols.

Mastigocladus laminisus, a hotspring cyanobacterial species, survives room dryness for several months and freezing for 1 year. This species was isolated from the volcanic island of Surtsey (80 km form Iceland), after about 7 years since its formation. The alga had probably reached the newly formed habitat via air. Other Icelandic hotspring species with high sensitivity outside their actual habitat have not been observed.

Man

Different kinds of human activities that lead to introduction of species to lakes where they had not been originally present are difficult to list. A summit would certainly involve large- and small-scale commerce (ballast water, shipments of tropical large, and ornamental fish), the overall increasing tourism that ranges from transport of large ships to small plastic toys of children, and, sad enough to say, the activities of energetic field naturalists carrying algae on their plankton nets from one lake to another.

Dispersal Distances

The probability of successful dispersal depends on the effectiveness (velocity, distance) of the carrier and the ability of algae to tolerate the transport conditions. Water beetles and aquatic mammals can be effective over a distance of a few kilometers. Dragonflies have been shown to be effective in a distance of almost 1000 km. In experiments, *Asterionella* remained viable in the feces of waterfowl for a maximum of 20 h, perhaps corresponding to 2200 km flight. Many ducks migrate up to 4800 km but with many stops. The longest recorded nonstop distance (3200 km in 48 h) was made by *Anser albifrons* from West Greenland to Scotland. Long-distance north–south transport can be done by N–S migrating birds like the south polar skua and the Antarctic tern. For airborne transport, it is difficult to give maximal distances. Since viable *Melosira* was found at 3000 m height, a considerable distance can be possible. Dust containing diatom frustules can certainly be blown across the Atlantic Ocean.

Dispersal Times

Dispersal times are also difficult to estimate and our knowledge is restricted to deductions and analysis of particular cases. The relatively poor flora of hot springs in the Azores (under glacial ice cap for a long time and becoming available for colonization only some 8000–10 000 years ago) compared to hot springs in other regions with similar temperatures and chemistry in other parts of the world probably reflects insufficient colonization times.

Because the time that elapsed since algal data were available at all has been quite short in comparison to evolutionary time of even such quickly multiplicating organisms like algae, it is difficult to say how long it might be before a species with nonuniform distribution populates many waters of the world. The recent expansion of some southern species (*Aphanizomenon aphanizomenoides*, *A. issatschenkoii*, *Cylindospermopsis raciborskii*) has been recorded through the last 30–50 years. In case of *C. raciborskii* it is demonstrated that it lasted for less than a century until the originally pantropical species invaded north up to the latitude of 53° N.

Biogeography and Speciation

In the discipline of biogeography, it is customary to consider the area richest in species or morphological diversity as the area of evolutionary origin of a given taxonomic group. The genus *Cylindrospermopsis* or desmids, in general, have their evolutionary origin in the tropics. Almost nothing is known about the times taken for a new phytoplankton species to evolve. For example, blue-green algae are a very old group, and even 3.5 Ma forms occurred that are morphologically similar to recent taxa. Other works based on methods of molecular genetics or phytogeography, however, often argue for the relatively recent evolutionary age of certain blue-green algal species.

Types of Geographic Distributions

Although the ease and quite long distances of dispersal would allow the majority of species to become cosmopolitan, this has seldom been tested and it is unlikely, especially in term of genetics, to be the case. The absence of certain species from whole continents cannot generally be explained by the lack of suitable habitats. In this respect, K-selected species that make higher demands upon their environments are more likely to occur less frequently or less widely than r-strategists. The establishment of a cosmopolitan, pantropic, circumpolar, regional, etc. (see later) flora depends on an efficient distribution in relation to the rate of evolution. If the dispersal rate of new genotypes is quicker than the rate of evolution then a pantropic, holarctic, or cosmopolitan distribution will result. In the opposite case (the rate of dispersal of genotypes is slower than the rate of evolution), floristic regionality will result and thus explain regional or endemic species.

Because of a paucity of sampling in many regions of the world, the question whether there is a nonuniform distribution of taxa is generally answerable only at a spatial scale of thousands of kilometers. The chance has been always open to find new or endemic species in remote locations. For example, the narrow Indo-Malayan endemism, *Peridinium baliense* Lindemann, became immediately pantropical when some 80 years later it was found in a South American lake.

Subcosmopolitan Taxa

It is reasonable to replace the term 'cosmopolitan' (strictly speaking, occurring almost anywhere and in many kinds of lakes therefore has to be ubiquist) by the term 'subcosmopolitan' where it applies to species occurring throughout the world but only in 'appropriate' habitats. The best examples are hot spring algae that flourish within a narrow range of environmental conditions and occur more or less wherever these conditions are fulfilled. Best examples include the cyanobacterium *Matigocladus laminosus* (Cohn ex. Born & Flah.) Kirchner in hot springs with 45–60 °C, or the enigmatic red algae, *Galdieria sulphurica* (Galdieri) Merola and *Cyanidium caldarium* (Tilden) Geitler that inhabit hot, acidic, sulfur springs with pH values from 0.5 to 3 and temperatures up to 56 °C.

The wide distribution of such species is, therefore, mosaic-like, reflecting the corresponding distribution of habitats. Among planktonic organisms, species like *Microcystis aeruginosa*, *M. wesenbergii*, *Planktothrix agardhii* and, indeed, a majority of planktonic algae have wide geographic distribution. The relative proportion of subcosmopolitan taxa in phytoplankton is certainly higher than in any other group of freshwater nonplanktonic algae. A comparison of typical summer assemblages in temperate lakes indicates a great overlap with year-round associations in tropical lakes (e.g., *Urosolenia* spp., *Aulacoseira granulata* var. *angustissima*, *C. raciborskii*). On the other hand, a detailed survey on the Indo-Malayan phycogeographical region has shown that altitudinal gradients also exist. Tropical lowland lakes are dominated by subcosmopolitan, pantropical taxa. The composition of algal communities changes markedly along an altitudinal gradient: tropical taxa are gradually replaced by taxa characteristic for cooler climates. The transition zone is between 1700 and 2500 m above sea level. For example, the typical temperate species, *Ceratium hirundinella*, have been reported only from high mountain lakes in the Indo-Malayan region.

Pantropic, Temperate, and Circumpolar (Holarctic) Taxa

Many of phytoplankton species are distributed in large latitudinal bands. Among them, species occurring only roughly between the two tropics are called pantropical. Numerous algae are found in this group like all *Cylindrospermopsis* species (except *C. raciborskii* and *C. cuspis*), *Anabaena fuellebornii*, *A. iyengarii*, *A. leonardii*, *A. oblonga*, *A. recta*, *Anabaenopsis tanganyikae*, *Aulacoseira agassizii*, *A. ikapoensis*, *Schroederia indica*, *Mallomonas bangladeshica*, *M. bronchoartiana*, *M. tropica*, *Synura australiensis*, *Ceratium bracyceros*, and *Peridinium gutwinskii*. Several well-known species (*Planktothrix rubescens*, *Limnothrix redekei*, *Anabaena solitaria*, *A. flos-aquae*, *A. lemmermannii*, *Anabaenopsis arnoldii*, *A. milleri*, *Aphanizomenon flos-aquae*, *A. issatschenkoi*) are restricted to temperate zones. Some of them are found at quite high latitudes, others (e.g., *A. issatschenkoi*) only in warmer temperate regions.

Some rare algae occur in wide latitudinal range; however, there are several phycogeographical regions where they do not occur. For example, *Asterionella formosa* has never been reported from the Indo-Malayan region even though there are many lakes where it, theoretically, could occur.

A marked bipolarity was found in geographic distribution of silica-scaled chrysoflagellates: a high degree of similarity is apparent between the floras of the climatically comparable regions of the northern and southern hemispheres. Typical taxa are *Mallomonas parvula*, *M. paxillata*, *M. transsylvanica*, *M. cristata*, *M. alata*, and *M. alveolata*. More examples on distribution of silica-scaled flagellates are discussed elsewhere.

A more restricted distributional type characterizes the occurrence of *Cyclotella tripartita* and *Stephanocostis chantaicus* which are restricted to latitudes higher than 50° but only in the northern hemisphere. Such species are northern circumpolar species. Desmids provide another example: in contrast to arctic regions, the desmid flora of Antarctica is extremely poor.

Regional Taxa, Endemics

Endemism may have two quite different kinds of origin: (1) the most pure form of endemism would be where a species evolve uniquely at a location and remains exclusive to that location; (2) endemism can also occur as a result of habitat fragmentation or destruction and a subsequent extinction from all localities except one. Distinction is difficult because, in most cases, no fossils remained. There is quite a number of species that, for a long time, have been known only from the type locality (local endemics). Others were found to be restricted to a certain region (regional endemics).

Tyler differentiates between fragile endemics with restricted distribution and robust endemics that have a wider distribution. For example, *Aphanizomenon. manguinii* and *Trichormus subtropicus* have so far been recorded only on several islands in the Caribbean region. *Cyclotella tasmanica* occurs exclusively in Tasmania, but is widely distributed there. Australia, and especially Tasmania, contains a considerable number of endemic species and genera, among them planktonic algae, called 'flagship' taxa. Such species are, for example, *Tessellaria volvocina*, *Dinobryon ungentariforme*, *Chrysonephele palustris*, *Prorocentrum playfairi*, *P. foveolata*, and *Thecadiniopsis tasmanica*.

See also: Chrysophytes – Golden Algae.

Further Reading

Atkinson KM (1980) Experiments in dispersal of phytoplankton by ducks. *British Phycological Journal* 15: 49–58.

Atkinson KM (1988) The initial development of net phytoplankton in Cow Green Reservoir (Upper Teesdale), a new impoundment in northern England. In: Round FE (ed.) *Algae and the Aquatic Environment*, pp. 30–40. Bristol: Biopress.

Borics G, Grigorszky I, Padisák J, Barbosa FAR, and Doma ZZ (2005) Dinoflagellates from tropical Brazilian lakes with description of *Peridinium brasiliense* sp. nov. *Algological Studies* 118: 47–61.

Castenholz RW (1983) Ecology of blue-green algae in hotsprings. In: Carr NG and Whitton BA (eds.) *The Biology of Blue-Green Algae. Botanical Monographs,* vol. 9, pp. 379–414. Oxford: Blackwell Scientific.

Coesel PFM (1996) Biogeography of desmids. *Hydrobiologia* 336: 41–53.

Dumont HJ (1999) The species richness of reservoir plankton and the effect of reservoirs on plankton dispersal (with special emphasis on rotifers and cladocerans. In: Tundisi JG and Straskraba M (eds.) *Theoretical Reservoir Ecology*, pp. 505–528. Leiden: Backhuys Publishers.

Haworth EY and Tyler PA (1993) Morphology and taxonomy of *Cyclotella tasmanica* spec. nov., a newly described diatom from Tasmanian lakes. *Hydrobiologia* 269: 49–56.

Hoffmann L (1996) Geographic distribution of freshwater blue-green algae. *Hydrobiologia* 33675: 33–40.

Komárek J (1985) Do all cyanophytes have a cosmopolitan distribution? Survey of freshwater cyanophyte flora of Cuba. *Algological Studies* 75: 11–29.

Kristiansen J (1996) Biogeography of freshwater algae. *Developments in Hydrobiology*, vol. 118. Dordrecht, Boston, London: Kluwer Academic.

Kristiansen J (1996) Dispersal of freshwater algae – A review. *Hydrobiologia* 336: 151–157.

Kristiansen J and Vigna MS (1996) Bipolarity in the distribution of silica-scaled chrysophytes. *Hydrobiologia* 336: 121–136.

Kristiansen J (2005) *Golden Algae. A Biology of Chrysophytes.* Koenigsten: A. R. G. Gantner Verlag.

Krienitz L and Hegewald E (1996) Über das Vorkommen von wa¨rmliebenden Blauaglgenarten in einem norddeutschen See. *Lauterbornia* 26: 55–64.

MacArthur RH (1972) *Geographical Ecology.* New York: Harper and Row.

MacArthur RH and Wilson EO (1967) *The Theory of Island Biogeography.* Princeton, NJ: Princeton University Press.

Overeem MA (1937) On green organisms occurring in the lower trophosphere. *Botanica Neerlandica* 34: 388–442.

Padisák J (1997) *Cylindrospermopsis raciborskii* (Woloszynska) Seenayya et Subba Raju, an expanding, highly adaptive blue-green algal species: geographic distribution and autecology. *Archiv für Hydrobiologie/Monographc Studies* 107: 563–593.

Round FE (1988) Algae and the Aquatic Environment. Bristol: Biopress.

Scheffler W and Padisák J (1997) *Cyclotella tripartita* Håkansson (Bacillariophyceae), a dominant diatom species in the oligotrophic Stechlinsee (Germany). *Nova Hedwigia* 65: 221–232.

Scheffler W and Padisák J (2000) *Stephanocostis chantaicus* (Bacillariophyceae): Morphology and population dynamics of a rare centric diatom growing in winter under ice in the oligotrófhic Lake Stechlin, Germany. *Archiv für Hydrobiologie/ Algological Studies* 133: 49–69.

Schlichtling HE (1960) The role of waterfowl in dispersal of algae. *Transactions of the American Microscopical Society* 79: 160–166.

Schlichtling HE, Speziale BJ, and Zink RM (1978) Dispersal of algae and protozoa by Antarctic flying birds. *Antarctic Journal of the United States* 13: 1147–1149.

Sekbach J (1999) *Enigmatic Microorganisms and Life in Extreme Environments*, pp. 687. Dordrecht, Boston, London: Kluwer Academic.

Stefaniak K and Kokociński M (2005) Occurrence of invasive Cyanobacteria species in polymictic lakes of the Wielkopolska region (western Poland). *Oceanological and Hydrobiological Studies* 34(Supplement 3): 137–148.

Symoens JJ, Kusel-Fetzmann E, and Descy JP (1988) Algal communities of continental waters. In: Symoens JJ (ed.) *Vegetation of Inland Waters*, pp. 183–221. Dordrecht: Kluwer Academic.

Talling JF (1951) The element of chance in pond populations. *The Naturalist* 1951: 157–170.

Tyler PA (1996) Endemism in freshwater algae. With special reference to the Australian region. *Hydrobiologia* 336: 127–135.

Vyverman W (1996) The Indo-Malaysian North-Australian phycogeographical region revised. *Hydrobiologia* 336: 107–120.

Algae of River Ecosystems

R J Stevenson, Michigan State University, East Lansing, MI, USA

Introduction

Algae in rivers occupy two distinct habitats, the benthos and water column. Benthic algae are algae attached to or associated with substrates in streams and rivers. (The term benthic algae is used to refer to microscopic and macroscopic algae on or associated with substrates. Thus, periphyton and biofilms (microphytobenthos), macroalgae (microphytobenthos), and metaphyton are benthic algae). Phytoplankton are suspended in the water column. The amount of these algae and their function and biodiversity varies greatly among different types of streams and rivers and with time. River ecosystems, broadly defined, include all flowing water sections within a watershed and may be bordered or interrupted by lakes, reservoirs, and wetlands. Groundwater connections are also important parts of river ecosystems that affect algae. Algal biomass is low on substrates in many shaded mountain headwater streams or when suspended in large rivers during high flow, but biomass can be very high in benthos of open canopy streams draining fertilized lands or as phytoplankton during slow flow, summer conditions. Correspondingly, photosynthesis, respiration, and nutrient transformation by algae increases on an areal or volumetric basis with increasing amounts of algae in the habitat, but cell-specific metabolism often decreases with increasing algal biomass due to competition.

Taxonomic and morphological diversity varies among habitats and seasonally within rivers as well as among rivers with different environmental conditions. Species composition varies in most river ecosystems from diatom-dominated benthos and phytoplankton during spring to a varying diversity of green algae and Cyanobacteria in warmer waters. However, a wider diversity of almost all kinds of algae, euglenoids, reds (Rhodophyta), crytomonads, chrysophytes, xanthophytes, dinoflagellates, and even browns (Phaeophyta) can be found in rivers. Size of these organisms varies from less than 5 μm for small diatoms and chrysophytes in the plankton to meters in length for some filamentous green benthic macroalgae. Morphologically, unicellular, colonial, and filamentous algae occur in both benthic and planktonic habitats. Both unicellular and colonial flagellates use flagella to move vertically in the plankton. Unicellular diatoms may use their raphe, and filamentous cyanobacteria move through sheaths in benthic habitats. Many planktonic species rely simply on the water current or periodic resuspensions to maintain sufficient exposure to light to survive.

Despite the great variation in algae of rivers and the complexity of factors that affect them, we have learned much about the environmental factors that regulate the processes that control algal biomass, diversity, and function. For example, floods and grazing invertebrates can reduce algal biomass, reduce primary production, and shift species composition to flood- or grazer-tolerant organisms. Whether or not floods and grazing invertebrates are important in river ecosystems is determined by the climate, geology, and resulting hydrology of the ecosystem. In this article, I will use a hierarchical conceptual framework for the factors that affect algae in rivers and streams. This framework will help organize the interrelationships among factors that regulate the complex patterns of algal biomass, function, and biodiversity, and thereby, will help predict patterns of spatial variation in these algal attributes among ecosystems and seasonal variation within these ecosystems. Ultimately, this broader conceptual model will help us predict longer-term changes in algae of river ecosystems and how human activities can be managed to solve and prevent environmental problems.

Algae are important elements of river ecosystems and important determinants of the goods and services provided by rivers. Algae are important in the food web for primary production, biogeochemical cycling, and habitat formation and alteration. Algae support human well-being by producing food and cleaning water for drinking. Excess algae in rivers cause many problems by altering physical habitat and depleting dissolved oxygen supplies. This affects biodiversity and productivity of rivers. Algae can also cause taste, odor, and toxicity problems in drinking water supplies and foul the pipes and filters of water users. So understanding algal ecology is important for managing river ecosystems to protect the goods and services that rivers provide.

Factors Affecting Algae in Rivers

Both benthic algae and phytoplankton in rivers live in highly variable habitats. This makes relation of pattern and process difficult. If we focus our attention on algae at a specific location, five fundamental

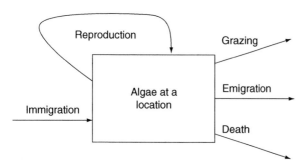

Figure 1 Five fundamental processes determine the biomass and species composition of algae at location.

processes affect how much algae occur in that location (**Figure 1**). Immigration of cells or groups of cells into the space provides the initial colonists as well as ongoing replacement for cells that are continuously leaving the habitat via loss factors, such as grazing, emigration (drifting and sinking), and death. Reproduction by cells within that location, as well as immigration, affects accrual positively. These fundamental processes determine the structure of algal communities in rivers, the biomass of individual species as well as all algae at a location, and therefore species composition. Many other biological processes, such as photosynthesis, respiration, and nutrient uptake, are considered as functions. Many physical processes, such as eddy mixing and diffusion, affect nutrient availability and sinking rates (equivalent to emigration from plankton and immigration to benthos). Processes such as eddy mixing and diffusion are considered as indirect processes because they regulate fundamental processes, in this case, reproduction and emigration rates. Finally, both abiotic and biotic factors either directly or indirectly regulate the five fundamental processes. These include light intensity and duration, nutrient concentration, and density of grazers, bacteria, and viruses that have direct effects and flood frequency, climate, and geology that have indirect effects. Many patterns are possible because they depend on the relative magnitude of each of the biological processes and their many direct and indirect environmental determinants. Relating patterns in algal biomass and species composition to the biological processes and environmental factors in rivers require, in most cases, application of nonequilibrium models in algal ecology.

Factors that regulate algae in rivers operate at different spatial and temporal scales (**Figure 2**). The processes of accrual and losses of cells at a specific location operate at the cellular or local spatial scale and reflect the biological responses to other ecosystem factors that are regulated at larger spatial and temporal scales. Abiotic factors that directly affect

algae include both resources and stressors, such a nutrients and light versus pH, salinity, shear stress, and heavy metals. Biotic factors can also be classified as positive or negative, and include commensalistic and mutualistic interactions as well as competition, predation, disease, and allelopathic interactions. At the habitat scale, riparian canopy, current velocity, and substrate presence and size affect algae, but primarily by mediating the direct biotic interactions, resources, and stressors. At watershed and regional scales, climate and geology ultimately regulate land use, hydrology, and geomorphology of rivers. They also regulate species biogeography and their availability to colonize rivers. The spatial and temporal hierarchy of these factors will regulate the complexity of possible interactions and facilitate prediction of local conditions and algal structure and function in rivers.

Local Abiotic Factors

Resources

Algal reproduction can be regulated by light, nutrients, and space. A variety of metabolic processes have similar functional responses to light intensity and nutrient concentration. Metabolic rates increase relatively rapidly in response to light and nutrient increases at low levels of resource availability, but eventually saturate at high levels of resources and respond little to further increases (**Figure 3**). Light has the possibility of having an additional negative effect at very high intensities because of photoinhibition processes. Light intensity strongly regulates photosynthetic rates and carbohydrate synthesis. Nutrient concentrations regulate their uptake rates and use in protein and lipid synthesis. The differential effects of light and nutrients on algal metabolism have been used to explain the paradox of how algae in thin biofilms can reproduce in very low light as fast as in very high light ($10–1000\,\mu m$ quanta $m^{-2}\,s^{-1}$), yet photosynthetic rates of algae are known to respond greatly over the same range in light conditions.

Variation in light and nutrients does affect algal biomass, metabolism, and species composition in streams and rivers. High light intensity and nutrient concentrations are required for accrual of high algal biomass on substrata or in the water column. Although reproduction of algal growth saturates at low light and nutrient levels, self-shading and reduced nutrient transport (because eddy mixing and diffusion are less than nutrient uptake rates) limit resource supply at the local scale as algae accumulate on substrates. For example,

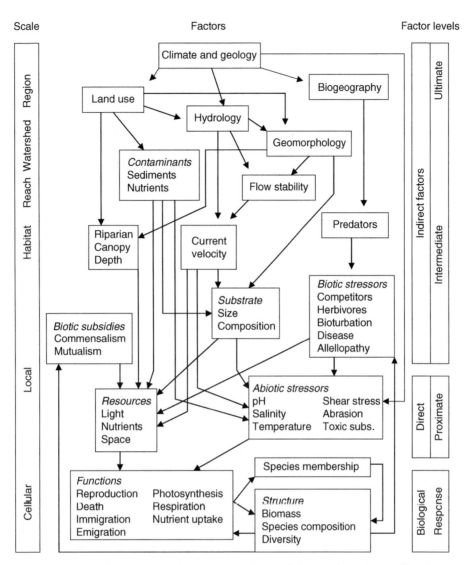

Figure 2 Hierarchical interrelationships among proximate, intermediate, and ultimate determinants of benthic algal structure and function. The relative spatial scale is shown at the left. Modified from Stevenson RJ (1997) Scale-dependent causal frameworks and the consequences of benthic algal heterogeneity. *Journal of the North American Benthological Society* 16: 248–262.

phosphorus limitation of thin biofilms of diatoms requires concentrations lower than $5\,\mu g\,l^{-1}$ PO_4–P, whereas $30\,\mu g\,l^{-1}$ is required to saturate growth of thicker mats. As algal biomass increases, more light and nutrients are required to produce growth and accrual of high biomasses in relatively short periods of times. However, high algal biomasses can accumulate over longer periods of times at lower light and nutrient concentrations in hydrologically stable habitats, such as some springs, where disturbance does not interrupt accumulation processes. Thus, the process-based approach helps us understand how we can have high algal biomass at intermediate and high nutrient concentrations.

Experiments using nutrient diffusing substrates and dosing in experimental streams show that both nitrogen and phosphorus can limit algal growth in streams.

There is also some evidence that micronutrients can limit growth of some algae that need large amounts of an element for enzymes in critical processes, such as Fe availability for nitrogen fixation by cyanobacteria. In several published meta-analyses of research with benthic algae, the trend is for about half of all streams being nutrient limited; and of those streams that are nutrient limited, a quarter are limited by P, a quarter are limited by N, and about half are colimited by both N and P. In streams with high accumulations of diatoms, Si may become limiting for both benthic and planktonic algae. Nutrient ratios have been used successfully to predict which lakes are limited by N or P, but this approach has not been as successful in streams where nutrient concentrations in the water column are highly variable, supply rates may differ with time, groundwater supplies are not often accounted for, and

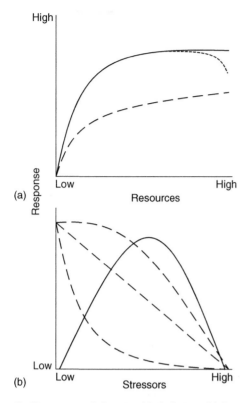

Figure 3 Responses of algae to abiotic factors. (a) Resources have asymptotic responses with rapid increases to saturating levels (solid line). Light can have a negative effect at very high levels (dotted line). Density of algae reduces supply rates and response of algae to given resource levels (dashed line). (b) Stressors negatively affect algal processes either nonmonotonically like temperature (solid line) or monotonically (family of dashed lines), when there is no positive effect at low levels of stressors.

differential leakage of N versus P over time may increase relative N demand.

Much less is known about space limitation than nutrients and light. Space limitation has not been evaluated well because it is difficult to isolate space versus density-dependent interactions. Space limitation applies most directly in the case of algae that must attach directly to substrates, such as the diatom *Cocconeis*. Space becomes an indirect factor or resource when algal biomass accumulates on substrata and reduces light and nutrient supply rates to cells in benthic algal mats. Space limitation can be relieved for algae in general when epiphytic microalgae can grow on macroalgae, such as *Cladophora*; but some species may be excluded from this extra space because they can attach to rocks but not well to *Cladophora*.

Species composition as well as biomass and function varies among rivers corresponding to differences in light and nutrient concentrations. Species require different amounts of nutrients and light to survive.

Thus, low levels of nutrients and light constrain species membership to those species that can survive in these resource-stressed habitats. High nutrients or light enables species with requirements for high resource levels to colonize a habitat with high nutrient or light. Some experiments, theory, and now, some field evidence suggest that tradeoffs for species do exist between being able to grow fastest in low and high nutrient concentrations, which would help explain the dramatic changes in species composition along nutrient gradients.

Diversity is complexly related to resource availability. As resources increase, more algal species can invade and successfully reproduce in the river based on physiological requirements. Evenness of species abundances change nonmonotonically with increasing nutrients, from low to high to low evenness as nutrient concentrations increase. Evenness is low in low nutrient conditions because only a few species are adapted to grow well in low nutrient concentrations. As nutrients increase, more species can invade the habitat and their growth rates increase faster than species adapted to low nutrients; therefore, evenness of species abundances is highest when evenness of species reproduction rates is highest. At high nutrient concentrations, growth rates and species abundances become uneven again as some species can grow faster than others in high nutrient concentrations. Many models predict that numbers of species (richness) would follow the same pattern along resource gradients. These models predict that richness would increase as the habitat became more available to more species and could decrease in high nutrients conditions because one or more resources became depleted, competitive relationships shift, and some species are competitively excluded. Observations of this pattern have not been satisfactory because estimating species richness of organisms is difficult in habitats where they are so abundant and hard to see.

Stressors

Stressors are environmental factors that can have negative effects on metabolism and other attributes of species performance. Stressors affect reproduction according to one of two groups of response functions (**Figure 3(b)**). One group consists of negative linear and nonlinear responses in which there is no or very little positive effect of a stressor on species performance. Fine sediments that bury cells and toxic inorganic and organic substances are examples of factors with monotonic negative effects. Suspended sediment has strong indirect effects on phytoplankton, by reducing light transparency. Temperature and the ionic factors (salinity, pH, and alkalinity) produce nonmonotonic effects on species performance. Algal performance is optimal at some intermediate level of these factors because high and low levels of temperature, salinity, pH, and alkalinity limit species performance for one reason or another. Temperature, for example, stimulates metabolism as temperature increases from relatively low to intermediate levels as kinetic energy increases. However, high temperatures denature enzymes and reduce function. Salinity, pH, and alkalinity probably affect enzyme-mediated processes, and thereby create an intermediate condition which is optimal for species.

Because almost all algal species are negatively affected by sediments and toxic substances, biomass of algae can be negatively affected by these stressors. Similarly, the negative effect of high temperatures (above 30 °C) is 'toxic' for most algae, except for some cyanobacteria that tolerate temperatures as high 55 °C in hot springs. So, high temperatures may reduce algal biomass, as low temperatures may. However, algal species are adapted to an unusually wide range of pH, alkalinity, and salinity, so that biomass is only affected by the ionic factors in extreme conditions. Throughout much of the range of the ionic factors, species composition differs with varying ionic factors, but functional redundancy is able to maintain community-level reproduction and biomass.

Local Biotic Factors

Biotic factors can have positive or negative effects on algae in rivers. We know relatively little about commensalistic and mutualistic interactions, but some examples do exist. One example of commensalism would be the attachment of some benthic diatoms on stalks of other diatoms, which provides an advantage to species that can attach to stalks and has little negative effect on the stalked diatoms. Diatoms with endosymbiotic cyanobacteria provide an example of mutualistic interactions. Of the negative interactions, much more is known about the herbivory and competition versus allellopathic interactions and disease-like effects of fungi, bacteria, and viruses. Many have hypothesized that the latter two biotic stressors should have great effects in dense microbial assemblages like benthic algae. They are known to be important in lake and ocean phytoplankton. Unusual white circles in periphyton with high numbers of bacteria and fungi indicate 'disease,' but little investigation has pursued this line of research.

Competition is probably a more important determinant of algal biomass, function, and species composition for benthic algae than for phytoplankton. Phytoplankton seldom accumulate to sufficiently high densities to deplete nutrient concentrations in rivers because of the relatively short residence times of these organisms in their habitat. However, competition may be important in very slow flowing, lake-like rivers where residence time is sufficiently high to deplete nutrients or light by biological uptake or shading. These processes are thought to be important in benthic algal communities. If species membership is constrained to diatom-dominated communities, peak biomass of communities may be constrained by light and nutrient depletion. We know that light and nutrient availability within benthic algal mats decreases with increasing density; less light penetration and decreasing nutrient transport rates and cell nutrient content has been documented as diatom biomass increases on substrata. Per capita rates of metabolism and reproduction decrease with increasing benthic algal biomass. Species composition changes with increasing biomass of diatoms on substrata. All indicate autogenic changes in environment that are consistent with strong competitive regulation of benthic algal communities.

Herbivory is also an important determinant of algae in rivers. It is more frequently important for benthic algae than for phytoplankton because of the lack of time for zooplankton to accumulate in the water column of rivers. Low disturbance frequency by floods and drought is important for determining whether herbivory is important for benthic algae, too. When river conditions allow herbivores to accumulate to high densities, they can regulate biomass, function, and species composition of benthic algae in rivers. The importance of zebra mussels in some rivers is a good example of herbivores affecting river phytoplankton, but examples of zooplankton regulation are not common. Filter feeding invertebrates like blackflies and net-spinning caddisflies may also be important regulators of suspended algal abundance in streams, but this is not well understood.

Aquatic insects, snails, and some fish consume benthic algae, but aquatic insects and snails are the most important in most situations. Protozoa have also been shown to consume benthic algae, but their importance does not seem to be as great as cased caddisflies, mayflies, and snails that graze algae from substrata. These invertebrates can constrain diatom biomass to very low levels. Many of them can consume filamentous green algae during early stages of growth, but not after they have exceeded the size that can be controlled. They seem to avoid consumption of filamentous Cyanobacteria, but push it back from actively grazed areas or may knock it off the substrate. Invertebrate herbivores can reduce algal biomasses from 10 to 0.5 µg chl a cm^{-2}. In addition, they selectively graze overstory versus understory diatoms (stalked and filamentous forms versus tightly adnate and prostrate forms). Although grazers consume algae, not all are killed. Estimates of algae passed alive and viable through guts of aquatic invertebrates often exceed 50%. In a sense, grazing of cells removes cells from the substratum and either causes death or emigration of the cells downstream when they are egested alive. Because of the importance of grazing as a process of removing algae from a location and conceptually within food webs, I have included it as one of the five fundamental processes.

Bioturbation is another process that affects benthic algal biomass. Invertebrates, fish, and terrestrial animals are common sources of disturbance of benthic algae as they move through streams. Diel patterns in algal drift are observed in some stream that correspond to the dawn and dusk activity patterns of aquatic invertebrates. Paths of disturbed benthic algae in shallow riffles can be observed in deeper upstream–downstream channels where fish have moved from pool to pool. Movement of fish in pools disturbs the development of periphyton and clouds the water. Raccoons and larger animals, such as manatees, crocodiles, hippopotamus, and humans, probably have great effects on benthic algae when moving, but these effects have not been quantified.

Habitat-Scale Factors

Riparian canopy, depth, substrate, and current velocity vary among habitats within a reach and have indirect effects on the five fundamental algal processes. Covariation among depth, substrate, and current velocity are regulated by the riffle-pool structure carved in most rivers. Coarser substrates are found in the shallow, high-velocity riffles. Finer substrates are found settling in the deeper, low-velocity pools. Riparian canopy and depth regulate light availability, whereas current velocity and substrate have relatively complex effects that warrant further discussion.

Substrate is very important for algae as a stable surface for attachment and growth and potentially as a source of nutrition. Some algae have special morphological adaptations for attaching to substrata, such as the raphe and mucilaginous stalks and tubes of diatoms. Many filamentous green algae produce specialized cells that attach to a substrate and then form filaments when they divide. Thus, a major ecological division occurs in the algae about which we know relatively little, except for these morphological adaptations. Benthic algae have morphological adaptations that enable their attachment to substrata, and phytoplankton do not. Given the great differences in habitat conditions when suspended in the water versus attached to substrata, such as greater ranges

in current velocity and denser packing of cells for benthic algae, great physiological differences must accompany these morphological adaptations.

For benthic algae, substrate size and nutritional value may have great effects. The smallest substrates, silt and organic sediments, are smaller than the smallest algal cells. So organisms that live in this habitat tend to be large and motile, such as the diatoms *Nitzschia* and *Surirella* or the flagellated euglenoids. Sands, slightly larger and inert, are large enough for attachment by small diatoms in streams. During hydrologically stable periods, other algae can colonize sands, but only small diatoms can survive in the crevices of individual sand grains when the sands tumble across the bottom of streams. As rock substrates become larger, they tend to become more resistant to hydrologic disturbance. Almost all pebble and larger substrates can support diverse diatom assemblages. As substrates become larger and more stable, they can support luxurious growths of filamentous green algae that require longer stable periods for colonization by spores and growth of filaments. Woody debris and plants are other common substrates in rivers and both may have nutritional properties that affect algae. Plants leak nutrients that become available to epiphytic algae attached to them. Wood provides nutrition to bacteria, which may actually have a negative effect on benthic algal colonization because of competition with bacteria for nutrients.

Current velocity has watershed- and habitat-scale effects on algae. For phytoplankton, current velocity determines residence time of water and algae in the river and the time for cells to accumulate. For benthic algae, increases in current velocity during rain events may be sufficient to scour algae from stable substrates. Higher velocities can disturb the substrate in riffles and cause severe scouring of algae from substrates. The time between scouring events provides the time for recovery of benthic algal communities. Thus, variations in current velocity at the watershed scale determine the frequency and intensity of disturbance and are important, ultimate determinants of nonequilibrium ecological dynamics of algae in rivers.

The habitat-scale effects of current velocity on benthic algae are relatively well understood compared to their effects on river phytoplankton. Eddy mixing of the water column surely slows sinking of phytoplankton and shear in the water column may reduce nutrient depletion in waters surrounding cells. One of the important premises of current effects on algae is that algae, in still waters, develop nutrient-poor shells of waters around cells. This shell develops because nutrient uptake rates from waters around cells exceed diffusion rates of nutrients into those waters. Water near surfaces, whether cell or substrate surfaces, has different physical properties than water away from these surfaces. Although measurements of nutrient concentrations in these 1–2-µm thick layers around cells has not been practical, many observations between current velocity, metabolic rates, and transport of nutrients through periphyton mats indicate that these shells do exist. Thus, as current velocity around cells increases, the shearing of layers of water around cells increases and disrupts this layer of nutrient-poor (and potential waste-rich) water from around cells.

Habitat-scale effects of current velocity on benthic algae have both positive and negative effects. Although the shear stress of current velocity decreases immigration rates and likely increases emigration rates from habitats with moderate and fast current velocities, the increased physical mixing of water through attached masses of microalgae or macroalgae stimulates metabolic rates. Concentration gradients in micro- and macroalgal communities can cause severe nutrient depletion within this microhabitat. Assuming a common 10–30 cm s^{-1} range in velocities from slow (pool) to fast (riffle) current habitats, profound differences in current effects on immigration and reproduction help relate the patterns of benthic algal colonization after flood disturbance to these processes. Initially, algal biomass and accrual rates will be slower in riffles than in pools because immigration is the dominant process affecting colonization when algal biomass on substrates is low, for example $(5–10) \times 10^3$ cells cm^{-2} (**Table 1**). As more cells accumulate, reproduction becomes more important than daily accrual of cells and the positive effects of current outweigh the negative effects. So we eventually observe higher algal biomasses in riffles than in pools, even though that pattern may not be evident immediately after flood-related disturbances.

Table 1 Illustrating the changing importance of immigration versus reproduction during algal community development on substrates

Day	Density	Immi	Repro
1	1100	1000	100
2	2419	1000	319
3	4001	1000	582
4	5897	1000	896
5	8168	1000	1271
6	10887	1000	1719
7	14139	1000	2252
8	18025	1000	2886
9	22663	1000	3638
10	28190	1000	4527
11	34766	1000	5576
12	42573	1000	6807
13	51820	1000	8247
14	62741	1000	9921
15	75595	1000	11854
16	90663	1000	14068
17	108242	1000	16579
18	128636	1000	19394
19	152140	1000	22504
20	179023	1000	25883
21	209499	1000	29476
22	243699	1000	33200
23	281636	1000	36937
24	323171	1000	40535
25	367984	1000	43813
26	415561	1000	46577
27	465193	1000	48632
28	516004	1000	49811
29	567001	1000	49997
30	617146	1000	49145

Accumulation of algal cells (density, cells cm^{-2}) on substrates for 30 days is modeled with a spreadsheet. In this model, only immigration (Immi) and reproduction (Repro) are processes affecting algal cell accumulation. Grazing, death, and emigration are assumed to be 0. Immigration is assumed to stay at 1000 cells cm^{-2} day^{-1} throughout the colonization period. Twenty percent of all cells from the last day and 10% of the immigrating cells reproduce each day. A carrying capacity 1 000 000 cells cm^{-2} was included to slow reproduction as density dependent competition increased on substrates.

Reach to Regional Scale Factors

Although reach or segment scale factors (such as hydrology, channel geomorphology, flow stability, and stream size) interact to shape habitat structure and local abiotic conditions, regional processes associated with climate, geology, and biogeography regulate the more general spatial and temporal patterns of algae in rivers (**Figure 2**). Climate and geology also affect human activities in watersheds, which have profound effects on contaminants in streams, channel geomorphology, and hydrology. Extensive discussion of human effects of algae in rivers is beyond the scope of this article and is covered elsewhere in this encyclopedia. I want to synthesize the review of algal ecology in rivers by describing three common

patterns: seasonal and longitudinal patterns of phytoplankton in rivers, algae and the river continuum hypothesis, and the effects of disturbance on benthic algal–grazer interactions.

Phytoplankton in rivers have a distinct longitudinal pattern that varies in magnitude with seasonal variations in discharge. Suspended algal densities are usually low in headwater sections of rivers and largely composed of algae that have emigrated from substrates and are entrained in the water column. As water flows downstream, species better adapted to life in the water column immigrate into a mass of water and stay in it as water moves downstream. Gradually, fewer benthic species occur in the water than do planktonic species. The longer that mass of water resides in the river before reaching a lake, estuary, or ocean, the higher the biomass of algae in the water column will become. More phytoplankton accumulates in rivers of geologic regions where rivers have low gradients versus those with high gradients. Climate determines the time of year and the extent of hydrologically stable periods when algae can accumulate in streams. In climates with periodic storms during the dry season, high discharge events may disrupt the development of phytoplankton communities. During long dry and hot periods, high and often problematic biomasses of algae can develop with sufficient quantities that deoxygenation and accumulation of toxic algae occurs.

The river continuum hypothesis proposes that the ecology of rivers is regulated by upstream–downstream patterns in the hydrogeomorphology and connectivity within rivers. Although originally defined for free-flowing rivers in temperate climates with forested landscapes, the model has been adapted for dammed rivers and many ecological regions. The original model described how narrow streams in forests would be covered by a riparian canopy that would limit light, and therefore, algal production. As streams became wider downstream, light would become more available and benthic algae would be able to accumulate. Correspondingly, the base of the food web was predicted to shift from allochthonous detritus in the headwaters to autochthonous primary production in the mid-reaches of the river. Farther downstream as algae became entrained in the water column and depth increased, autochthonous productivity in rivers would shift from benthic to planktonic algae. Although this upstream–downstream pattern in hydrogeomorphology and connectivity may vary greatly among regions, it tends to vary with climate and geologic conditions of a region, and to some extent with local landscape conditions. Most differences occur in headwater regions where the river may not have accumulated sufficient erosional power to carve channels and broad floodplains. Headwaters vary from springs arising in deserts to high gradient channels in mountains or low gradient channels emerging out of wetlands. Each of these provides a different starting point in downstream regulation of the ecology of rivers and very different environments for algae.

One of the challenges for understanding the ecology of algae in streams has been understanding the relationship between resource availability and disturbance intensity. This challenge underpins: (1) the prediction of top-down or bottom-up regulation of river food-webs, (2) relationships between nutrients and algal biomass for management of these important contaminants of streams, and (3) biodiversity of algae in streams. Increases in nutrient concentrations should result in an increase in algal biomass of streams. Increases in hydrologic stability also should result in increases in algal biomass of streams because algae have, on average, longer periods of time to accumulate between disturbances. Thus, in streams that are relatively hydrologically stable, we should have greater responses of algae to nutrients than in frequently disturbed streams. This is what we see in New Zealand streams where hydrologic stability ranges from great to average on the global scale. However, in the middle of North America, we see the lower end of the disturbance continuum. Low disturbance streams respond relatively little to increases in nutrient concentrations compared to relatively high disturbance streams in this region. Here, accrual of algae is constrained by high grazing pressure in hydrologically stable streams. Flood and drought disturbances in relatively hydrologically variable streams of this region constrain grazer abundances, so grazers no longer regulate algal biomass in the most hydrologically disturbed regions of central North America – which compared to the global range of possibilities, is only intermediate on the hydrologic disturbance scale. Globally, we find complexity arising from the nonlinear and multitrophic level effects of disturbance on algal-resource interactions. Herbivores and disturbance constrain algal response to nutrients in the most hydrologically stable habitats and the most hydrologically variable habitats, respectively; and algae respond most to nutrients at intermediate levels of disturbance when they have sufficient time to recover from disturbance, but not so much time that invertebrates can also recover.

Summary

Interactions between algae and their environment are complex, yet predictable based on accurate linkage of processes, scale of determining factors, and algal attributes. Although much is yet to be understood about algae in rivers, we have learned much by a marriage of three basic scientific approaches: large-scale field observations from environmental monitoring projects, experiments, and process-based models. Study of algae in rivers provides a model for how to understand complex systems that are tightly coupled to human effects on the environment. Future work in rivers should address the importance of biodiversity and ecosystem function, how ecological systems respond to environmental change, and the importance of algae in ecosystem goods and services. The lessons that we have learned in the study of algae in rivers should be applied to studies of algae in other habitats as well, such as wetlands, terrestrial habitats in rainforests, and the arctic tundra.

See also: Algae; Photosynthetic Periphyton and Surfaces.

Further Reading

Biggs BJF (1996) Patterns in benthic algae of streams. In: Stevenson RJ, Bothwell ML, and Lowe RL (eds.) *Algal Ecology: Freshwater Benthic Ecosystems*, pp. 31–56. San Diego, CA: Academic Press.

Biggs BJF, Duncan MJ, Jowett IG, Quinn JM, Hickey CW, Davies-Colley RJ, and Close ME (1990) Ecological characterisation, classification, and modeling of New Zealand rivers: An introduction and synthesis. *New Zealand Journal of Marine and Freshwater Research* 24: 277–304.

Desy JP (1987) Phytoplankton composition and dynamics in the River Meuse (Belgium). *Archiv für Hydrobiologie* 78(Suppl): 225–245.

Hauer FR and Lamberti GA (2006) *Methods in Stream Ecology*, 2nd edn. Amsterdam: Elsevier.

Reynolds CS Descy JP (1996) The production, biomass, and structure of phytoplankton in large rivers. *Archiv für Hydrobiologie*. 113(Suppl): 161–187.

Reynolds CS, Descy JP, and Padisák J (1994) Are phytoplankton dynamics in rivers so different from shallow lakes? *Hydrobiologia* 289: 1–7.

Stevenson RJ (1997) Scale-dependent causal frameworks and the consequences of benthic algal heterogeneity. *Journal of the North American Benthological Society* 16: 248–262.

Stevenson RJ, Bothwell ML, and Lowe RL (1996) *Algal Ecology Freshwater Benthic Ecosystems*. San Diego: Academic Press.

Stevenson RJ, Rier ST, Riseng CM, Schultz RE, and Wiley MJ (2006) Comparing effects of nutrients on algal biomass in streams in 2 regions with different disturbance regimes and with applications for developing nutrient criteria. *Hydrobiologia* 561: 140–165.

Vannote RL, Minshall GW, Cummins KW, Sedell JR, and Cushing CE (1980) The river continuum concept. *Canadian Journal of Fisheries and Aquatic Sciences* 37: 130–137.

Chrysophytes – Golden Algae

J Kristiansen, University of Copenhagen, Øster Farimagsgade, Copenhagen K, Denmark

Introduction and Taxonomy

Under the broad designation 'chrysophytes' there are two main classes – the Chrysophyceae and the Synurophyceae. They have a common pigment composition (chlorophylls *a* and *c*, and the yellow xanthophyll and fucoxanthin), the heterokont flagellation, and the storage product chrysolaminaran. Resting stages are endogenous silicified cysts, called stomatocysts. Some Chrysophyceae and all Synurophyceae bear silica scales. The separation in these two classes is based on differences in the chlorophyll *c* composition in the photosynthetic system, and in the ontogeny of the silica scales. About 1200 species of chrysophytes have been described, mainly flagellates. Most of these are solitary, but many are united in globular or branched colonies (**Figure 1**). Cells are naked or located in an envelope, called lorica, or they are surrounded by an armor of silica scales with species-specific structures. Many immotile forms have a cell wall. Reproduction takes place by longitudinal cell division. Sexuality has been observed in several genera, e.g., *Synura* and *Dinobryon*: vegetative cells from different colonies act as gametes: they fuse and form a zygote, which then encysts for later germination.

In the broader sense, sometimes the haptophytes are also included among chrysophytes. They have similar pigments, but their cell construction is quite different, e.g., in the possession of a third flagellar-like appendage, the haptonema. They play an important role in the sea (e.g., the coccolithophorids), only a few genera, e.g., *Chrysochromulina*, are found in fresh water, often in gravel pits, where they can attain enormous cell concentrations (*see* **Other Phytoflagellates and Groups of Lesser Importance**).

The vast majority of the chrysophytes are living in fresh water; in the marine environment they play a minor role. The silicoflagellates have a few extant species only (e.g., the genus *Dictyocha*), but mass developments sometimes occur. In such blooms, the genus *Dinobryon* is only represented by *D. balticum*. Further there are pedinellids and several species of *Paraphysomonas*.

The freshwater chrysophytes occur mainly in the phytoplankton. Common genera are *Dinobryon, Uroglena, Synura,* and *Mallomonas*. Some species may occur in mass development in the plankton. Other species are benthic, epiphytic, or epilithic. The latter are mainly confined to running waters and will not be discussed further.

Occurrence

Most chrysophytes occur in ponds, often slightly humic and with moderate concentrations of nutrients, and often slightly influenced by human impacts. A typical temperate chrysophyte locality is a pond situated in a rural landscape, with cattle in adjacent fields and with vegetation around providing some shade. They occur most abundantly in spring, right after ice-break, examples are *Chrysolykos planktonicus, Mallomonas akrokomos,* and *M. teilingii*. They are also found in sheltered forest ponds, where they may form a golden surface layer, neuston, consisting of Chromophyton. Epiphytic chrysophytes such as *Epipyxis* and *Chrysopyxis* grow on filamentous algae.

In lakes they may occur as regular elements in the seasonal succession, in temperate lakes mainly in early spring (e.g., species of *Mallomonas*) before the diatom maximum. However, species of *Dinobryon*, especially *D. divergens*, may also be common in the clear water phase or appears in sudden peaks during the summer stagnation. These species are important members of the food web, as they are mixotrophic, ingesting bacteria, and they are thus both primary and secondary producers.

In Arctic and Antarctic regions, submerged moss communities harbor a rich chrysophyte flora, but mainly occurring in the encysted stage, with hundreds of stomatocyst morphotypes.

Very clear lakes such as those fed by melt-water from glaciers are almost sterile and contain only few chrysophytes. In moderately oligotrophic waters chrysophytes thrive abundantly and may attain 90% of the phytoplankton biomass. Such waters are poor in phosphates, and chrysophytes can only thrive there because they can compensate P-deficiency by ingesting bacteria or by having symbiotic associations with them, since bacteria, at low ambient concentrations, have a more effective mechanisms for phosphate uptake than chrysophytes. In such waters, mixotrophic chrysophytes may constitute more than 50% of the total phytoplankton biomass.

In mesotrophic waters, chrysophytes may constitute half of the phytoplankton biomass.

In eutrophic lakes, Chrysophytes normally play a minor role. In such lakes, Chrysophytes will have their main occurrence prior to the vernal diatom maximum. In highly eutrophic lakes, dominated by chlorococcalean green algae or by blue-green algae, chrysophytes are of minor importance. The blue-greens

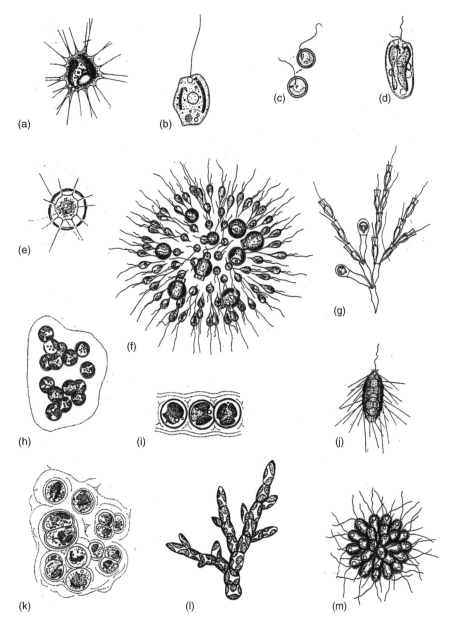

Figure 1 Morphological diversity within the chrysophytes. (a) *Chrysamoeba*, (b) *Chromulina*, (c) *Chrysococcus*, (d) *Ochromonas*, (e) *Chrysothecopsis*, (f) *Uroglena*, (g) *Dinobryon*, (h) *Chrysocapsa*, (i) *Sphaeridiothrix*, (j) *Mallomonas*, (k) *Gloeochrysis*, (l) *Chrysoclonium*, (m) *Synura*. From Kristiansen 2005, compiled from Nygaard 1945, Huber-Pestalozzi 1941, Starmach 1985. With permission from Koeltz Scientific Books, Koenigstein, Germany.

produce massive surface blooms in summer and autumn, shading other phytoplankton. However, chrysophytes may occur in spring in minor quantitative importance, but often with high species diversity.

As mentioned earlier, the best chrysophyte localities are slightly humic, somewhat shaded ponds, with some organic addition. More than 60 species can been found in such a locality during a one-year investigation.

Humic, especially mesohumic, waters host a special chrysophyte flora. In such brown waters there is a pronounced stratification, because of the limited penetration of light. This favors chrysophyte flagellates, which can perform vertical diurnal movements, and their share increases with the coloration of the water.

Mass developments of chrysophytes occur in stagnant waters, often in oligotrophic lakes, after some addition of nutrients, e.g., fertilizers from agriculture. *Synura* often shows mass development in drinking water reservoirs. Such blooms are not toxic, but they give the water a very unpleasant smell and taste – fishy-cucumber-like – mostly because of unsaturated aldehyde exudates.

Other types of mass developments occur as neuston formation. This is a community, mostly consisting of Chromophyton, that lives attached to the water surface. The algae locate themselves on the water surface (epineuston) and adjust their chloroplasts uniformly in relation to the sun. They reflect the sunlight, therefore if observed from a certain angle the water surface looks like a golden mirror.

Seasonality

Seasonality, the variation in occurrence during the year, is governed by several factors, both internal and external.

In the first place, the species have individual strategies, as regards their adaptation to stress (lack of nutrients) and to disturbance (water mixing). C-strategists are small, rapidly growing species, sensitive to grazing and disturbance. They occur in early spring. R-strategists are dependent on water mixing – occur in spring and autumn. The S-strategists are large and good swimmers, tolerant to nutrient stress and stagnation; they are summer forms. In chrysophytes, it is often difficult to distinguish clearly between these life-strategy types.

Temperature variation, combined with variation in light, is the main cause for seasonality. As temperature and light maxima do not coincide, but are shifted, there are e.g., strong-light cold-water forms (most chrysophytes) in spring and low-light warm-water forms in autumn.

In tropical regions these distinctions are not valid as there are only moderate temperature variations. Here seasonality may be governed by rainy and dry seasons as seen in chrysophyte studies from Brazil.

Most seasonal studies have been made in temperate countries. The majority of chrysophytes have their maxima in early spring, often starting to grow under the ice before it breaks, and then they disappear and do not reappear until next spring, or they have a second occurrence in the autumn. Many species are adapted to slightly higher temperatures and occur during summer also, one example is *M. caudata*.

In very productive water bodies, chrysophyte occurrence is restricted to spring, when competition with other groups of algae, especially blue-greens, is less severe.

In arctic lakes, chrysophytes have only one occurrence period, during summer.

It is important to note that species of the same genus may exhibit very different temporal patterns. In *Synura*, *S. petersenii* may occur the whole year. The other species are more or less characteristic to spring and autumn, except *S. sphagnicola*, which is restricted to the summer period. Similar variation has been observed within the very large genus *Mallomonas*, which can be divided into five species groups according to their temperature optima. Similarly, the genus can also be divided in species groups according to seasonality.

Grazing by zooplankton may influence seasonality. Species with long bristles are less endangered than species without and can thus survive for longer periods. Parasitism by chytrids and other organisms may also influence occurrence periods.

Seasonality may include morphological changes called cyclomorphoses. Species of *Dinobryon* may develop longer cell envelopes at summer temperatures. Moreover, the ratio between bristle types in *M. crassisquama* (serrated bristles versus bristles with helmet-shaped tips) is affected by temperature.

Formation of resting stages, called stomatocysts, takes place during the occurrence period or at the end of it. In *Synura*, sexual cysts are formed at high cell concentrations after fusion of gametes. After maturation, stomatocysts, either vegetative or sexual, sink to the bottom sediment and germinate only if favorable conditions arise, e.g., next spring. At germination the protoplast escapes from the cyst through the apical porus and divides.

Environmental Factors

Environmental factors determine the occurrence of chrysophytes in the different types of water bodies. On of the most important factors for distribution is pH, such as expressed in the system by Hustedt. Even species within the same genus can show distinct differences as regards distribution in relation to pH. Most *Dinobryon* species prefer neutral to slightly alkaline waters – they are alkaliphilic. However, one species, *D. pediforme*, has been exclusively found in acidic, often humic waters. Other acidobionts are some minute species of *Ochromonas*, which exclusively grow in extremely acid mining lakes.

Species of the genus *Synura* also show variation in pH preference. *S. petersenii* can be found at almost any pH (but most frequently at 6–7); it is indifferent. *S. uvella* occurs only in alkaline waters (alkalibiontic), and *S. sphagnicola* only in humic, acidic waters (acidobiontic). In the genus *Mallomonas* there is a similar pattern. A few species are definitely acidobiontic (e.g., *M. canina*) but most are alkaliphilic or alkalibiontic, occurring in a broader pH range.

Minor taxonomic differences may be reflected in pronounced differences in pH ranges. Of the two varieties of *M. acaroides* one (var. *acaroides*) is alkalibiontic, and the other (var. *muskokana*) acidobiontic.

Soil conditions are decisive for the distribution of chrysophytes, whether acid or alkaline (calcareous). In Denmark, the western part of the country was covered during the last glaciations ending 11 000 years ago, is still characterized by sandy diluvial soil and oligotrophic waters with low pH and conductivity. In the contrary, the rest of the country has younger moraine deposits, with alkaline lakes and with high conductivity.

Salinity is also an important environmental factor. As mentioned, there are only few marine chrysophytes. Some species of *Paraphysomonas* are apparently indifferent; they can be found both in fresh water and in the sea. However, some freshwater chrysophytes, although not occurring in the sea, have some preference for salty waters, such as saline lakes, e.g., *M. salina*. *M. tonsurata* also has some affinity to traces of salt in the water. It can suddenly appear in lakes if roads in the neighborhood are salted in winter.

Availability of nutrients, such as phosphates, has already been addressed briefly. Chrysophytes in phosphate poor waters have to rely on bacteria, which have more efficient P-uptake mechanisms, either as food or as symbionts. In general, chrysophytes are auxotrophic: in addition to their photosynthesis they depend on uptake of organic compounds, e.g., vitamin B, from the water.

A special issue is the relation of chrysophytes to pollution. Slight pollution of a lake may enhance the growth of chrysophytes, both qualitatively and quantitatively. However, under heavier pollution, in cyanophyte lakes, the chrysophytes will suffer from the shading effects of the blue-greens. Some species are, however, directly benefited from high concentrations of nutrients. This applies to *M. heterospina*, which is found, *inter alia*, in pools below bird cliffs.

The occurrence of chrysophytes in relation to nutrients and pollution can be registered in the modified Kolkwitz–Marsson saprobic system, as made usable not only for self-purification of streams but also for nutrient contents of stagnant waters. Most chrysophytes can then be classified as beta-mesosaprobic.

Chrysophytes as Indicators

The distinct occurrence spectra of many chrysophytes make them useful as indicator organisms. They can be used to monitor ongoing changes in nutrient loading, and in sediment investigations they can serve as markers of long-term environmental changes.

Species number in itself may indicate water quality: few species, but often in high cell numbers, may indicate oligotrophy. Higher species numbers may indicate eutrophy, but with increasing pollution, the chrysophytes will decrease or disappear.

Well defined species, which have a narrow environmental spectrum and are easily recognizable, can serve as indicator species. Among them, previously mentioned are *M. canina*, *D. pediforme*, and *S. sphagnicola* as indicators of acidic and *D. divergens* and *S. uvella* of alkaline environment.

Indices based on ranking of the species according to their occurrence in relation to pollution can help in comparing localities and to monitor changes resulting from altered nutrient loading or other factors.

Occurrence or disappearance of indicator species shows altered influence of pollution, acidification, or increase in salinity. The restoration of Lake Trummen in Sweden could be monitored by the establishment of a rich chrysophyte flora, notably with *S. petersenii* and *M. eoa*.

Chrysophytes as Indicators in Paleoecology

The main practical value of chrysophytes lies in their importance as markers in paleoecology, as seen from their occurrence in sediments where they can serve as indicators for reconstruction of lake history.

Two main types of chrysophyte remnants in sediments can be used: chrysophyte cysts and chrysophyte scales. Both consist of silica and can be preserved in, best in undisturbed, sediments.

Stomatocysts have thick silica walls and preserve well in sediments. They are more or less globular and possess spines and other specific ornamentation, and a multitude of morphotypes has been described by means of SEM. The trouble is that they can hardly be identified at species level especially because knowledge of extant cysts and their species affiliation is catastrophically low. The genus *Mallomonas* has more than 180 species, but cysts are only known and adequately described from less than one-third. Likewise, the variation range of the individual morphotypes is not known.

Still, these morphotypes can be used, in two ways.

As a group, they can be compared to other algal remains. A shift in the ratio diatoms/stomatocysts indicates a trophic change. Also a change in the composition of the stomatocyst assemblage indicates environmental changes. This is seen in sediments from regions in the United States from the time when settlers arrived, and can thus show a change from oligotrophic to more eutrophic conditions.

Further, it has become possible to calibrate the occurrence of some morphotypes with the occurrence of diatoms with known ecological spectra. In this way, individual morphotypes can also be given indicator values.

The use of chrysophyte cysts in paleolimnology was initiated fifty years ago in studies of sediments from an acidic brown-water lake, Lake Gribsø north of Copenhagen, dating back to 3000 BC. By light microscopy 77 morphotypes were found. As they could not be identified to known taxa, they were equipped by provisional descriptive names (such as *cysta spinosa*), and their individual stratigraphic occurrences in relation to, e.g., diatoms and pollen diagrams could be ascertained.

Now hundreds of stomatocyst (statospore) morphotypes from all over the world have been studied by means of SEM. To prevent chaos, a special statospore working group was established at the First International Chrysophyte Symposium (1983) to ensure uniform nomenclature and documentation.

An important step in the utilization of stomatocyst morphotypes was the comparison of fossil morphotypes with those from recent sediments with known environmental parameters. In this way fossil morphotypes can be given indicator values.

In varved (laminated) sediments, even the seasonal variation in the deposition of the stomatocysts can be followed, because color of sediment bands represent different seasons of the year. Thus, also temperature variations in morphology of individual morphotypes can be followed, without having to be affiliated to actual taxa.

In Arctic regions, changes in stomatocyst assemblages can show changes in salinity and can thus be correlated with progress and retraction of glaciers, thus providing information about climate change.

In oceanic islands, studies of stomatocysts in sediments may be rewarding. Sediment investigations in the crater lake of Rana Raraku (Easter Island) show how changes in stomatocyst assemblages correlated with the invasion of people from South America about 1400, and later with the visits in the seventeenth century by European sailors accidentally introducing chrysophytes from their water barrels.

The other chrysophyte remains in sediments – the silica scales – provide opportunities for a much finer differentiation and resolution. Studied by electron microscopy, their detailed ultrastructure gives absolutely certain identification to extant species with known environmental ranges. Unfortunately, these delicate structures are easily damaged and can best be obtained from undisturbed, laminated sediments.

The scale assemblages can then directly be compared with chrysophytes in recent phytoplankton samples. For example pH values can then be directly transferred from the recent to the fossil samples.

Studies of silica scales in sediments have yielded enormous amounts of information about lake history. One of the earliest examples is the study of Hall Lake, Washington, showing changes in the chrysophyte flora as related to the homesteading and the related agriculture, to the building of a sawmill, and to the construction of a new highway along the lake. This was reflected in successive maxima of *M. caudata*, *S. lapponica*, *M. crassisquama*, and *S. petersenii*. Simultaneously, this could be related to changes in the diatom flora and the appearance of weeds accompanying agriculture, such as ragweed (*Ambrosia*).

Similar studies in Finland have dated the introduction of agriculture.

In other places, *M. tonsurata* has indicated the start of winter-salting of the local roads.

Studies of acidification history based on chrysophyte scales in sediments have been even more important. Because of acid rain caused by industrialization many soft-water lakes have become acidified. Several acidobiontic chrysophytes have been used as indicators for the start of this impact, notably *M. canina*, *M. hindonii*, and *S. sphagnicola*.

For instance, it can be seen from such studies that the mysterious disappearance of trout and the resulting breakdown of tourism in the thirties in some regions in the USA, in fact, was due to acidification caused by the growing industry as revealed by the acidobiontic chrysophytes found in the sediments. Numerous studies of this kind show the importance of silica-scaled chrysophytes, in combination with diatoms, in documenting acidification progress and possible decline.

In recent years and in the future, the study of chrysophyte scales in sediments will become important in monitoring global climate change.

Biogeography of Chrysophytes

The study of chrysophyte biogeography falls into two categories: the study of dispersal mechanisms and the study of the resulting distribution patterns (*see* **The Phycogeography of Freshwater Algae**).

The dispersal of algae from one water body to another involves change between elements, and thus exposes the organisms to considerable stress. The dispersal capacity depends on the ability of the cells to tolerate this challenge. The most common transport vectors are wind, birds (and other animals), and man.

Transport via wind requires in most cases that the cells, which are whipped from the water surface, can tolerate desiccation. Algae with thick or gelatinous cell walls may tolerate such conditions and can potentially be dispersed in that way. The wall-less chrysophyte cells cannot, but the successful transport of stomatocysts may be a possibility but it has not been directly observed yet. The observations of chrysophytes spending most of their life history in the cyst stage, but with very few free-swimming flagellates, in isolated islands may support this idea.

The most obvious dispersal method is by birds, either in the intestine (endozoic) or on feathers or feet (epizoic). Many algae with thick cell walls, e.g., desmids can tolerate endozoic dispersal and this may also be possible for chrysophyte stomatocysts. Many other algae with thick or gelatinous walls have been observed to be dispersed in the epizoic way, but very few chrysophytes. *Synura* can be dispersed on feathers of water-fowl, perhaps as gelatinous palmella stages.

Some distribution patterns coincide with the routes of migrating birds. Examples have been demonstrated from Romania, and from the southeast United States, where species richness and the occurrence of certain species in some places has been explained by coincidences of chrysophyte occurrence with several migratory routes of birds. Bipolar distribution in North and South America has also been explained this way.

Many experiments on dispersal by birds have been made, however, with very few positive results as regards chrysophytes.

Dispersal by man has been postulated in a few cases. The earlier-mentioned change in stomatocyst assemblage composition in Easter Island in the seventeenth century might be due to European visitors who washed and refilled their water barrels. Another case is the apparently new occurrence of *M. vannigera* in the Hudson Bay area in Canada, proposed to have taken place by ballast water in ships from the Baltic area, where the species, until a few years ago, was thought to be endemic.

The dispersal capacity varies very much between the species. Some are apparently well adapted and are found almost everywhere, except on isolated islands. Other species have apparently great difficulties. *M. pseudocoronata*, very common in North America, has apparently not been able to cross the Davis Strait to West Greenland. Endemicity of many species can be due to low dispersal capacity.

Chrysophyte Distribution

Beyerinck's old statement as to the distribution of microorganisms – 'that everything is everywhere, only the environment selects' – has been revived during the recent past. This is mainly due to the demonstration of the occurrence of almost all species of *Paraphysomonas* in a pond in Northern England, as a proof of their putative worldwide distribution, ignoring the fact that almost all of them had been originally described from vicinity of Cambridge, and the dispersal of cells over that distance does not prove anything about ubiquitous occurrence. Investigations on the occurrence patterns of chrysophytes have been done on the silica-scaled forms, because the ultrastructure of the scales

guarantees correct identifications, which is obligatory for all distribution studies. There are about 250 species of Synurophyceae and Chrysophyceae with such scales, and these identifications can be treated with absolute confidence, because all records are documented with electron micrographs of the scales.

Such investigations started in the fifties in Europe, and simultaneously in Alaska and Japan. In the seventies, extensive investigations started in North America and Russia. In the eighties, Africa and the Southern Hemisphere (South America, Australia, and New Zealand) were included, and in the nineties, studies in China and Central Asia have been initiated.

Remaining are large areas in Asia and Africa. Also Antarctica is in serious lack of investigations, whereas there are several investigations in the Arctic.

These investigations now cover so large areas of the Earth that it is possible and reasonable to distinguish several distribution types, and it is evidently far from the 'everything is everywhere' hypothesis, even if many species are found in a high number of sites in all continents and must be termed cosmopolitans. *S. petersenii* and *P. vestita* have been found almost everywhere where investigations have taken place and in almost every type of environment. Many of the small species of *Paraphysomonas* are certainly also cosmopolitans. Another group contains those species that are widely distributed, but not found in all parts of the world. Together, these account for only 20% of the Synurophyceae, which, because of their silica scales, is the best known group as regards biogeography.

Species with arctic-northern temperate distribution constitute a large group, including, *inter alia*, 59 *Mallomonas* species. A special arctic flora cannot be distinguished. Apart from one or two species it is mainly a diluted version of the temperate flora, decreasing in species number towards North.

A related group consists of those with bipolar distribution, occurring both in northern and southern temperate regions. This is similar to many higher plants. But a special circumpolar (Gondwanaland) chrysophyte flora does not exist, unlike the case of both higher plants and marsupial animals.

The early concept was that chrysophytes were restricted to temperate or cold regions. However, also the tropics have a very rich and special flora. Some species are pantropic (e.g., *S. australiensis*), occurring in tropical parts of all the continents. Others are more restricted.

A very large percentage (e.g., almost half of the Synurophyceae) have only been found in one place or in a very restricted region. They are endemics. Such species are not only found in islands and other isolated regions, but just as well in thoroughly investigated areas such as eastern North America.

Endemic species are vulnerable, i.e., they lose their endemicity if they are found in other places. Some appear to be genuine, having remained endemic for many years, and are thus also still endangered. Others have been endemic for some years, and are then found elsewhere. In fact, almost every species described starts as an endemic, but is later found in other places. For example, many species first described from South America were soon found in other parts of the world, some of them with bipolar distributions.

It must be concluded that the distribution of chrysophytes depends on many factors, both environmental and historical. Necessary conditions are the dispersal capacity of the species, and available relevant vectors. Time is an important factor, varying from species to species. Cosmopolitan species have positive scores in each of these parameters.

Concluding Remarks

The chrysophytes play a special role in several ways. Their mixotrophy gives them a key role in the food web. Their role as indicator organisms in paleoecology, together with diatoms, make them useful also in other disciplines, such as climatology and archeology.

See also: Other Phytoflagellates and Groups of Lesser Importance; The Phycogeography of Freshwater Algae.

Further Reading

Kristiansen J (2002) Phylum Chrysophyta (golden algae). In: John DM, Whitton BA, and Brook AJ (eds.) *The Freshwater Algal Flora of the British Isles*, pp. 214–244. Cambridge: Cambridge University Press.

Kristiansen J (2005) *Golden Algae. A Biology of Chrysophytes*, 165 pp. Liechtenstein: A.R.G. Gantner Verlag.

Kristiansen J and Preisig HR (eds.) (2000) Encyclopedia of chrysophyte genera. *Bibliotheca Phycologica* 110: 1–260.

Kristiansen J and Preisig HR (2007) *Synurophyceae. Süsswasserflora von Mitteleuropa/Freshwater Flora of Central Europe*. Heidelberg: Spektrum Akademischer Verlag Vol. I, Part 2, VIII + 252 pp.

Nicholls KH and Wujek DE (2002) Chrysophyte algae. In: Wehr JD and Sheath RD (eds.) *Freshwater Algae of North America*, pp. 471–503. San Diego: Academic Press.

Sandgren CD (1983) Survival strategies of chrysophyte flagellates: reproduction and formation of resistant resting cysts. In: Fryxell G (ed.) *Survival Strategies of the Algae*, pp. 23–48. Cambridge: Cambridge University Press.

Sandgren CD (1988) The ecology of chrysophyte flagellates: Their growth and perennation strategies as freshwater phytoplankton. In: Sandgren CD (ed.) *Growth and Reproductive Strategies of Freshwater Phytoplankton*, pp. 9–109. Cambridge: Cambridge University Press.

Siver PA (2002) Synurophyte Algae. In: Wehr JD and Sheath RD (eds.) *Freshwater Algae of North America*, pp. 523–557. San Diego: Academic Press.

Cyanoprokaryota and Other Prokaryotic Algae

A Sukenik and T Zohary, The Yigal Allon Kinneret Limnological Laboratory, Migdal 14950, Israel
J Padisák, University of Pannonia, Veszprém, Egyetem, Hungary

Introduction

The Cyanobacteria (from the Greek word κυανos; = blue), also known as Cyanophyta, Cyanoprokaryota, Chloroxybacteria or blue-green algae, constitute the largest, most diverse, and most widely distributed group of photosynthetic prokaryotes. Cyanobacteria are among the oldest organisms on Earth. Fossil traces of cyanobacteria have been found in Archaean rocks of Western Australia, dated 3.5 billion years old. As soon as they evolved they played a major role in raising the level of free oxygen in the atmosphere of early Earth.

The Division Cyanobacteria belongs to the Kingdom Monera, which, together with the Eubacteria ('true' bacteria) and the Archaeobacteria, make up the Prokaryota. Like all other prokaryotic organisms cyanobacteria lack cellular organelles; their DNA lies free in the center of the cell and is not enclosed within a nucleus. Cyanobacteria contain chlorophyll and other photosynthetic pigments, which give them blue-green or other, often strong colors. They acquire their energy through oxygenic (oxygen evolving) photosynthesis, thus are often referred to as algae (blue-green algae), although their prokaryotic characteristics are well defined and differentiate them from eukaryotic algae (Protista).

From the earliest observation and recognition by botanists, cyanobacteria were a source of fascination and attraction. They soon became model organisms for studying fundamental biological processes and environmental phenomena as reflected by upsurge of publications, and more recently web sites (see Further Reading and Relevant Websites at the end of this chapter). Learning the intricacies of these organisms and the recent elucidation of the detailed structure of their genome could lead to breakthroughs in the understanding of biological carbon sequestration, UV protection, and hydrogen production, processes which are essential for coping with current global changes.

This chapter briefly covers the unique structural, physiological, and biochemical features of cyanobacteria, focusing on characteristics that support their proliferation in aquatic ecosystems. A detailed account of their adaptation to inland water environments is portrayed to cover the important role they play in a wide range of ecosystems.

Origin and Diversity

Role of Cyanobacteria in Earth's Evolution

The cyanobacteria are ancient organisms appearing in fossil records in sedimentary rocks deposited in shallow seas and lakes. Fossilized cyanobacterial colonies in rock formations, called stromatolites, have been described from various geological strata while living examples of stromatolites can be found in the Caribbean and Western Australia (**Figure 1**). The cyanobacteria have been tremendously important in shaping the environment and the course of evolution throughout Earth's history. The current oxygenic atmosphere was generated by cyanobacterial photosynthesis during the Archaean and Proterozoic Eras, which led to the gradual increase in the concentration of free oxygen in the earth's atmosphere from 1% to 21%, supporting the development of aerobic respiration and life as we know it today, and leading to the formation of a protective ozone layer.

Cyanobacteria (including the prochlorophytes) are also claimed to contribute to the evolution of plants. Sometime in the late Proterozoic, or in the early Cambrian, cyanobacteria began to take up residence within certain eukaryote cells, providing organic compounds for the eukaryote host via photosynthesis in return for a home. This event, known as endosymbiosis, is proposed to have occurred several times independently during the evolution of protists and higher plants. With time, the endosymbionts became the eukaryotic chloroplast.

Morphological Forms

Cyanobacteria are limited in form to relatively simple shapes (**Figure 2**) that have not changed over billions of years. These include simple unicells (as in the genera *Synechoccocus* and *Chroococcus*), colonies (e.g., *Microcystis*), and filaments, also known as trichomes (e.g., *Oscillatoria*, *Aphanizomenon*). More complex forms, which are variations on the original simple forms, also occur such as branched filaments (*Calothrix*, *Fischerella*), filament aggregates (e.g., *Gloeotrichia*), and sheets or mats (*Merismopedia*, *Hydrococcus*). Some of the filamentous forms evolved to have specialized cells – heterocytes

Figure 1 Natural and man-induced scenes involving cyanobacteria. (a) Stromatolites in Numbung National Park, Western Australia. Cyanobacteria that construct the stromatolites move toward the light and always remain on the surface of the mineral sediments deposited during their growth over thousands of years. (b) Purnululu National Park, Western Australia. Layers of sandstone with high clay levels support the growth of cyanobacteria. Cyanobacteria inhabit the top few millimeters of the sandstone, coating it with a dark color layer that protects it from erosion. (c) A surface scum of *Microcystis* in Dianchi Lake, Kunming, China August 2005. (d) *Microcystis* scum is collected from Biyanchi Lake, (Kunming, China) by on board screening to be used for land fertilization. (e) A scum of the cyanobacterium *Microcystis aeruginosa* from Hartbeespoort Dam, near Johannesburg, South Africa, 1985. (f) An infrared aerial photo of the Hartbeespoort Dam shore, showing the surface scum in pink. The strong red areas are trees on the shore. (g) A hypersaline gypsum crust deposited at the bottom of a saltern evaporation pond of the Israel Salt Company in Eilat. The upper orange layer is dense communities of unicellular *Aphanothece-Halothece* type cyanobacteria. The dark-green layer below contains *Phormidium*-type filamentous cyanobacteria. Photograph courtesy of Aharon Oren, The Hebrew University of Jerusalem. (h) A thick (*c.*50 cm) 'hyperscum' of the cyanobacterium *Microcystis aeruginosa* accumulated against the wall of Hartbeespoort Dam, South Africa, with photooxidized surface layer, mid-1980s.

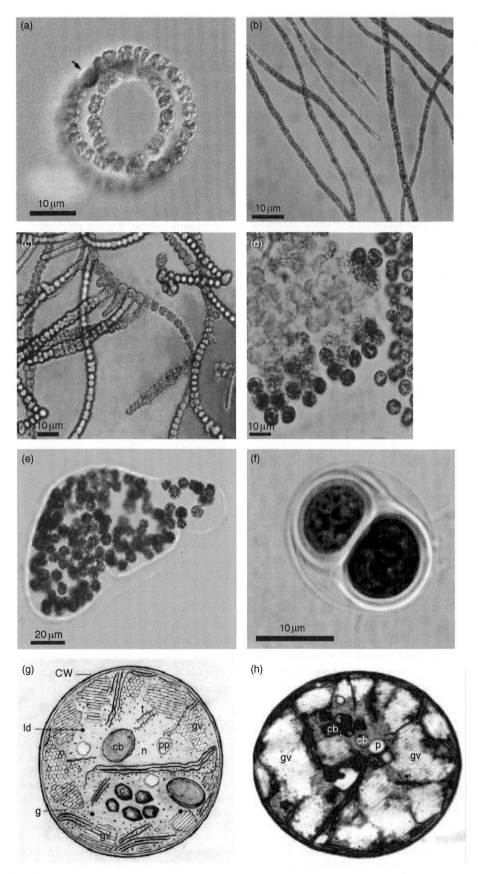

Figure 2 Continued

(old term: heterocysts) – for nitrogen fixation and akinetes as resting stages (see later). There are no flagellated forms or life stages.

The simplest cyanobacteria grow as single cells, which may be enclosed by a sheath or gelatinous envelope or may be clustered into colonies. Cell sizes range from <1 μm diameter (in the unicellular marine picocyanobacteria *Prochlorococcus* and *Synechococcus*) to >50 μm in length (e.g., *Cyanothece*, specialized cells in a cluster). Cells vary in form: spheres, rods, oval, fusiform, and irregular shapes. Some planktonic colonial forms produce massive accumulations at the surface of the water (e.g., *Microcystis*) whereas benthic colonial species form thick attached macroscopic mats (e.g., *Lyngbya*). In some forms the cells are heteropolar, i.e., their apical and basal parts are morphologically different and they possess specific patterns of cell division (*Clastidium*). In filamentous species the cells form one physiological entity as neighboring cells are connected through pores in their crosswalls. Trichomes are capable of fragmentation or complete disintegration into separate cells.

Classification and Phylogenetics of Cyanobacteria

The division Cyanobacteria contains about 150 genera and more than 2000 species, occurring in a wide variety of morphological and ecological forms (see later). Their morphological characteristics were the basis for the early taxonomy of the Cyanophyta (by Geitler in the 1930s, by Desikachary in the 1950s) according to the Botanical Code. In parallel, bacteriologists (e.g., Stanier, Castenholz) have applied physiological and biochemical properties for those species that existed in culture, and named the division Cyanobacteria. Over the last decades, ecological characteristics, ultrastructure, and molecular evidence have substantially influenced our knowledge of this group. As a result, species concepts and classification are undergoing radical changes (see Oren 2004, http://ijs.sgmjournals.org/cgi/reprint/54/5/1895). Further classification changes can be expected in the future, but at present the traditional morphological approach is still needed as it cannot yet be replaced satisfactorily by other approaches.

Traditionally, the cyanobacteria were classified based on morphology into five sections/orders, referred to by the numerals I–V (**Table 1**). The first three – Chroococcales, Pleurocapsales, and Oscillatoriales – are of coccoid and filamentous species that are nonheterocytic. The allocation into these taxonomic groups is not fully supported by phylogenetic studies: unicellular and simple filamentous forms evolved in a polyphyletic fashion and are scattered throughout several clades. Therefore, organisms of similar morphology are not necessarily closely related. On the other hand, the heterocytic cyanobacteria – Nostocales and Stigonematales – are monophyletic. Recent phylogenetic studies based on DNA sequencing have supported the notion of monophyletic origin of the division Cyanobacteria.

Recently, a sixth order was added to the Cyanobacteria – the prochlorophytes, formerly designated as division Prochlorophyta. Phylogenetic analyses revealed that these organisms do not represent a separate monophyletic lineage but they belong to different clades of coccoid and trichal Cyanobacteria. So far only three genera belonging to three families known. Two coccoid shaped genera (*Prochloron*, *Prochlorococcus*) are marine. The third, containing the only known freshwater species (*Prochlorothrix hollanica*), is filamentous (unbranched), gas vacuolated, abundant in the shallow eutrophic lakes of the Netherlands.

Habitats

Cyanobacteria are found in a diverse range of habitats, from oceans to fresh water, from bare rock to soil, from hot springs and hydrothermal vents to the arctic and under ice. It is actually difficult to identify environments in which Cyanobacteria do not occur, one such environment lacking Cyanobacteria is highly acidic (pH < 3) water bodies. Despite covering a wide range of environmental factors as a group, individual species of cyanobacteria are narrowly adapted organisms, which makes them sensitive to sudden changes in prevailing conditions (see later).

Figure 2 Light and electron microscope photographs of various cyanobacteria species. Photographs B–F courtesy of Dr. Alla Alster Kinneret Limnological Laboratory, IOLR. (a) A trichome of *Anabaena circinalis* (Nostocales) from Lake Texoma, USA. Arrow points to a heterocyte. (b) Trichomes of *Aphanizomenon ovalisporum* (Nostocales) culture isolated from Lake Kinneret, Israel. (c) A mass of culture-grown trichomes of the true-branching *Fischerella* sp., (Stigonematales). (d) Light microspore photograph of a *Microcystis aeruginosa* (Chroococcales) colony showing darker cells in the periphery having more gas vacuoles and chlorophyll than interior cells. (e) A colony of *Microcystis wesenbergii* (Chroococcales), from Lake Kinneret, Israel. A characteristic colorless delimited mucilage enveloping the cells. (f) A dividing cell of *Chrococcus turgidus* (Chroococcales), surrounded by colorless mucilagenous envelopes. Cell division is by binary fission. (g) Schematic presentation of a thin section of *Microcystis aeruginosa* showing c, carboxysome; cw, cell wall; g, glycogen granule; gv, gas vacuole; ld, liipid droplet; n, nucleoplasm region; p, plasmalemma; pp, polyphosphate body. Arrows point to thylakoid membranes. Hand drawn based on electronmicrograph by Alla Alster, Kinneret Limnological Laboratory, IOLR. (h) A thin section of *Microcystis aeruginosa* from a thick surface scum. Symbols as in (g). P stands for poly-β-hydroxyburate granule; Cell diameter: c.4 μm.

Table 1 Classification of Cyanobacteria into morphological orders as suggested by Castenholz & Waterbury (1080), but the prochlorophyte group included as Order Prochlorales

Group	Features	Examples of genus
Nonfilamentous Cyanobacteria		
I. Order Chlorococcales	Unicellular or nonfilamentous aggregates of cells held together by outer wall or gel-like matrix. Binary division in one, two, or three planes, symmetric or asymmetric or by budding.	Chamaesiphon, Chroococcus[a], Cyanobacterium, Cyanobium, Cyanothece, Dactylococcopsis, Gloeobacter, Gloeocapsa, Gloeothece, Microcystis[a], Synechococcus, Synechocystis
II. Order Pleurocapsales	As mentioned in the previous entry but reproduce by internal multiple fission with production of daughter cells smaller than parent, or by a mixture of multiple fission and binary fission. Rarely form akinetes.	Cyanocystis, Dermocarpella, Stanieria, Xenococcus, Chroococcidiopsis, Myxosarcina, Pleurocapsa
Filamentous Cyanobacteria		
III. Order Oscillatoriales	Binary division in one plane forms 1-seriate trichomes, sometime with 'false' branches; Neither heterocysts nor akinetes are formed.	Arthrospira, Borzia, Crinalium, Geitlerinema, Halospirulina, Leptolyngbya, Limnothrix, Lyngbya, Microcoleus, Oscillatoria, Phormidium, Planktothrix, Pseudanabaena, Schizothrix, Spirulina, Starria, Symploca, Trichodesmium, Tychonema
IV. Order Nostocales	Binary division as mentioned in the previous entry. One or more cells per tricome differentiate into heterocyst under a low concentration of combined nitrogen. Some species produce akinetes.	Anabaena[a], Anabaenopsis, Aphanizomenon[a], Calothrix, Cyanospira, Cylindrospermopsis, Cylindrospermum, Nodularia, Nostoc, Rivularia, Scytonema, Tolypothrix
V. Order Stigonematales	Binary division periodically or commonly in more than one plane forms multiseriate trichomes or trichomes with true branches or both. Apparently always posses ability to form heterocysts; Some species also form akinetes.	Hapalosiphone, Chlorogloeopsis, Fischerella[a], Geitleria, Iyengariella, Nostochopsis, Stigonema
Prochlorophytes	Coccoid shaped from marine environment.	Prochloron, Prochlorococcus
VI. Order Prochlorales (prochlorophytes)	Comprises two families: Prochloraceae and Prochlorococcaceae with a single genus in each one.	
	Filamentous (unbranched), gas vacuolated from fresh water lakes. Comprises a single family Prochlorothrichaceae with a single genus.	Prochlorothrix

Source
Whitton BA and Potts M (2000) *The Ecology of Cyanobacteria: Their Diversity in Time and Space*, 259pp. Boston: Kluwer Academic.
[a]See photographs in **Figure 2**.

We limit our discussion to cyanobacteria of inland water, but also mention the oceanic environment.

Fresh and Brackish Water

Cyanobacteria contribute to the plankton and benthos, epiphyton and epilithon (see later) of the various types of natural freshwater and brackish systems, including lakes, rivers, streams, bogs, wetlands, salt pans, permanent and temporary ponds, and reservoirs. The bloom-forming genera *Microcystis, Anabaena, Aphanizomenon, Cylindrospermopsis* often dominate the summer-fall assemblages of these types of water bodies, especially if eutrophic.

Oceanic Waters

Cyanobacteria also constitute a considerable proportion of the marine phytoplankton, mainly in the picoplankton ($0.2–2\,\mu m$) size group, with two main genera – *Synechococcus* and *Prochlorococcus*. *Prochlorococcus*, the most abundant photosynthetic organism in the oceans and presumably on Earth, is ubiquitous between 40° S and 40° N latitude, where it occupies the uppermost 100–200 m layer, forming a 'deep chlorophyll maximum'. The filamentous *Trichodesmium* and *Nodularia* are widely known as marine bloom-forming genera. *Trichodesmium* can fix atmospheric nitrogen and flourish in the open ocean while *Nodularia* inhabits shallow water sediments in brackish environments.

Extreme Environments

Cyanobacteria are the main and often sole photoautotrophic organisms in extreme environments, including thermal, saline, arid, and endolithic habitats. Hot springs constitute an extreme environment with high temperature, elevated concentrations of H_2S and inorganic ions such as calcium, magnesium, chloride, bicarbonate, sulphate, and high pH (8–10). Cyanobacteria are among the relatively few groups of organisms that are adapted to life under such extreme conditions, they thrive in hot spring environments of up to ∼70 °C. Typical hot spring genera are the coccoid *Aphanocapsa*, *Chroococcus*, *Cyanobacterium*, and *Synechococcus*, and the filamentous *Mastigocladus*, *Oscillatoria*, and *Phormidium*. These cyanobacteria may form mats several centimeters thick (**Figure 1**) and may have extremely high rates of primary production. Cyanobacteria dominate the microalgal communities of hot as well as cold deserts, both in soil crusts and endolithic communities. Their mucilaginous growth form enables them to adhere to the surfaces and retain sparse water resources from dew, intermittent rain, or snow and melt-water in cold regions.

Microbial Mats

Microbial mats are formed by filamentous entangled organisms that produce macroscopically mat-like structures covering water sediments. Cyanobacteria are often the key organisms in microbial mats because of unique physiological characteristics of various species and genera, such as adaptation to extreme light conditions and to UV radiation, and abilities to maintain metabolic activity under anoxic conditions, to optimize photosynthesis at a wide range of irradiance levels, and to fix atmospheric nitrogen.

Life Mode

All Cyanobacteria are autotrophs, having the capacity to synthesize organic carbon by reduction of inorganic carbon. Most cyanobacteria use for this synthesis light energy (photosynthesis), but a few species in specialized dark environments use instead chemical energy (chemosynthesis).

The more abundant life form of cyanobacteria is planktonic – as microorganisms that are suspended in water and move with the motions of the water. But also abundant are benthic forms, attached to the bottom sediments or submersed rocks, where cyanobacteria often form mats. In standing water these mats may later float. Other, epiphytic forms colonize submerged and emergent plants and larger (usually filamentous) algae. Still other life forms are epipelic (colonizing sediments) or epipsammic (colonizing sand), growing on wet or moist soil. Some cyanobacteria are symbionts in lichens, hornworts, cycads, Gunnera, corals, and water plants (e.g., *Azolla*).

Cellular Structure

General Organization

Cyanobacteria are characterized by simple morphology (**Figure 2**). They lack internal organelles such as chloroplasts, mitochondria, vacuoles, or a discrete nucleus as well as structural mechanisms for movement such as flagella or cilia. The DNA lies bundled up in the center of the protoplast ('nucleoplasm') but is not enclosed by a membrane. The thylakoids (photosynthetic membranes) are also dispersed within the cytoplasm. The cyanobacterial DNA lacks the histone proteins associated with eukaryotic chromosomes. Recent studies indicated that Cyanobacteria have circular chromosomes of 2.5–6.4 Mb, as well as a number of plasmids. This correlates with about 3000 genes in the unicellular *Synechococcus*, and with more than 7400 genes in the large filamentous *Nostoc punctiforme*.

Cell Envelopes, Cell Wall, Sheath

Like all eubacteria, the cell wall of cyanobacteria consists of peptidoglycan (murein) often perforated by tiny holes (c.70 nm in diameter). This rigid layer is covered by an outer membrane, associated with a layer of Glycocalyx (a network of polysaccharides). These layers define outer and inner periplasmic spaces, each of which serves different functions in transport and surface metabolism. Cells are often embedded in sheaths of mucilage (hydrated polysaccharide). In many species (e.g., *Microcystis wesenbergii*, **Figure 2**) the sheaths develop into a large common mass of mucilage to form three dimensional structures.

The Photosynthetic Apparatus

Cyanobacteria have an elaborate and highly organized system of photosynthetic lamellae, the analogs of the eukaryotic thylakoid membranes. The cyanobacterial thylakoids (**Figure 2**) contain chlorophyll *a* but lack chlorophyll *b* or *c*, which serve the light harvesting antennae of higher plants and eukaryotic algae. Instead they contain phycobilisomes, structures composed of several chromoproteins, or phycobiliproteins (phycoerythrin, phycocyanin and allophycocyanin), that are connected by linker peptides. The phycobilisomes are organized in rows on the outer surface of the thylakoids and optimize light harvesting and energy transfer to the photosynthetic reaction centers. Phycobilisomes also occur in eukaryotic algae belonging to Rhodophyta and Glaucophyta. The prochlorophytes are exceptional; they do not contain phycobiliproteins and phycobilisomes but do have chlorophyll *b* as an accessory pigment though both their chlorophyll forms (*a* and *b*) are divinyl derivatives.

The Photosystem II (PSII) reaction center of cyanobacteria is a large complex of numerous proteins. It consists of the reaction center core and core antennae but lacks any chlorophyll–xanthophyll binding proteins that are typical as peripheral antennae in plants and algae. The Photosystem I (PSI) reaction center of cyanobacteria is comparable to the core complex of higher plants. Most of the gene products comprising the cyanobacterial reaction center I and II are highly conserved in higher plants. The carboxylation enzyme (Ribulose bisphosphate carboxylase/oxygenase – RubisCO) of Cyanobacteria is a large protein complex with a molecular mass of 500–600 KDa (Type I RubisCO) similar to that of higher plants and chlorophytes. RubisCO and other Calvin-cycle enzymes are localized in polyhedral bodies (carboxysomes) forming distinct structures of 200–300 nm in diameter (**Figure 2**).

Pigmentation

The photosynthetic pigments of cyanobacteria include chlorophyll *a*, β-carotene, zeaxanthin, echinenone, myxoxanthophyll, and other xanthophylls in addition to an array of water soluble chromoproteins, organized in the phycobilisomes. These pigments create a rainbow of possible colors – yellow, red, violet, green, deep blue, and blue-green – based on the relative abundance of each chromophore complex. The core of a phycobilisome is formed by the chromoprotein allophycocyanin, transferring energy via a terminal acceptor to chlorophyll *a* in the reaction centers. Light energy is funneled to the core protein via rod like antenna complexes consisting of the light-harvesting chromoproteins phycocyanin and phycoerythrin or phycocyanin and erythrophycocyanin, depending on the cyanobacterial strain. The amount and composition of phycobilisomes vary in a well-controlled mechanism in response to quality and quantity of irradiated light.

Carotenoids have a dual role in cyanobacteria; they are active photosynthetically and provide protection against photooxidation. β-Carotene is associated with PSII and PSI reaction centers whereas other xanthophylls provide photoprotection from high light and near UV irradiation. In contrast to algae and plants, cyanobacteria do not have a xanthophyll cycle, although they contain ample quantities of zeaxanthin.

Aerotopes (gas vacuoles)

Many planktonic cyanobacteria contain aerotopes (formerly known as gas vacuoles). These protein structures are aggregates of elongated gas-filled chambers, or gas vesicles, each having a hexagonal cross-section, organized like a bee-hive (**Figure 2**). The proteinaceous vesicle wall has a hydrophilic outer surface and a hydrophobic inner surface, making it permeable to gases but not to water. Hence they contain gases in equilibrium with the surrounding protoplast. The gas vesicles have a density of about one tenth that of water; their presence reduces the overall cell density and provides cyanobacteria with a mechanism to control their vertical position in the water column. The aerotopes can be seen with the light microscope as irregular, highly refractile dark-colored bodies. Gas vesicles are found in other bacteria but not in eukaryotes.

Storage

Under conditions of excess supply of particular nutrients cyanobacteria store those in specialized bodies (**Figure 2**). The reserve polysaccharide of cyanobacteria is cyanophycean starch (composed of α-1,4 glucan), which is similar to glycogen and to the amylopectin fraction of starch found in higher plants. The cyanophycean starch is deposited in tiny granules between the thylakoids. Lipids are stored in lipid droplets. Cyanobacteria often accumulate granules of the storage protein cyanophycin (a polymer of the amino acids aspartate and arginine). Polyphosphate (volutin) bodies are storage granules of polymerized phosphate (linear polymers of tens to hundreds of phosphate units linked by high-energy phosphoanhydride bonds); these can be viewed under the light microscope, especially after toluidine blue staining. Cyanobacteria can store enough phosphorus to perform two to four cell divisions, which corresponds to a 4–32-fold increase in biomass. A few cyanobacteria species accumulate poly-β-hydeoxybutiric acid (PHB) when exposed to stress conditions.

Heterocyte

Filamentous genera of the orders Nostocales and Stigonematales growing in oxygen rich environments conduct the obligatory anaerobic nitrogen fixation in specialized cells called heterocytes. This lineage includes, for example, *Anabeana*, *Nostoc*, *Scytonema*, and *Fischerella*. Heterocytes differentiate from common filament cells; they are larger, empty appearing cells interspersed among the vegetative cells, or at the end of the trichome (terminal heterocytes). Heterocytes lack the ability to produce oxygen, and reduce the permeability of oxygen relative to nitrogen by a surrounding barrier consisting of laminae of long chain lipids. Movement of fixed carbon metabolites from vegetative cells to the heterocytes and of fixed nitrogen from the heterocytes to the vegetative cells takes place via open pores in the membrane barrier between these cells.

Akinete

Under deteriorating environmental conditions, some cyanobacteria form larger, thick-walled cyst-like cells called akinetes, which can persist until conditions favorable for growth return. Akinetes contain high amounts of storage compounds: glycogen and proteins (cyanophycin). They are resistant to adverse environmental conditions such as desiccation, extreme temperatures, and phosphate-deprivation. During their dormancy, akinetes maintain low level of metabolic activities. Their germination is triggered by environmental factors such as temperature, light, and nutrients and provides a well acclimated inoculum for population proliferation. In most cases they do not have an obligate requirement for a dormant period.

Primary Metabolism and Physiological Responses

Acclimation to Light Intensity and Quality

Cyanobacteria, as many other photosynthetic organisms, can acclimate to the quantity and quality of light present in their environment. Under different irradiance levels cyanobacteria regulate antenna pigment complexes, photochemical reaction centers, and enzymes for CO_2 fixation to optimize absorption and utilization of light energy. Furthermore, via an integrated regulatory process, cyanobacteria tune light harvesting and photosynthetic functioning to both light intensity and nutrient availability. In a process termed complementary chromatic adaptation, cyanobacterial cells respond to changes in wavelength of light by regulating the synthesis of specific polypeptides of the phycobilisome.

Carbon Acquisition

Cyanobacteria can grow under a wide range of CO_2 concentrations, from several percent CO_2 in air to as low as nanomolar dissolved CO_2 concentrations in aqueous media. To compensate for the relatively low affinity of the carboxylating enzyme – RubisCO – for its substrate CO_2, cyanobacteria possess an inorganic-carbon-concentrating mechanism (CCM). This mechanism involves an energy-dependent influx of inorganic carbon, accumulation of this carbon pool in the form of HCO_3 in the cytoplasm, and the generation of CO_2 at carbonic anhydrase sites within the carboxysomes. The CCM is induced under low CO_2 conditions to facilitate an internal inorganic carbon (Ci) concentration 1000-fold higher than the external one.

Nitrogen Assimilation

Nitrogen can constitute up to 11% of the dry weight of a cyanobacterial cell. Nitrogen sources are usually ammonium, nitrate, nitrite, and elementary nitrogen in N_2-fixers; organic-N (e.g., urea, amino acids) can be assimilated by some species. When ammonium is available, it is preferentially used over other nitrogen sources; its presence imposes a prompt cessation of the uptake of nitrate. Ammonium also acts to repress the synthesis of proteins involved in nitrate assimilation. When nitrate is taken up, intracellular nitrate is reduced to nitrite by nitrate reductase (Nar), the resulting nitrite is then reduced to ammonium by nitrite reductase (Nir). In some cyanobacteria either Nar or both Nar and Nir have been found firmly bound to thylakoid membranes. Heterocytic cyanobacteria and also some nonheterocytic ones can also convert N_2 directly into ammonium using the enzyme nitrogenase.

Phosphate Uptake

Phosphate is an essential nutrient; its abundance promotes cyanobacterial blooms. Under conditions of ample phosphate supply, cyanobacteria accumulate inorganic phosphate in excess of their immediate requirements in polyphosphate bodies. In low-phosphate environments many cyanobacteria can induce the formation of high-affinity phosphate scavenging systems and upregulate enzymes to hydrolyze dissolved inorganic phosphate molecules. A high-affinity phosphate transport system encoded by the gene cluster *pstSCAB* has been identified in several cyanobacteria and is upregulated by P deficiency. When starved for phosphorus, many cyanobacteria increase their phosphate uptake rate several-folds concomitant with the induction of periplasmic alkaline phosphatases.

Assimilation of Trace Metals

Various trace metals are essential for normal functioning and development of cyanobacteria. Iron is an essential component of electron transport in almost all living organisms, 22–23 iron molecules are required for a complete functional photosynthetic apparatus in cyanobacteria. N_2-fixing species need iron, as well as molybdenum, for their nitrogenase enzyme. Iron is considered limiting in many limnetic environments in which the soluble ferrous (Fe^{2+}) is rapidly oxidized to ferric (Fe^{3+}), which is practically insoluble under most natural pH conditions. Cyanobacteria, like other bacteria, maintain two basic responses to support growth under iron-limited conditions. They greatly improve their ability to scavenge iron by synthesizing siderophores, which solubilize iron as siderophore-iron complexes. In addition, the cellular demand for iron is reduced by substituting proteins that contain iron or whose synthesis requires iron by other compounds, e.g., the mobile electron carrier flavodoxin replaces iron–sulfur-containing ferredoxin.

Buoyancy Regulation

Cyanobacteria have developed mechanisms for regulating their position in the water column. Their aerotopes may provide positive buoyancy, i.e., the ability to float to the surface of the water column where light is abundant, and there is access to CO_2 via atmospheric dissolution despite high pH conditions. Under low or no wind conditions, positively buoyant cyanobacteria accumulate at the surface to form surface scums (**Figure 1**). As a result of photosynthesis at the surface, the cells accumulate high-density storage materials, which provide the ballast needed for sinking through the water column. Another mechanism for sinking is accumulation of osmotically active sugars and uptake of ions, which increase the cellular turgor pressure causing the collapse of gas vesicles.

Gas vesicle production is induced by low light conditions. Under surface scum condition, the cells underneath the ones at the air–water interface are shaded; this promotes the accumulation and thickening of the scum. Calm winds further push the floating scum to lee-shores, ultimately forming massive 'hyperscums,' decimeters in thickness (**Figure 1**).

Motility

Most cyanobacteria lack an active mechanism for motility. However, some filamentous forms lacking sheaths, like *Oscillatoria*, exhibit a gliding motility, rotating or gliding, and often bending rhythmically back and forth. Such motility is created by cellular secretion of dehydrated polyelectrolytic gel from specialized barrel-shaped pores in the cell membrane. The gel swells rapidly upon contact with the water, generating a force that pushes the cells.

Reproduction

The only means of reproduction in cyanobacteria is asexual. Cyanobacteria are able to reproduce through binary fission, budding, or fragmentation. These forms of asexual reproduction explain the various appearances that cyanobacteria take on: patches, slimy masses, strings, filaments, and branched filaments, for example. Binary fission is reproduction that involves merely duplicating the DNA and dividing in half. Budding, or cell fission, involves the formation of smaller cells from larger ones. Fragmentation is breaking into fragments, each of which then regenerates into a complete organism. Coccoid forms usually reproduce by binary fission. Cell division occurring in 1, 2, or 3 planes in successive generations is diagnostic in determining families. Filamentous forms reproduce by trichome fragmentation, or by formation of special hormogonia. Hormogonia are distinct reproductive segments of the trichomes. They exhibit active gliding motion upon their liberation and gradually develop into new trichomes.

Asexual reproduction and strain-selection led to the success of cyanobacteria as universal branch of biota, which managed to turn the atmosphere from reductive to oxidative environment. Nevertheless many species are strictly adapted to given conditions and need a stagnant environment to be successful.

Bloom Formation

Cyanobacteria proliferate under favorable conditions of nutrient abundance, warm water temperature, calm weather conditions, and presence of light, often to the extent that they color the water. High phytoplankton density leads to high turbidity and low light availability, conditions under which cyanobacteria have competitive advantages. Their proliferation is further enhanced by low grazing pressure, as many of the colonial and filamentous cyanobacteria are not grazed by crustacean zooplankton. Furthermore, because of their positive buoyancy, many species of planktonic cyanobacteria float through stable layers of water and accumulate at the surface where they form floating scums. This surface accumulation gives cyanobacteria access to CO_2 (when its concentrations in the water are low) and light, and at the same time they shade the water column below them. Under the same stable water column conditions, most other algae sink. Allellopathy also plays a role (*see* Chloride). Thus, by competitive exclusion, blooming cyanobacteria tend to dominate the planktonic algal assemblage.

Cyanobacterial bloom formation is particularly widespread, in both marine and fresh waters, among several key genera, including *Microcystis*, *Aphanizomenon*, *Anabaena*, *Planktothrix*, *Gloeotrichia*, *Trichodesmium*, *Coelosphaerium*, *Gomphosphaeria*, and in recent years also *Cylindrospermopsis*. Blooms of these genera tend to dominate water bodies of all types, especially if they are eutrophic. Examples are:

blooms of *Microcystis* in fishponds and wetlands in Israel, in the Dnieper reservoirs in Russia, in eutrophic lakes Suwa and Kasumigaura in Japan, in South African reservoirs, in Biyanchi Lake, China (**Figure 1**) and in many other locations worldwide. Blooms of *Aphanizomenon* species in the Baltic Sea, in Lake Kinneret Israel and in newly constructed lakes in Queensland Australia; *Cylindrospermopsis raciborskii* blooms in Lake Balaton, Hungary and water reservoirs in southeast Australia, and *C. cuspis* in Lake Kinneret, Israel. Metalimnetic *Planktothrix* blooms in Lake Lucerne, Switzerland. *Gloeotrichia* blooms in Scottish reservoirs and small lakes in Maine, USA. More details on cyanobacteria blooms are discussed elsewhere. In the temperate zone cyanobacterial dominance is usually pronounced during late summer-fall. In tropical and subtropical regions it can continue year-round. In many cases these blooms are undesirable, ranging from being a nuisance (due to accumulation and rotting of stagnant scums and as a result of taste and odor problems) to being a health hazard, in the case of toxin-producing species.

Cyanobacterial blooms collapse, gradually or abruptly, when conditions are no longer suitable for growth and when the standing stock biomass can no longer be maintained. Common reasons for bloom collapse include exhaustion of nutrients, increased vertical mixing, cooler temperature, pathogens such as cyanophages and enhanced photooxydation due to exposition of floating scum to high light.

The increasing frequency of cyanobacteria bloom events and the expanded geographical coverage of such blooms demonstrate the wide range of adaptive characteristics of this ancient group of organisms. Nevertheless, cyanobacterial blooms frequently collapse when exposed to a small but rapid change in their environment. This 'cyanobacterial paradox' was phrased by Paerl (1988), who stressed that both environmental conditions (in many cases extreme conditions) and spatial–temporal stability should be considered in any attempt to predict cyanobacterial blooms and dominance in lakes.

Mass Production

Cyanobacteria are the source of many valuable products and carry physiological processes that have a potential use for mankind, including light-induced hydrogen evolution by biophotolysis. Cyanobacteria may be used for food or fodder because some strains have a very high content of proteins and vitamins. The best example is the use of naturally occurring *Arthrospira* (known as '*Spirulina*') blooms as food by indigenous populations in Africa and Central America. In recent years this cyanobacterium has become an important commercial source for health food. Several production plants of various sizes and sophistication are producing near 2000 t dry weight biomass per annum at different parts of the world. The mass production of *Spirulina* opens markets for by-products such as vitamins and other neutraceticals (food additives of health value), ingredients for the cosmetics industry, and fluorescence dyes (based on phycobiliproteins) for medical and biological research and diagnostics. Another cyanobacterium marketed on a large scale as a health food is *Aphanizomenon flos-aquae*, collected from natural blooms in Klamath Lake, Oregon, USA.

Nitrogen fixing cyanobacteria were evaluated as biofertilizers for rice fields in various geographical locations and this low cost fertilization approach has become a common practice to enhance rice production. Other more advanced biotechnology applications, currently investigated in cyanobacteria, are related to recent global changes and the demand for alternative sustainable energy sources and atmospheric CO_2 sequestering. Under certain conditions various cyanobacterial species, instead of reducing CO_2, consume biochemical energy to produce molecular hydrogen. Hydrogen production has been identified to occur in a wide variety of cyanobacterial species under a vast range of culture conditions via the activity of several unique enzymes (nitrogenase, hydrogenases). Recruiting the efficient carbon assimilation process of cyanobacteria for atmospheric CO_2 sequestering was proposed as potential solution for reducing atmospheric CO_2 concentration.

Furthermore, the vast genetic data and the availability of several cyanobacteria genome sequences open challenges for biotechnology application and innovations for years to come, from waste water purification processes and environmental biotechnology (bioremediation) to advanced biosensors and production of fine chemicals.

Prospects for the Future

Under the current scenarios of climate changes (global warming and increase in UV radiation and atmospheric CO_2 concentration), cyanobacteria are likely to become even more abundant than what they already are. Warm temperatures give cyanobacteria a competitive advantage over many other algal groups. Furthermore cyanobacteria are well equipped to cope with UV radiation and high CO_2 concentrations. Thus, cyanobacteria are likely to proliferate in more habitats and form more toxic blooms. As the availability of drinking water is globally diminishing, cyanobacteria present a growing threat to the quality of the remaining water, especially those coming from above ground sources (lakes and reservoirs). Furthermore, cyanobacteria blooms in many water bodies jeopardize their potential use for recreational activities and affect their biodiversity and stability. Current research attempts should lead to better understanding of cyanobacterial responses to environmental cues, and point to suitable controlling measures to regulate and reduce events of toxic blooms.

At the same time, the biotechnological potential of cyanobacteria should be further explored. Efficient metabolic processes such as photosynthesis, nitrogen fixation, and production of special secondary metabolites can be engaged into energy production facilities as well as fine chemical and pharmaceuticals manufacturing. The genomic data of more than six different cyanobacteria species is currently revealed, which opens new directions to fundamental studies of evolutionary processes, regulatory mechanisms, and unique metabolic pathways.

See also: Cyanobacteria; Phytoplankton Productivity; Protists.

Further Reading

Badger MR and Price GD (2003) CO$_2$ concentrating mechanisms in cyanobacteria: Molecular components, their diversity and evolution. *Journal of Experimental Botany* 54: 609–622.

Bryant DA (ed.) (1994) *The Molecular Biology of Cyanobacteria*, 654 pp. Dordrecht, Boston: Kluwer Academic Publishers.

Carr NG and Whitton BA (eds.) (1973) *The Biology of Blue-Green Algae*, 676pp. Oxford: Blackwell Scientific.

Chorus I and Bartram J (eds.) (1999) *Toxic Cyanobacteria in Water: A Guide to their Public Health Consequences, Monitoring, and Management*, 456 pp. London/New York: E & FN Spon.

Dutta D, De D, Chaudhuri S, and Bhattacharya SK (2005) Hydrogen production by Cyanobacteria. *Microbial Cell Factories* 4: 36–44.

Eldridge DJ and Green RSB (1994) Microbiotic soil crusts—A review of their roles in soil and ecological processes in the rangelands of Australia. *Australian Journal of Soil Research* 32: 389–415.

Falconer IR (2005) *Cyanobacterial Toxins of Drinking Water Supplies: Cylindrospermopsins and Microcystins*, 279 pp. Boca Raton, FL: CRC Press.

Fay P (1983) *The Blue-Greens: Cyanophyta-Cyanobacteria*, 88 pp. London/Baltimore, MD: E. Arnold.

Fay P and Van Baalen C (eds.) (1989) *The Cyanobacteria*, Amsterdam/NY/Oxford: Elsevier.

Huisman J, Matthijs HCP, and Visser PM (2005) *Harmful Cyanobacteria*, 241 pp. Dordrecht, Netherlands: Springer.

Johnson CH and Golden SS (1999) Circadian programs in cyanobacteria: Adaptiveness and Mechanism. *Annual Review Of Microbiology* 53: 389–409.

Kehoe DM and Gutu A (2006) Responding to Color: The Regulation of Complementary Chromatic Adaptation. *Annual Review of Plant Biology* 57: 127–150.

Komárek J (2003) Coccoid and colonial cyanobacteria. In: Wehr JD and Sheath RG (eds.) *Freshwater Algae of North America. Ecology and Classification*. London: Academic Press.

Komárek J and Angnostidis K (1998) Cyanoprokaryota 1. Teil: Chroococcales. In: *Süsswasserflora von Mitteleuropa, Band 19/1*, 548 pp. Stuttgart: Fisher Verlag.

Komárek J and Angnostidis K (2005) Cyanoprokaryota 2. Teil: Oscillatoriales. In: *Süsswasserflora von Mitteleuropa, Band 19/2*, 759 pp. München: Elsevier/Spektrum Akademischer Verlag.

Komárek J, Kling H, and Komárková J (2003) Filamentous cyanobacteria. In: Wehr JD and Sheath RG (eds.) *Freshwater Algae of North America. Ecology and Classification*. London: Academic Press.

Paerl HW (1988) Growth and reproductive strategies of freshwater blue-green algae (cyanobacteria). In: Sandgreen CD (ed.) *Growth and Reproductive Strategies of Freshwater Phytoplankton*. Cambridge University Press.

Rai AK (ed.) (1997) Cyanobacterial Nitrogen Metabolism and Environmental Biotechnology, 299 pp. New Delhi: Narosa.

Reynolds CS (1987) Cyanobacterial water blooms. *Advances in Botanical Research* 13: 68–143.

Reynolds CS and Walsby AE (1975) Water blooms. *Biological Reviews* 50: 437–481.

Rippka R, Deruelles J, Waterbury JB, Herdman M, and Stanier RY (1979) Generic assignments, strain histories and properties of pure cultures of cyanobacteria. *Journal of General Microbiology* 111: 1–61.

Stanier RY and Cohen-Bazire G (1977) Phototrophic prokaryotes: The cyanobacteria. *Annual Review of Microbiology* 31: 225–274.

Schaeler1 DJ and Krylov VS (2000) Anti-HIV activity of extracts and compounds from algae and cyanobacteria. *Ecotoxicology and Environmental Safety* 45: 208–227.

Stal LJ (1995) Physiological ecology of cyanobacteria in microbial mats and other communities. *New Phytologist* 131: 1–32.

Schwarz R and Forchhammer K (2005) Acclimation of unicellular cyanobacteria to macronutrient deficiency: Emergence of a complex network of cellular responses. *Microbiology* 151: 2503–2514.

Whitton BA and Potts M (2000) *The Ecology of Cyanobacteria: Their Diversity in Time and Space*, 259 pp. Boston: Kluwer Academic.

Relevant Websites

http://www-cyanosite.bio.purdue.edu – Cyanosite.

http://www.kazusa.or.jp/en/plant/ – CyanoBase.

http://www.people.vcu.edu/~elhaij/cyanonews/ – CyanoNews (The Cyanobacteriologist's Newsletter).

http://141.150.157.117:8080/prokPUB/chaprender/jsp/showchap.jsp?chapnum=504 – The Prokaryote – Cynaobacteria plant Symbiosis.

http://141.150.157.117:8080/prokPUB/chaprender/jsp/showchap.jsp?chapnum=403 – The Prokaryote – The genus *Prochlorococcus*.

http://www-cyanosite.bio.purdue.edu/cyanotox/cyanotox.html – The Toxic Cynaobacteria Home Page.

Cyanobacteria

W F Vincent, Laval University, Quebec City, QC, Canada

Introduction

Cyanobacteria are photosynthetic prokaryotes that capture sunlight for energy using chlorophyll *a* and various accessory pigments. They are common in lakes, ponds, springs, wetlands, streams, and rivers, and they play a major role in the nitrogen, carbon, and oxygen dynamics of many aquatic environments. The exact timing of appearance of the first cyanobacteria-like microbes on Earth is still unclear because of controversy over the interpretation of Precambrian fossils; however, much of their present diversity was achieved more than 2 billion years ago, and they likely played a major role in the accumulation of oxygen in the Earth's early atmosphere. In addition to their remarkably long persistence as free-living organisms, cyanobacteria also form symbiotic associations with more complex biota; for example, the nitrogen-fixing species *Anabaena* (or *Nostoc*) *azollae* forms a symbiotic association with the floating fern *Azolla*, which is widely distributed in ponds and flooded soils. The chloroplasts in plants and algae appear to be originally derived from an endosymbiosis in which a cyanobacterium was engulfed and retained within a colorless eukaryotic cell.

Modern-day cyanobacteria include some 2000 species in 150 genera and 5 orders (**Table 1**), with a great variety of shapes and sizes. Ecologically, there are three major groups in the aquatic environment: mat-forming species, which form periphytic biofilms over rocks, sediments, and submerged plants; bloom-formers, which create a wide range of water quality problems and that are most common in nutrient-rich (eutrophic) lakes; and picocyanobacteria, which are extremely small cells (<3 μm in diameter) that are often abundant in clear water lakes. Additional groups include colonial non-bloom-formers, which are common in a variety of aquatic habitats, including mesotrophic lakes, wetlands, and saline waters; metaphytic species that form aggregates that are loosely associated with emergent macrophytes (water plants such as rushes and reeds that extend out of the water into the air above); certain species that grow as periphyton but that can also enter the plankton; and various symbiotic associations.

Cellular Characteristics

Cyanobacteria were formerly classified as blue-green algae (*les algues bleues* in French, *las algas azules* in Spanish) because of their algal-like appearance, their possession of chlorophyll rather than bacteriochlorophyll, and their photosynthetic production of oxygen by a two-photosystem process as in algae and higher plants. The most widely used taxonomic schemes for these organisms are largely based on the International Botanical Code, with separation according to classic morphological criteria. Ultrastructural studies, however, clearly show that the Cyanobacteria are prokaryotic; that is, they lack nuclei and other organelles and they have a peptidoglycan cell wall that is typical of gram-negative Eubacteria. They also possess several features that set them apart from other bacteria, especially their photosynthetic apparatus and oxygen production. The term 'blue-green algae' is still widely used by the media and in the water quality management area. The current taxonomic separation of species, especially the coccoid and nonheterocystous (see definition later) taxa, is believed to be artificial and not reflective of evolutionary relationships, and will likely be substantially revised as more genetic data become available.

All cyanobacteria contain chlorophyll *a* and most contain the blue phycobiliproteins phycocyanin and allophycocyanin, giving the cells their characteristic blue-green color. Many taxa also contain the phycobiliprotein phycoerythrin, making the cells red, or sometimes black. The phycobiliproteins are located in structures called phycobilisomes on the thylakoid (photosynthetic) membranes, and these are highly efficient 'light guides' for the transfer of captured solar energy (excitation energy) to the reaction centers of photosynthesis, specifically photosystem II. A group of photosynthetic microbes formally classified as a separate prokaryotic phylum, the Prochlorophyta, contains chlorophyll *b* in addition to *a*, but lacks phycobiliproteins and thus phycobilisomes. On the basis of genetic data (specifically, the gene sequence for 16S ribosomal RNA) this latter group is now placed within the Cyanobacteria. This group includes the species *Prochloroccus marinus*, one of the most abundant phototrophs in the sea, and a filamentous freshwater phytoplankton species, *Prochlorothrix*.

Although cyanobacteria lack membrane-bound organelles, they have a variety of cellular structures and inclusions that have specialized functions and that contribute to their ecological success. These include the photosynthetic thylakoid membranes containing the phycobilisomes, and the nucleoid region or centroplasm in the center of the cell,

Table 1   The five orders of cyanobacteria recognized in the classic botanical taxonomic scheme

Order	Characteristics	Illustrative genera
1. Chroococcales	Coccoid cells that reproduce by binary fission or budding	*Aphanocapsa, Aphanothece, Gloeocapsa, Merismopedia, Microcystis, Synechococcus, Synechocystis*
2. Pleurocapsales	Coccoid cells, aggregates or pseudo-filaments that reproduce by baeocytes	*Chroococcidiopsis, Pleurocapsa*
3. Oscillatoriales	Uniseriate filaments, without heterocysts or akinetes	*Lyngbya, Leptolyngbya, Microcoleus, Oscillatoria, Phormidium, Planktothrix*
4. Nostocales	Filamentous cyanobacteria that divide in only one plane, with heterocysts; false branching in genera such as *Scytonema*	*Anabaena, Aphanizomenon, Calothrix, Cylindrospermopsis, Nostoc, Scytonema, Tolypothrix*
5. Stigonematales	Division in more than one plane; true branching and multiseriate forms; heterocysts	*Mastigocladus* (*Fischerella*), *Stigonema*

In the bacterial classification scheme, the orders are referred to as subsections of Phylum BX: Cyanobacteria.

which contains the complex folded, circular DNA, often in multiple copies. The cells also contain various storage bodies, including glycogen (polyglucose) granules, which store carbon; cyanophycin granules, which are nitrogen stores composed of arginine and aspartic acid; carboxysomes composed of ribulose 1,5-bisphosphate carboxylase/oxygenase, which act as a store of this photosynthetic enzyme as well as of nitrogen; and polyphosphate granules. These inclusions allow cells to accumulate energy and nutrients far in excess of their present requirements when they are under favorable conditions, and to subsequently use these reserves for maintenance and growth when they encounter resource-poor conditions. The cells of several planktonic genera contain up to several thousand gas vacuoles (**Figure 1**), which are hollow, water-impermeable cylinders made up of protein subunits. These fill with gases that diffuse in from the surrounding medium. They provide buoyancy to the cells and colonies, allowing the cyanobacteria to float towards the surface where the light conditions are improved for photosynthesis. Some species may undergo diurnal migration up and down the water column by varying the amount of dense carbohydrate inclusions that act as ballast in their cells.

Many filamentous cyanobacteria produce different cell types that play specific physiological, reproductive, or ecological roles. The most well known of these is the heterocyte (often called a heterocyst, although it is not a cyst). This thick-walled cell (**Figure 1**) is formed by members of the Nostocales and Stigonematales and is the location of the enzyme nitrogenase for nitrogen fixation, the conversion of nitrogen gas into ammonium and then amino acids. This cell type is not a strict prerequisite for nitrogenase activity, however, because several nonheterocystous taxa in other orders are also known to fix nitrogen. Another specialized cell type is the akinete, a structurally reorganized cell

that is formed under unfavorable conditions, and that allows cyanobacteria to overwinter in the sediments. Some genera such as *Nostoc* produce hormogonia, a motile series of cells formed for reproduction. In one order of coccoid forms, the Pleurocapsales, reproduction is via the production of up to several hundred minute cells called baeocytes.

Ecophysiology

Cyanobacteria photosynthesize using water as the electron donor and produce oxygen as in algae, although a small number of strains can also use hydrogen sulfide (H_2S) and convert it to elemental sulfur. In general, cyanobacteria can tolerate low oxygen conditions and concentrations of H_2S that are toxic to eukaryotic algae. This tolerance may contribute to their ability to survive in anoxic, eutrophic lake sediments as well as in certain mat environments.

Cyanobacteria are highly tolerant of UV radiation and have four major strategies to eliminate or mitigate the toxic effects of this most reactive waveband of underwater solar radiation. Some species avoid UV exposure by their choice of habitat, for example, by growing in dense macrophyte beds, in sand beneath the surface, or in the metalimnion (region of the thermocline) of lakes. Many species have effective UV-screens that include scytonemin (peak absorption in the UV-A waveband) and mycosporine-like amino acids (peak absorption in the UV-B). Some benthic (bottom-dwelling) species such as *Nostoc*, *Calothrix*, and *Gloeocapsa* have such high concentrations of scytomenin that they are black. All cyanobacteria produce carotenoids, for example, β-carotene, myxoxanthophyll, echinenone, oscilloxanthin, and canthaxanthin, which are effective scavengers of reactive oxygen species, the damaging photoproducts of UV exposure.

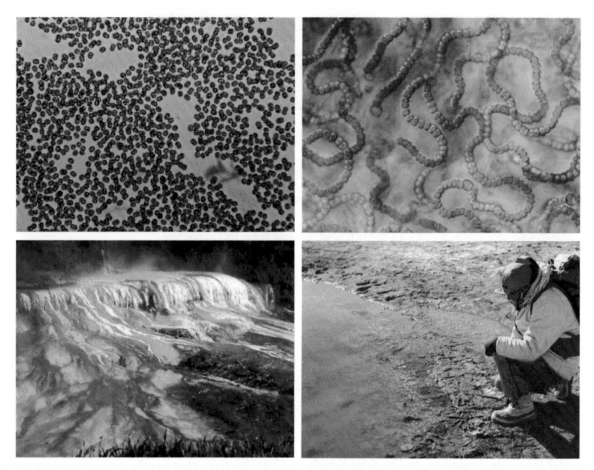

Figure 1 Cyanobacteria in inland water ecosystems. Top left: Photomicrograph of *Microcystis aeruginosa* from a eutrophic lake. The bright areas inside each cell are due to the scattering of light by the gas vacuoles. Top right: Photomicrograph of *Nostoc* from a high Arctic lake. The larger, more spherical cells are heterocysts, the sites of nitrogen fixation. Bottom right: A carotenoid-rich cyanobacterial mat in a pond on the McMurdo Ice Shelf, Antarctica. Bottom left: Cyanobacterial mats in a geothermal spring, New Zealand.

Cyanobacteria also have various enzymatic defences against reactive oxygen species, such as superoxide dismutase. Finally, cyanobacteria have a variety of damage repair mechanisms to identify and repair UV-damaged proteins, for example, in photosynthetic reaction centers, and DNA. This combination of mechanisms and high tolerance to UV radiation, perhaps also combined with the multiple 'back-up' copies (six or more identical 'chromosomes') of the cyanobacterial genome in each cell, likely contribute to the ecological success of cyanobacteria in many types of habitats exposed to bright sunlight, for example, as surface blooms, as benthic layers in shallow waters, and in the picoplankton of oligotrophic, transparent lakes.

Cyanobacteria occur over a wide temperature range; however, most tend to have warm temperature optima for growth. They occur in thermal springs, with the highest temperature limit of 74 °C recorded in Yellowstone National Park, USA. Bloom-forming cyanobacteria prefer temperatures over 15 °C, and even the mat-forming species that occur throughout the Polar regions in frigid waters or on ice are not psychrophilic (optimal growth at low temperatures). Instead they are psychrotolerant (also termed psychrotrophic); that is, they tolerate the cold and grow very slowly relative to their optimal growth rates at much higher temperatures.

In general, cyanobacteria prefer alkaline conditions, and during a bloom, the pH may rise to more than 9. Under these conditions, most of the inorganic carbon is in the form of carbonate and is unavailable to many eukaryotic algae. Cyanobacteria have excellent inorganic carbon concentrating mechanisms, with separate membrane transport systems for CO_2 and for bicarbonate (HCO_3^-). Once inside the cell, the carbon is largely present as bicarbonate, and the enzyme carbonic anhydrase catalyzes its dissociation into CO_2, the substrate for ribulose 1,5-bisphosphate carboxylase/oxygenase.

Some cyanobacteria are highly tolerant of salt and grow in media with high osmolarities. Saline and hypersaline lakes often contain picocyanobacteria, mat-forming species such as *Microcoleus chthonoplastes*, bloom-forming, nitrogen-fixing species such as *Nodularia spumigena*, and the salt-tolerant colonial species *Aphanothece halophytica*. Certain taxa are obligate halophiles, and can grow in environments with salt concentrations of up to 360 g of salt per liter, 10 times the salinity of seawater. Salt-tolerant cyanobacteria produce a range of osmolytes, that is, solutes that maintain their high internal osmolarity and turgor pressure without causing toxic effects on proteins and other cellular constituents. The most important of these are glucosylglycerol (which can account for up to 30% of dry weight in cells grown in hypersaline media) and glycine betaine.

The ability of some species of cyanobacteria to fix nitrogen gives them a competitive advantage in low nitrate, low ammonium waters, and may also contribute substantial quantities of new nitrogen to aquatic ecosystems. One particularly important habitat for cyanobacteria is the flooded rice fields, where the cyanobacteria increase the nitrogen content and fertility of the soil because of their nitrogen-fixing capability.

Taste, Odor, and Toxin Production

Cyanobacteria produce a variety of compounds that strongly affect water quality. These include molecules that affect the taste and odor of water, notably geosmin and 2-methylisoborneol, both of which impart an earthy or musty odor to the water. An additional cyanobacterial group of compounds is cyclocitrals, which are carotenoid breakdown products that impart a grassy odor to the water. Of greater concern to water resource managers is the production of three classes of toxins: hepatotoxins, which attack the liver; neurotoxins, which attack the nervous system; and dermatotoxins, which cause skin irritation. These toxins are especially produced by certain planktonic species. *Microcystis aeruginosa*, a bloom-forming species that is common in eutrophic lakes and reservoirs throughout the world, produces microcystin, a cyclic peptide that is a hepatotoxin. Two other bloom-forming genera *Anabaena* and *Aphanizomenon* often occur in association with *Microcystis* and produce the alkaloid neurotoxin anatoxin. *Cylindrospermopsis*, a tropical species that has been increasingly observed in temperate lakes, contains the potent hepatotoxin cylindrospermopsin, also an alkaloid. Several benthic species produce anatoxins as well as the dermatotoxins lyngbyatoxin and aplysiatoxins. Cyanobacterial toxins have been implicated in the death of animals, including birds, farm stock, dogs, and a small number of humans. The presence of toxic cyanobacteria in drinking reservoirs and recreational lakes is of particular concern, and often results in the temporary closure of such water supplies. The toxins are water-soluble and are not destroyed by boiling the water prior to drinking. The difficulty of managing such blooms is compounded by the large variability in toxin production among strains of the same species, and with environmental conditions.

Ecology

Mat-Forming Species

Cyanobacteria are a common component of the periphyton (the ensemble of microorganisms attached to submerged surfaces), forming crusts and films over rocks (epilithon), plants (epiphytes), sand (epipsammon), sediments (epipelon), and other substrates. In many environments, these biofilms accumulate from millimeters to centimeters in thickness as vertically structured, microbial mats that form a benthic layer at the bottom of the water column, or that detach and float at the surface.

Mat-forming cyanobacteria have a particularly long history. Some of these organisms form laminated, lithified structures called stromatolites that are well represented in the fossil record all the way back to the Precambrian. The fossils contain layered, filament-like inclusions and bear striking resemblance to living stromatolites that occur today in a small number of marine habitats, and also in certain lakes. The stromatolite communities are thought to have been the main primary producers on Earth for more than 1 billion years throughout the Proterozoic, and a major contributor to atmospheric oxygen that in turn set the stage for the rise of oxygen-requiring microbes and animals. They are less successful today, in part, because of the presence of grazing animals, including crustacean zoobenthos and zooplankton.

Cyanobacterial mats and films are common in wetland systems, especially in alkaline habitats where they are often encrusted in calcium carbonate. The common genera in swamps include *Chroococcus*, *Leptolyngbya*, *Lyngbya*, *Phormidium*, *Microcoleus*, *Schizothrix*, and *Scytonema*. Benthic cyanobacteria are especially important in rice-field wetlands because of nitrogen fixation capabilities, and include the heterocystous genera *Anabaena*, *Calothrix*, and *Tolypothrix*. Mats and films are common in many extreme environments, including cryoconites holes on glaciers, in pools on Arctic and Antarctic ice shelves, in saline and hypersaline lakes, and in geothermal springs.

The nitrogen-fixing cyanobacterium *Nostoc* is a frequent member of benthic lake, pond, stream, wetland, and semiaquatic communities throughout the world (**Figure 1**), and is well known for its ability to survive prolonged desiccation. It can form flattened mats, lobes, and sheets that coat the rocks and sediments, or discrete spherical colonies that range in size from less than a millimeter to 25 cm. These grape- or pearl-like colonies are harvested from lakes and wetlands in several parts of the world as a food source and for their medicinal properties. In China, *Nostoc commune* is known as *Tian-Xian-Mi* ('rice of heavenly immortals'), and has been used as a medicine for more than 1500 years.

Mat-forming cyanobacteria create their cohesive structure by the excretion of extracellular polymeric substances that bind together the individual cells or filaments. This polysaccharide-rich matrix likely confers desiccation and freeze tolerance to the mats in extreme environments such as semiaquatic habitats that periodically dry out. The extracellular polymeric substance traps sediment particles and forms the physical medium for a great variety of other organisms, including viruses, heterotrophic bacteria, protists such as diatoms, flagellates, and ciliates, and sometime microinvertebrates such as turbellarians (flatworms), rotifers, and nematodes. Detailed analyses of these mats and films have shown that they have very different chemical properties relative to the overlying water; for example, they may have much higher pH and concentrations of inorganic nitrogen and dissolved reactive phosphorus that are orders of magnitude greater than the bulk properties of the surrounding lake or river water. They are also highly differentiated vertically. There are typically strong gradients in oxygen concentration, ranging from supersaturated near the surface to (sometimes) anoxic, H_2S-rich conditions at the base of the mats, where photosynthetic sulphur bacteria may reside.

Microbial mats dominated by cyanobacteria are often brightly colored (**Figure 1**), with large changes in pigment composition down the mat profile. The surface layer is rich in photoprotective pigments, especially carotenoids (orange and red) but also sometimes scytonemin (black or brown), overlying a deeper blue-green layer rich in light-harvesting phycobiliproteins and chlorophyll *a*. Certain mat-forming species of cyanobacteria are motile and are able to change their position in the mats by a gliding motion through the extracellular polymeric substance matrix.

Some of the most spectacular examples of such mat-forming cyanobacteria occur in lakes, ponds, and streams in the Arctic and Antarctic (**Figure 1**).

Cyanobacterial mats are typically 1–2 mm in thickness, but in some lakes and ponds they form mucilaginous mats up to several tens of centimeters in thickness. Similar mats and films also occur in abundance at the other thermal extreme, in hot pools and geothermal springs (**Figure 1**). Apart from their physiological adaptation to high temperatures, these communities have been of interest to microbial ecologists because of their biogeography. A long-standing theory of microbial distribution is that 'everything is everywhere' and that the local environment selects for a particular microbial flora that is globally distributed. Analyses of hot spring cyanobacteria indicate that certain taxa are indeed cosmopolitan and occur at many sites throughout the world. For example the species *Mastigocladus* (or *Fischerella*) *laminosus* occurs at temperatures of up to 58 °C, and is found in hot springs worldwide. However, the high-temperature form of *Synechococcus* that grows at temperatures of up to 74 °C appears to have a more restricted global distribution, suggesting some biogeographical separation of strains.

Bloom-Forming Species

Bloom-formers tend to be found in warm, stable, nutrient-rich lakes and are largely absent from the polar and alpine regions. Three genera are particularly common and often co-occur: *Anabaena*, *Aphanizomenon*, and *Microcystis*. Some lakes contain a metalimnetic bloom of *Planktothrix* (*Oscillatoria*). The factors causing the dominance of bloom-forming cyanobacteria are of great interest to water quality managers because of the production of toxins and other secondary metabolites by these organisms (see earlier), and the most important factors have been vigorously debated by researchers for several decades. No single factor has been identified, but rather a combination of several conditions appears to lead to cyanobacterial blooms. First, the overproduction of biomass requires large quantities of its elemental constituents, and in inland water environments the factor limiting biomass production is often phosphorus availability. The increased enrichment of lakes by phosphorus, leading to eutrophic conditions (high nutrients and dense phytoplankton concentrations), is generally accompanied by the development of cyanobacterial dominance and blooms. However, this does not occur in all seasons, nor even every year, implying that secondary factors also play a role. Nitrogen availability tends to be of lesser importance, and although many studies have pointed to low N:P ratios as a factor associated with cyanobacterial dominance, this is often a correlate rather than the cause of eutrophic water conditions. High pH and low CO_2 concentrations were also considered causative factors; however, these conditions are also the consequence of a large photosynthetic biomass. Iron availability has also been considered a secondary limiting factor for cyanobacterial growth development in some lakes.

Temperature clearly plays a role in tipping the balance towards cyanobacteria, with blooms becoming more likely as the water column warms above 15 °C. In part this is because cyanobacteria tend to have high temperature optima for maximum growth. However, it is also related to the increased frequency and strength of diurnal thermoclines (near-surface temperature and density gradients) that accompany warming and result in a more stable water column. The absence of mixing allows the cyanobacterial colonies to adjust their position in the water column via their gas vacuoles. This responsiveness to temperature suggests that cyanobacterial blooms will become increasingly common as the world heats up because of greenhouse gas emissions and global climate change. These effects may be exacerbated in regions that have reduced rainfall, and therefore longer lake residence times that favor eutrophication.

The positive buoyancy of bloom-forming cyanobacteria can lead to the sudden appearance of surface scums, which create aesthetic, health, and other water quality problems. These are blown by the wind and can accumulate from throughout the lake to spectacular concentrations along the shoreline and in bays. The shading of phytoplankton lower in the water column, and the rise in pH that accompanies the production of such surface blooms, are two of several mechanisms that continue to favor the ongoing growth and dominance of cyanobacteria once they become established. The production of resting spores (akinetes) and their resistance to grazing pressure by zooplankton are additional strategies that contribute to the ecological success of these bloom-forming organisms.

Picocyanobacteria

Picocyanobacteria are the smallest of cyanobacteria cells, often around one thousandth of a millimeter in diameter; however, they typically occur in high concentrations (the term pico refers to cells smaller than 2 μm). Two of the genera *Synechococcus* and *Prochlorococcus* are so widely distributed in the open ocean that they are probably the most abundant photosynthetic cell types in the biosphere. Freshwater picocyanobacteria are usually ascribed to the genus *Synechococcus*, although this is an artificial taxonomic category that encompasses diverse genotypes and phenotypes. They are readily observed by epifluorescence microscopy,

and increasingly by flow cytometry, and comparative studies have shown that they contribute a large proportion of total photosynthetic biomass in oligotrophic lakes. Their high surface-to-volume ratio confers an advantage in low nutrient conditions, and their small size can also be an advantage for light capture because of a lack of internal shading ('package effect'). These minute cells provide a nutrient and energy source to flagellates and ciliates in the microbial food web, and they are also subject to loss by viral attack (cyanophage).

Picocyanobacteria can reach very high concentrations in nutrient-rich saline lakes where the competition for light may be severe, and where their superior light-capturing ability confers an advantage. An interesting dichotomy is found in the Polar regions, where although picocyanobacteria are typically in low abundance or absent from the marine environment, they are often common in cold lakes and rivers. For example, picocyanobacteria are largely absent from the coastal Southern Ocean, but phycoerythrin-rich *Synechococcus* populations occur in the coastal, meromictic (incompletely mixed because of their salinity gradients) lakes of the Vestfold Hills, Antarctica, coloring the water red with cell concentrations of up to 10 billion per liter.

Other Ecological Groups

Some small-cell species of coccoid cyanobacteria form colonies, and although these rarely produce dense blooms, they are often abundant in the phytoplankton of mesotrophic and eutrophic waters. The most common genera are *Aphanocapsa*, *Aphanothece*, *Coelosphaerium*, *Gomphosphaeria*, and *Merismopedia*. The relationship between these genera and free-living picoplankton is still not clear, and it may be that some are simply an aggregate life-form of solitary cells. On the basis of genetic analyses, Bergey's Manual of Systematic Bacteriology does not recognize many of these classic genera, and places them within genera containing solitary cells. For example, *Aphanocapsa* and *Merismopedia* are assigned to the form genus *Synechocystis*, and *Aphanothece* to *Cyanothece* or to *Synechococcus*. There is evidence from marine food web analyses that picocyanobacterial aggregates may provide a direct food source for higher trophic levels such as copepods, but little is known about their trophic role in inland waters, despite their frequent presence in the freshwater plankton.

Certain cyanobacterial taxa are often found in loose association with emergent plants in ponds and wetlands. This community is referred to as the metaphyton, and includes genera such as *Chroococcus*,

Leptolyngbya, *Merismopedia*, and *Phormidium*. In peat bogs, genera such as *Aphanothece*, *Chroococcus*, *Hapalosiphon*, *Merismopedia*, and *Tolypothrix* are found in association with the *Sphagnum* moss, and occur at acidic pH's as low as 4.

Some cyanobacteria occupy both benthic as well as planktonic habitats. For example, the saline lake species *Aphanothece halophytica* commonly forms mucilaginous, benthic films, but it is also often found in the phytoplankton of such lakes, sometimes at high concentrations. One common freshwater species, *Gloeotrichia echinulata*, develops on sediments and plants in eutrophic lakes and ponds in spring and early summer, and then in late summer can become gas vacuolated, producing an abundant population of large (from submillimeter to several centimeters) spherical colonies in the plankton.

The symbiotic interactions of cyanobacteria include close associations with fungi to form lichens, sometimes found in semiaquatic habitats but usually in the surrounding terrestrial environment; on or inside mosses; as an endosymbiont within the semiaquatic angiosperm *Gunnera*; and within the floating fern *Azolla*, which is often cultivated in rice fields for its nitrogen content. Certain cyanobacterial species are also cultivated in artificial or naturally eutrophic ponds, lagoons, and lakes as a protein source for human consumption, animal feed, fertilizer, or health food. The most important of these is *Arthrospira* ('Spirulina'), which is commercially exploited because of its nutritional value, lack of toxins, high growth rates (doubling times less than 12 h), and tolerance to environmental stress.

See also: Cyanoprokaryota and Other Prokaryotic Algae; Harmful Algal Blooms; Microbial Food Webs; Phytoplankton Productivity; Viruses.

Further Reading

Castenholz RW (2001) Phylum BX. Cyanobacteria: Oxygenic photosynthetic bacteria. In: Boone DR and Castenholz RW (eds.) In: *Bergey's Manual of Systematic Bacteriology*, 2nd edn, vol. 1, pp. 473–599. Springer: New York.

Chorus I and Bartram J (1999) *Toxic Cyanobacteria in Water: A Guide to Their Public Health Consequences, Monitoring and Management* 432 pp. U.K.: Taylor & Francis, London.

Kaebernick M and Neilan BA (2001) Ecological and molecular investigations of cyanotoxin production. *FEMS Microbiology Ecology* 35: 1–9.

Knoll AH (2005) *Life on a Young Planet: The First Three Billion Years of Evolution on Earth*, 277 pp. Princeton, NJ: Princeton University Press.

Komárek J and Anagnostidis K (2001) Cyanoprokaryota I. Teil. *Chroococcales*. In: Ettl H, Gartner G, Heynig H, and Mollenhauer D (eds.) *Susswasserflora von Mitteleuropa*, vol. 19, part 1, 548 pp. Jena: G. Fischer.

Komárek J and Anagnostidis K (2005) Cyanoprokaryota II. Teil. Oscillatoriales. In: Büdel B, Gärtner G, Krienitz L, and Schagerl M (eds.) *Susswasserflora von Mitteleuropa*, vol. 19, part 2, 759 pp. Heidelberg: Elsevier.

Matthijs HCP, Petra M, Visser PM, and Huismann J (eds.) (2005) *Harmful Cyanobacteria*, 243 pp. Springer.

Rautio M and Vincent WF (2006) Benthic and pelagic food resources for zooplankton in shallow high-latitude lakes and ponds. *Freshwater Biology* 51: 1038–1052.

Tomitani A, Knoll AH, Cavanaugh JM, and Ohno T (2006) The evolutionary diversification of cyanobacteria: Molecular-phylogenetic and paleontological perspectives. *Proceedings of the National Academy of Sciences of the USA* 103: 5442–5447.

Vincent WF (2007) Cold tolerance in cyanobacteria and life in the cryosphere. In: Seckbach J (ed.) *Algae and Cyanobacteria in Extreme Environments*, pp. 287–301. Heidelberg: Springer.

Wehr JD and Sheath RG (eds.) *Freshwater Algae of North America Ecology and Classification*, 918 pp. San Diego, CA: Academic Press.

Whitton B and Potts M (eds.) (2000) *The Ecology of Cyanobacteria: Their Diversity in Time and Space*, 669 pp. The Netherlands: Kluwer Academic.

Wiedner C, Rücker J, Brüggemann R, and Nixdorf B (2007) Climate change affects timing and size of populations of an invasive cyanobacterium in temperate regions. *Oecologia* 152: 473–484 doi: 10.1007/s00442-007-0683-5.

Diatoms

S Sabater, University of Girona, Girona, Spain

The Diatom Cell and its Taxonomical Entity

Diatoms are silicified algae between 5 and 200 μm in diameter or length, although sometimes they can be larger than 1 mm. Diatoms are functionally single cells. However, they can appear in nature in many different forms: associated into filaments, forming cell chains, arranged within mucilage tubes, living freely in the water column, or attached to any single substratum.

The siliceous cell wall encloses the organs of the cell and has ornamented structures (**Figure 1**). The structure (the *frustule*) has two distinct parts (outer – *epitheca*; and inner – *hypotheca*), which are adapted to each other similar to that of a shoe's box or a Petri dish. The remarkable properties of this siliceous wall are its transparency, which allows the entrance of the light to the chloroplasts of the cell, and its perforated structure, which makes possible the transport of gases and solutes.

Both of the two different thecas are composed by the *valve* (a flattened plate, the most commonly visible part of the diatom cell under the light microscope) and the *connecting band*, which is attached to the edge of the valve. The connecting bands of the two thecas are one attached to the other and constitute the *girdle* (also called *girdle bands*), which in many taxa are complemented by other siliceous bands (the *intercalary bands*). The girdle bands often do have minute teeth, which are useful to hold the valves together by their edges. Under natural conditions, the frustule is covered by muccopolysaccharide materials, which have been excreted by the same diatom cell through the different openings. When the frustules are acid digested (in order to eliminate the organic matter and achieve a rigorous microscopical determination), very often the bands and valves tear apart from each other.

Throughout the valves there are several structures. Insome pennate diatoms (Monoraphidineae and Raphidineae), there is a *raphe* (longitudinal slot in the theca) in one or the two valves respectively. It has been suggested that this structure is designed to prevent the longitudinal splitting of the valve under turgor. The raphe can be continuous from pole to pole (e.g., *Nitzschia*), but in most cases it is separated in two parts by the central nodule (e.g., *Navicula*). In a few cases, the raphe is reduced to the final part of the valve (e.g., *Amphipleura*). In some genera

(e.g., *Navicula*, *Cymbella*), there exist polar nodules in the respective ends. In the pennate diatoms that do not have a raphe system (Araphidineae), an ornamented area commonly appears in the central position, which is named as the *pseudoraphe*. In some groups of raphid diatoms, the raphe is subtended internally by bridges of silica, termed *fibulae*. In this case, there is a tendency for the raphe to rise above the level of the valve, forming a *keel*, which can be shallow or deep.

Besides the raphe, there are two different types of wall perforations: the simple *pore* or hole, and the *areola*, which are more complex. Pores are homogeneously silicified and are spatially defined within striae or costae. Areolae are hexagonal structures reminding a honeycomb, where the different areas are separated by vertical spacers that often have pores to allow for communication between them. Sieve membranes are placed at the ends of the areolae, either inside or outside. When pores or areolae are arranged in rows, the structure is referred as a *stria*. Other special pores are specially placed in the extreme of one or the two poles of the pennate diatoms, and are the ones that excrete materials that allow the attachment of the cells to the substratum.

The valve surface can have extensions (*processes*) whose main function is to assist colony integration between contiguous cells. These are hollow extensions of tube processes (fultoportulae or rimoportulae). *Fultoportulae* occur in the Thalassiosirales, and consists of a tube which penetrates the silica structure and is internally supported by two or more buttresses. These structures are the point of exit of synthesized chitin or other products. *Rimoportulae* are simpler and more commonly found structures that consist of a tube that opens to the inside of the valve. In other cases, there are silicified *spines* that are holding the cells (e.g., *Fragilaria*).

Cell chains can also be formed when valve faces remain in contact after cell division with the aid of mucilage pads. Cells may be linked by polysaccharide pads at cell corners, eventually leading to colonies (*Tabellaria*, *Asterionella*, *Synedra*). Mucilage produced from special pores in either pole of the cell may produce long stalks (*Gomphonema*). Diatom cells show passive movement, completed through the excretion of mucilage by its many holes and slots. This movement ability is especially important when diatoms live on sediments, where the continuous displacement of this material could imply the

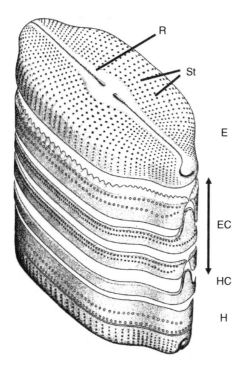

Figure 1 The structure of the diatom frustule, exemplified in a pennate taxon. The drawing shows a exploded view of the frustule, with the epivalve (E), the girdle bands of the epicingulum (EC) and hypocingulum (HC), and the hypovalve (H). The drawing also shows the raphe (R) and the striae (St) on the valve surface. Adapted from Round FE, Crawford RM, and Mann DG (1990). *The Diatoms. Biology and Morphology of the Genera*, 747 pp. Cambridge, UK: Cambridge University Press.

burying (and ultimate death) of the cells unless they could move again to the light. Although motility is less common than passive gliding in diatoms, some examples are remarkable. *Bacillaria paradoxa* forms colonies in which adjacent cells are linked by interlocking raphe ridges so that the cells move synchronously, causing the colonies to expand and contract.

The number of diatom species is estimated to be around 10 000, but some argue that this number could be much higher. To determine the taxonomic specificity of diatoms, one should consider the shape of the body of the valve, the length-to-breadth ratios, and the forms of the apical, axial, and central areas of the cell. The extent, position, and structure of the raphe are important taxonomic characters. The different ornaments on the valve face of the diatom frustule (pores, areolae, processes, spines, hyaline areas, and so on) are distinguishing features used to classify and describe diatoms. Stria densities are traditionally measured as the number in 10 μm, and also their arrangement in the centre and the ends of the cell can be differential. Fibula and *costae* (distinctly thickened ribs on the valve face) densities also need to be determined in many species, while

differences in fibula width or length, together with the evenness of spacing may be useful.

There is a large degree of variation in the diatom cell, which compromises the identification based uniquely in morphological characters of the specimens. Culture experiments have shown that described taxa are simply part of a single gradient of forms along one or more environmental factors such as temperature, salinity, or nutrients. It is also true that there is a given degree of uncertainty in some taxa because naming has been done on single specimens or in narrow populations, therefore not accounting for the total variability of them. Modern molecular techniques combined with traditional microscopical observations need to be combined in order to advance in the classification of these algae. These synergistic approaches could answer clearly about the total species number as well as on to the phylogenetic relationships between them.

Diatoms are easy to collect and to preserve. Their siliceous wall allows for natural preservation in sediments when reducing conditions are not extreme. Even though the arrangement of the chloroplasts has been also indicated as a taxonomic parameter, in order to clearly observe the ornaments of the siliceous wall, a digestion with oxidants is required. Diatoms can be mounted permanently using synthetic resins (e.g., Naphrax) and therefore stored, exchanged, and reexamined as many times as required.

Diatoms are classified into two Orders (**Figures 2** and **3**). The Centrales (Biddulphiales) have valve striae arranged basically in relation to a point, an annulus or a central areola, and tend to appear radially symmetrical. The Pennales (Bacillariales) have valve striae arranged in relation to a line and tend to appear bilaterally symmetrical. Currently, there is a certain debate in the taxonomic classification of diatoms, between the classical classification and a newer system aiming to reflect natural classification. This latter system has split many of the classical genera into new genera. However, it is difficult to select between any of the two classifications because of lack of supporting data (genetic, morphological) and a certain inconsistency of the two.

The Biology of the Diatoms

Diatoms live as diploid cells that progressively increase in volume as they accumulate photosynthetic products. When they reach a threshold volume, they divide vegetatively. This requires to have doubled their various organelles that will be shared between the two daughter cells after mitosis. This type of reproduction allows the diatoms to achieve (under

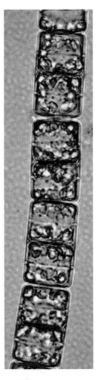

Figure 2 Centric diatoms. Left, *Cyclotella meneghiniana*, scanning electron microscopy (SEM), 6000×. Right, *Melosira varians*, living filament, light microscopy (LM), 400×.

no resource limitation) large clonal populations, and therefore the ability to conquer a new space and to control the available resources. According to species, doubling times can last from about 0.3 to 5 days. Because the siliceous valves are rigid and, once laid down, do not increase in size, cell volume increases by expansion in the girdle region between mitotic divisions. At the completion of mitosis, each offspring possesses one parent valve with its girdle elements (the epitheca) plus a newly secreted valve and its girdle elements (the hypotheca). Because the new valves are deposited within the confines of the parent cell they cannot be larger than the parent valves and gradual diminution in cell size may occur over several generations. The decrease of valves size causes the progressive change in the relative width to height and the morphology of the cell, that is, proportions may be altered through this process. Even though not all diatoms undergo size reduction throughout their vegetative division, those which do show this reduction must also have a stage in the life history when maximum size is restored. This often involves *sexual reproduction* and the production of *auxospores*. The auxospores are formed by the fusion of two gametes, flagellated or not depending on the taxonomic group. Auxospores are swelled cells where a new cell wall of maximum dimensions is produced. This is possible because the cell wall of the auxospores is only lightly silicified with bands or scales. This cell will later divide to produce a new cell, where the original dimensions are reestablished. Some genera (such as *Melosira*) are able to produce *dormant cells*, cells with condensed protoplasm and which are able to survive for longer periods when they settle to the sediments. They might persist until resuspended to the photic zone, when they are quickly able to differentiate into vegetative cells.

Diatoms respond to a certain number of environmental and biological variables. Light, water temperature, substratum type, water velocity, mineral composition and content, nutrient availability, and grazing are amongst the most important factors. Their architectural growth and community composition is a direct response to the overall factors that mould their growth.

The type of attachment as well as the growth form is associated to the *substrata* type to which the cells grow. Those on rocks or cobbles (*epilithic*) are well adapted to resist the drag force of waters and produce prostrate forms capable of a tight attachment. Solid current-exposed surfaces favor the establishment of flat cells or those that have strong adherence to the substratum (*Cocconeis*, *Ceratoneis*, *Tabellaria*), while under low flow rates stalked and filamentous forms may grow into tufts (*Synedra*). Those attaching to plants (*epiphytic*) are less tightly adhered as they use

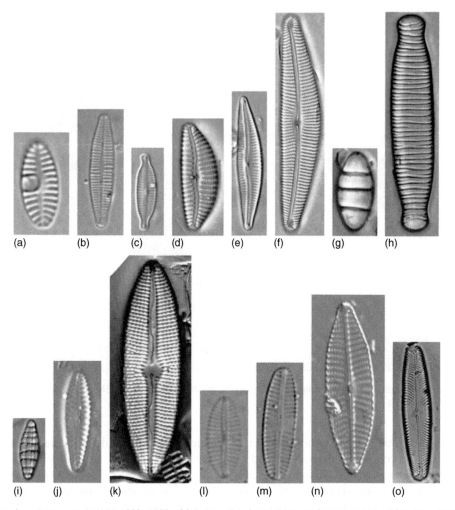

Figure 3 Pennate river diatoms at the LM; 1000–1500× (a) *Achnanthes lanceolata* var. *frequentissima*; (b) *A. biasolettiana*; (c) *Cymbella microcephala*; (d) *C. affinis*; (e) *C. delicatula*; (f) *C. helvetica*; (g) *Diatoma mesodon*; (h) *D. vulgaris* forma *linearis*; (i) *Denticula tenuis*; (j) *Gomphonema angustum*; (k) *Mastoglia smithii*; (l) *Navicula subminuscula*; (m) *N. seminulum*; (n) *N. gregaria*; (o) *Pinnularia microstauron*.

the hydrodynamic protection of the host. Common epiphytes are *Gomphonema*, *Cocconeis*, or *Tabellaria*. Others living on sediments (*epipelic*) do not need to be attached but must have a particular motility. *Gyrosigma*, *Diploneis*, *Amphiprora*, *Amphora*, and *Mastogloia* posses this ability and are common in this habitat. Finally, those living attached to sand grains (*epipsammic*) are firmly attached by their entire valve to resist the movement and related abrasion of the substrata in relation to water. Typical examples are *Cocconeis* and *Amphora*.

Most diatoms are adapted to a wide range of *water temperature* and *light*. However, some diatoms are well adapted to dim light environments (*Aulacoseira roseana*, *Campylodiscus*, *Surirella*), while others take advantage of full light, especially when water temperatures are not excessively high, to produce massive growths (e.g., *Melosira varians*). Most diatoms have

their temperature preference at 15–25 °C. However, a few are able to resist more than 30 °C (*Nitzschia palea*, but photosynthesis declines at 33 °C) or to effectively make photosynthesis at 5 °C (*Aulacoseira islandica*).

Water pH is a summarizing factor of the geochemistry of the waters, and as such its different values create preferences amongst the diatom taxa. Calcium (their abundance being associated to the carbonic buffer system, and therefore to pH) seems to be a key element in determining the distribution of the taxa. Under extreme pH values (both low and high) the diversity of the diatoms (and of the algal community) may be expected to be lower than in circumneutral waters. In the highly eutrophic lakes of East Africa (water pH of 9.5–10.5), a few diatoms coexist with filamentous cyanobacteria. In acidic environments (pH values between 6 and 2.5), diatoms are commonly the dominant algal group. Genera such as *Eunotia*, *Frustulia*, *Stenopterobia*, *Pinnularia*, and *Tabellaria* are widespread in acidic waters. In moderate-to-highly alkaline waters, species of *Denticula*, *Epithemia*, *Nitzschia*, and *Rhopalodia* are the most frequent.

Salinity summarizes the overall ionic composition and hardness of the waters. Salinity might affect the osmotic pressure within the diatom cell, as well as the nutrient uptake and other physiological processes. As such, salinity it is a key factor in the distribution of the diatom communities. Some taxa are able to thrive under salinity of 3–4% (but rarely above this), and are recognized as brackish water species: *Cyclotella quillensis*, *Chaetoceros elmorei*, *Navicula salinarum*, *Diploneis interrupta*, *Amphora coffeaeformis*. Other taxa are able to tolerate moderate salinities (0.5–2%), and named as halophilic: *Anomoeoneis sphaerophora*, *Navicula cincta*, *Cyclotella meneghiniana*. Finally, the most are purely freshwater species, as they require low mineral content to have viable growth.

Diatoms are sensitive to changes in *nutrient* concentrations and ratios. As such, they respond rapidly to changes, both towards an increase of nutrients (eutrophication) or because of their recovery. Benthic (periphytic) communities may respond quicker to eutrophication than the planktonic community. Nutrient excess may favor a certain (low) number of tolerant diatom taxa (e.g., *Navicula gregaria*, *Nitzschia palea*), but inhibit the growth of others specialized in lower nutrient availability (*Achnanthes biasolettiana*, *Cymbella microcephala*). Nitrogen limitation favors the growth of *Epithemia* and *Rhopalodia*, since they may have endosymbiotic nitrogen-fixer cyanobacteria. There is a good knowledge on the optima and tolerances of diatom species to nitrogen and phosphate, even though these values may be conditioned by other factors such as light and temperature.

Grazing by herbivores affect not only the biomass of the diatom community (high grazing intensity depletes algal biomass), but it also may determine its preferential growth form. There is a tight relationship between the diatom communities and the herbivore density and morphology of the mouthparts. Depending on the mouthpart morphology of the grazers, the selected growth form to be ingested may change. Mayflies tend to feed on the outer parts (filaments, stalked forms), while snails, caddisflies, or tadpoles may feed on tightly attached forms (encrusting or prostrate forms). The type of grazers therefore favors one or the other growth forms, depending on their abundance.

Diatom Distribution in Inland Waters

Diatoms are abundant in nearly all habitats where water is at least occasionally present. Therefore, they occur throughout the whole range of freshwater habitats. These microscopic algae are common (and sometimes achieve overwhelming numbers) in standing and flowing waters. Being autotrophic cells mean that they are restricted to areas with light availability, i.e., either surface waters or water depths up to the light compensation point in lake systems. While some grow in the water column itself, as members of the phytoplankton (especially centric and araphid pennates), some others are associated with different types of substrata (**Figure 4**).

Those in the water column are commonly centric diatoms, but some pennate species with extended colonies that resist sinking may be also abundant. In the planktonic environment, diatoms form blooms during spring and autumn. Even though some taxa may exhibit pulses at the two periods, most often the dominant species are different in each of them. Hence, in the spring bloom usually the pennate taxa occur, in particular *Asterionella formosa*, and also several species of *Fragilaria* and *Synedra*. In autumn, the most typical are the centric species, including the genera *Aulacoseira*, *Stephanodiscus*, and *Cyclotella*. Their time of dominance is being defined by their different abilities towards the use of the resources (nutrients, light) as well as by their differential buoyancy.

Some diatoms subsist in subaerial environments, such as springs, illuminated caves, as epiphytes of mosses, or on periodically dried areas of streams and lakes. Under the most extreme cases, these organisms only receive splashing waters. These aerophils usually store large amounts of oils and build inner plates to resist desiccation and osmotic variations. *Ellerbeckia arenaria* is a superb example of this. Other taxa common under these situations are *Melosira roseana*, *Pinnularia borealis*, *Navicula contenta*, *Fragilaria construens*, *Achnanthes lanceolata*, and *Denticula tenuis*.

Pennate diatoms are the most common in the littoral areas of lakes. Amongst the most common attached to cobbles and plants are *Diatoma vulgare*, *D. hiemale*, *Melosira varians*, several *Synedra* and *Fragilaria*, as well as some *Cymbella* and *Gomphonema*. Others common in the superficial lake sediments are *Caloneis*, *Diploneis*, *Fragilaria*, *Frustulia*, *Gyrosigma*, *Cymatopleura*, *Navicula*, *Neidium*, *Pinnularia*, *Nitzschia*, *Stauroneis*, and *Surirella*.

In fast-flowing rivers, taxa need to tolerate low water temperatures and the drag effect of these waters. Only a few thrive under these situations. Among them are *Ceratoneis arcus*, *Diatoma hiemale*, *Gomphonema minutum*, *Meridion circulare*, and *T. flocculosa*. In calcium-richer headwaters, *Denticula tenuis* as well as *Cymbella* and *Gomphonema* taxa are replacing some of the previously existing taxa.

In middle reach waters with increasing mineral content, numerous species of the genera *Navicula*, *Fragilaria*, *Nitzschia*, and *Gomphonema* are dominant and form a highly diverse diatom community. When the nutrient content of the water increases, or organic pollution is relevant, the diatom community decreases its diversity and becomes dominated by a few tolerant taxa, in particular of the genera *Nitzschia* (e.g., *N. palea*) and *Navicula* (e.g., *N. accomoda*). In the lower river stretches both benthic (i.e., *Gomphonema parvulum*, *Navicula gregaria*) and planktonic taxa (*Thalasiossira*, *Cyclotella*) are common.

Using Diatom Communities in Environmental Assessment

Diatoms are used as reliable indicators of environmental (or stress) factors because they respond quickly to altered chemical, physical, and biological parameters. Because of their small size and rapid growth, they may be very useful as 'early warning system' for ecosystem changes. The changes in diatom communities are, however, rarely linear to the degree of change, and the response shows a feedback on external parameters. Whole assemblage indicators include diversity indices, community similarity indices, and inferential or autoecological descriptors. Disturbances trigger not only direct responses, but also indirect responses, including adaptive changes of the diatom community, which need to be beard in mind for an accurate use of this information.

The diversity of a diatom assemblage reflects the number of species (species richness) and the evenness of species abundance. Diversity indices (e.g., Shannon-Wiener, Margalef's, or Simpson's) have been effectively used as indicators of changes in community structure when comparing impacted and reference sites, but did not reliably accounted for a measure of disturbance across systems. This apparent inconsistency is related with internal structuring factors, which affect species abundance and evenness and renders the diversity parameter not suitable for monitoring water quality. However, because of their particular sensitivity, changes in environmental factors are reflected in the composition of the diatom community and may be summarized by means of transfer functions, multivariate approaches, or indices.

Transfer functions have been applied for monitoring pH, salinity, and nutrient trends. These expressions look for the regression between the autoecological value of diatom communities and a master environmental parameter. To be reliable, these model

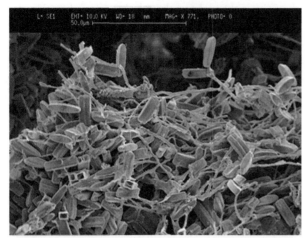

Figure 4 Growth forms and attachment strategies in diatoms. Images are SEM photographs. From top to bottom: Epilithic pennate diatoms (mostly *Achnanthes minutissima*) amongst muccopolysaccharide filaments; 770×. Epiphytic pennate diatoms (mostly *Cocconeis placentula*, attached, and *Rhoicosphaenia abbreviata*, stalked) on *Cladophora glomerata*; 300×. Epipelic pennate (mostly *Navicula* spp.) and centric (*Melosira*, bottom left) diatoms mixed with clay particles; 1000×.

expressions require to have been calibrated (tested) in a number of real cases. In the case of pH, local and regional transfer expressions have been developed and successfully used to predict present and past water pHs. In other cases, transfer functions of nitrogen and phosphorus concentration using diatom communities have been applied successfully. These types of expressions have been particularly applied in lake ecosystems, where diatoms preserved in the sediments can be used to reconstruct the history of the lake. However, similar transfer functions have been used to detect current trends in acidification or eutrophication in river waters.

Diatom indices are one of the most common methods to account for the indicator value of the organisms. Indices have been developed to reflect the abilities of diatom communities in detecting variations in water pH, salinity, nutrient content, and total phosphorus. The widespread use of diatoms as indicators for water quality originates from several facts. Diatom taxa may account up to 80% of the species present in streams, rivers, littoral of lakes and wetlands. The most tolerant taxa are favored while the sensitive taxa are depleted or eliminated, following the rules of interspecific competition, and this might be reflected by accurate diatom indices.

Diatom indices constitute a way of summarizing the information provided by the autoecological preferences of the taxonomic composition of the diatom community. Diatom indices can infer general or specific environmental conditions (i.e., water quality or phosphorus loading). Diatom indices have been developed and widely used mostly in European inland waters, some of them being based in a detailed account of the species present while others (with lower precision power) rely on the genera present.

Most of the diatom indices currently used are inspired on the saprobic index of Kolkwitz and Marsson. However, the indices take into consideration the structure of the community, and therefore consider not simply the presence of the taxa but also their proportion in the community. Most of these indices are calculated according to the formula designed by Zelinka and Marvan, which consider the sum of the different species abundance influenced by their *sensitivity* to the described disturbance, and by their *indicator* value (the latter being opposite to the unspecificity for any situation).

$$\text{ID} = \frac{\sum_{j=1}^{n} a_j s_j v_j}{\sum_{j=1}^{n} a_j s_j}$$

where a is the relative abundance; s, sensitivity value; v, indicator value; n is the number of species observed in the periphyton community.

Diatom characteristics have been included in *multimetric indices*, mostly used in the United States. These indices are composite (average or sum) from different parameters (growth form, motility, sensitivity, and so on). These multimetric indices are designed for the assessment of the biological integrity of the system and the causes of degradation. Comparisons between diatom made indices are made with other trophic levels (invertebrates, fishes) and geomorphological situations, therefore providing an ecological integrated assessment of the ecosystem.

The use of diatom indices is now a question of routine, but they are not exempt of limitations. Uncertainties in their application might occur because of different factors. Sometimes the lack of information about the autoecology of some species may reduce the precision of the index. The tougher the stressor, the more reliable is the information provided by the index; it has been observed that inaccurate prediction of phosphorus loading by diatom communities systematically may occur at lower phosphorus concentrations. Other causes for misuse may occur when there is a mixture of human-produced influences such as nutrient enrichment, habitat alteration, or toxic inputs, where the diatom communities might reflect the overall effect of the disturbance and not the effect of a particular stressor.

Indices can be biased by factors other than human-induced components, including biogeographical or biogeochemical factors. A correct diagnosis has to consider which part of the information available is related to human influences and which part is related to characteristics defining the ecoregion. Accurate knowledge of the autoecological characteristics of the species is necessary to identify the correct reference conditions.

Multivariate techniques (principal component analysis, canonical correspondence analysis, or cluster analysis, amongst the most common) are used to assess the multidimensional patterns of ordination, as well as to relate environmental factors to the ordination patterns. Therefore, these techniques make a useful complement to indices, since they allow determining the ecological factors that are responsible for most of the variation. Often, the information extracted from multivariate approaches may be later used to refine the value provided by the indices. In particular, variance partitioning applied to some of the aforementioned analyses may help to separate the regional versus general factors in the determination of the distribution of diatom assemblages and might be therefore very useful in refining the information provided by the indices.

Further Reading

Bennion H, Johnes P, Ferrier R, Phillips G, and Haworth E (2005) A comparison of diatom phosphorus transfer functions and export coefficient models as tools for reconstructing lake nutrient histories. *Freshwater Biology* 50: 1651–1670.

Cox EJ (1996) *Identification of Freshwater Diatoms From Live Material*, 158 pp. London: Chapman & Hall.

DeNicola DM (2000) A review of diatoms found in highly acidic environments. *Hydrobiologia* 433: 111–122.

Descy JP (1979) A new approach to water quality estimation using diatoms. *Nova Hedwigia* 64: 305–323.

Lange-Bertalot H (1979) Pollution tolerance of diatoms as a criterion for water quality estimation. *Nova Hedwigia* 64: 285–304.

Margalef R (1983) *Limnología*, 1120 pp. Barcelona: Omega.

Potapova MG and Charles DF (2002) Benthic diatoms in USA rivers: Distributions along spatial and environmental gradients. *Journal of Biogeography* 29: 167–187.

Round FE (1965) *The Biology of the Algae*, 269 pp. London: Edward Arnold.

Round FE, Crawford RM, and Mann DG (1990) *The Diatoms. Biology and Morphology of the Genera*, 747 pp. Cambridge, UK: Cambridge University Press.

Sabater S, Gregory SV, and Sedell JR (1998) Community dynamics and metabolism of benthic algae colonizing wood and rock substrata in a forest stream. *Journal of Phycology* 34: 561–567.

Stoermer EF and Smol JP (1999) *The Diatoms: Applications for the Environmental and Earth Sciences*, 469 pp. Cambridge, UK: Cambridge University Press.

Werner D (1977) *The Biology of Diatoms*, 498 pp. Berkeley: University of California Press.

Green Algae

L Naselli-Flores and R Barone, Università di Palermo, Palermo, Italy

Generalities

Green algae constitute the most heterogeneous group of photoautotrophic protoctists inhabiting the biosphere and show an enormously wide variability of shape, size, and habit. Even color may be highly variable and range from grass-green to orange and purple. Taxonomy is attempting to put order in this group of organisms since at least a couple of centuries but the puzzle still has not been completely resolved and new information is added day by day through the development of new biomolecular tools and observation techniques. According to the most recent taxonomic trees, the green algae have evolved in two major lineages: the chlorophytes clade (e.g., *Cladophora*, *Chlamydomonas*) and the charophyte clade (e.g., *Chara*, *Spirogyra*). The first one includes the majority of what have been traditionally called 'green algae,' whereas the second one contains a smaller number of green algal taxa, although some are widespread and as familiar as any other green alga.

From a taxonomic point of view, green algae constitute a paraphyletic group since they likely have a common ancestor with plants: they have the same type of pigments and produce the same kind of carbohydrates during photosynthesis as do land plants. Nevertheless, the poor fossil record of early land plants makes it difficult to tell exactly how the transition from water to land occurred.

Green algae are all photoautotrophic (even though some *Chlamydomonas* species were demonstrated to be mixotrophic under particular conditions) and their growth depends on nutrients and underwater light availability, which, in turn, is linked to physical parameters such transparency, hydrodynamics, temperature, conductivity, and so on. Nevertheless, the ability of single species in exploiting resources may be strongly different as concerning nutrient uptake, light harvesting, and resistance to flushing; the different combinations of these parameters, which in inland water ecosystems may strongly fluctuate in short time scales, give raise to a variety of assemblages that may show an amazingly high biological diversity. As primary producers, green algae have an importance in the biosphere comparable to that of rainforests.

These organisms have dimensions ranging from a few micrometers to several centimeters and show substantial differences in their morphological organization: from swimming or nonmotile unicells and colonies to filaments and various levels of tissue organization (pseudoparenchymatous, parenchymatous, or thalloid) and branching morphologies. These, to some extent, reflect their different functional adaptation to the highly variable aquatic environment. Actually, green algae colonize all types of inland water, from fresh water to hypersaline, from very acidic to strongly alkaline, lentic or lotic. Green algae may proliferate in very small ecosystems as 'phytotelmata' (tree holes filled with water in dead or living trunks, but also axils of bromeliads) and 'kamenitza' (small pools formed in the carbonatic rocks by the combined action of rain and endolithic algae) as well as in very large lakes, and are worldwide distributed in all climatic regions from the poles to the equator. They can also grow in wet subaerial environments like the moss-covers or on the ice and snow and even in dry conditions as lichen phycobionts. Their ubiquity makes rather difficult to summarize their ecology in a few words since these organisms are archetypical of the incredibly vast possibility of adaptation shown by Life on our planet. Nevertheless, from an ecological point of view, a broad, even not always net and clear, distinction can be made between benthic and planktic way of living. The lack of a net boundary is due to the fact that the benthic algae may be suspended into the water column by turbulence (tychoplankton) and the planktic ones may sink to the bottom; both these groups can keep on growing actively in spite of their allocation in a different habitat.

The Benthos

Benthic green algae spend most of their vegetative life attached or in contact to a substratum. They include both macroscopic filamentous or plant-like forms and microscopic unicellular or colonial forms and inhabit flowing as well standing waters.

In eutrophic rivers, *Cladophora* grows firmly attached to the stones and to the rocks on the riverbed by a disk-like holdfast or downward growing rhizoids. It is a major component of the primary production and its filaments are branched and form dense tufts (**Figure 1(a)**), offering refuge and nutrition to several species of aquatic animals. *Cladophora* shows a siphonaceous or coenocytic habit, where the thallus grows in size without any septa being formed between adjacent cells (**Figure 1(b)**). Other green algae, both

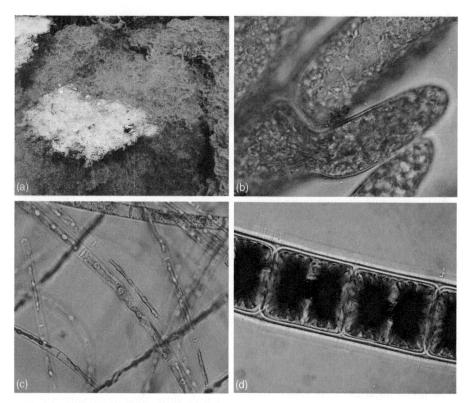

Figure 1   Periphytic and metaphytic green algae. (a) *Cladophora* and *Mougeotia* tufts; (b) *Cladophora* branching; (c) *Mougeotia*; and (d) *Zygnema* (Magnification ×400).

filamentous (e.g., *Oedogonium*, *Spirogyra*) and unicellular (e.g., *Characium*), may grow attached or lying on *Cladophora* and create peculiar epiphytic assemblages, which may include bacteria, fungi, protozoa, and sessile animals, altogether called 'periphyton.' Nevertheless, periphythic algae are also capable of growing on other surfaces as stones (epilithic), wood, and mud (epipelic) or even on (epizoic) or within (endozoic) hard parts of aquatic animals as valves of molluscs, crustacean carapaces, or turtle shields. They produce an adhesive extracellular matrix of polymeric substances (EPS) that provides the attachment of the microorganisms to the surfaces they are growing on. These phototrophic biofilms gain important applications in wastewater treatment, antifouling coatings for ship hulls, and CO_2 sequestration.

In those environments where attaching to a substrate is not strictly necessary and there is no risk of being flushed away, benthic green algae may grow unattached and form large tufts lying on the bottom of ponds and along the coastal zone of lakes. The algae living unattached but in contact with the bottom are generally referred as 'metaphyton' (**Figure 1**). Metaphyton is also formed by filamentous and unicellular species. *Spirogyra*, *Zygnema*, and *Mougeotia* are three common genera of filamentous metaphytic green algae commonly colonizing shallow

environments, even temporary; they also colonize slow-flowing rivers but, contrary to *Cladophora*, do not resist well to shear stress and are easily flushed away when floods occur. Moreover, they can resist complete desiccation, thanks to sexual reproduction, since their zygote stages act as resistant resting form. Thus, when their environment is drying, two adjacent filaments (or two adjacent cells in the same filament) conjugate by forming a tubular outgrowth that fuse to provide a pathway for amoeboid gamete transfer. The resulting zygote (zygospore) produces a thick wall made up of several layers and can remain dormant and resistant to adverse condition for a considerable time. An analogous strategy is adopted by some common unicellular metaphytic green algae like *Closterium* and *Micrasterias*. More in general, green algae reproduce vegetatively through cell division or fragmentation, and many of them reproduce asexually through production of motile zoospores (**Figure 2**) or nonmotile aplanospores.

In some lakes, a rare growth form of metaphytic algae has been observed: small tufts of unattached *Cladophora*-like filaments (*Aegagropila linnaei*), thanks to a gentle wave movement, combined with the muddy bottom substrate, take a ball form with a velvety appearance. These balls are called as 'marimo' in Japan, where they are protected and considered a natural treasure, and are also known as Lake ball or Moss ball. The species achieving this growth form are adapted to low light conditions and are formed by densely packed filaments. The internal part of the ball is also green and the cells packed inside have dormant chloroplasts, which become active if the ball collapses or breaks apart.

Another group of benthic green algae, called rhizophytes, live attached to the muddy substratum by creeping rhizoidal branches. These algae, belonging to the Order Charales, may reach 1 m in length and are very similar to higher plants because, contrarily to all the other green algae, these have female reproductive structures (oosporangia) protected by five spirally twisted sterile cells and bearing a crown of 5 or 10 cells. Male gametangia (antheridia) are also protected (**Figure 3**) and according to their location on the same or on a different individual, as well as to their lifespan, four different types of life histories can be recognized for the species belonging to this group: monoecious annuals, monoecious perennials, dioecious annuals, and dioecious perennials. The oospores resulting from sexual reproduction may be dormant for long times and form dense seed bank on the bottom of ephemeral water bodies. Individuals are formed by an erect branch bearing whorls of

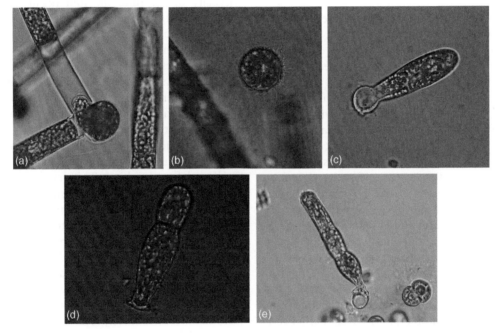

Figure 2   *Oedogonium* asexual reproduction. (a) zoospore release; (b) multiflagellate (stephanokont) zoospore; (c) germlings; (d) juvenile stage with holdfast; (e) young filament with rhizoids attached to a small centric diatom. (Magnification ×400).

Figure 3   *Chara* branchlets with gametangia: (a) magnification ×6 and (b) magnification ×12.

secondary branchlets, where reproductive structures can occur. Some of the species belonging to this group are lime-encrusted and have a 'crunchy' aspect which contributed to their common name of 'stoneworts.' *Chara* generally inhabits clear, hard waters or alkaline ponds (even temporary) and lakes where it can form meadows extending up to 25 m in depth in very clear environments. Occasionally, it may be found in the shallow of slow-flowing rivers. Another genus of this order, *Nitella*, is only slightly encrusted or not at all and may inhabit softer or even acid water bodies.

The Plankton

Planktic green algae include a large number of microscopic organisms, unicellular or colonial, adapted to spend part or all of their lives in apparent suspension in the open water of lakes, ponds, and rivers. The size spectrum covered by the different species of planktic green algae is comparable to that occurring between the trees of a forest and the grasses growing at their foot. In particular, the smallest green algae, forming in large part the so-called eukaryotic picoplankton, have dimensions of less than 2 μm and a spherical or elliptic shape. The species (e.g., *Choricystis minor*, *Pseudodictyosphaerium jurisii*) belonging to this dimensional group do not have swimming representatives and inhabit water bodies with a trophic status ranging between oligotrophy to hypertrophy. Nevertheless, they are most abundant in the less productive environments where they can contribute considerably (up to 60%) to primary production.

Larger green algae include both swimming, flagellated forms and nonmotile ones, whose existence is linked to the degree of entrainment into the circulation movements occurring in the water column (**Figure 4**). Actually, planktic green algae are generally denser than water and have the tendency to sink downwards. Only in one case (*Botryococcus*), they produce lipid droplets which decrease their relative density and make them floating upwards (**Figure 4(c)**). Thus, they are never isopycnic with the medium and the existence and level of abundance of single species largely depends on their own ability in traveling along the water column to exploit nutrients and harvest light. To balance their inability to control horizontal position or to swim against significant currents, planktic green algae have developed a wide variety of adaptive strategies. In addition, since they represent an important source of nutrition for the zooplankton, they also have to counteract the grazing pressure exerted by these 'herbivores.'

The adaptive strategies shown by planktic green algae act on multiple levels. The species involved differ significantly in their physiological abilities (i.e., in exploiting nutrients and light), as well as in their shape and size. The possibility to be grazed by zooplankton and the sinking velocity of planktic algae depend on the shape and size; sinking can be described by the Stokes' equation:

$$V_s = \frac{2\,gr^2(\rho' - \rho)}{9\eta\Phi}$$

where v_s is the sinking velocity (m s^{-1}), g is the gravitational acceleration (m s^{-2}), r is the radius of the sinking spherical particle (m), ρ' is the specific gravity of the sinking particle (kg m^{-3}), ρ is the specific gravity of the fluid medium (kg m^{-3}), and η is the viscosity of the fluid medium (kg m^{-1} s^{-1}). Φ is a species-specific form resistance factor by which the sinking velocity of the alga differs from that of a sphere of identical volume and density. Thus, the variety of shapes exhibited by green algae represents an evolutionary compromise directed both to an increased grazing resistance and to an optimization of sinking velocity. It seems important to underline that the modification of sinking velocity allowed by the form resistance factor is not a way to merely reduce sinking and in some cases it can increase the velocity itself and favor sinking. This has been observed to happen to desmids (e.g., *Staurastrum*, *Staurodesmus*, *Cosmarium*) inhabiting the epilimnia of tropical lakes where diurnal stratification (atelomixis) can occur. Actually, the higher water temperatures experienced by tropical lakes as compared to temperate ones decreases the density of the water. A faster sinking due to the lower density is counteracted by the hydrodynamic properties of such tropical lake where water column overturn takes place daily. In these environments, night-time convection is stronger than in temperate lakes, and the low resistance form helps these organisms in optimizing their traveling into the water column and in the exploitation of nutrient and light resources. Thus, optimizing sinking enhances the physiological adaptations of these algae in relation to the different physical and chemical conditions of the water column. Shapes and sizes displayed by planktic green algae are thus as different as the wide environmental variability showed by aquatic ecosystems. Moreover, very dynamic environments support a large diversity of different sizes and shapes of species either present in the assemblages or as propagules 'ready to develop' as the environmental template changes. Moderately eutrophic lakes and ponds very often show an amazingly high diversity of green algae. They may represent 50–90% of the species present in the phytoplankton assemblage and almost the whole of the phytoplankton biomass. A typical assemblage is generally constituted by one or a few dominating

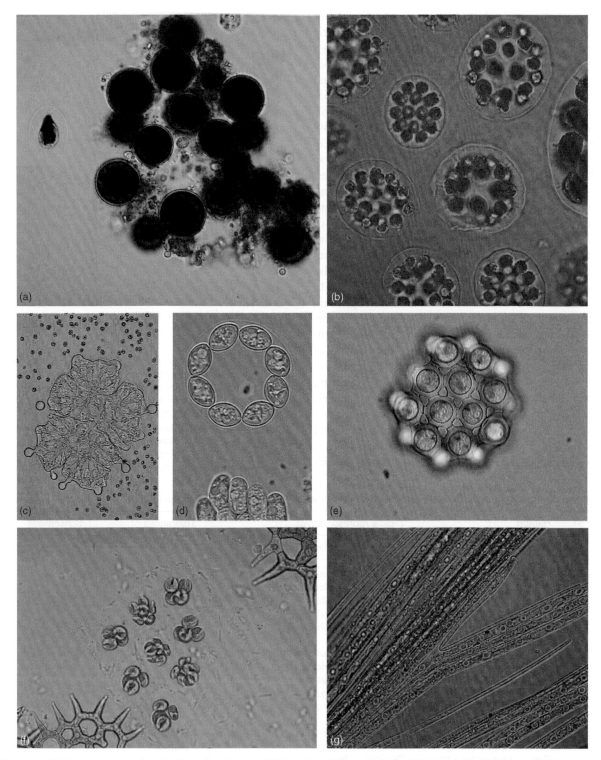

Figure 4 (a) *Haematococcus* (vegetative cell and cysts); (b) *Eudorina*; (c) *Botryococcus*; (d) *Oocystis*; (e) *Coelastrum*; (f) *Kirchneriella*; (g) *Closterium*. (Magnification ×400).

and codominating species and a procession of subdominating and rare species which, in some cases, can comprise several tens of taxonomic units. This specific richness, due to the inherent variability of the aquatic environments, allows the persistence of many rare species which represent 'the ecological memory' of the ecosystem.

The attention paid by plankton ecologists to the size and shapes of these algae has its origin in the need to categorize them not only on the basis of their phylogenetics, still subject to massive reappraisal, but also on the functional basis of their roles in aquatic ecosystems. It has been observed that algae forming a single functional group also have similar morphologies, as quantified by the dimensions of the algal 'units' (cells or colonies, as appropriate, together with any peripheral mucilage): surface area (s), volume (v), and maximal linear dimension (m) are powerful predictors of optimum dynamic performance. Thus, grouping the planktic green algae, as well as other phytoplankton, on the basis of these selected morphological traits allows to distinguish among habitats with a different accessibility to light and nutrient resources. By using this method, the morphological traits distribute the algae into a triangular portion of the Cartesian plane, whose vertices represent three different environmental conditions in relation to nutrients and light availability (**Figure 5**). In accordance with their different abilities in exploiting these resources, planktic algae are positioned on this plot and the three corners, named C, S, and R, summarize the three main strategies they may adopt. In particular, the corner C is occupied by the more invasive species, i.e., those investing in rapid growth; around the corner S, the more acquisitive species, i.e., those investing in resource conservation, are displaced; the corner R receives the more acclimating species, i.e., those investing in efficient light conversion. It is possible to verify that those species

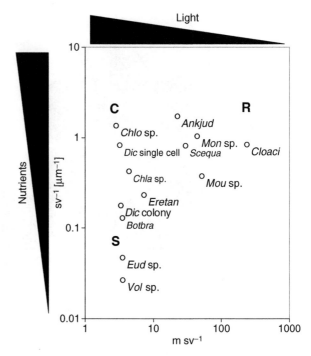

Figure 5 C–S–R strategic ordination of some selected green algae performed according to morphological traits. *Ankjud: Ankyra judayi; Botbra: Botryococcus braunii; Chlo sp.: Chlorella sp.; Chla sp.: Chlamydomonas sp.; Cloaci: Closterium aciculare; Dic: Dictyospaherium sp.; Eretan: Eremosphaera tanganyikae; Eud sp.: Eudorina sp., Mon sp.: Monoraphidium sp.; Mou sp.: Mougeotia sp.; Scequa: Scenedesmus quadricauda; Vol sp.: Volvox sp.; s: individual or colony surface; v: individual or colony volume; m: individual or colony maximal linear dimension.*

Table 1 Examples of functional groups of some representative planktic green algae along with their tolerances and sensitivities

Codon	Habitat	Typical representatives	Tolerances	Sensitivities
F	Clear epilimnia	Colonial non-motile greens e.g., *Botryococcus, Kirchneriella, Pseudosphaerocystis, Radiococcus, Coenochloris, Oocystis lacustris, Gemellicystis, Gloeocystis, Paulschulzia*	Low nutrients high turbidity	?CO_2 deficiency
G	Short, nutrient-rich water columns	*Eudorina, Pandorina, Volvox*	High light	Nutrient deficiency
J	Shallow, enriched lakes, ponds and rivers	*Actinastrum, Ankistrodesmus, Crucigenia, Lagerheimia, Tetrastrum, Tetraedron, Pediastrum, Coelastrum, Scenedesmus, Golenkinia, Oocystis borgei*		Settling into low light
N	Mesotrophic epilimnia	*Cosmarium, Staurodesmus*	Nutrient deficiency	Stratification, pH rise
P	Eutrophic epilimnia	*Closterium aciculare, Staurastrum pingue*	Mild light and Carbon deficiency	Stratification, Si depletion
T	Deep, well-mixed epilimnia	*Geminella, Mougeotia*	Light deficiency	Nutrient deficiency
X1	Shallow mixed layers in enriched conditions	*Chlorella, Ankyra, Monoraphidium*	Stratification	Nutrient deficiency, filter feeding
X3	Shallow, clear, mixed layers	Picoplankton	Low base status	Mixing, grazing
W1	Small organic ponds	*Gonium pectorale*	High BOD	Grazing

Source

Reynolds CS, Huszar V, Kruk C, Naselli-Flores L, and Melo S (2002) Towards a functional classification of the freshwater phytoplankton. *Journal of Plankton Research* 24: 417–428.

characterized by elongated, needle shape (e.g., *Closterium aciculare*) will be positioned by their morphological traits in the vicinity of the R corner; those showing spherical shape and small dimensions (e.g., picoplankton) will tend to accumulate in the vicinity of the C corner and those with a larger spherical (e.g., *Pseudosphaerocystis* spp. and *Botryococcus* spp.) or slightly ellipsoidal shape (e.g., *Eremosphaera tanganykae*) will concentrate around the S corner. Grouping green algae according to their ecological performance is the first step to attempt a functional classification of these organisms. One of the most successful functional classification, addressed toward an ordering of all phytoplankton taxa, distributed green algae in nine ecological groups (**Table 1**). This ordination well approximates the results obtained by dividing the algae according to their morphological descriptors.

It has been observed that planktic green algae may show a high degree of phenotypic plasticity (**Figure 6**). This may be regarded as an attempt performed by a population to counteract adverse environmental conditions or, more correctly, as a mechanism that tend to select those specimens bearing morphological characteristics that best fit the environmental template. Common examples are those offered by green algae typically forming colonies or coenobia (e.g., *Scenedesmus, Coelastrum, Dictyosphaerium, Micractinium*) that under high nutrient availability and low grazing pressure start growing as unicells till forming dense populations of 'small green balls' that may create problems in their correct taxonomic identification. Other environmental constraints, as grazing pressure, may produce morphological responses: an example can be spine formation in *Scenedesmus* after exposition to infochemicals released

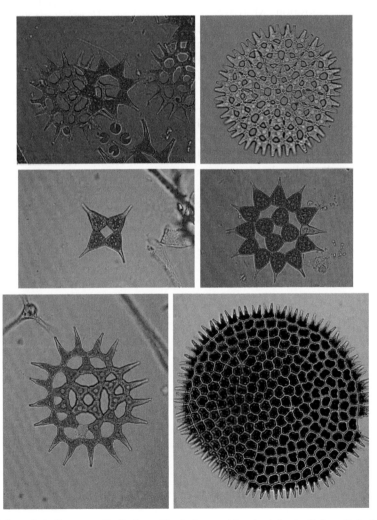

Figure 6 Morphological diversity in the genus *Pediastrum*. The last coenobium is stained with Lugol's iodine solution. (Magnification ×400).

by daphnids. Spines increase the greatest axial linear dimension (GALD) of these organisms and thus reduce the probability to be eaten by planktic herbivores. An analogous mechanism is adopted by *Eremosphaera tanganyikae*, a green alga typical of African Great Lakes. *Eremosphaera* being a big, nonmotile chlorococcalean species, can sustain abundant populations only in the season with a deep mixing. During wet season, larger specimens are gradually selected by copepods' grazing pressure on smaller ones. Thus, the final effect can be seen as an increase in size performed by this species to minimize grazing pressure. A further mechanism adopted by planktic green algae to modify their shape and size involves the production of mucilage that may bind both unicells and colonies. Mucilage is formed by hygroscopic lattice-like polymers of carbohydrate and substances resembling acrylic, analogous to EPS produced by periphytic algae, and seems to have several functions such as density reduction, nutrient sequestration, and processing, as well as a source of defense against grazing and digestion, and heavy metal exposition. Moreover, it may maintain a reducing microenvironment around the cells. Although the reasons of mucilage production are still not completely understood, it is apparent that the thickness of this mucilaginous investment is itself variable and responsive to environment.

Green algal assemblages, both benthic and planktic, well integrate physical, chemical, and biological parameters of freshwater ecosystems. Actually, they contain a high number of species, making data useful for statistical and numerical applications to assess water quality. Their response time is rapid, as is recovery time, with recolonization after a disturbance often more rapid than for other organisms. They thus could provide a powerful monitoring tool, once the processes governing their relative importance are fully understood.

See also: Algae; Comparative Primary Production.

Further Reading

Hepperle D and Krienitz L (2001) Systematics and ecology of chlorophyte picoplankton in German inland waters along a nutrient gradient. *International Review of Hydrobiology* 86: 269–284.

Lewis LA and McCourt RM (2005) Green algae and the origin of land plants. *American Journal of Botany* 91: 1535–1556.

Luo W, Krienitz L, Pflugmacher S, and Walz N (2005) Genus and species concept in *Chlorella* and *Micractinium* (Chlorphyta, Chlorellaceae): Genotype versus phenotypical variability under ecosystem conditions. *Internationale Vereinigung für Theoretische und Angewandte Limnologie: Verhandlungen* 29: 170–173.

Padisák J (1992) Seasonal succession of phytoplankton in a large shallow lake (Balaton, Hungary) – A dynamic approach to ecological memory, its possible role and mechanisms. *Journal of Ecology* 80: 217–230.

Padisák J, Soróczki-Pintér É, and Rezner Z (2003) Sinking properties of some phytoplankton shapes and the relation of form resistance to morphological diversity of plankton – An experimental study. *Hydrobiologia* 500: 243–257.

Reynolds CS (1997) *Vegetation Processes in the Pelagic: A Model for Ecosystem Theory.* Oldendorf/Luhe: Ecology Institute.

Reynolds CS (2006) *The Ecology of Phytoplankton.* Cambridge: Cambridge University Press.

Reynolds CS (2007) Variability in the provision and function of mucilage in phytoplankton: Facultative responses to the environment. *Hydrobiologia* 578: 37–45.

Reynolds CS, Huszar V, Kruk C, Naselli-Flores L, and Melo S (2002) Towards a functional classification of the freshwater phytoplankton. *Journal of Plankton Research* 24: 417–428.

Sabater S, Gregory SV, and Sedell JR (1998) Community dynamics and metabolism of benthic algae colonizing wood and rock substrata in a forest stream. *Journal of Phycology* 34: 561–567.

Wehr JD and Sheath RG (eds.) (2003) Freshwater Algae of North America. Ecology and Classification. Amsterdam: Academic Press.

Wetzel RG (2001) *Limnology. Lake and River Ecosystems,* 3rd edn. Amsterdam: Academic Press.

Other Phytoflagellates and Groups of Lesser Importance

N Salmaso and M Tolotti, IASMA Research Center, S. Michele all'Adige, (Trento), Italy

Introduction

The dominant algae capable of colonizing a large diversity of freshwater environments belong to the Cyanoprokaryota, Chlorophyta, Bacillariophyceae (diatoms), and Chrysophyceae. However, under specific environmental conditions, or when the competition of the other algae lessen, other phytoflagellates may become important constituents of the algal communities, often developing with high biomasses. This is the case of the Dinophyta, Cryptophyta, and Euglenophyta. Other species of lesser importance in the structuring of the algal communities, but having notable implications in the maintaining of biodiversity and exploitations of specific niches include the Haptophyta, Xanthophyceae, Eustigmatophyceae, and Raphidophyceae.

Dinophyta

The phylum Dinophyta (dinoflagellates, dinophytes) is composed of a diverse assemblage of unicellular organisms. Most of them are biflagellated and free-swimming (e.g., among the Peridiniales, *Ceratium*, *Peridinium*, *Gymnodinium*) (**Figure 1**), whereas a few are nonmotile unicells reproducing by motile specialized cells (zoospores and gametes) with features such that to permit their inclusion among the Dinophyta. Nonmotile cells (Dinococcales) are free-floating (e.g., *Cystodinium*), or attached either with stalks (e.g., *Tetradinium*) or directly (*Cystodinedria*) to a substrate.

All the dinoflagellates have a multilayered covering (the amphiesma). Traditionally, species belonging to this division have been separated into thecate (or armored) and naked (athecate or unarmored) forms. Thecate species have cellulose plates arranged underneath the outer membrane of the amphiesma. The arrangement of plates is species-specific and is used to classify the armored forms. Athecate cells may be active swimmers (gymnodinoid forms) or nonmotile.

Typical thecate and athecate motile cells – which represent, with the Peridiniales, the most prevalent vegetative forms in pelagic environments – have transverse (cingulum) and longitudinal (sulcus) furrows connecting and hosting two ventrally inserted flagella (**Figure 2**).

In motile dinoflagellates, life cycles include an alternation of a growing, assimilative (vegetative) cell with a nonmotile resting cyst stage. Vegetative cells are haploid and reproduce asexually by mitosis to produce other daughter cells, which will produce later new cell walls and, in the thecate forms, new thecae. Life cycles described in a number of dinoflagellates reveal a diverse spectrum of sexual modes of reproduction. Vegetative cells undergo mitosis to produce isogametes (gametes with the same size and shape, as in *Scripsiella*) or anisogametes (gametes of different size, as in *Ceratium*). Isogamic and anisogamic species may be homothallic or heterothallic (requiring only one or two different clones, respectively, for sexual reproduction). In some taxa, the triggering of sexual reproduction may be induced in culture by nitrogen deficiency or by changing the temperature. The fusion of the gametes originates a motile zygote (planozygote) characterized by two longitudinal flagella and two or one transverse flagellum. In several species the planozygote presumably may represent the only zygotic stage undergoing eventually meiosis. In few other species, the planozygote may remain motile for a variable period of growth (even for weeks) before losing flagella and transforming into a nonmotile zygote (hypnozygote), which finally settles to the sediments (resting cysts) (**Figure 3**). Planozygotes may form again following germination of nonmotile hypnozygotes. Cysts appear to require an obligatory dormancy period (from days to several months) before they are able to establish new populations. Apparently, the excystment is controlled by a complex set of factors (temperature, internal biological clocks, oxygen levels). Cysts with thick membranes and abundant food reserves are also produced by asexual reproduction (temporary cysts). Explicit evidences of cyst formation by sexual reproduction (zygotes) have been easily and extensively obtained in laboratory conditions. However, more field research needs to be carried out to verify the fraction of cysts obtained by sexual reproduction.

Nonmotile taxa have multifaceted, heteromorphic life cycles that may include asexual reproduction in cysts, parasitic or photosynthetic stages, amoeboid stages, gymnodinoid swarmers that can also behave like gametes.

Dinoflagellates may rely on different nutrition modes including autotrophy, auxotrophy (require exogenous vitamins), mixotrophy (a combination of autotrophy and heterotrophy), and organotrophy (heterotrophy in organisms without chloroplasts, including phagotrophy, i.e., ingestion of particles). It is noteworthy that many marine and freshwater dinoflagellates have no chloroplasts, living exclusively on organic compounds (e.g., the freshwater naked species *Gymnodinium helveticum*). Even among the armored species, rapid progresses in the clarification of feeding strategies have demonstrated that the theca does not represent an insurmountable or even a significant barrier to phagotrophy.

Dinoflagellates are typical constituents of the algal communities in marine and brackish waters. Freshwater members of this group are less common both quantitatively and as number of species. Typical representatives of pelagic environments include various motile forms (Peridiniales); nonmotile members (Dinococcales), which are represented only by a few genera, may be found in small water bodies such as bogs and peat pools. However, owing to the insufficient number of specific taxonomic works, complete information about the geographic distribution of many species is lacking. In this section, a brief account of the ecological properties of the freshwater motile species will be given.

Though present generally in low numbers, a few species are sometimes able to develop with large biomasses, such that to discolor the standing waters. Most of the common motile dinoflagellates show defined seasonal patterns, being present as benthic stages and with rare or no representatives during the winter months in the temperate zones. Benthic resting cysts may provide an important pool of inocula for the development of dinoflagellates at the beginning of the growing season. In this case, early stages of population growth can benefit from higher concentrations of nutrients at the mud–water interface. In several lakes (e.g., lakes Garda, Balaton, and Esthwaite Water), the growth of *Ceratium hirundinella* begins in spring, followed by exponential growth until July and rapid decline in mid autumn (**Figure 4**). The development of populations appears correlated with the warming of lake water and, in the deeper lakes, with the timing and duration of the mixing period.

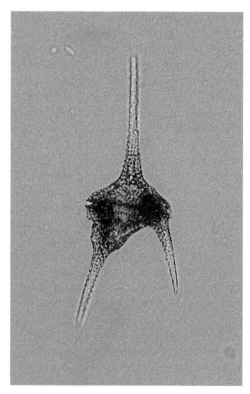

Figure 1 *C. hirundinella*. Cell length: 190 µm. A long flagellum is visible near the long horn.

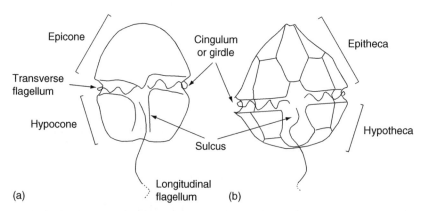

Figure 2 Typical (a) athecate (e.g., *Gymnodinium*) and (b) thecate (e.g., *Peridinium*) dinoflagellate cells in ventral view. The cingulum encircles the cell and divides it into an anterior (epitheca) and a posterior (hypotheca) portion (epicone and hypocone, respectively, in unarmored species). The transverse flagellum, which provide a whirling movement, extends to a minimum distance of two-third the cell's perimeter and remains always positioned in the furrow as the cell is actively moving in the water. The sulcus extends into the hypotheca/hypocone (sometimes into the epitheca/epicone) and hosts the propelling flagellum.

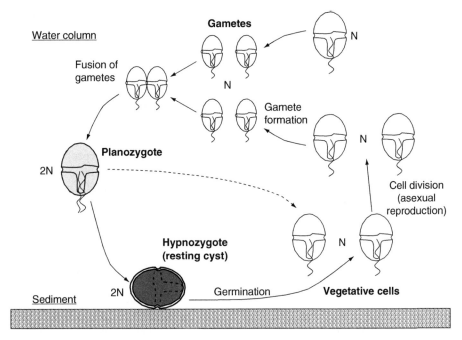

Figure 3 A generalized and simplified life cycle of motile, photosynthetic dinoflagellates. N and 2N: haploid and diploid stages.

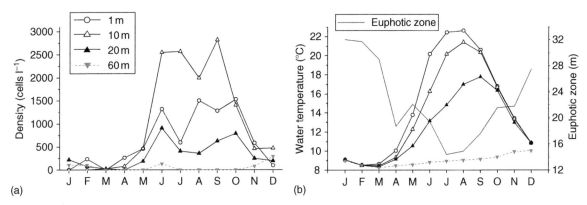

Figure 4 (a) Example of seasonal dynamics of *C. hirundinella* in a deep (z_{max} = 350 m) temperate lake (Lake Garda, Northern Italy; 45° 40′ N, 10° 42′ E). (b) The maximum abundance and vertical segregation of *Ceratium* are correlated with the water temperature and the timing and duration of the mixing period, and the extent of the euphotic layer (monthly averages for the years from 1998 to 2005).

Studies carried out since the 1960s on *Peridinium gatunense* (formerly identified as *P. cinctum*) in the Lake Kinneret showed that in the subtropical warm belt benthic stages could be produced in summer to withstand severe conditions, leading to a strong decrease of the vegetative stages. Young cells of *P. gatunense* emerging from cysts were documented along the littorals from November (around 20 °C) to February (around 14 °C); this phase was followed by exponential growth until March–April and drastic decrease as of June. This stable and predictable annual pattern was lost since the mid 1990s, when absence of the prevailing spring *Peridinium* blooms in some years was observed, in parallel with the appearance in summer and autumn of nitrogen-fixing cyanobacteria and unusual winter blooms of *Microcystis aeruginosa*. These modifications took place during a period of continuous reduction of the lake's water level because of shortage of water in Israel (−0.7 m year⁻¹ between 1993 and 2002).

Large dinoflagellates generally have low specific growth rates and long generation times compared with smaller nanoplankton. Doubling times *in situ* typically range from 2–3 to more than 30 days (corresponding to specific growth rates between <0.02 and 0.2–0.3 ln units day⁻¹). However, low growth rates, which may represent a negative competitive factor, can be balanced by competitive advantages, making them able to establish blooms in some lakes. Dinoflagellates can move vertically within the water column up to several meters daily to contrast sinking and in response to shifting light and nutrient concentrations. Nonmotile algae may experience severe survival problems in thermally stratified and stable pelagic environments because of susceptibility to sinking losses, which may be particularly high in heavy silicified diatom cells. During thermal stratification, cells concentrating near the bottom of the euphotic zone may contain higher concentrations of phosphorus than those located at the surface. In this respect, dinoflagellates may take advantage from luxury consumption of phosphorus. For example, when ambient phosphorus is in excess, *Peridinium* is able to accumulate it in internal pool of polyphosphates. Cells of *Peridinium* rich in internal stores of phosphorus have the capacity to divide up to three to four times in media deficient in this element. The diverse heterotrophic capabilities of dinoflagellates may provide alternative ways to obtain nutrients. Moreover, in larger species, further competitive ability is obtained from higher resistance to grazing. These characteristics may contribute to explain why many dinoflagellates are able to grow even when nutrient concentrations reach their seasonal minima. More specifically, the largest slow growing species (e.g., *Ceratium, Peridinium*) are able to develop substantial biomass because loss rates (either due to grazing or sinking) also are moderate.

Marine dinoflagellates are able to form localized 'red tides', which can sometimes cause, with the production of specific toxins, various human illnesses. Though less frequent, dense aggregations of dinoflagellates are also known in freshwater habitats. One of the best dinoflagellate bloom former is *C. hirundinella*, which can develop in huge quantities in small or leeward basins. In such occasions, high densities (up to several thousands cells per milliliter) are associated with marked brown water coloration and, often, with unpleasant smell. Summer blooms causing extended reddening of the surface waters due to huge development of *Tovellia sanguinea* (*Glenodinium sanguineum*) were reported until the first half of the 1960s in Lake Tovel (Northeast Italian Alps).

Cryptophyta

Cryptophytes include both pigmented and colorless species. Colored and photosynthetic species have one or two chloroplasts; in this group, common genera are represented by *Cryptomonas, Plagioselmis* (**Figures 5** and **6**); and *Chroomonas*; the genus *Rhodomonas* has been eliminated because it was considered later an unnecessary name for *Pyrenomonas* (the two generic names are synonyms). Colorless

species include *Chilomonas* (possess a reduced chloroplast without pigments, the leucoplast) and *Goniomonas* (lacks plastids and a nucleomorph). The nucleomorph, a vestigial nucleus originating from a secondary endosymbiosis event, is a distinctive structure located in the periplastidial compartment between the two outer and the two inner plastid membranes. A further group of colorless and heterotrophic flagellates that was historically incorporated in the cryptophytes includes the genera *Katablepharis* and *Leucocryptos* ('katablepharids,' both with marine and freshwater representatives). These genera, except for the presence of ejectosomes and the site of flagellar insertion, do not possess other diacritic structural features that can justify their inclusion in the phylum Cryptophyta. At present, the taxonomic position of katablepharids is not precisely determined.

The first evidence of dimorphism in cryptophytes was reported for the marine species *Proteomonas sulcata*. Recent morphological and molecular studies have also provided evidence for the presence of dimorphism in the freshwater *Cryptomonas*, which, in some strains, include *Campylomonas*-like cells in single-cells-derived cultures, advocating for a synonymization of the two genera and suggesting that dimorphism in the cryptophytes may be more widespread than is presently envisaged. More in general, the systematics of cryptophytes has experienced many reorganizations in the last 20 years and other modifications are expected as a result of the application of new phylogenetic information obtained from molecular sequencing studies.

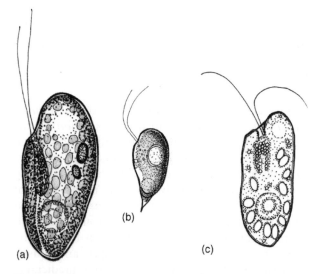

Figure 5 Examples of photosynthetic ((a): *Cryptomonas obovata* Skuja, cell length *c.*30 μm; (b): *Plagioselmis nannoplanctica* (Skuja) Novarino, cell length *c.*8 μm) and heterotrophic ((c): *Chilomonas paramecium* Ehrenberg, cell length 25 μm) cryptomonads.

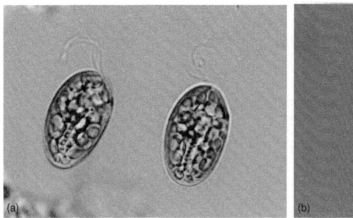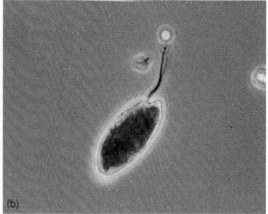

Figure 6   (a) *Cryptomonas* (living cells, 35 μm). (b) *Cryptomonas* cf. *reflexa* (cell length 30 μm, fixed with Lugol's solution).

The primary method of reproduction of cryptophytes is by longitudinal cell division. Sexual cycles have been documented only in a few cases (e.g., *Chroomonas* and the marine *Proteomonas*). Growth rates in culture conditions are high, even for the large species. *Cryptomonas erosa* and *C. ovata* are known to double their populations in less than a day; higher rates should be expected for species with smaller body sizes. Though cryptophytes are usually free-swimming, some species are able to produce palmelloid cells (single or irregular colonies) and resting stages (cysts) to survive unfavorable environmental conditions. The cells of *Cryptomonas* may assume a palmelloid stage becoming nonmotile and living in bundles kept together by mucilage.

Members of the cryptomonads have colonized almost any marine and freshwater habitats, from the arctic regions to the tropics. The largest diversity is reached in lakes of the temperate regions, where these organisms are found under very different environmental conditions. Contrary to many other phytoplankters, which may show a clear seasonality in response to specific factors (e.g., the spring bloom of the large diatoms during high water turbulence and nutrient availability in temperate lakes), cryptomonads are characterized by a permanent presence throughout the year. Populations present in low densities commonly increase intermittently, often following the decline of other previous dominant phytoplankters or following periods of water turbulence, when they are circulated with deeper waters rich in nutrients and when grazing may be reduced. This opportunistic behavior is favored by their r-selected behavior, i.e., the ability to respond quickly to the availability of environmental resources, taking advantage from potential high growth rates. Cryptomonads – and flagellates, except the biggest

ones, in general – are good or optimal food for herbivorous ciliates (e.g., *Coleps* and *Prorodon*), rotifers (e.g., *Keratella* and *Polyarthra*), and microcrustaceans (e.g., *Eudiaptomus* and *Daphnia*). Possibly, grazing has an important effect on the densities of the planktonic cryptomonads, and this could contribute to explain the generally low densities of cryptophytes in most pelagic environments.

In many lakes of different trophic status, photosynthetic cryptophytes may reach their maximum densities well below the surface: at depth where the balance between the nutrient availability and light required for sustaining photosynthesis provides growth rates able to balance losses due to zooplankton grazing. This behavior is favored by the presence of accessory pigments (phycobiliproteins) in the cells, which render them capable of photoadaptation and high efficiency light harvesting in poorly illuminated environments. Yet, even in this case, the deep algal maxima, when located around the metalimnion or at the bottom of the euphotic zone, and when subjected to high grazing pressures in oxygenic environments, are generally dominated by larger organisms, such as dinophytes, cyanobacteria, and chlorophytes. Deep chlorophyll maxima (DCM) due to cryptophyte populations belonging to *Cryptomonas* (including *C. phaseolus*, *C. undulata*, *C. rostratiformis*, *C. erosa*) are known to form in eutrophic or moderately productive lakes, developing strong physical and chemical gradients. In these environments, cryptophyte maxima of several thousands cells per milliliter localize near the oxic–anoxic boundary of the water column. Some of the species perform diurnal cycles, involving a downward migration into sulfide-rich waters during the night, and an upward migration in the morning toward oxygenated and sulfide-free waters. The amplitude of the vertical movements varies among the species and with the strength of the environmental gradients. Diel vertical migrations into the sulfide-rich chemocline could constitute a mechanism to reduce grazing and exploit waters rich in nutrients, though other factors such as light gradients and endogenous clocks could also be implicated. In these examples, the cryptophytes are entirely phototrophic, sustaining growth by photosynthetic adaptations to low light intensities rather than to phagotrophy. However, extreme environmental conditions, such as those met in the perennially ice-covered lakes of Antarctica, which are characterized by low levels of photosynthetically available radiation (PAR), can induce mixotrophy among photosynthetic cryptophytes. During the winter periods, the growth of these phytoflagellates in total darkness can only be explained by heterotrophy (ingestion of bacteria).

As in marine waters, cryptophytes are known to form occasional nuisance blooms in rivers and lakes, causing also taste and odor problems in the use of drinking waters.

Euglenophyta

Most of the species belonging to this group are colorless and nonphotosynthetic flagellated protozoans. Members of the heterotrophic euglenoids are bacterivorous (e.g., *Petalomonas* and *Ploeotia*) or capable of ingesting eukaryotic prey (e.g., *Dinema*, *Peranema*). Other heterotrophic euglenoids lacking the ability to phagocytosis are osmotrophs, acquiring nutrients by absorbing organic molecules from the environment (e.g., *Astasia*, *Hyalophacus*). However, the most known euglenoids in field samples are green autotrophs (e.g., *Euglena*, *Phacus*, *Trachelomonas*) (**Figures 7** and **8**). More precisely, green euglenoids are photoautotroph requiring, besides light and nutrients, one or more vitamins to synthesize organic matter. Some of the green species (e.g., *Euglena*, *Colacium*) are facultatively heterotrophic, i.e., they are able to grow in the dark if organic carbon is provided; other species (*Euglena gracilis*) are able to synthesize vitamin E (α-tocopherol). The present section will focus on photosynthetic genera.

The euglenoids have a rich variety of forms, from spindle-shaped (*Euglena*) to ovoid (*Trachelomonas*) or leaf-like (*Phacus*). Some genera (e.g., *Euglena*, *Colacium*) are able to change cell shape (metaboly), a property not shared by euglenoids possessing a rigid cell surface (e.g., *Phacus*). Other species possess a mucilaginous covering, the lorica, which, in some organisms (*Trachelomonas*), may be equipped with spines. A noticeable feature of pigmented euglenoids is the eyespot, an orange to red anterior stigma always located outside chloroplasts; the eyespot is

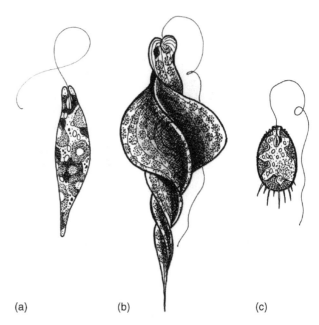

(a)   (b)   (c)

Figure 7 Examples of photosynthetic euglenophyta: (a) *Euglena gracilis* Klebs, cell length 50 μm; (b) *Phacus helicoides* Pochman, cell length 100 μm; (c) *Trachelomonas armata* var. *longispina* (Playfair) Deflandre, lorica length 40 μm.

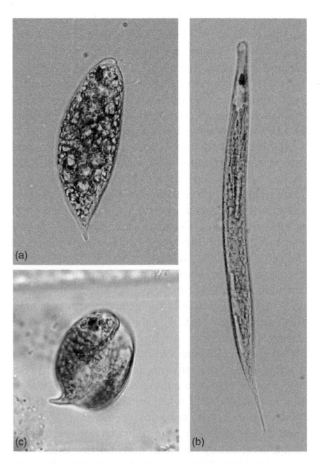

Figure 8 (a) *Euglena* sp. (cell length 70 μm); (b) *Euglena acus* (cell length 175 μm); (c) *Phacus* sp. (cell length 35 μm).

sometimes lacking in *Phacus*. The euglenoids are able to live in free open waters or attach themselves to surfaces. For example, species belonging to *Euglena* are pelagic or may form scums on the air–water interface and green layers above the sediments. Members of the genus *Colacium* have a free-swimming phase and a nonmotile phase, attaching to the exoskeleton of aquatic invertebrates. Some organisms have been proved to form cysts appearing in soil or atmospheric samples. Some genera (*Euglena*, *Colacium*) are known to form palmelloid stages. In this condition, the cells of *Euglena* loose their flagella, became nonmotile and covered with mucilage, and divide actively before becoming flagellated again.

Photosynthetic green euglenoids are often dominant in waters rich in dissolved organic matter (DOM). In these environments, the requirement of vitamins (e.g., B1, thiamine and B12, cyanocobalamin in some species of *Euglena*) by photoautotrophs may be supported by the activity of bacteria utilizing DOM. When conditions are favorable, the euglenoids can develop with large biomasses in open waters or in muds and sediments of eutrophic and hypertrophic lakes and reservoirs, ponds, and ditches exposed to animal excrement. Large bright green masses may also develop near heaps of dung. Numerous studies have documented a marked presence of euglenoids (e.g., *Euglena*, *Lepocinclis*) in sewage disposal sites and wastewater ponds; on the other side, striking decreases in the contribution of this group have been observed after sewage diversion from shallow lakes and ponds. Members of this group (*Lepocinclis*, *E. mutabilis*) are known to develop also at critically high acidity levels (pH < 3) in the microphytobenthos and plankton of lakes residuals of surface mining.

Some species can be bright red because of a pigment called the astaxanthin (e.g., *E. sanguinea*). With warm water and high solar irradiation, these species can develop in large numbers, forming red scums on the water surface.

Haptophyta

Most of these yellow-green to gold-brown heterokontophytes are euplanktic flagellates, although some coccal, palmelloid, and even filamentous benthic species are known. Some species (e.g., the marine genus *Hymenomonas*) show alternating nonmotile haploid and flagellated diploid stages in their life cycle. Asexual reproduction occurs by binary cell division, while mechanisms of sexual reproduction, observed in several species, remain unclear.

Cells normally bear two smooth equal or subequal flagella and a third filament (haptonema) arising close to the flagella, which moves autonomously and can be coiled and uncoiled according to various physical and chemical stimulus (as in *Chrysochromulina* species). The function of haptonema is still poorly known, although it is clear that it does not serve for cell movement. It seems to be involved in the heterotrophic nutrition (phagotrophy) observed in several marine species.

Haptophytes are covered with several layers of tiny polysaccharide scales, detectable by electron microscopy. In the large almost exclusively marine group coccolithophorids, organic scales are covered by an additional layer of calcified scales (coccoliths) formed by regularly and specifically arranged fibers. Present bloom-producing marine coccolithophorids are among the most important primary producers in oceanic environments. After cell death, coccolithophorids sink to the sea bottom, forming thick mud layers. Since skeletal remains of these algae are the major component of chalk deposits originated during successive explosion stages during the Earth's history (e.g., in Jurassic and Cretaceous), these algae have great stratigraphic value and represent the best known haptophytes. Only one freshwater coccolithophorid has been reported (*Hymenomonas roseola*) so far.

Although haptophytes are more common in marine waters, a few species are widespread in fresh waters. *Chrysochromulina parva* (**Figure 9**) is the most common species in lakes, ponds, and rivers of temperate regions. Higher densities (up to several thousands of cells per milliliter) are usually recorded in relatively nutrient rich waters, although this species represents a very common element of phytoplankton even in oligotrophic lakes.

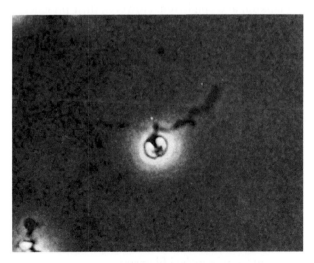

Figure 9 *Chrysochromulina parva* Lackey (cell diameter *c*.6 µm, fixed in Lugol's solution). One flagellum and the uncoiled haptonema (on the right) are visible.

An economically important aspect is represented by the toxin production by some species of Haptophytes. *Prymnesium*, an euryhaline genus, has been intensively studied because of its production of extracellular hemolytic toxins (prymnesins), which are responsible, through damaging the epithelial cells of the fish gills, for fish-killing in coastal lagoons, estuaries, and high salinity aquaculture ponds. In addition, *Prymnesium* toxins induce cell lyses in many aquatic protist species. Recent studies presented evidences that toxin excretion in *P. parvum* could represent a means to immobilize and ingest small motile preys. Though massive blooms of marine *Chrysochromulina* species (e.g., *C. polylepis*) are well known to induce extensive mortalities among marine biota, toxic episodes associated with production of toxins by freshwater species need to be analyzed thoroughly with further studies and observations. Expression of toxicity is still poorly known but laboratory studies suggest that it may be controlled by growth conditions, such as phosphorus deficiency.

Xanthophyceae

Xanthophyceae are in general less frequent in marine than in freshwater environment, although several benthic species are common in brackish waters. They occur often under dystrophic or mesotrophic conditions, showing their highest diversity in acidic waters enriched with dissolved organic matter.

Xanthophytes include a variety of cell organizations. Single flagellated cells (e.g., *Chloromeson*, **Figure 10 (a)**) are relatively rare, except in brackish waters, where they are usually metaphytic or epiphytic among submersed vascular plants. Solitary and coenobial coccal types are much more common in freshwater euplankton and epiphyton. However, their presence is often overlooked since they are easily mistaken for coccal green algae because of the similar morphology and their yellow-greenish color, which is due to the lack of fucoxanthin, an accessory pigment typical of other heterokontophytes. Sessile coccal genera, e.g., *Ophiocytium* (**Figure 10(b)**) and *Mischococcus*, live attached to other algae, submerged macrophytes, and detritus in dystrophic and oligotrophic small water bodies.

A trichal xanthophyte common in freshwater plankton is *Tribonema* (**Figure 10(c)**), which forms straight filaments where cells are held together by overlapping H-shaped cell walls. Several epiphytic species prefer cold, nutrient rich waters, where they can be found on reeds and other vegetal substrate especially in spring and autumn. Other trichal xanthophytes (i.e., *Heterodendron*) form branched filaments.

The most common and best known xanthophytes have siphonal organization, characterized by

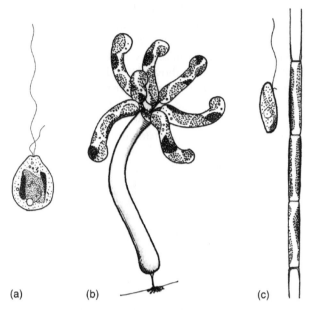

(a)   (b)   (c)

Figure 10 Examples of xanthophyte organization. (a) monadoid *Chloromeson agile* Pascher, cell length 10 μm; (b) First generation empty cell of coccal *Ophiocytium ilkae* (Istvánffi) Heering, attached to the substratum by a short stalk with a group of daughter cells (length *c*.40 μm) originated by germination of zoospores liberated by the cap-like top of the mother cell; (c) unbranched filament *Tribonema angustissimum* Pascher, cell length 15 μm, with a zoospore; H-shaped cells are clearly visible at the ends of the filament.

globose (*Botrydium*) or tubular (*Vaucheria*) thalli, where numerous nuclei and chloroplasts are dispersed in the protoplast and not separated by cell walls. Ecology of both genera is very well known because of their peculiar morphology and physiology. *Botrydium* is a very common epipelic alga living on humid mud of lakes and pond shores, where it can produce very large clusters. Its single-celled, multinucleate vesicles are attached to the substratum by a sort of rhizoid apparatus entering the mud. The thallus of *Vaucheria* shows a zonation of nuclei, chloroplasts, and vesicles near the filament apex and a peculiar light adaptation, which consists in changing chloroplast distribution and orientation according to changing light intensity and spectral composition. Chloroplast displacement occurs by protoplast currents and is aimed at optimizing light harvesting and reducing light induced pigment damages. Several *Vaucheria* species live in brackish and sea waters, while others are very common on damped ground and in well-aerated ditches, pools, and slow-flowing small streams, where they can form mats and cushions.

For the majority of Xanthophyceae only vegetative reproduction is known, which can occur by vegetative cell division, filament fragmentation, nonmotile spores (aplanospores) or flagellated spores (zoospores). Although free swimming xanthophytes are very few, reproduction by flagellated spores is very common in the whole class. Several taxa (e.g., *Tribonema* and *Botrydium*) can also produce akinetes, modified vegetative cells with thickened cell wall that can survive adverse environmental conditions.

Sexual reproduction is less common and has been reported only in *Tribonema*, where isogamous union of gametes produces a dormant zygote, *Botrydium*, with isogamous or anisogamous gametes, and *Vaucheria*, which presents a very characteristic oogamy (**Figure 11**), with male and female reproductive structures arising near the apex of vegetative filaments on monoecious or dioecious thalli.

Eustigmatophyceae

Eustigmatophytes represent a small group of poorly known heterokonts, which have been moved from Xanthophyceae in the 1970s because of the evidence of some exclusive characteristic, such as the presence of chlorophyll *a* only and of violaxanthin as major accessory and harvesting pigment. Moreover, motile cells (with one or two flagella) bear an eyespot, which is formed by a cluster of carotenoid granules close to the chloroplast but never enveloped by membranes. Such a stigma has a parallel only in the euglenophytes. Polyhedral pyrenoids are present only in nonmotile cells. Only asexual reproduction, by autospores or naked zoospores, has been observed.

Vegetative eustigmatophytes are typically spherical, elliptical (*Ellipsoidion*), or irregular (*Monodopsis*) coccoid cells, single or grouped in regular aggregates enveloped in layered mucilages, as in *Chlorobotrys* (**Figure 12**). They have sparse and limited distribution in euplankton and tichoplankton of lakes, ponds, pools, bogs, and streams and they seem to prefer dystrophic and oligotrophic cold waters (*Pseudotetraëdriella kamillae*). However, their diffusion might be underestimated because of both their small dimensions (2–32 μm) and their previous classification as xanthophytes or even Chlorococcales, as is the case of the sessile *Pseudocharaciopsis*.

Raphidophyceae

These small groups of large (30–100 μm), dorsiventral and naked heterokonts bearing two unequal flagella arising from an apical pit show a separation of the cytoplasm into a central, dense endoplasm containing mitochondria and a large nucleus, and a superficial ectoplasm containing numerous discoid yellow-green chloroplasts, vacuoles, and trichocysts. Autotrophic marine and freshwater taxa contain accessory pigments similar to those of chrysophytes and xanthophytes, respectively. Saprophytic and phagotrophic raphidophytes are also known, such as the marine genus *Reckertia*. Asexual reproduction occurs by longitudinal cell division, while sexual reproduction has been revealed and investigated only very recently.

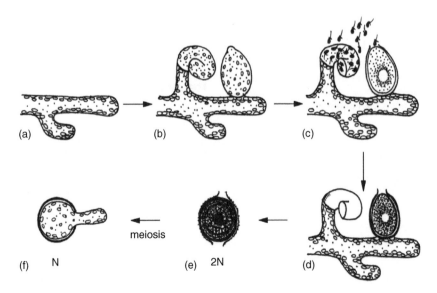

Figure 11 Sexual reproduction (oogamy) of monoecious *Vaucheria* sp. (a) multinucleate filament apex; (b) antheridium and oogonium (male and female gametangia, respectively) separated from the filament by a setting cell wall; (c) extrusion of zoogametes and fertilization of the oocyte; (d) zygote maturation; (e) resting zygote; (f) zygote germination and arising of a new thallus. N and 2N: haploid and diploid stages.

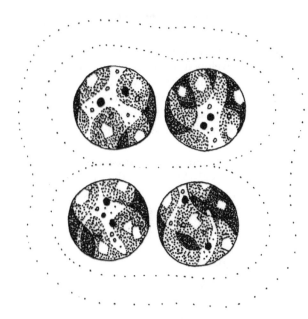

Figure 12 Eustigmatophyte *Chlorobotrys regularis* (W. West) Bohlin, cell diameter *c*.10 μm.

Freshwater raphidophytes, e.g., *Gonyostomum*, *Merotrichia* and *Vacuolaria*, are relatively common in dystrophic and eutrophic acidic water bodies. *Gonyostomum*, in particular, is widespread in small humic lakes, ponds, floodplains, and bogs, in association to *Sphagnum*, in Scandinavia, Finland, United States, and Central Europe. Its abundance has markedly increased, possibly in relation to freshwater acidification and eutrophication, in the last decades in Fennoscandia, where nuisance summer blooms are relatively frequent. Blooming marine genera (e.g., *Chattonella* and *Fibrocapsa*) are reported to produce fish toxic red tides along the Japanese coasts and, in lesser intensity, also in the Mediterranean lagoons.

See also: Algae; Phytoplankton Nutrition and Related Mixotrophy; Phytoplankton Population Dynamics: Concepts and Performance Measurement.

Further Reading

Canter-Lund H and Lund JWG (1995) *Freshwater Algae. Their Microscopic World Explored.* Bristol: Biopress Limited.

Cronberg G (2005) The life cycle of *Gonyostomum semen* (Raphidophyceae). *Phycologia* 44: 285–293.

Elbrächter M (2003) Dinophyte reproduction: Progress and conflicts. *Journal of Phycology* 39: 629–632.

Ettl H (1978) *Süßwasserflora von Mitteleuropa. Vol. 3/1. Xanthophyceae.* Jena: Gustav Fisher Verlag.

Green JC and Leadbeater BSC (eds.) (1994) *The Haptophyte Algae.* Oxford Scientific Publications Systematic Association, Special Volume No. 51. Oxford: Clarendon Press.

Hansen G and Flaim G (2007) Dinoflagellates of the Trentino Province, Italy. *Journal of Limnology* 66: 107–141.

Hegewald E, Padisák J, and Friedl T (2007) *Pseudotetraëdriella kamillae*: Taxonomy and ecology of a new member of the algal class Eustigmatophyceae (Stramenopiles). *Hydrobiologia* 586: 107–116.

Hoef-Emden K and Melkonian M (2003) Revision of the genus *Cryptomonas* (Cryptophyceae): A combination of molecular phylogeny and morphology provides insights into a long-hidden dimorphism. *Protist* 154: 371–409.

Jacobson DM (1999) A brief history of dinoflagellate feeding research. *Journal of Eukaryotic Microbiology* 46: 376–381.

Matsuoka K and Fukuyo Y (2000) *Technical Guide for Modern Dinoflagellate Cyst Study.* Tokyo: WESTPAC-HAB.

Novarino G (2003) A companion to the identification of cryptomonad flagellates (Cryptophyceae = Cryptomonadea). *Hydrobiologia* 502: 225–270.

Popovský J and Pfiester LA (1990) *Süßwasserflora von Mitteleuropa. Vol. 6. Dinophyceae (Dinoflagellida).* Jena: Gustav Fisher Verlag.

Rieth A (1980) *Süßwasserflora von Mitteleuropa. Vol. 4/2. Xanthophyceae.* Jena: Gustav Fisher Verlag.

Sandgren CD (ed.) (1988) *Growth and Reproductive Strategies of Freshwater Phytoplankton.* Cambridge: Cambridge University Press.

Santos LMA (1996) The Eustigmatophyceae: Actual knowledge and research perspectives. *Nova Hedwigia* 112: 391–405.

Taylor FJR (ed.) (1987) *The Biology of Dinoflagellates.* Oxford: Blakwell Scientific Publications.

Van den Hoek K, Jahns HM, and Mann DG (1993) *Algen.* Stuttgart/New York: G. Thieme Verlag.

Wehr JD and Sheath RG (eds.) (2003) *Freshwater Algae of North America. Ecology and Classification.* Academic Press.

Relevant Websites

http://www.algaebase.org/.
http://www.nhm.ac.uk/research-curation/projects/algaevision/.
http://protist.i.hosei.ac.jp/.
http://megasun.bch.umontreal.ca/protists/protists.html.
http://ucjeps.berkeley.edu/INA.html.

Photosynthetic Periphyton and Surfaces

M E Azim, University of Toronto, Toronto, ON, Canada

Definition and Synonyms

In classical limnology, the term 'periphyton' refers to the microfloral community living attached to the surfaces of submerged objects in water. This definition does not include fungal, bacterial, protozoan, and other attached animal components, which are included in the German word 'Aufwuchs.' Aufwuchs includes all the organisms those are attached to, or move upon, a submerged substrate, but which do not penetrate into it. Sometimes, the terms 'euperiphyton' (immobile organisms attached to the surfaces by means of rhizoids, gelatinous stalks, or other mechanisms) and 'pseudoperiphyton' or 'metaphyton' are used. Metaphyton forms cohesive floating and subsurface mats, usually composed of filamentous green algae. The mats originate as epiphyton or, in some cases, as edaphic mats, which are detached by water turbulence and float, in the water column or on the surface, because of trapped gases within them. Some authors prefer to call periphyton as attached algae or attached microorganisms, but these terms disregard the many other forms that live in periphyton assemblages. The scope of this review is limited to photosynthetic periphyton and therefore throughout this review, the term periphyton is used to refer to the attached algae. Depending on the size, periphyton can be categorized as micro- and macroperiphyton. Microperiphyton is microscopic algae growing on submerged substrates and integral to the aquatic food webs, representing the primary source of plant material to feed aquatic animals, including insects, other invertebrates, and fish. Macroperiphyton is mostly filamentous algae, attached by one end to substrates and visible to naked eye. In general, these morphologically large attached algae cannot be eaten by aquatic animals and therefore they can develop very dense populations. However, macroperiphyton can also serve as the surface for attachment of microperiphyton.

The 'Surfaces' refer to the submerged physical objects on which periphyton colonizes. They are also called substrata or substratum or simply substrates. They can be inert, providing only physical surfaces, and can also be nutrient-diffusing, providing both surfaces and nutrients for periphyton. Research findings have shown that in some cases, periphyton communities especially associated with macrophytes show substrate specificity and therefore the use of artificial substrates for ecological experiments would not be recommended. However, this dependency often cannot be separated from other habitat conditions such as the age of the substrate during colonization, time in the season when a substratum is colonized, and position in the littoral or stream. In general, substrate specificity is prominent when nutrient availability is low and less important at high nutrient availability.

Periphyton communities that develop on natural substrates (e.g., macrophytes) may have detrimental effects on the host substrates. Epiphytic coating on the leaf can interfere with the light availability and carbon uptake, hampering the photosynthesis and causing a reduction in oxygen diffusion rate. Periphyton could potentially be beneficial to the host macrophyte too: periphyton may form a protective UV-B shield on the substrate, which can minimize the attenuation of harmful UV-B.

There are different types of surfaces/substrates and periphyton has different synonyms depending on the surfaces on which it grows (**Table 1**). Epiphyton is composed of prostrate, erect, and heterotrichous algae, growing on the external surfaces of submersed and emergent vascular and nonvascular macrophytes. Epipelon includes motile algae inhabiting soft sediments. A related algal type is plocon, which includes nonmigratory algal crusts, typically composed of cyanobacteria and diatoms that form on the surface of exposed or submersed sediments. These crusts occasionally detach because of buoyancy of accumulated gas bubbles and float at the water surface. Plocon may correspond to 'epibenthic' mats or 'edaphic algae' studied in estuarine marshes where they can become several centimeters thick. Other surfaces on which periphytic algae are commonly found in lentic environments are sand grains (epipsammon), rocks (epilithon), wood (epixylon), and animals (epizoon).

Taxonomic Composition and Morphological Structure

Algae are very diverse and cosmopolitan in nature. Many of them are represented in freshwater periphyton, if they are attached to submerged substrate surfaces throughout or partially during their life cycles. The taxonomic diversity and abundance of periphyton depend on a range of factors such as habitat and surface types, light intensity, grazing pressure, seasonality, nutrient availability, acidity, and physical disturbances. The most frequently encountered groups are the diatoms (Bacillariophyceae), Cyanobacteria

(Cyanophyceae), and green algae (Chlorophyceae). Some taxa of red (Rhodophyceae) and brown (Phaeophyceae) algae are also common in periphyton especially in brackish/saline waters. Some representative periphyton taxa in different habitats are listed in **Table 2**.

Table 1 Category of surfaces and different synonyms of periphyton based on the surfaces

Surfaces/substrates	Synonyms
Aquatic plant	Epiphyton
Sediment/mud/silt	Epipelon/plocon
Wood/tree	Epixylon
Rock	Epilithon
Sand	Epipsammon
Animals	Epizoon

Table 2 Some periphyton taxa that dominate in different habitats

Habitats	Group/genera
Ponds	Bacillariophyceae: *Achnanthes, Cocconeis, Fragilaria, Gomphonema, Microphora, Navicula, Nitzschia, Synedra* Chlorophyceae: *Chlorella, Gonatozygon, Oedogonium, Scenedesmus, Stigeoclonium, Tetraspora, Ulothrix, Zygnema* Cyanophyceae: *Anabaena, Chroococcus, Lyngbya, Microcystis, Oscillatoria, Rivularia* Euglenophyta: *Difflugia, Trachelomonas*
Lake	Bacillariophyceae: *Cymbella, Diatoma, Gomphonema, Navicula, Nitzschia, Stephanodiscus, Synedra* Chlorophyceae: *Scenedesmus, Monoraphidium*
Marsh/wetland	Bacillariphyceae: *Achnanthes, Amphora, Navicula, Nitzschia, Diatoma, Fragilaria* Chlorophyceae: *Spirogyra, Mougeotia* Cyanophyceae: *Lyngbya, Oscillatoria*
Stream/river	Bacillariophyceae: *Achnanthes, Cocconeis, Cymbella, Diatoma, Fragilaria, Meridion, Navicula, Gomphonema, Rhoicosphenia, Synedra* Chlorophyceae: *Closterium, Oedogonium, Scenedesmus, Spirogyra, Stigeoclonium, Ulothrix* Cyanophyceae: *Phormidium, Oscillatoria, Calothrix*
Estuary	Bacillariophyceae: *Berkeleya, Cocconeis, Fragilaria, Gomphonema, Melosira, Navicula, Nitzschia, Proschkinia, Synedra* Cyanophyceae: *Anabaena, Hormothamnion, Oscillatoria* Chlorophyceae: *Bryopsis, Cladophora* Rhodophyceae (red algae): *Ceramium, Polysiphonia* Phaeophyceae (brown algae): *Ectocarpus, Sphacelaria*

An individual alga might be found free swimming in the water column or attached to a surface, or both, depending on when the observation is made and also on the habitat. Algae need some adaptations that help them to attach to a surface and cope with the periphytic environment. There is a broad diversity of habitats and surfaces to which algae adapted, showing a wide range of morphological structures. Morphological characteristics differ among groups of algae and therefore, different taxa show a variety of mechanisms for adapting to a sedentary life attached to surfaces. Morphology ranges from unicellular through amorphous colonies to filamentous and multicellular forms. Some of the more obvious morphological characteristics of periphyton include stalks with sticky ends, sticky capsules, cushions or filaments, muscular suction pads, glue, or simply clinging to the substrate. Attachment to sediment can also be achieved by rooting rhizoids and with a muscular foot.

The size of periphytic algae ranges from that of a single cell of about one thousandth of a millimeter in diameter to a large alga (*Cladophora*) up to many tens of centimeters in length. The shape varies from simple nonmotile single cells to motile, multicellular, filamentous structures. Nonmotile forms of unicellular, colonial, and filamentous algae attach to substrates by specially adapted cells and mucilaginous secretions. Some taxa (e.g., *Stigeoclonium*) have morphologically distinct basal and filamentous cells. The basal cells form broad horizontal expanses of cells across the substratum surface, and filaments develop vertically from the basal cells. Mucilaginous secretions can be amorphous for unicellular Cyanobacteria and green algae or organized into special pads, stalks, or tubes for diatoms.

Most of the unicellular and colonial cyanophytes are nonmotile, whereas filamentous cyanophytes comprise both motile and nonmotile forms. Gliding motility is performed using sheaths of mucilage. In contrast, unicellular and colonial chlorophytes and chrysophytes comprise both motile and nonmotile forms, and the filamentous species are entirely nonmotile. Flagella and pectin affect motility in the chlorophytes while chrysophytes exploit flagella and pseudopods for movement. Unicellular bacillariophytes have both motile and nonmotile form and colonial and filamentous species are nonmotile. The pennate diatoms have characteristic slits or pores on both sides of their frustules called raphes. Polysaccharide mucilage is excreted through the pores that sticks to the substrate and can be used by the diatom to move. Rhodophytes and xanthophytes consist of only filamentous nonmotile forms. Euglenoids and cryptomonads are unicellular motile organisms that rely on flagella for movement. There are both motile and nonmotile unicellular dinoflagellates, with colonial dinoflagellate species being exclusively nonmotile. As their name indicates, the means of motility among dinoflagellates are through flagella.

Colonization and Detachment Process

In general, the development of a periphyton layer on a clean substrate/surface is initiated by the deposition of a coating of dissolved organic substances (mainly amino acids and mucopolysaccharides) by electrostatic forces. Within hours, a coating of bacteria begins to form by hydrophobic reactions. The dissolved and nonliving particulate organic matters serve as support for the attachment and as substrate for the metabolism of bacteria. Therefore, the presence of free-floating organic particles in eutrophic waters stimulates this process. Bacteria actively attach to the substrates using mucilaginous strands. In a primarily physical process, the mucilage produced by bacteria offers potential binding sites for a

variety of colloidal, organic and inorganic elements. Bacteria also produce extracellular enzymes that make a significant contribution to the processing of dissolved organic substances especially for degrading larger fractions of organic molecules into assimilable lower weight molecules as well as inorganic substances. Whether bacterial colonization is a prerequisite for subsequent attachment of other organisms, or indeed what the exact role of the bacteria is in this process is not fully known. Provided that sufficient light is available, algae will subsequently attach and form the major structure of the developing periphyton. After a few days of bacterial colonization, small pennate diatoms (often *Cocconeis*, *Navicula*) can adhere to the organic matrix secreted by bacteria. These are followed by erected or stalked and long-stalked species and then by diatoms with rosettes and mucilage pads. During the climax stages of development, green or red algae with upright filaments or long strands can grow, forming a layered community within weeks (**Figure 1**).

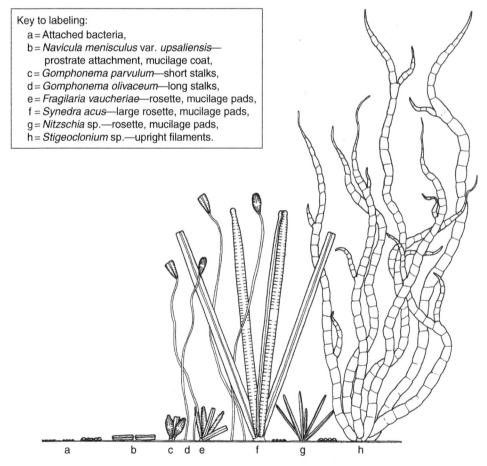

Key to labeling:
a = Attached bacteria,
b = *Navicula menisculus* var. *upsaliensis*—prostrate attachment, mucilage coat,
c = *Gomphonema parvulum*—short stalks,
d = *Gomphonema olivaceum*—long stalks,
e = *Fragilaria vaucheriae*—rosette, mucilage pads,
f = *Synedra acus*—large rosette, mucilage pads,
g = *Nitzschia* sp.—rosette, mucilage pads,
h = *Stigeoclonium* sp.—upright filaments.

Figure 1 The sequence of colonization processes in periphyton communities. Reproduced from Hoagland KD, Roemer SC, and Rosowski JR (1982) Colonization and community structure of two periphyton assemblages with emphasis on the diatoms (Bacillariophyceae). *American Journal of Botany* 69: 188–213.

Initially, periphyton biomass increases exponentially through colonization and growth and reaches a peak which is termed as the 'accrual phase.' Then there is a shift to dominance of loss processes through death, emigration, sloughing, and grazing, which is termed the 'loss phase' (**Figure 2**). The time required to reach the peak biomass (PB) varies from a few days to several months, depending on the availability of light and nutrients and on grazing intensity. An equilibrium of periphyton biomass is reached, where immigration (incoming movement) equals emigration (outgoing movement). The possible factors influencing this equilibrium are those controlling biomass development, such as nutrients, light, and temperature; and those regulating biomass losses, including disturbances resulting from substratum instability, current velocity, suspended solids, grazing by invertebrates and fish, disease, parasitism, and age. A distinction can be made between removal of periphyton by sloughing and by other disturbances related to water motion. As the community increases in size, algae close to the substrate may senesce as a result of light and or nutrient limitation and become detached from the substrate. This process is called sloughing (autogenic). By contrast, abrupt water motion in the form of spates can dislodge the outer parts of living periphyton layers from their base.

Dense growths of filamentous green algae (such as *Cladophora*) may become several centimeters in thickness, particularly on gravel and rocks in eutrophic rivers and wave-swept littorals. Such dense growths may effectively outshade the lower layers, which deteriorate and may become susceptible to sloughing. Under well-lit, eutrophic conditions and at higher temperatures ($>15\,°C$), dislodgement of dense periphyton may occur within 4 weeks since first colonization. Maximum densities generally have been reached over similar time spans of several weeks but these densities are highly variable. In streams, dislodgement is often initiated by sudden increases in velocity, associated with discharge peaks. Where wave- or flow-related turbulence is intensive, periphyton communities often remain less dense and smaller adnate diatoms dominate. The variation in accumulation, growth, and loss leads to a substantial range in productivity estimates (0.1–20 g ash-free dry matter, $AFDM\ m^{-2}\ d^{-1}$). Losses of periphyton communities by natural senescence and cellular death constitute a major fate of production. A significant portion of death is genetically programmed and estimates indicate at least 30% and often $<50\%$ of cellular death is viral in origins.

Disturbance to the substrates and surrounding habitats of periphyton communities occurs over a large spectrum from severe physical disruption of substrate, such as during river flooding, to much more gradual but nonetheless severe shifts in water level, water depth, and associated physical and chemical factors. Although much study has been directed to disturbances of substrates and periphyton in streams, much greater periphyton productivity and ecosystem constituents occur in quiescent habitats (flood plains, littoral areas, and especially wetlands). Very little is known about disturbance events, periodicities, and hydrology in these environments in relation to periphyton development, growth, productivity, and trophic interactions.

Periphyton Biomass and Productivity

The most useful methods for determining photosynthetic periphyton biomass are the chemical extracts of photosynthetic pigments, especially of chlorophyll *a* and cell counts. Other potential methods include determinations of dry matter weight or AFDM weight, which indicate not only the photosynthetic biomass but also other associated organic or inorganic particles, including bacteria. However, the photosynthetic part of the periphyton assemblage can be separated by autotrophic index (AI), a ratio of chlorophyll *a* to AFDM, assuming that on average 1.5% chlorophyll *a* contains in photosynthetic periphyton dry matter. Pigment concentrations can also be converted to carbon units using assessments of the C:Chlorophyll *a* ratio of 50:1 for periphytic algae, which leads to concentrations of several grams of algal carbon per square meter in mesotrophic and eutrophic systems. In general, high proportions of AFDM (i.e., low ash contents) in periphyton are positively correlated with high chlorophyll concentrations, but chlorophyll

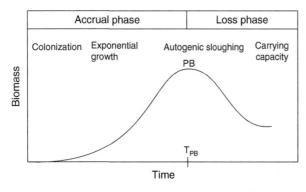

Figure 2 An idealized periphyton growth curve for streams showing the accrual and loss phases. Reproduced from Biggs BJF (1996) Patterns in benthic algae of streams. In: Stevenson RJ, Bothwell ML, and Lowe RL (eds.) *Algal Ecology: Freshwater Benthic Ecosystems.* pp. 31–56. San Diego, California: Academic Press.

appears to saturate around $10 \, \mu g \, cm^{-2}$, where AFDM may still continue to increase.

The capacity of chlorophyll a to reemit sorbed light energy as fluorescent radiation has also been exploited to monitor algal biomass in periphytic mats. Pulse Amplitude-Modulated (PAM) fluorescence (F_0) under dark conditions correlates well with chlorophyll a content. The use of F_0 as a surrogate of chlorophyll a has the advantage of allowing for continuous monitoring without the need to perform the chemical extraction of the sample. Changes in fluorescence detected by PAM methodology has been used extensively as an ecotoxicological tool for the detection of toxicants on periphytic algal biomass. In spite of these advantages, the optical conditions in periphyton impose some restrictions to the application, especially in very dense algal mats.

Chlorophyll a is a main photosynthetic pigment, but taxonomic groups of microalgae contain characteristic sets of accessory pigments. High performance liquid chromatography (HPLC) analysis of pigment extracts provides pigment fingerprints that give clues to the contribution of diatoms, green algae, and cyanobacteria. Multiwavelength excitation has become available on recent PAM fluorometers and the deconvolution of the fluorescent signal of mixed samples potentially reveals the contribution of algal groups with different sorption spectra. While such measurements have been applied to monitor the parallel growth of diatoms and Cyanobacteria in dual species-cultures, the technique may need further development to be effective in monitoring Cyanobacteria and diatoms in complex natural periphyton.

Several methods have been adapted to determine the productivity of photosynthetic periphyton. Measurement of oxygen variations has been used traditionally because it can provide accurate estimates of net production and respiration. Chambers or bell-jars under light and dark conditions have been used in a variety of environments. Open systems' measurements use variations in dissolved oxygen between day and night (one station) or between two different stations in order to determine the net metabolism and respiration of the overall system. Such measurements require an accurate estimate of the reaeration coefficient between the water and the atmosphere.

At a smaller scale, determination of the productivity of periphyton has also been approached by means of microelectrodes. The use of microelectrodes allows the determination of vertical profiles of oxygen concentration at the appropriate spatial scale, affecting the microbial processes. The applications of very short light-dark shifts enable the determination of the instantaneous rate of oxygen production for each layer. Consequently, these stratified production rates are vertically integrated into aerial rates similar to the routines in phytoplankton. Depth-integrated, aerial rates of production may therefore be considered an appropriate measure of productivity.

In addition to oxygen-based methods, the photosynthetic activity of periphyton has been measured by means of the incorporation of stable isotope-labeled carbon. The addition of the label requires that the mixing or diffusion of the label is sufficient and there is an associated risk of changing isotope ratios in complex natural samples. The stable ^{13}C-labeled bicarbonate marker can be added in periphyton and the conversion of this labeled inorganic carbon in organic fractions can be determined with isotope ratio mass spectrophotopmeter (IRMS). The specific degree of labeling (ratio $^{13}C/^{12}C$) of different fractions of algae, bacteria, and grazers can therefore be used to derive the turnover rates that are responsible for the production rate of the different trophic groups.

PAM has also been used as a measure of photosynthetic electron transport. The technique has recently been compared with the oxygen technique with good results. Optical techniques are particularly useful for monitoring the productivity of periphyton in managed systems or in systems with a more or less constant optical condition of the substrate. Applications of PAM to detect nutrient limitations have proved useful in planktonic samples, but the method still needs to be developed for periphyton samples.

Biomass and productivity of periphyton communities vary over several orders of magnitude. The causes of this high variability can be external as well as internal to the developing periphyton community, and biotic as well as abiotic in nature. Light availability is a prime abiotic factor that determines the periphyton abundance and its taxonomic dominance (**Figure 3**). Whenever sufficient light is available, periphyton communities will be dominated by algae but in low light condition, the communities will be dominated by heterotrophic organisms. Nutrient availability especially N and P play significant role on periphyton biomass and productivity. In general, enhanced nutrient availability led to shifts in taxonomic composition and increases in periphyton density and thickness. Since periphytic communities are often dominated by diatoms, the availability of Si should not be ignored besides nitrogen and phosphorus. Inorganic carbon has received substantially less attention, but has been shown to strongly govern periphyton development in interaction with nutrients and grazing pressure, particularly at high nutrient availability, increases of dissolved inorganic carbon above 2 mM (a median level for moderately hard fresh waters) led to fourfold increases in periphyton density provided grazing pressure would not decimate

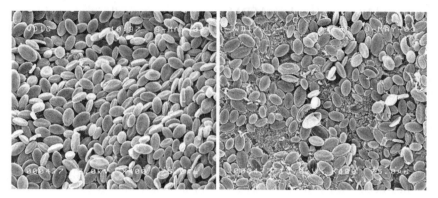

Figure 3 SEM pictures of diatom dominated periphyton in high light (*left*) and low light (*right*) conditions showing the differences in abundance and community composition.

this potential growth capacity. Heavy grazing of periphyton by animals (insect larvae, crayfish, and certain fishes) can result in reductions of periphyton biomass and productivity. In some cases, moderate grazing of periphyton helps regeneration of clonal vegetation and stimulates growth rates by means of improved light availability within the periphyton communities and perhaps enhanced nutrient availability from the water as well as from the feeding activities of the animals. Supporting evidence for the latter, however, is poor. There is some evidence that grazing mortality of periphyton is greater in streams than in standing waters.

In many water bodies, the contribution of the periphyton community to productivity is greater than that of the phytoplankton. Some earlier studies reported that periphyton contributed 42–97% of the total annual productivity, especially in shallow lakes (2–3 m) with large littoral zones. In some cases, the contribution of periphyton to annual primary productivity can be as high as $1\,kg\,C\,m^{-2}$. In general, high values for periphyton productivity are found on coral reef systems, with values typically in the range of 1–$3\,g\,C\,m^{-2}\,day^{-1}$. The highest value for periphyton productivity of $7.9\,g\,C\,m^{-2}\,day^{-1}$ was reported in acadja systems (artificial brushparks), which was 4.5 times higher than that of the lagoon phytoplankton community. In ponds provided with periphyton substrates, half of the primary productivity was contributed by periphyton (1.0–$1.7\,g\,C\,m^{-2}\,day^{-1}$).

Role of Periphyton in the Aquatic Food Web

Periphyton is immensely important in aquatic systems as it provides community structure and primary productivity that supports a wide range of aquatic organisms. Periphyton is easily grazed upon by small

invertebrates, fishes, and other aquatic animals and hence contributes considerably to the productivity of aquatic ecosystems, natural or human-made. However, despite early recognition that the littoral regions of aquatic ecosystems represent some of the most productive communities on Earth, there is still debate about the origin of this productivity. Although there are extensive literatures on aquatic plant productivity and pelagic phytoplankton productivity, less is known about the contributions of periphyton to aquatic systems.

Studies of algal productivity in shallow water environments demonstrate that periphytic algae provide an abundant, rapidly renewed, easily assimilated food resource that can be as or more important than macrophytes. Moreover, the abundance of submersed surfaces in lakes and wetlands argues that periphytic algae may be at least as important as phytoplankton in these habitats. Recent studies using stable C, N, and S isotopes to establish trophic linkages between primary producers and their prospective consumers have demonstrated that in many cases, the isotopic composition of periphyton matches that of grazers more closely than other potential primary producers. Research has also shown that the grazing on a two-dimensional layer of periphyton is mechanically more efficient than filtering algae from a three-dimensional planktonic environment.

Bottom-up and top-down (BU:TD) hypothesis of periphyton argued that bottom-up processes determine maximum attainable biomass at each trophic level whereas top-down processes are responsible for observed deviations from predicted relationships. In streams, algal biomass is mainly limited by nutrients, grazing pressure, current velocity, and variation of discharge. Nutrient enrichment experiments often demonstrate significantly higher grazer biomass than in controls, suggesting that primary production is channeled efficiently into secondary production. Grazers and nutrients interact on a fine scale, since grazing can physically disrupt the periphytic mat and allow for greater nutrient availability. The physical structure of periphyton communities is such that gaseous and ionic exchanges with the surrounding aqueous environment occur at rates in orders of magnitude slower than occur in the surrounding water. As a result, much of the metabolism within the periphyton community is coupled mutualistically. Because of this intimate physiological interdependence, internal nutrient and energetic recycling within the relatively closed community is intense. Metabolism, growth, and productivity within periphyton communities rely heavily on internal recycling and conservation of resources – the result is an unusually high efficiency of utilization and retention of captured external resources and a very high productivity.

The fate of the nutrients and organic energy of periphyton communities within aquatic communities is highly diverse spatially and temporally. Most of the organic matter produced by algae/Cyanobacteria of periphyton is degraded to CO_2 microbially by bacteria and protists within the periphytic community. Many of the nutrients released during decomposition are actively sequestered and retained by the viable components of the periphyton community. The fate of available living or dead organic matter of the periphytic communities that is not utilized and respired microbially is very complex, and often depends on the changing conditions of the supporting substrates. Although most of the organic production is heterotrophically metabolized by microbes and relatively little of the periphytic production passes to higher trophic levels, the impacts of higher trophic levels on the periphytic development and its ecosystem roles can be significant at certain periods.

Utilization of periphyton by protists such as sessile flagellates and ciliates is very poorly known as they feed within the periphytic matrix and microchannels. Utilization of released dissolved organic substrates from cells is largely metabolized to CO_2, much of which is recycled and utilized, along with organic and inorganic nutrients, within the periphyton. As in the pelagic planktonic system, most of the periphyton production enters the pool of detrital organic matter in both lake and river ecosystems and is rapidly metabolized and recycled in an efficient community mutualism.

Evaluation of the rates of energy fixation by primary producers in terrestrial and aquatic communities, and the rates of transfer of this energy to higher trophic levels, are foundations of trophic dynamic analyses. In the pelagic zone of lakes, trophic structure and energy fluxes considered almost solely particulate organic matter (POM). Flows of energy within the trophic structures focused on predation by ingestion of particles of organic matter. Many particulate filtering and predatory organisms consume variable amounts of particulate detritus (dead POM) that clearly dominates over living POM. Total annual budgets of carbon fluxes in lakes and especially rivers demonstrated the overriding importance of dissolved organic matter (DOM) derived from terrestrial and littoral-floodplain production of higher plants and periphyton. Much of the heterotrophic respiration of organic matter occurs in sediments, particularly in shallower waters, with net evasions of CO_2 to the atmosphere in quantities many times in excess of the rates of carbon fixation by the phytoplankton. The productivity of allochthonous organic matter and littoral productivity of wetland-littoral aquatic plants and associated periphyton can and often does constitute a major or the dominant input of organic carbon to the aquatic ecosystems.

Great emphasis has been given to periphyton as a food supply for animals and movement of carbon, nutrients, and energy within the food web. A number of processes contribute to mortality, losses, and interference competition, including predatory mortality (largely by grazing invertebrates and fishes), natural physiological senescence, death (often genetically programmed death), and population turnover, diseases and viral mortality, and physical disturbances from water movements and substrate instabilities. However, although the role of periphyton in aquatic food webs is understood in lotic systems, there remains a pressing need for research on all aspects of lentic periphyton ecology, especially on the role of periphyton in the biogeochemical cycles of nutrients and its responses to the climate change and anthropogenic chemical exposures.

Application of Periphyton

Periphyton is of growing interest worldwide. It is immensely important for various reasons: (1) it is a significant, and often the dominant, contributor to the carbon fixation and nutrient cycling in aquatic ecosystems, (2) it is an excellent indicator for changes occurring in the aquatic environment and the usefulness of lentic periphyton for biomonitoring of environmental change is increasingly recognized, (3) it is used to improve water quality in lakes and reservoirs, (4) it increases food availability and improves environment in aquatic production systems and therefore, is used in aquaculture, (5) it can provide specialty foods for fish and shellfish larvae culture, and (6) it can be used for waste water treatment.

See also: Comparative Primary Production; Cyanoprokaryota and Other Prokaryotic Algae; Diatoms; Green Algae; Phytoplankton Productivity.

Further Reading

Aloi JE (1990) A critical review of recent freshwater periphyton field methods. *Canadian Journal of Fisheries and Aquatic Sciences* 7: 656–670.

Azim ME and Asaeda T (2005) Periphyton structure, diversity and colonization. In: Azim ME, Verdegem MCJ, van Dam AA, and Beveridge MCM (eds.) *Periphyton: Ecology, Exploitation and Management*, pp. 15–33. Wallingford, UK: CABI Publishing.

Azim ME, Verdegem MCJ, van Dam AA, and Beveridge MCM (2005) Periphyton and aquatic production: an introduction. In: Azim ME, Verdegem MCJ, van Dam AA, and Beveridge MCM (eds.) *Periphyton: Ecology, Exploitation and Management*, pp. 1–13. Wallingford, UK: CABI Publishing.

Biggs BJF (1996) Patterns in benthic algae of streams. In: Stevenson RJ, Bothwell ML, and Lowe RL (eds.) *Algal Ecology: Freshwater Benthic Ecosystems*, pp. 31–56. San Diego, California: Academic Press.

Goldsborough LG, McDougal RL, and North AK (2005) Periphyton in freshwater lakes and wetlands. In: Azim ME, Verdegem MCJ, van Dam AA, and Beveridge MCM (eds.) *Periphyton: Ecology, Exploitation and Management*, pp. 71–89. Wallingford, UK: CABI Publishing.

Hoagland KD, Roemer SC, and Rosowski JR (1982) Colonization and community structure of two periphyton assemblages with emphasis on the diatoms (Bacillariophyceae). *American Journal of Botany* 69: 188–213.

Rosemond AD, Mulholland PJ, and Brawley SH (2000) Seasonally shifting limitation of stream periphyton: Response of algal populations and assemblage biomass and productivity to variation in light, nutrients and herbivores. *Canadian Journal of Fisheries and Aquatic Sciences* 57: 66–75.

Sabater S and Admiraal W (2005) Periphyton as biological indicators in managed aquatic ecosystems. In: Azim ME, Verdegem MCJ, van Dam AA, and Beveridge MCM (eds.) *Periphyton: Ecology, Exploitation and Management*, pp. 159–177. Wallingford, UK: CABI Publishing.

Stevenson RJ, Bothwell ML, and Lowe RL (eds.) (1996) *Algal Ecology: Freshwater Benthic Ecosystems*. San Diego, California: Academic Press.

Van Dam AA, Beveridge MCM, Azim ME, and Verdegem MCJ (2002) The potential of fish production based on periphyton. *Reviews in Fish Biology and Fisheries* 12: 1–31.

Vermaat JE (2005) Periphyton dynamics and influencing factors. In: Azim ME, Verdegem MCJ, van Dam AA, and Beveridge MCM (eds.) *Periphyton: Ecology, Exploitation and Management*, pp. 35–49. Wallingford, UK: CABI Publishing.

Wetzel RG (2005) Periphyton in the aquatic ecosystem and food webs. In: Azim ME, Verdegem MCJ, van Dam AA, and Beveridge MCM (eds.) *Periphyton: Ecology, Exploitation and Management*, pp. 51–69. Wallingford, UK: CABI Publishing.

Wetzel RG (1983) *Periphyton of Freshwater Ecosystems*. The Hague, The Netherlands: Dr. W. Junk Publishers.

ZOOPLANKTON

Contents

Rotifera

R L Wallace and H A Smith, Ripon College, Ripon, WI, USA

Introduction

Although comprising a small phylum (*c.* 2000 species) of tiny animals traditionally described as pseudocoelomate, rotifers are significant to freshwater systems. Because of their high reproductive rates, they often reach population densities well over 1000 individuals per liter, occasionally numerically dominating zooplankton communities. In sewage ponds, exceedingly large populations can develop ($>10^4$ individuals per liter), and aquaculture systems regularly achieve densities nearing 10^6 individuals per liter. Rotifers may represent 25% or more of total zooplankton biomass in natural systems; and they are an important connection between the microbial loop and the classical food web. Thus, rotifers link microbes and protists to species in higher trophic levels, such as arthropods (copepods, predatory cladocerans, and insect larvae) and small fishes. Despite a typical lifespan of one to a few weeks, their rapid reproduction enables rotifers to be grown in mass quantities in aquaculture systems, where they are readily consumed by commercially valued species. Rotifers also serve as bioindicators for ecotoxicological studies.

Few, if any, places on earth are devoid of these tiny animals as long as unfrozen water is present for at least a few days. They occupy lakes, ponds, streams, marshes, and bogs. Moreover, their ability to tolerate harsh conditions permits colonization of sewage lagoons, birdbaths, ephemeral desert pools, arctic ponds, and glacial melt-waters. Rotifers also populate the water film covering mosses and liverworts, as well as the interstitial water of soils and sediments (limnoterrestrial). Most rotifers live in fresh water, but they also are found in brackish waters; perhaps fewer than 100 species are exclusive to marine habitats. Rotifers may be present in tide pools, and they inhabit ice–water interstices of marine and freshwater systems.

Because research typically has focused on free-swimming forms, we know much more about them; however, only about 100 species are truly planktonic. Most rotifers move by crawling or swimming. Nearly all are benthic or associated with plants in the inshore, crawling over them or forming temporary attachments to their surfaces. About 80 species (*c.* 4%) are sessile; 20 species taxonomically related to the sessile forms have abandoned sessile life to return to the plankton (e.g., *Collotheca pelagica*, *Sinantherina semibullata*). Though predominantly solitary, a few species (*c.* 25) form colonies ranging from an adult with 1–3 offspring to groups of thousands of individuals.

Morphology

Rotifers are tiny animals (*c.* 50–2000 μm) possessing a wormlike to saccate shape with several notable characteristics (**Figure 1**). Chief among these are an apical field called the corona, a muscular pharynx termed the mastax, and a syncytial body wall strengthened by a filamentous layer of special proteins. While none of these elements is unique among invertebrates, together they represent the three principal features of the phylum.

The corona, an apical ciliated region used for locomotion and gathering food, is eye-catching because in many species the cilia beat in a way that creates the illusion of a rotating wheel. Although this feature serves as the basis for the phylum Rotifera, or wheelbearer, the appearance is an artifact of metachronous ciliary action. Arrangement of cilia varies considerably

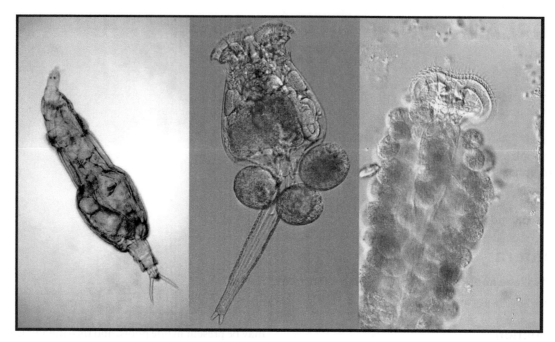

Figure 1 Three examples of common rotifers. Left, *Rotaria* sp. (crawling-swimming; animal *c.* 400 μm) (Photomicrograph courtesy of RLW); middle, *Brachionus* sp. (swimming; animal *c.* 500 μm) (Photomicrograph courtesy of Dr. Elizabeth Walsh, University of Texas-El Paso, TX), right, *Ptygura pilula* (sessile; animal *c.*1000 μm) (Photomicrograph courtesy of Dr. Don Ott, University of Akron, OH).

among species, but often consists of two concentric bands, the trochus, and cingulum (**Figure 2**). However, not all rotifers possess a corona that resembles a rotating wheel. Many have a simple, flat, ciliated field, and others have a corona resembling a funnel or bowl. In the latter, the mouth is located at the bottom of the funnel and its rim is ringed by long setae (**Figure 3**). Rotifers with this coronal form (Families Atrochidae and Collothecidae) capture prey one at a time using a process like that of a Venus flytrap.

The mastax contains a set of jaws called trophi (**Figure 4**). Morphology of the corona and trophi determine the way a rotifer collects and processes its food. These structures, especially the trophi, can be so specialized that they gained taxonomic importance, permitting one to characterize rotifers to the level of family, genus, and even species. The basic rotifer trophi is composed of three units: the incus and paired mallei, each of which has its own component parts. On the basis of morphological variations in these components, nine types of trophi are recognized. These function in grinding, grasping, piercing prey, or creating a suction action. Complex variations on the jaw structure represent both delight and bane to those who study rotifers.

The body wall (integument) of rotifers often is transparent and therefore overlooked, but it still is an important character. This layer is syncytial, a feature seen in other invertebrates and certain tissues of vertebrates. However, within the integument is a deposit of filamentous proteins. The proteinaceous layer is referred to as the intracytoplasmic lamina (ICL). Rotifers possessing a thick ICL have a rigid body wall and are termed loricate; those with a thin ICL and less rigid body wall are called illoricate. Intermediates between these two extreme are seen, but stiffness of the ICL is of no taxonomic consequence. In addition, all rotifers exhibit consistency in the total number of cells in the body of adults. This feature, termed eutely, is important because of its significance in wound healing; eutelic animals damaged by disease or predation can heal wounds, but they cannot regenerate lost parts.

Additionally, rotifers possess sensory, secretory, and digestive organs. Some rotifers have one or two eyespots; these may be frontal, lateral, or associated with a large saccate structure called the retrocerebral organ. This organ is located near the brain. Rotifers also possess bands of muscles, but lack circulatory and respiratory systems.

Classification

Rotifers fall into three basic groups that are divided into two classes: Pararotatoria and Eurotatoria. Pararotatoria comprises a small class of marine rotifers

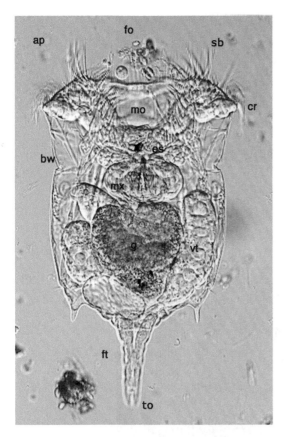

Figure 2 *Epiphanes brachionus* under bright field microscopy. (Symbols: ap, apical field; bw, body wall; cr, corona; es, eyespot; fo, food; ft, foot; g, gut; mo, mouth; mx, mastax; sb, sensory bristles; to, toes; vt, vitellarium.) (Animal *c*. 500 μm.) (Photomicrograph courtesy of Dr. Don Ott, University of Akron, OH).

Figure 3 Funnel shaped corona of *Collotheca campanulata*, a sessile rotifer of Family Collothecidae. (Animal *c*. 700 μm.) (Photomicrograph courtesy of RLW).

with only three species in two genera (Order Seisonidea), all of which appear to be epizoic on a marine crustacean. Seisonids possess a corona and trophi unlike that of other rotifers; they also have paired gonads and are obligatorily sexual. Bdelloids and monogononts together form Class Eurotatoria (the focus of this chapter), but they differ greatly in their morphology and life history. Subclass Bdelloidea consists of about 400 species of worm-like animals, most of which inhabit the benthos or crawl along surfaces. Movement in many of these species resembles that of a leech. Male bdelloids have never been seen and reproduction is parthenogenetic. In Subclass Monogononta, more than 1500 species are recognized in three orders (Collothecaceae, Flosculariaceae, Ploima). All have a single gonad. Monogononts may be sessile, sedentary, or planktonic, and reproduction involves both asexual and sexual phases.

A complicating factor throughout the phylum is the phenomenon of cryptic speciation, whereby two or more morphologically similar taxa actually constitute distinct species. Subtle morphological or physiological variations often suggest that populations recognized as a single species represent a species complex. However, detailed genetic analysis is necessary to confirm the occurrence of cryptic speciation.

Another major unresolved issue in rotifer classification is their relation to the Acanthocephala or spiny-headed worms, a parasitic group traditionally considered a distinct phylum. Acanthocephalans are obligatorily sexual and possess a complex life cycle involving a vertebrate as the definitive host, and an arthropod as the intermediate host. Although genetic analyses suggest acanthocephalans are highly evolved rotifers, further research is needed to clarify the phylogenetic relationship of these two taxa.

Life History Strategies

Reproduction

As a phylum, rotifers possess a complex yet intriguing reproductive scheme (**Figure 5**). Except for the obligatorily sexual seisonids nearly all rotifers are female. Bdelloid reproduction is by asexual parthenogenesis, while monogonont reproduction involves asexual (amictic) and intermittent sexual (mictic) cycles. In both bdelloids and monogononts, asexual eggs

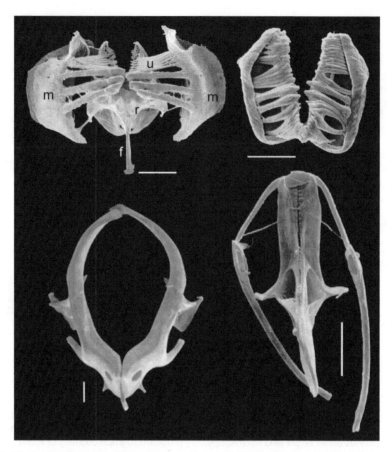

Figure 4 Examples of rotifer trophi. Upper left, *Sinantherina semibullata* (microphagous); upper right, *Adineta ricciae* (microphagous); lower left, *Asplanchna brightwellii* (predator); lower right, *Dicranophorus robustus* (predator). (Scale bars = 10 μm, except for *A. ricciae* = 5 μm). (Symbols: f, fulcrum; m, manubrium; r, ramus; u, uncus.) (SEM photomicrographs courtesy of Dr. Hendrik Segers, Belgian Biodiversity Platform, Belgian Institute of Natural Sciences, Brussels, Belgium.)

usually hatch within a few days, and thus are termed subitaneous (**Figure 6**).

Because monogonont reproduction is predominantly amictic, populations usually are all female. Nevertheless, in the typical monogonont life cycle asexuality occasionally is interrupted by bouts of sexual reproduction (mixis), when certain signals (e.g., photoperiod, food, population density) stimulate a switch from the amictic to the mictic cycle. While amictic eggs are diploid (as are all bdelloid eggs), those made during the mictic cycle are haploid (**Figure 6**). These relatively small eggs hatch into haploid males, or if fertilized form a diploid zygote. Production of male offspring may occur for several weeks or be so brief – days or perhaps a week – that males are not collected during sampling. The males of most species lack a functional gut and are much smaller than females, a phenomenon termed male dwarfism. Upon contact with a female of their own species, males initiate specific mating behaviors. Mating pheromones present on the female's body surface induce these reactions. Altogether, the combination

of amictic and mictic phases in monogononts is termed cyclic parthenogenesis. However, some species are capable of simultaneously producing amictic and mictic eggs; this condition is termed amphoteric. Then again, no males have been found in populations of some monogononts, although it is assumed that they are capable of sexual reproduction.

During development the embryos produced by mixis are encased within a multilayered shell; this encysted embryo is called a resting egg (**Figure 7**). Resting eggs sink to the bottom of the sediments, where they may remain for decades, analogous to a seed bank. This dormancy is referred to as diapause, so another term for the resting egg is a diapausing embryo. The tough cyst wall of resting eggs confers resistance to conditions present in the sediments, including hypoxia or the drying of ephemeral pools. Environmental cues such as changes in light and temperature initiate hatching.

In all, mixis provides a pool of potential colonists for the future while fostering diversity through genetic recombination. Ability of these structures to withstand desiccation may enable resting eggs to serve as dispersive agents, potentially explaining many monogononts' widespread distribution.

Coloniality

An interesting life history feature of rotifers is the formation of colonies (**Figure 8**). While most rotifers are solitary, about 25 monogononts (Families Conochilidae and Flosculariidae) and one bdelloid form colonies. All species that form colonies are filter feeders that ingest small particles (microphagous feeding). It appears that coloniality enhances feeding efficiency in some species. Also, the gelatinous sheaths that envelop the colonies of certain species provide protection from some predators.

Sessile Rotifers

Thought mobile as juveniles, within a day of hatching the young of some species attach to a surface (usually an aquatic plant), undergo a kind of metamorphosis, and become permanently sessile (**Figure 3**). Sessile species are found in three monogonont families: Atrochidae, Collothecidae, and Flosculariidae.

Parasitism

As with all multicellular organisms, rotifers are parasitized by viruses, bacteria, and fungi (**Figure 9**). Unfortunately, we know relatively little of how these infectious agents affect rotifer populations. Yet, in some instances a population can be annihilated within a few days. A small number of rotifers are parasites themselves. Perhaps more accurately termed specialized herbivores, these species act as endoparasites of freshwater algae, such as *Vaucheria* (**Figure 10**) and *Volvox*. Animals parasitized by rotifers include freshwater oligochaetes, eggs of pulmonate snails, and arthropods. Rotifers also act as vectors in the transmission of viruses to shrimp in aquaculture.

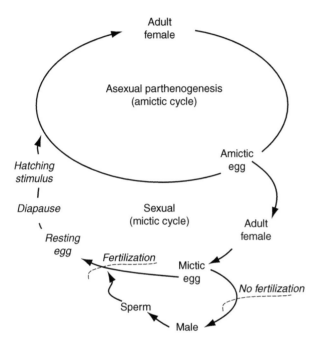

Figure 5 Simplified schematic of the typical rotifer life cycle showing amictic (asexual) and mictic (sexual) phases. Bdelloids reproduce using only asexual parthenogensis. (RLW and HAS.)

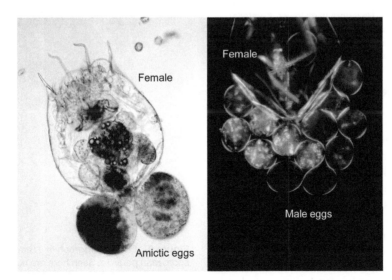

Figure 6 Female *Brachionus* sp., a planktonic rotifer carrying eggs. Left, female with two amictic eggs. Bright field microscopy. (Eggs, *c*. 90 × 70 µm.); right female with many male eggs. Dark field microscopy. (Eggs *c*. 50 µm.) (Photomicrographs courtesy of RLW.)

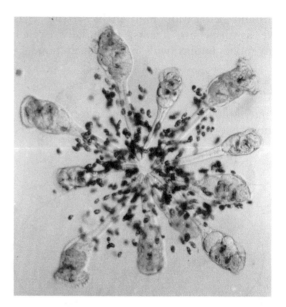

Figure 8 *Conochilus unicornis*, a planktonic colonial rotifer. Embedded in the gelatinous matrix of this colony are many small, unicellular algae. (Colony size approximately 500 μm in diameter.) (Photomicrograph courtesy of RLW.)

Figure 7 Resting eggs of three free-swimming rotifers. Upper, *Brachionus durgae*; middle, *Hexarthra mira*; lower, *Keratella valga*. (Scale bars = 10 μm.) (SEM photomicrographs courtesy of Dr. Hendrik Segers, Belgian Biodiversity Platform, Belgian Institute of Natural Sciences, Brussels, Belgium.)

Practical Considerations

Sampling

If one is interested only in the general composition of a community, elaborate collection equipment is not required. However, the tools and techniques needed differ according to habitat.

Planktonic rotifers may be collected by slowly dragging a fine-mesh net (*c.* 30–50 μm) through the water. Nets with coarser mesh sizes will miss small-bodied species. However, a different strategy is required when a quantitative assessment is desired. In that case, plankton net tows can be integrated over the water column sampled. Also, a pump equipped

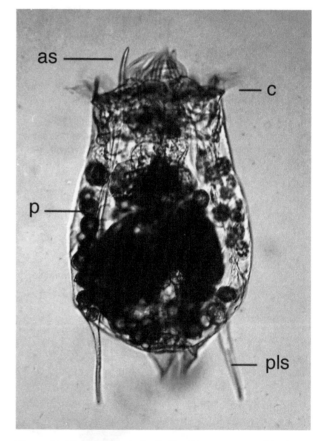

Figure 9 Photomicrograph of a *Brachionus* sp. parasitized by a sporozoan parasite. Seen here are numerous spherical cells of the parasite within the body cavity. (Symbols: as, anterior spines; c, corona; p, parasite; pls, posteriolateral spines.) (Animal *c.* 250 μm.) (Photomicrograph courtesy of RLW.)

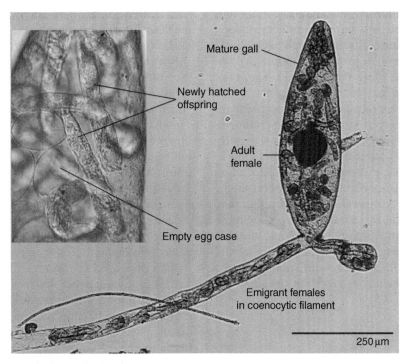

Figure 10 *Proales werneckii*, a small monogonont rotifer that parasitizes the coenocytic alga *Vaucheria*. (Photomicrograph courtesy of Dr. Don Ott, University of Akron, OH.)

with a filter to concentrate the sample, a discrete sampler (e.g., Van Dorn bottle, Kemmerer bottle, or Schindler-Patalas trap), or a closing net (e.g., Clarke-Bumpus sampler) is commonly employed.

Plants of the littoral zone provide habitat for the phytal community, an assemblage of species that have a propensity to stay close, and often attach, to vascular macrophytes, mosses, and filamentous algae. Of course, nets function poorly in regions choked with plants. In this case, a flexible, plastic tube (water-core) may be maneuvered between plants to collect several liters of water. Dipping a jar into a pond next to an unattached plant so that it slides into the jar with its surrounding water permits collection of local free-swimming rotifers, as well as sessile species on the plant. To extract free-moving species from aquatic or terrestrial mosses a small clump of wet moss is gently squeezed over a jar, rewetted with filtered pond water, and then squeezed again.

Because they attach permanently to plants, sessile species can be challenging to collect and study. Small plants (e.g., *Lemna*) and those with highly dissected leaves (e.g., *Elodea*, *Myriophyllum*, and *Utricularia*) may be cut into small pieces for examination under a dissection microscope. Broad-leaved plants (e.g., *Nuphar*, *Nymphaea*) must be cut into thin strips and examined on the leaf's edge.

Sampling frequency is an important consideration, and it is best to sample as often as possible. A rigorous sampling schedule of once or twice a week is not realistic for most studies; especially given that many species must be examined live for identification. However, less frequent sampling (e.g., monthly) could give a misleading picture trends if highs or lows in population density are missed. On the other hand, collection of rotifers does not require traditional sampling strategies or even that adults of the target species be present. Resting eggs may be induced to hatch when small amounts of sediments are incubated at temperatures about 10–20 °C warmer than the bottom of the lake. This technique also works for dry sediments from ephemeral ponds.

To prevent high mortality due to oxygen depletion samples should not be concentrated too much, and field samples generally should be kept cool during transport to the laboratory. However, cooling can kill warm-water species. Similar precautions need to be taken once the samples are in the laboratory. Fixatives (e.g., formalin, alcohol, Lugol's iodine) are often used to preserve specimens. However, centrifugation should not be used to concentrate samples as it damages rotifers, thus creating artifacts that can lead to misidentification.

Culturing

Numerous species of rotifers have been cultured, but most efforts have concentrated on a few species, often of the genus *Brachionus*. For many species determining the optimal food is a serious problem, making them difficult to culture. Routine practice sometimes requires that fresh food be provided almost daily and medium be refreshed a few times per week. However, bdelloids have been grown in infusions of various grains, manure, soil extracts, and even dilute suspensions of dry commercial pet food. These cultures require much less maintenance. Another low maintenance system is that of greenhouse pools and laboratory aquaria. Yet, species composition will depend on the presence of small zooplanktivorous fishes, snails, flatworms, or small crustaceans, and insects that may consume rotifers. Adding a small amount of sediment or aquatic plants can enrich the planktonic and sessile rotifer community in an aquarium. Containers range in size from small vessels with volumes of less than 10 ml to aquaculture systems the size of swimming pools.

Identification

Rotifer identification typically begins with an assessment of external characteristics such as the shape of the body and corona. Illoricate species should be examined live because fixation can cause them to contract into an unrecognizable lump. Identification is easier in loricate species, but trophi may need to be analyzed for both loricate and illoricate monogononts. Hydrolysis of specimens with bleach dissolves the surrounding tissues, thus freeing the trophi for detailed study. However, most bdelloid trophi are not sufficiently distinct to be useful for taxonomic purposes.

Regrettably, while several good printed keys are available, nearly all are limited in geographic coverage. Some Internet sites contain identification keys, but these generally are inadequate or do not reflect recent advances in rotifer taxonomy. The most comprehensive resource is the series *Guides to the identification of microinvertebrates of the continental waters of the world*.

Ecology

Habitats

Rotifers may be found in nearly every conceivable wet place – ranging from bogs to rivers, lakes to birdbaths, and hot springs to glacial melt-waters. The physical structure of the habitat is significant in determining the composition of the rotiferan community. For example, the community in the pelagic region differs greatly from that of the weedy littoral zone. There is a strong relationship between development of littoral vegetation and rotifer species richness, even when sessile forms are not considered.

Rotifers are not restricted to the water column; in the sandy region just above the waterline is a diverse array of interstitial species. These rotifers live in water held by capillary action between sand grains. This habitat extends into the littoral zone of lakes; however, low oxygen concentrations make the deeper sediments unsuitable for rotifers. Many psammic (sand) species tenaciously grasp the sediments and are not easily dislodged for study, even when treated with anesthetic agents.

Like aquatic sediments, terrestrial soils also are home to rotifers, including many bdelloids. Unfortunately, this habitat has not received much study, even though population density can be quite high (up to 2×10^6 individuals per square meter). Yet, population levels and species composition vary widely, probably due to differences in soil moisture, pH, and organic content. Although depauperate in species richness and offering only miniscule populations, subterranean ground waters and springs welling up at the surface serve as additional habitats.

Distribution

Our knowledge of rotifer biogeography appears to reflect the distribution of researchers and the places they have sampled as much as the distribution of the species themselves. Moreover, what we know is based mainly on studies of monogononts, with relatively little known about the bdelloids. Thankfully the number of sites and species studied is increasing, albeit not fast enough.

Early researchers were confident that rotifers could disperse to every water body, but the environment would permit survival of only those preadapted to the ambient conditions. This idea was based on the assumption that monogonont resting eggs and bdelloid xerosomes (see Anhydrobiosis) could be easily transported (e.g., by wind or by birds). Indeed, recent studies have confirmed that many species possess cosmopolitan distributions. Yet, endemism is an important feature in many species, and the true extent of cosmopolitanism may be overestimated because of difficulties in distinguishing among closely related species (i.e., cryptic speciation).

Community Structure

Rotifers are only one component of freshwater invertebrate communities and their relative importance varies by habitat (i.e., benthic, phytal, or planktonic). Other members include numerous protists (amoebas, flagellates, ciliates), hydroids, flatworms, gastrotrichs, tardigrades, mollusks, freshwater annelids, nematodes, aquatic mites, microcrustaceans (cladocerans and copepods), and some insects. In certain instances, jellyfish (*Craspedacusta*) and the swimming stages of parasitic flukes may be present. Typically, the planktonic community is dominated by protists in number of individuals (**Figure 11**) and by microcrustaceans in biomass. However, rotifers occasionally dominate in one or both categories, and even at smaller population densities comprise an important portion of the zooplankton community. They play a critical role in

community dynamics by serving as an intermediate in the transfer of energy and nutrients from the microbial loop to the classical food web. Though almost always an important part of the plankton, little data is available regarding their relative significance in benthic and phytal habitats.

Population Dynamics

The influence of environmental conditions on population growth has intrigued researchers for decades. Factors that appear to be important include temperature, pH, salinity, alkalinity, food, and season, as well as the presence of competitors, predators, and parasites.

The dynamics of rotifer population growth have been studied in a variety of ways, with early work simply correlating food levels to egg production. However, knowledge of the individual species, including its reproductive potential, can be combined with information about the environment to model population levels. Such models employ growth rate equations of varying complexity. The best predictions of growth are made from laboratory cultures. Perhaps one of the most frequently studied genera in this regard is *Brachionus*, especially the species *B. calyciflorus* and *B. plicatilis*.

Nevertheless, in natural systems population growth and community structure are much more complex and difficult to predict. Population dynamics may vary among similar habitats, as well as between seasons and years in the same habitat. This can be illustrated by examining the population density of rotifers in a lake over several years (**Figure 12**).

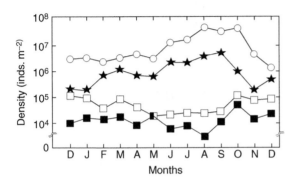

Figure 11 Numbers of protists and micrometazoans in the plankton of fresh waters. Lake Oglethorpe (Georgia, USA) is typical of many inland waters with respect to the relative abundance of protists (open circles), rotifers (stars), nauplii (open squares), and micrometazoans (closed squares). Reproduced from Figure 2 in Pace ML and Orcutt JD, Jr. (1981). The relative importance of protozoans, rotifers, and crustaceans in a freshwater zooplankton community. *Limnology and Oceanography* 26: 822–830, with permission from the authors. Copyright 1981 The American Society of Limnology and Oceanography, Inc.

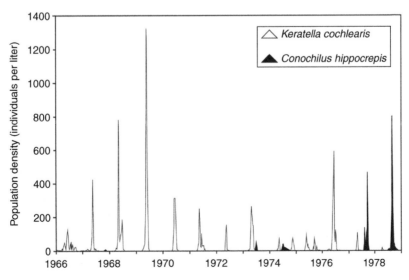

Figure 12 Rotifer population dynamics. Tracked here are two species (*Keratella cochlearis* and *Conochilus hippocrepis*) of planktonic rotifers from Lake Washington (Washington, USA) for a 14-year period. (Adapted from the Lake Washington database of Dr. W. 'Tommy' Edmondson, as funded by the Andrew Mellon Foundation; data courtesy of Dr. Daniel Schindler, University of Washington, WA).

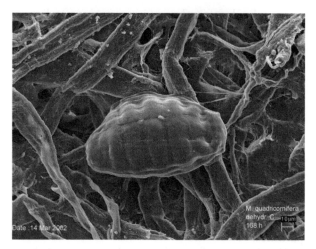

Figure 13 Xerosome of *Macrotrachela quadricornifera*. Filter fibers appear in the background. (Length *c.*125 μm.) (Photomicrograph courtesy of Dr. Giulio Melone, University of Milano, Italy.)

Even lakes in close proximity may differ greatly in species composition and population levels.

Anhydrobiosis

If dehydrated slowly, some bdelloids desiccate, shrinking into a small mass (**Figure 13**). At this point, they have entered a state of anhydrobiosis and may remain desiccated for years, becoming active again after rehydration. Although not completely understood, this process also is seen in juveniles of some nematodes and adult tardigrades, resulting in structures referred to as dauer larvae and tuns, respectively. The bdelloid anhydrobiont sometimes is called a tun; however, to differentiate between rotifers and tardigrades, herein it is termed a xerosome. Like monogonont resting eggs, xerosomes can form an egg bank that remains viable for decades, thereby permitting survival during unfavorable times. Thus, resting eggs and xerosomes represent future communities of rotifers. Similar to monogonont resting eggs, the enhanced resistance of the xerosome to desiccation is thought to increase the group's capacity for dispersal, and may help to explain the cosmopolitan distribution of many bdelloids.

Feeding

Rotifers use a variety of methods to procure their food. In filter or suspension feeding an animal (microphage) consumes tiny algae, bacteria, yeast, and protists by creating filtering currents that bring small particles to its mouth (e.g., *Brachionus*, *Keratella*). Many rotifers that crawl over surfaces feed by a scraping action that removes food from the surface

Figure 14 An example of phoresis: *Daphnia* sp. with more than 20 attached *Brachionus rubens*. (Photograph courtesy of RLW.)

(e.g., certain bdelloids). Some large rotifers such as the planktonic genus *Asplanchna* and the sessile genera *Collotheca* and *Cupelopagis* feed by grasping and swallowing their prey whole. A small number of rotifers are detritus feeders and a few are parasitic. Some consume a variety of foods, while others are extremely specific. *Acyclus inquietus* feeds on the eggs and young of the sessile rotifer *Sinantherina socialis*; *Trichocerca cylindrica* sucks out the contents of the eggs of other rotifers.

Competition

For the most part rotifers are not strong competitors, especially when compared with large cladocerans (>1 mm) such as *Ceriodaphnia* and *Daphnia*. These microcrustaceans are capable of feeding on the same size food as the filter feeding rotifers but with greater efficiency; in this way the arthropods act as exploitative competitors. However, the filtering currents of large cladocerans (e.g., *Daphnia* and *Scapholeberis*) are so strong that they can sweep slow-swimming rotifers (e.g., *Brachionus*, *Keratella*) into their branchial chambers. Because of the swift motions of the arthropod's thoracic legs, the rotifers are seriously injured or killed. This is termed interference competition. Both forms of competition can greatly affect rotifer population levels. In fact, rotifers can be extirpated from laboratory cultures containing *Daphnia*. Protists such as *Paramecium* also compete with rotifers, but in general protists are thought to be inferior competitors. Herbivorous copepods consume some of the same food resources, but we know less about their competitive abilities.

Phoresis may represent a specialized form of interference competition in which one organism attaches to, and colonizes, another. For instance, certain algae and protists colonize the surface of many rotifers, while others embed themselves in the gelatinous sheaths secreted by some colonial species (**Figure 8**). The impacts of colonization on rotifer populations are unknown. In addition, certain rotifers (e.g., *Brachionus rubens*) will attach to cladocerans (e.g., *Daphnia*) (**Figure 14**). This could inhibit the cladoceran's fitness in a variety of ways – adding mass that interferes with its ability to maintain itself in the water column, competing for food particles, and obstructing the arthropod's filtering current.

Predator–Prey Interactions

Rotifers are caught between the two millstones of size-selective predation. Tactile predators such as amoebas (heliozoans), ciliates (*Stentor*), hydrozoans (*Hydra*), copepods, and large, predatory rotifers readily consume small species. On the other hand, large rotifers are vulnerable to visual predators such as small fishes and certain aquatic insects.

Rotifers have evolved a variety of features to reduce predation pressure (**Table 1**). Many possess permanent spines (**Figure 15**), while others can undergo changes in morphology from one generation to the next, a phenomenon termed phenotypic plasticity. In *B. calyciflorus*, for example, the change in form is manifest by the production of posteriolateral spines that articulate with the animal's body (**Figure 16**). While swimming these spines point backward. However, when disturbed the animal retracts its corona, which increases the internal hydrostatic pressure of the body cavity, this in turn extends the spines laterally. In the presence of the predatory rotifer *Asplanchna*, offspring of unspined *B. calyciflorus* are born with spines. This embryonic induction is a

Table 1   Examples of evolutionary developments of rotifers to reduce predation pressure

Strategy	Rotifer species	Predator	Comments
Fixed spines	*Brachionus satanicus*, *Kellicottia longispina*, and *Keratella cochlearis*	*Asplanchna* spp.	Spines act as foils, making it more difficult for the predator to consume the rotifer
Moveable spines	*Brachionus calyciflorus*, *Filinia longiseta*, and *Polyarthra dolichoptera*	*Asplanchna* spp.	As mentioned in the previous entry. In *B. calyciflorus* spines are induced by presence of *Asplanchna*. This phenomenon is known as phenotypic plasticity
Setous, arm-like appendages and paddles	*Hexarthra mira* and *Polyarthra dolichoptera*, respectively	*Asplanchna* spp. and predatory copepods	Contraction of powerful internal muscles move these appendages, thus permitting rapid jumps away from the predator
Passive sinking	Most species	Predatory copepods	Rotifers stop swimming, thus ceasing the creation of currents that might be detected by tactile predators
Spherical shape	*Trochosphaera equatorialis*	*Asplanchna* spp.	The spherical shape of this rotifer makes it difficult for small predators to grasp
Transparent mucus sheaths in certain colonial species	*Conochilus unicornis*, *Lacinularia flosculosa*, and *Octotrocha speciosa*	*Asplanchna* spp.	Individuals retract into the gelatinous matrix, making it impossible for the predator to extract them
Tubes and mucus sheaths in many sessile rotifers	*Limnias melicerta*, *Collotheca gracilipes*, *Floscularia ringens*, *Ptygura crystallina*, *Octotrocha speciosa*, and *Stephanocerus fimbriatus*	Various predators (e.g., insect larvae and snails) that forage on the surfaces of aquatic plants	These tubes and sheaths are believed to afford the adult rotifer refuge from grazing invertebrates
Chemical deterrence	*Sinantherina semibullata* and *Sinantherina socialis*	Small zooplanktivorous fishes and certain aquatic insects	Specialized structures called warts are located at the anterior end of each animal in the colony; they appear to hold chemicals that render the animals unpalatable

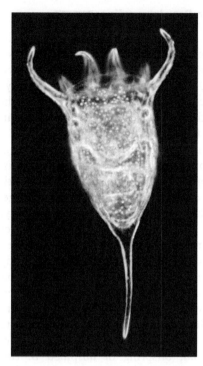

Figure 15 *Keratella taurocephala* in dark field microscopy. Note permanent spines at the anterior as well as the single large posterior spine. (Animal *c*. 225 μm.) (Photomicrograph courtesy of Dr. Robert Moeller, Miami University of Ohio, Oxford, OH.)

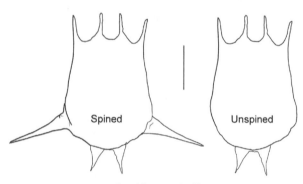

Spined   Unspined

Brachionus calyciflorus

Figure 16 Spined and unspined forms of *Brachionus calyciflorus*. In both animals the corona has been retracted (Bar *c*.100 μm.) (RLW & HAS.)

response to the presence of chemicals released by *Asplanchna* (a kairomone). Spined forms of *B. calyciflorus* are protected from ingestion, whereas unspined forms are nearly always eaten when captured by *Asplanchna*.

Applied Studies

Aquaculture

Many freshwater and marine fishes feed on rotifers, especially those taking their first meal after exhausting their yolk supply. Aquaculture industries utilize this relationship in creating large-scale, commercial systems. They maintain production systems with dense rotifer populations (i.e., kg quantities of biomass produced per day) as a food source for commercially valuable fishes. Larval stages of shrimps and crabs are also fed with rotifers.

Ecotoxicology

Rotifers are exploited for ecotoxicological studies because they are easily cultured, mature and reproduce rapidly, and have short generation times; these features facilitate studies of multiple cohorts. Moreover, toxic agents appear to be readily taken up across their integument, and through the gut while feeding. Some studies are straightforward analyses that expose cohorts of newborns to different levels of a potential toxicant, and then examine the number surviving over a specific time interval. Other common endpoints include changes in reproductive potential, feeding rate, and swimming behavior.

In conclusion, rotifers not only play a critical role in the food web of natural and artificial systems, but also serve many purposes in fundamental and applied research, particularly as a bioindicator for monitoring ecosystem health.

Acknowledgments

We thank Drs. Elizabeth J. Walsh, Thomas Schröder, and Winfried Lampert for their helpful comments on the manuscript and our colleagues who provided original artwork.

Glossary

Acanthocephala – Traditionally classified in their own phylum, these endoparasitic metazoans share the ancestral characteristic of an intracytoplasmic lamina with Rotifera, indicating a close phylogenetic relationship.

Amictic – The asexual portion of the life cycle of rotifers in which populations are exclusively female, and all eggs hatch into diploid females.

Amphoteric – The term used for females that produce both female and male eggs.

Anhydrobiosis – Literally life without water, a state in the lives of certain animals (e.g., nematodes, bdelloid rotifers, and tardigrades) in which the tissues have become dehydrated; see xerosome.

Aquaculture – Industrial production of commercially important aquatic animals.

Atrochidae – A family of monogonont rotifers possessing a funnel-shaped corona with long cilia.

Bdelloid (Bdelloidea) – A subclass of Class Eurotatoria comprising about 400 species; members lack males and reproduce exclusively by asexual parthenogenesis.

Benthic – The region of aquatic habitats that includes the interface between the water and sediments.

Bioindicator – An organism that serves as a miner's canary for conditions in an ecosystem (e.g., pollution or eutrophication).

Cingulum – In rotifers with a well-developed corona, the second circlet of cilia posterior to the mouth; compare trochus.

Classical food web – In aquatic systems, the construct used to describe trophic links and thus flow of energy and nutrients from the basal level of algae to top predators (often fishes); compare microbial loop.

Collothecaceae – One of three orders of Subclass Eurotatoria; includes Families Atrochidae and Collothecidae; compare Flosculariaceae and Ploima.

Collothecidae – A family of monogonont rotifers possessing a funnel-shaped corona with long cilia; most are sessile.

Conochilidae – A family of colonial, free-swimming monogonont rotifers that feed on tiny food particles.

Corona – The apical field of rotifers, typically possessing cilia or setae or both. A well-developed corona is composed of a ciliated cingulum and trochus.

Cryptic speciation – A term used to describe the phenomenon whereby two or more similar forms once assigned to a single taxon have been determined by genetic analysis to be distinct species; see species complex.

Cyclic parthenogenesis – The overall reproductive scheme of monogonont rotifers, in which asexual reproduction predominates but is interspersed with episodes of sexual reproduction; see amictic and mictic.

Diapause – A stage in the life cycle of some invertebrates in which metabolism is greatly reduced; occurs in monogonont resting eggs.

Ecotoxicology – The study of how chemicals act as toxic agents.

Eurotatoria – The class of rotifers comprising the bulk of the phylum; compare to Pararotatoria.

Eutely – A feature in some small invertebrates in which the number of cells of each individual remains constant. Cell number is determined early in embryonic development and subsequent increases in body size reflect increased cell size, not cell number.

Flosculariaceae – One of three orders of Subclass Eurotatoria, which includes the families Conochilidae and Flosculariidae.

Flosculariidae – A family of monogonont rotifers with diverse morphology that feed on tiny food particles. Many are sessile; a few are colonial.

Flukes (Platyhelminthes) – A type of parasitic flatworm; their larval stages are occasionally present in fresh waters.

Fulcrum – An unpaired piece of the trophi of monogonont rotifers (fulcrum and paired rami comprise the incus).

Illoricate – A term used to describe the body wall of a rotifer with a thin intracytoplasmic lamina.

Incus – An element of the rotifer trophi consisting of the fulcrum and paired rami.

Intracytoplasmic lamina – A deposit of special filamentous proteins within the integument of rotifers.

Kairomone – A chemical substance produced by an organism that causes a beneficial change in the behavior, physiology, or development of another organism to the disadvantage the producer.

Limnoterrestrial – An aquatic habitat found in a terrestrial setting, such as the film of water on terrestrial plants or within the interstices of soil.

Loricate – The body wall of a rotifer with a thick intracytoplasmic lamina; these body walls are somewhat rigid.

Littoral zone – The inshore region of a lake that is usually dominated by vegetation.

Male dwarfism – The condition of structural reduction in males of certain species of monogonont rotifers, often includes the loss of feeding capacity.

Malleus – Paired elements of the rotifer trophi consisting of the manubria and unci.

Manubria (um) – An elongate pair of club-shaped pieces of the rotifer trophi that articulate with the uncus (manubrium and uncus comprise the malleus).

Mastax – The muscular pharynx of rotifers, which controls movements of the trophi.

Metazoans – All multicellular animals.

Microbial loop – In aquatic systems, the construct used to describe trophic links and thus flow of energy and nutrients among viruses, bacteria, yeast, protists, and rotifers; compare classical food web.

Microphagous – Feeding that involves the consumption of tiny particles such as small algae, bacteria, and yeast.

Mictic – The sexual phase of the life cycle of monogonont rotifers in which eggs develop into males if unfertilized and resting eggs if fertilized; seen only in the Eurotatoria.

Mixis – The term for sexual reproduction in monogononts.

Monogonont (Monogononta) – A subclass of class Eurotatoria comprising about 1600 species that reproduce by cyclic parthenogenesis (i.e., asexual reproduction with episodes of sexuality).

Nematode (Nematoda) – A worm characterized in part by a cuticle that is molted; may be parasitic or free-living.

Pararotatoria – A tiny class of obligatorily sexual marine rotifers (seisonids) with only three species; compare to Eurotatoria.

Parthenogenesis – The term for asexual reproduction, as seen in bdelloid and monogonont rotifers.

Phylogeny (phylogenetic) – The study of the evolutionary relatedness of organisms.

Phytal – A freshwater habitat dominated by algae or vascular plants.

Ploima – One of three orders of Subclass Eurotatoria, which includes the bulk of rotifer families; compare Collothecaceae and Flosculariaceae.

Psammic – Belonging to the wet sandy region extending from the water's edge a few meters into the terrestrial habitat.

Rami (us) – Paired, toothed pieces of rotifer trophi that articulate with the fulcrum at their base and move like forceps.

Resting egg – A diapausing embryo enclosed in a tough, cyst-like coat; the outcome of a male monogonont rotifer fertilizing a young mictic female.

Retrocerebral organ – A complex, saccate organ of unknown function located near the brain of some rotifers, with ducts opening to the apical field.

Seisonids – See Pararotatoria.

Species complex – A taxon currently defined as a single species that may represent two or more sibling species that have undergone speciation; see cryptic speciation.

Subitaneous – Eggs that hatch without a delay in development.

Syncytium – In animals, a group of cells that have lost the plasma membrane separating cells, thus forming a multinucleate mass.

Tardigrade (Tardigrada) – A small phylum of eight-legged, tiny, arthropod-like animals; commonly called water bears in reference to the pawing movement of their appendages.

Trochus – In rotifers with a well-developed corona, the preoral circlet of cilia; compare cingulum.

Trophi – The jaws of rotifers, consisting of seven elements: incus (three pieces) and mallei (four pieces); sometimes used in species identification of monogononts. Morphology and the movement of trophi often differ among the taxa.

Tun – A desiccated animal of Phylum Tardigrada; equivalent to the xerosome in bdelloids.

Unci (us) – A pair of toothed pieces of the rotifer trophi that articulate with the manubrium (uncus and manubrium comprise the malleus).

Xerosome – The dormant body of bdelloids that have undergone desiccation; potentially viable for decades.

See also: Algae; Chrysophytes – Golden Algae; Cladocera; Competition and Predation; Copepoda; Cyanoprokaryota and Other Prokaryotic Algae; Cyclomorphosis and Phenotypic Changes; Diatoms; Diel Vertical Migration; Egg Banks; Eutrophication of Lakes and Reservoirs; Fungi; Green Algae; Microbial Food Webs; Other Phytoflagellates and Groups of Lesser Importance; Other Zooplankton; Phytoplankton Population Dynamics: Concepts and Performance Measurement; Protists; Viruses.

Further Reading

Brandl Z (2005) Freshwater copepods and rotifers: Predators and their prey. *Hydrobiologia* 546: 475–489.

Dumont HJ and Segers H (1996) Estimating lacustrine zooplankton species richness and complementarity. *Hydrobiologia* 341: 125–132.

Fussmann GF, Ellner SP, and Hairston NG Jr. (2003) Evolution as a critical component of plankton dynamics. *Proceedings of the Royal Society of London B* 270: 1015–1022.

Galkovskaya GA and Mityanina IF (2005) Structure distinctions of pelagic rotifer plankton in stratified lakes with different human impact. *Hydrobiologia* 546: 387–395.

Hampton SE (2005) Increased niche differentiation between two *Conochilus* species over 33 years of climate change and food web alteration. *Limnology and Oceanography* 50: 421–426.

Lair N (2005) Abiotic vs. biotic factors: Lessons drawn from rotifers in the Middle Loire, a meandering river monitored from 1995 to 2002, during low flow periods. *Hydrobiologia* 546: 457–472.

MacIsaac HJ and Gilbert JJ (1991) Discrimination between exploitive and interference competition between Cladocera and *Keratella cochlearis*. *Ecology* 72: 924–934.

Ricci C and Capriolo M (2005) Anhydrobiosis in bdelloid species, populations and individuals. *Integrative and Comparative Biology* 45: 759–763.

Ruttner-Kolisko A (1974) Planktonic rotifers: Biology and taxonomy. *Die Binnengewässer (Supplement)* 26: 1–146.

Segers H (1996) The biogeography of littoral *Lecane* Rotifera. *Hydrobiologia* 323: 169–197.

Stelzer C-P (2005) Evolution of rotifer life history strategies. *Hydrobiologia* 546: 335–346.

Stemberger RS and Gilbert JJ (1987) Defenses of planktonic rotifers against predators. In: Kerfoot WC and Sih A (eds.) *Predation: Direct and Indirect Impacts on Aquatic Communities*, pp. 227–239. Hanover: University Press of New England.

Wallace RL (2002) Rotifers: Exquisite metazoans. *Integrative and Comparative Biology* 42: 660–667.

Wallace RL and Snell TW (2001) Rotifera. In: Thorpe J and Covich A (eds.) *Ecology and Classification of North American Freshwater Invertebrates*, 2nd edn., pp. 195–254. New York: Academic Press.

Wallace RL, Snell TW, Ricci C, and Nogrady T (2006) Rotifera, Vol. 1: biology, ecology and systematics. In: Segers H and Dumont HJ (eds.) *Guides to the Identification of the Microinvertebrates of the Continental Waters of the World, No. 23*. 2nd edn. Leiden, The Netherlands: Backhuys Publishers.

Relevant Websites

http://dmc.utep.edu/rotifer/proj.html
http://jbpc.mbl.edu/wheelbase/
http://users.unimi.it/melone/trophi/
http://www.ansp.org/research/biodiv/rotifera_home.php
http://www.micrographia.com/index.htm
http://www.microscopy-uk.org.uk/mag/wimsmall/extra/rotif.html

Cladocera

J E Havel, Missouri State University, Springfield, MO, USA

Introduction

Cladocerans are common inhabitants in a variety of inland waters. Attracted by their large swarms, naturalists described and depicted cladocerans as early as the 1600s. Extensive sampling from natural habitats has provided detailed information about distributions, as well as population and community structure. Transparency of the body allows observing clutch sizes and thus estimating birth rates of natural populations. Their small size and ease of culture have made cladocerans the favorite subjects in experimental work, which provides detailed understanding of the life cycle, physiology, and interactions of cladocerans with their physical environments and with other species.

Taxonomy

Cladocerans belong to the crustacean class Branchiopoda, most of which have flattened, leaf-like appendages. Besides the cladocerans, other branchiopods include the phyllopods: fairy shrimp and brine shrimp (order Anostraca), tadpole shrimp (Notostraca), and clam shrimps (orders Spinicaudata and Laevicaudata). In contrast to the widespread cladocerans, phyllopods are restricted to fishless habitats, such as temporary ponds and salt lakes.

Cladocerans show a fantastic range of morphological variation (**Figures 1** and **2**), and are classified into 16 families within four distantly-related orders (**Table 1**). There is currently disagreement among systematists as to whether these four orders are monophyletic or polyphyletic in origin. All agree that the term 'Cladocera' is no longer a taxonomically accepted designation. Nevertheless, the term is still widely used for convenience. Most cladocerans (orders Anomopoda and Ctenopoda) have a carapace that covers the body below the head and also encloses a brood chamber (**Figure 3**). The orders Onychopoda and Haplopoda lack such a complete carapace, which instead only covers the brood chamber. The Haplopoda, represented by a single widespread species, *Leptodora kindtii*, has a long, cylindrical head (**Figure 1**) and a body size much larger than other cladocerans have (**Figure 4**). The Onychopoda have a more-rounded head and include *Polyphemus*, the marine and Ponto-Caspian Podonidae, and Ponto-Caspian Cercopagidae (**Figure 5**).

Currently, 86 genera (**Table 1**) and 400–500 species of cladocerans are recognized worldwide. These numbers are approximate, as taxonomy in several families is in a state of flux. Complicating the taxonomy is the occurrence of clonal lineages, interspecific hybrids, and phenotypic plasticity in response to environmental factors. The expanding use of molecular tools (e.g., DNA sequencing) has contributed to some taxonomic reassignments, discovery of further species, and provides clues for better understanding of evolutionary relationships. Nevertheless, most limnologists use morphological features (**Figure 3**) for routine identifications. Widely-used taxonomic keys and some recent monographs are listed below (see Further Reading), and a more extensive listing of monographs and detailed keys to specific groups is available through the 'Zooplankton Project' (see Relevant Websites).

Where Cladocerans Live

Along with rotifers and copepods, a variety of cladocerans are common in the plankton of lakes and ponds (**Figure 1**). Other groups, particularly the chydorids, are diverse and abundant in the near-shore vegetation and sediments (**Figure 2**). Cladocerans live in virtually all aquatic habitats, ranging from temporary ponds and wetlands to deep lakes. Although most cladocerans live only in fresh water, a variety of taxa can tolerate high salinities. Four genera occur in marine habitats (**Table 1**) and many of the onychopods are euryhaline, having evolved in the brackish Caspian Sea. A few anomopod taxa live in salt lakes (*Moina salina* and *Alona salina*). Cladocerans also live in flowing water. Chydorids (e.g., *Alona phreatica*) live with harpacticoid copepods and ostracods in the meiofauna of streams. Zooplankton are common in the water column of large rivers and can be surprisingly diverse. Here, the zooplankton assemblage appears to be an amalgam of species from upstream impoundments, near-shore slack areas (e.g., natural and artificial dikes), immigrants from floodplain wetlands, and those capable of growing and reproducing in the main channel.

Roles in Food Webs

Cladocerans are important components of food webs. Most species eat algae, supplemented with detritus

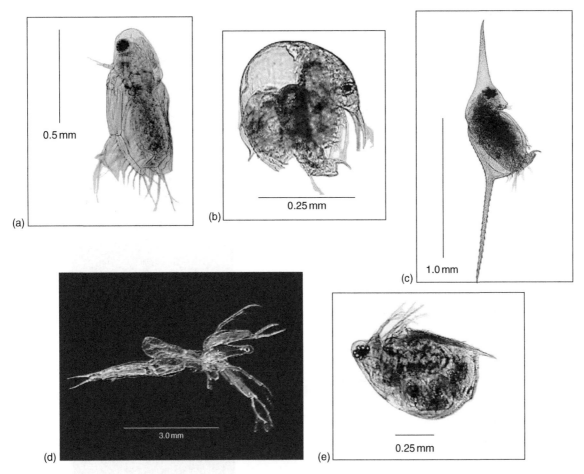

Figure 1 Some species of cladocerans living in pelagic habitats: (a) *Diaphanosoma birgei*, (b) *Bosmina longirostris*, (c) *Daphnia lumholtzi*, (d) *Leptodora kindtii*, (e) *Scapholeberis armata*. Photos B, D, and E are from the Zooplankton Project (see web site listed at end of this article). Photos A and C courtesy of Dave Hambright.

and bacteria. Filter feeding allows concentrating particles dispersed in the water column (see mechanism in Feeding and Digestion section). The exceptions are predaceous *Leptodora* (**Figure 1**) and the onychopods (**Figure 5**), which make their living by capturing other zooplankton. Littoral-benthic species of cladocerans (**Figure 2**) and those living in wetlands and rivers consume large quantities of detritus and associated bacteria. Cladocerans, in turn, serve as food for copepods and predaceous cladocerans, mysids, insects (e.g., *Chaoborus, Notonecta*), small fish species, and larval and juvenile stages of larger fish. In fishless habitats, salamanders and some water birds may also be important predators. In littoral habitats, cladocerans are also exposed to a variety of benthic predators, such as odonate nymphs and flatworms.

High grazing rates allow some cladocerans, such as *Daphnia*, to be keystone species in the pelagic zone of lakes. During late spring, their high numbers can control the abundance of algae, leading to a dramatic 'clearwater phase' that ends when *Daphnia* are depleted

by planktivorous fish. Managing the abundance of fish may thus have cascading effects through this food chain, affecting water clarity. However, the story is not quite so simple, as noxious and colonial algae that are difficult for *Daphnia* to eat may replace the edible forms, leading to increased turbidity even when grazers are abundant.

Structure and Function

Body Size and Molting

Cladocerans are generally small animals, with the maximum size of most species less than 1 mm. Nevertheless, the range in body size among cladoceran species is quite large (**Figure 4**). In the pelagic zone, 0.4 mm (adult) length *Bosmina longirostris* may coexist with 18 mm *L. kindtii*. In a single family (Chydoridae), body size ranges from 0.3 mm *Alonella nana* to 6.0 mm *Eurycercus glacialis*. Such differences in body size affect food selection, susceptibility to predators, and habitat use in benthic environments.

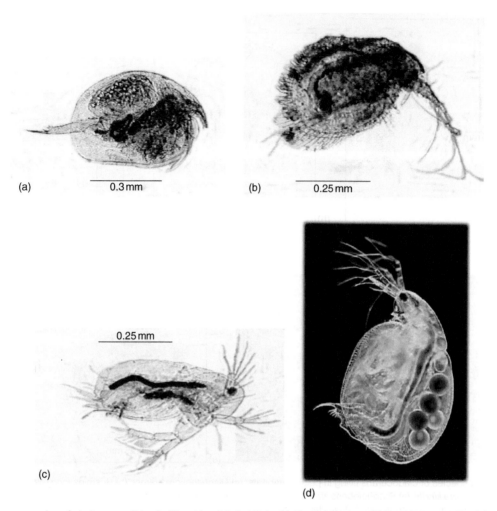

(a) ⎯⎯⎯⎯⎯ 0.3 mm

(b) ⎯⎯⎯⎯⎯ 0.25 mm

0.25 mm

(c)

(d)

Figure 2 Some species of cladocerans living in littoral-benthic habitats: (a) *Kurzia latissima*, (b) *Ilyocryptus spinifer*, (c) *Latonopsis occidentalis*, (d) *Simocephalus vetulus*. Photos A, B, & C are from the Zooplankton Project. Photo D taken by Rudolf Ruediger.

Cladocerans, like other crustaceans, have a chitinous exoskeleton that covers the entire surface of the body, including the setae on the limbs. The exoskeleton serves as points of attachment for the muscles and also protection against enemies. At each molt, the entire exoskeleton is shed and a new exoskeleton secreted. While the body is still soft, the animal imbibes water and swells to become as large as possible before the new exoskeleton hardens. Juveniles may double their body volume with each molt. The old shed exoskeleton (called an exuvium) bears a stunning resemblance to the animal that shed it. These exuviae are rapidly degraded by chitinases in the environment. Nevertheless, some hard parts, such as *Daphnia* mandibles, *Bosmina* rostrums, and chydorid head shields, persist and accumulate in the sediments, where they can later be used to describe ancient lake communities.

In contrast to the copepods, which stop growth at maturity, cladocerans continue to grow and molt throughout life. For instance, observations of laboratory isolates indicate *Daphnia schoedleri* to have up to 28 instars. Growth in size provides a larger brood chamber, which can accommodate larger clutches of eggs.

Carapace, Helmet, and Tailspine

The most common and abundant cladoceran species in freshwater environments (anomopods and ctenopods) have a carapace (valves, shell) that covers the body below the head (**Figure 3**). The carapace is hydrophobic, putting cladocerans near the air-water interface at risk of being trapped in the surface film. During certain periods, the carapace may be coated with epibionts, which must recolonize after each molt.

The carapace resembles an overcoat, open at the front and closed at the back. It is attached to the rest of the body at the neck. Water and suspended food can pass inside the carapace, entering the food groove between the legs. The posterior end of the carapace may be elaborated into a tailspine (**Figure 3(a)**) that is present in some species but not in others (**Figure 1**). In

Table 1 The taxonomy of 'Cladocera'

Orders and families	Genera
Anomopoda	
Acantholeberidae	Acantholeberis
Bosminidae	Bosmina, Bosminopsis, Eubosmina
Chydoridae	Acroperus, Alona, Alonella, Anchistropus, Archepleuroxus, Australochydorus, Bryospilus, Camptocercus, Chydorus, Dadaya, Disparalona, Dunhevedia, Ephemeroporus, Estatheroporus, Euryalona, Graptoleberis, Indialona, Karualona, Kozhowia, Kurzia, Leberis, Leydigia, Ledigiopsis, Monope, Monospilus, Nicsmirnovius, Oxyurella, Paralona, Planicirclus, Pleuroxus, Plurispina, Picripleuroxus, Pseudochydorus, Rak, Rhynchochydorus, Rhynchotalona, Spinalona, Tretocephala
Daphniidae	Ceriodaphnia, Daphnia, Daphniopsis, Megafenestra, Moina, Moinodaphnia, Scapholeberis, Simocephalus
Eurycercidae	Eurycercus
Ilyocryptidae	Ilyocrypsis
Macrothricidae	Bunops, Cactus, Drepanothrix, Grimaldina, Guernella, Lathonura, Macrothrix, Onchobunops, Pseudomoina, Streblocerus, Wlassicsia
Neothricidae	Neothrix
Ophryoxidae	Ophryoxus, Parophryoxus
Sayciidae	Saycia
Ctenopoda	
Holopedidiidae	Holopedium
Sididae	Diaphanosoma, Latona, Limnosida, Penilia, Pseudosida, Sarsilatona, Sida
Onychopoda	
Cercopagidae	Bythotrephes, Cercopagis
Podonidae	Caspievadne, Cornigerius, Evadne, Pleopsis, Podon, Podonevadne, Pseudoevadne
Polyphemidae	Polyphemus
Haplopoda	
Leptodoridae	Leptodora

Orders are based on Fryer (1987) and families and genera are based on Dumont and Negrea (2002) (see Further Reading). Genera that are underlined are restricted to the oceans; *Evadne* is common in both the ocean and the Caspian Sea; and all other genera in the Podonidae are restricted to the Ponto-Caspian region.

some species, the tailspine may be longer than the rest of the body (e.g., **Figure 1(c)**). The caudal appendage (posterior spine) of *Bythotrephes* and *Cercopagis* is extremely long and includes a complex mix of bends, spines, and claws (**Figure 5**). Such a long spine may reduce sinking rates or deter size-limited predators, but how it evolved is unclear.

In some daphniid species, the head above the compound eye may be extended to form a helmet. Depending on species, the helmet may take various shapes, ranging from an elongated head spine (**Figure 1(c)**), expanded recurved helmet (e.g., *Daphnia retrocurva*), and expanded dorsal crest (*Daphnia cephalata*) or spines (e.g., 'neck teeth' of induced *Daphnia pulex*). The sizes of the helmet and tailspine can be highly variable within and among populations, and show seasonal polymorphism (cyclomorphosis) as well as developmental responses to chemicals from predators.

Appendages

The primitive crustacean body plan is segmented, with a pair of jointed appendages on each somite. Cladocerans show this segmentation during the embryonic period, but the somites fuse by the time the newborns (neonates) emerge from the brood chamber. The head bears (from anterior to posterior) two pairs of antennae, a pair of mandibles, and one pair of first maxillae (maxillules) (**Figure 3(a)**). The second maxillae are either restricted to a small protrusion or they are absent. In some groups the first maxillae are also either reduced (onychopods) or absent (haplopods).

The first antennae are typically small and unbranched, and function as chemosensory organs. The second antennae are prominent, have two branches (**Figure 3(a)**) (except in *Holopedium*), and function in locomotion. For pelagic species (**Figure 1**), the antennae are used for swimming and have long plumose setae. Since water tends to flow around and not through the intersetule spaces under low Reynolds number (Re) conditions, these antennae behave hydrodynamically as paddles. Each second antenna is served by three large muscles, which allow strong sustained swimming and contribute to the ability of cladocerans to vertically migrate For benthic species, like *Ilyocryptus* (**Figure 2(b)**), the antennae are short and stout and are used for crawling and burrowing.

The trunk bears the thoracic appendages (legs), which vary in number: four pairs in onychopods, five to six in anomopods, and six pairs in haplopods and ctenopods. The legs perform a variety of functions, including grasping mates or prey and filtering or scraping food.

Feeding and Digestion

Most cladocerans (anomopods and ctenopods) are herbivorous, feeding on small particles (particularly algae). Their legs are flattened and setose. In pelagic species, beating of the legs creates water currents for moving water into the filtering chamber (food

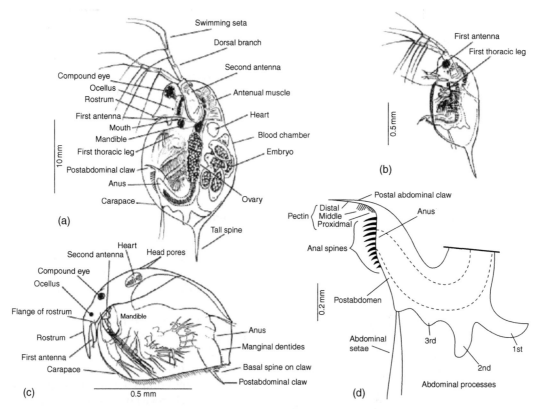

Figure 3 Anatomy of cladocerans: (a) *Daphnia pulicaria* female, (b) *D. pulicaria* male, (c) *Pleuroxus trigonellus* female, (d) detailed view of postabdomen from *D. pulicaria* female. A, B, and C reprinted from *Ecology and Classification of North American Freshwater Invertebrates*, J.H. Thorp and A.P. Covich (eds.), Dodson, S. I., and D. G. Frey, Cladocera and other Branchiopoda, pages 850–914, Copyright 2001, with permission from Elsevier.

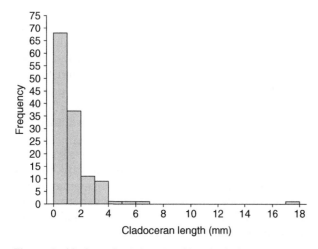

Figure 4 Maximum body lengths of female cladocerans from North America. Frequency indicates the number of species that fall within each size class. Body length does not include tailspine, but does include helmet. Data from Brooks (1959), see Further readings, plus other sources for exotic species *Bythotrephes longimanus, Cercopagis pengoi,* and *Daphnia lumholtzi.*

groove) between the legs. Littoral species scrape food off surfaces. The legs are armed with numerous hair-like setae and on each seta are fine setules that ensnare the particles. Electrostatic attraction also aids in this process. Because of the low *Re* hydrodynamic environment, water flows and particle movements are reversible, and the animal must expend considerable energy to squeeze water through the filters.

Other legs are used to scrape the filters and move the food toward the mouth, where it is manipulated by the maxillules and tasted by the first antennae. Food that is acceptable is ground by the mandibles, which are stout and have large tough surfaces. The labrum (upper lip) secretes mucous, which is mixed with the food before the bolus enters the mouth. If the food contains noxious algae, the entire bolus is rejected by flicking the postabdominal claws (**Figure 3(d)**) through the food groove. Because all food is collected and processed, filter-feeding clado-cerans waste considerable energy in environments where food density is low or quality is poor.

1.0 mm

Figure 5 Two onychopod invaders of the Laurentian Great Lakes, *Cercopagis pengoi* (top) and *Bythotrephes longimanus* var. *cederstroemi* (bottom). From Vanderploeg HA, *et al.* (2002). Dispersal and emerging ecological impacts of Ponto-Caspian species in the Laurentian Great Lakes. *Canadian Journal of Fisheries and Aquatic Sciences* 59: 1209–1228. With permission.

In contrast to these filter feeders, predaceous cladocerans (onychopods and haplopods) are raptorial feeders. Their legs are tubular and armed with stiff spines (**Figure 5**). Moving prey (smaller zooplankton) are captured with the legs, bitten with stoutly-toothed mandibles, and soft tissues sucked in by esophageal contractions.

Cladocerans have a one-way digestive tract (gut), ending in the anus on the postabdomen (**Figure 3**). The foregut and hindgut are lined with chitin. The foregut may have outpockets (hepatic caece) in the head that increase processing time of the food bolus. Animals feeding on algae show guts that are bright green, becoming brown toward the hindgut. Nutrients are absorbed in the midgut. Peristaltic contractions cause water to flow into the hindgut ('anal drinking'), allowing further processing of nutrients.

Circulation and Respiration

The small body size and flattened body shape of cladocerans results in a high surface-to-volume ratio (S/V). The high S/V simplifies the problem of exchanging nutrients, wastes, and gases between their internal spaces and the external environment. Circulation is increased with a heart (**Figure 3**). Cladocerans have no blood vessels. The clear blood moves from the heart directly into an internal cavity (hemocoel) that extends throughout the body. Gas exchange and excretion of nitrogenous wastes (as ammonia) occur across the flattened limbs and the inside surface of the carapace. When exposed to low oxygen for extended periods, cladocerans can manufacture hemoglobin, but usually it is absent or in low concentration.

Nervous and Sensory Systems

Cladocerans have a primitive nervous system, with a ventral ladder-like nerve cord extending through the trunk and a concentration of ganglia in the head. For sensation, cladocerans rely to a large extent on mechanoreception (vibrations) and chemoreception (taste and smell) to learn about the environment: detect enemies, find and discriminate food, and locate mates. Mechanoreceptors are associated with setae on both pairs of antennae, as well as on the legs and abdomen. Chemoreceptors (aesthetascs) are located on the first antennae. Additional 'taste buds' are found on the trunk limbs in some species. Most cladocerans have a single compound eye and supplementary ocellus (**Figure 3**). Although these structures are capable of detecting only light intensity and direction, they provide sufficient cues for vertical and horizontal migration. Predatory onychopods have a large enough eye for image formation and thus can use sight for detecting prey.

Reproduction, Development, and the Life Cycle

Females dominate most cladoceran populations. In adults, a pair of ovaries can be seen lying adjacent to the midgut (**Figure 3**). In well-fed animals, the ovaries can be recognized by their dark green color and stored lipids in the eggs. The lipids appear as featureless yellow-orange droplets and may also be dispersed in the body cavity. During each molt cycle, eggs are extruded into the brood chamber. Here they complete embryonic development, to be released as free-swimming juveniles. Following extrusion of the eggs, the ovaries appear temporarily clear until a new batch of eggs develops, during which they become dark again. Following their release from the brood chamber, juveniles feed in the same way as adults, but are more sensitive to starvation. Lipids leftover from the egg (maternal investment) are important for early survival. Thus, like humans, well fed mothers tend to have healthy offspring. Depending on species, there may be two to seven juvenile instars before maturity. Culturing cladocerans in the laboratory has revealed that adults continue to grow and reproduce. As size increases, the number of eggs that can be produced (and stored in the brood chamber) also increases, up to a limit, after which fecundity decreases. Fecundity is highly variable among species: *B. longirostris* carries one to two eggs, whereas *D. cephalata* may have up to 295 eggs in a single brood. Several important life history rates depend on temperature, which include embryonic development time (e.g., in *D. pulex* 2 days at 25 °C and 11 days at 10 °C), juvenile instar duration, and lifespan (1 month at 25 °C and 3 months at 10 °C).

Most cladocerans have a life cycle ('cyclic parthenogenesis') that alternates between asexual and sexual reproduction and includes a quiescent resting stage (**Figure 6**). Most of the time, typically over multiple generations, the asexual (amictic) eggs develop without fertilization (parthenogenesis). During this asexual phase, the population may consist of 100% females. The mother and her offspring are diploid and, barring mutation, genetically identical to one another. Such a lineage is called an isofemale line or clone. Researchers can use this characteristic an advantage for experiments, by sorting out genetic

sources of variation and studying genotype-by-environment interactions.

Following an appropriate environmental cue, such as crowding, food shortage, or decreasing photoperiod, cladocerans typically enter the sexual phase of the life cycle (**Figure 6**). Mictic females develop haploid eggs that require fertilization. Females also produce diploid eggs that develop into males. Males are considerably smaller than females, and also differ from females in having enlarged first antennae and first thoracic legs, structures that are used in mating (**Figure 3(b)**). These males are genetically identical to the isofemale line from which they are descended. Hence, cladocerans use environmental sex determination; this can be contrasted with the rotifers, which also exhibit cyclic parthenogenesis, but with haplodiploid sex determination. Male cladocerans produce haploid sperm and mate with mictic females carrying the haploid eggs. The resulting zygote becomes a resting egg. In anomopods and most ctenopods, resting eggs are covered by a modified carapace, forming a darkly-pigmented ephippium, which is released when the mother molts.

Resting eggs allow persistence during unfavorable environmental periods (e.g., drying, abundant predators), and may lie dormant for many years before they receive the appropriate hatching cue Since they are resistant to freezing, drying, and digestion by fish and birds, resting eggs are also important for cladoceran survival during overland dispersal. Some ephippia have barbs and spines, processes that may aid in attachment to fur and feathers of migrating vertebrates. Resting eggs typically hatch into amictic

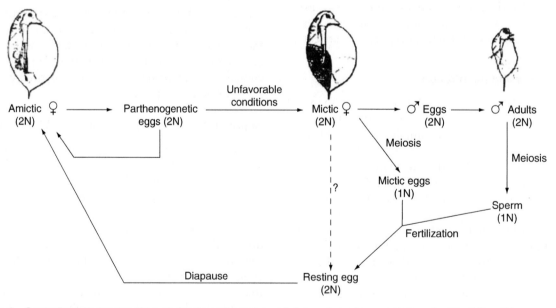

Figure 6 Generalized cladoceran life cycle (cyclic parthenogenesis). A few cladocerans are known to depart from this life cycle (see text).

females, although some anomopods from temporary ponds (*Moina* and *Daphniopsis*) hatch into both males and females. Other exceptions to the typical life cycle are some *Daphnia* that forgo the sexual phase and form resting eggs asexually (dotted line in **Figure 6**). These alternative life cycles are presumably adaptations to life in temporary ponds, where water may be available for only a few weeks.

Population Dynamics

Boom and Bust Dynamics

Cladocerans and rotifers are capable of explosive growth in population size, with numbers increasing by orders of magnitude over a few weeks (**Figure 7**). A typical cladoceran population in the water column of a dimictic lake has very low density during the winter. As light increases and waters mix in spring, resting eggs hatch, and the abundant food also allows overwintering individuals to increase their reproductive investment. The asexual subitaneous eggs develop without any pause. Since development time (and hence generation time) is shorter at warm temperatures, birth rate is higher. The population continues to grow until it is regulated by reduction in quantity or quality of food supply or other biotic interactions. Individuals in populations limited by food will typically carry very few eggs. Population declines can be rapid 'crashes.' Hence, the overall population dynamics of cladocerans is characterized by extreme oscillations in density.

In warm monomictic lakes (and reservoirs), the water column circulates all winter and food levels often remain high. Here, the harshest environmental conditions occur in summer, when temperatures can be high and cyanobacteria abundant. During this period native *Daphnia* species typically crash, whereas a tropical exotic (*Daphnia lumholtzi*) reaches maximal abundance (**Figure 7**). Nevertheless, the general pattern of rapid increases characterizes cladoceran populations in these lakes too.

Exploring Population Dynamics

During exponential growth (or decline), intrinsic growth rates (r) can be estimated between any two population estimates (N_0 and N_t) by rearrangement of the exponential growth curve [$r = (\ln N_t - \ln N_0 / t$]. Reproductive investment at the population level is typically reported by the egg ratio (E), calculated as the total number of eggs carried by adult females divided by the total number of females counted from the field. By including laboratory estimates of egg development time (D), birth rates can be estimated

using Paloheimo's formula [$b = \ln(E + 1)/D$]. Death rates (d) are then determined by difference ($r = b - d$). This approach has been very useful for better understanding the importance of biotic interactions in controlling seasonal succession.

However, there are two important limitations to this approach. First, in stratified lakes, temperature is quite different between the surface and deeper waters. As development time of eggs is strongly dependent on temperature, birth rates are highly sensitive to the depth where animals are spending most of their time. If the population is structured according position in the water column, some members of the population could be growing much faster than the rest of the population. Second, the simple formula ($r = b - d$) ignores the roles of immigration and emigration, which are sometimes substantial. During periods of high hatching rates of resting eggs, immigrants into the water column can cause population growth rate exceed birth rate. In reservoirs, advective losses via outflowing rivers may be very important, particularly during rainy periods when water renewal time is short.

Biogeography, Dispersal, and Exotic Species

Biogeographic Patterns

Cladocerans have a world-wide distribution, ranging from the Arctic to the tropics. Anomopods occur on all the continents, and include numerous widespread families (e.g., Bosminidae, Chydoridae, and Daphniidae). Some taxa have even been reported from Antarctica (e.g., *Daphniopsis studeri* and *Alona weinecki*). Ctenopods include holarctic species (*Holopedium gibberum* and *Sida crystalina*), widespread species of tropical origin (e.g., *Diaphanosoma*), and estuarine-marine species (e.g., *Penilia avirostris*). The sole haplopod, *L. kindtii*, is widespread in Europe, central Asia, and North America. Most of the onychopods are native to the Ponto-Caspian region, although several genera are widespread in the oceans (*Pleopsis*, *Podon*, and *Pseudoevadne*). The Polyphemidae includes one species that is holarctic (*Polyphemus pediculus*) and one species endemic to the Caspian Sea (*Polyphemus exiguous*).

Although freshwater ecologists long assumed that cladoceran species were cosmopolitan, we now know that most of these species have distributions that are more restricted geographically. Following detailed study of chydorid morphology, David Frey showed that many cases of apparent cosmopolitan distributions were due to a confused taxonomy and most species in fact have more-restrictive distributions. This is apparently the case in many other groups as well.

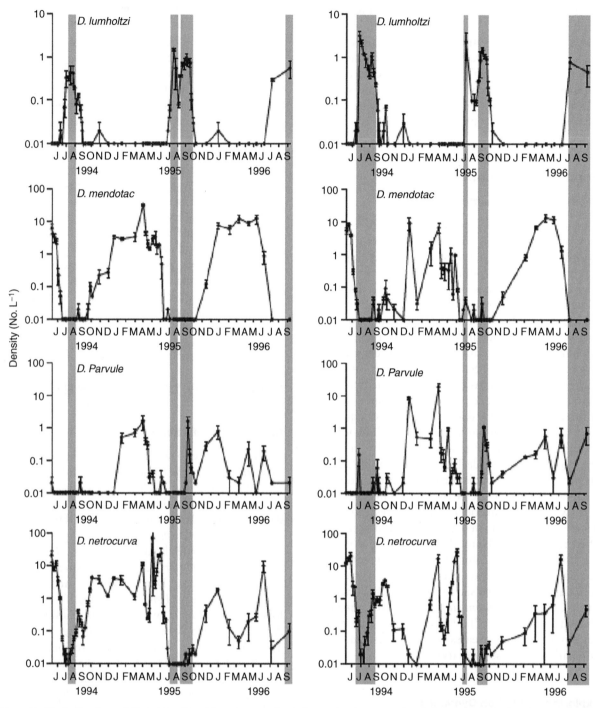

Figure 7 Similarity of population dynamics of four common *Daphnia* species from two sites (left and right panel) in Stockton Lake, Missouri USA, a warm monomictic reservoir. Shaded bars indicate periods when *D. lumholtzi* density exceeded 0.4 per liter. Adapted from Havel JE and Graham J (2006) Complementary population dynamics of exotic and native *Daphnia* in North American reservoir communities. *Archiv für Hydrobiologie* 167: 245–264. With permission (http://www.schweizerbart.de).

For example, nearly two-thirds of all *Daphnia* species are restricted to single continents, with rather few occurring worldwide (**Table 2**). In both the northern and southern hemispheres, richness of cladoceran species is greater in the temperate zone than in the tropics or near the poles.

Several explanations have been erected for explaining global distribution patterns of cladocerans. Fossil remains (e.g., *Archedaphnia* and *Propleuroxus*) have been detected in sediments of Permian age (225–280 Ma), and thus ancestors of modern species preceded breakup of ancient land masses (vicariance). An alternative explanation is that cladoceran species were once widely distributed across the latitudes when climates were similar, but became constricted in the tropics during the Tertiary, when temperatures became different. A simple explanation for low cladoceran diversity in the lowland tropics is the dominance of fish in even the smallest water bodies.

Dispersal and Its Scale

Resting eggs provide a convenient package for overland dispersal, and asexual reproduction allows a single hatchling to establish a population. Since a variety of potential vectors may carry resting eggs from sources to new habitats (wind, migrating vertebrates), one would assume that cladocerans should be capable of easily dispersing. However, the fact that many species have restricted distributions implies that successful colonization is generally a rare process. Furthermore, genetic studies have generally revealed strong spatial structure. Either overland dispersal is not as common as it appears or else most intact communities are not invasible. At small spatial scales (<1 km), communities are probably not limited by dispersal. Experiments have shown that open habitats can be readily colonized and that intact zooplankton communities tend to resist invasion. Invasibility may also be operating at the population level, as early-arriving clones develop a numerical advantage, both in the water column and in the egg bank, sufficient to out-compete later-arriving clones.

Larger-scales are difficult to manipulate experimentally, and so another approach is required. Time-series analyses of changing distributions of an invading exotic cladoceran (*D. lumholtzi*) reveal that distance to source populations affects risk of invasion in subsequent years. The likelihood of invasion decreases as distance to source populations increased, with the steepest drop at >20 km. In other words, at medium-large distances, dispersal can limit community structure. However, long-distance hops do occasionally occur and these can greatly increase the rate of range expansion by exotic species.

Exotic Species and Links to Humans

Human actions have undoubtedly extended the natural range of cladocerans, as well as other taxa. During the late twentieth century, ballast tanks of large ships introduced two Ponto-Caspian cladocerans from Europe to the Laurentian Great Lakes (*Bythotrephes longimanus* and *Cercopagis pengoi*, **Figure 5**). (A similar mechanism likely carried in zebra mussels and other invaders.) These cladocerans had first colonized the Black Sea, moving either with ships, via canals, or with invertebrates stocked in nearby reservoirs for fish forage. Other species have also been carried by humans; for instance, *D. lumholtzi* (**Figure 1**) was likely introduced to North America with stocked African fish. A listing of known species of exotic cladocerans is shown in **Table 3**. Since many

Table 2 Global distribution of *Daphnia* species, as frequency of occurrence on the continents

Distribution (number of continents)	Frequency (number of species)
1	49
2	8
3	11
4	4
5	2
6	0
7	0

Number of continents is based upon species distributions in Benzie (2005), see Further Reading. The six species occurring on four-five continents include: *D. magna*, *D. pulex*, *D. pulicaria*, *D. curvirostris*, *D. similis*, and *D. obtusa*.

Table 3 Documented examples of intercontinental invasions by cladoceran zooplankton

Species	Year invaded	Native range	Location where first population was reported
Alona weinecki	1774	Antarctica	Rano Raraku, Easter Island
Bythotrophes longimanus	1984	Eurasia	Laurentian Great Lakes (Lake Huron)
Cercopagis pengoi	1998	Eurasia	Laurentian Great Lakes (Lake Ontario)
Daphnia ambigua	1947	North America	Kew Gardens, England
*Daphnia galeata**	????	Eurasia	Laurentian Great Lakes (Lake Erie)
Daphnia lumholtzi	1990	Africa, Asia, & Australia	Joe Pool Lake, Texas
Daphnia obtusa	1994	Europe & North America	Australia
Daphnia parvula	1973	North & South America	Ponds and Lake Constance, Germany
Daphnia pulicaria	1988	North America	Fish pond in Macedonia
Eubosmina coregoni	1960s	Eurasia	Laurentian Great Lakes (Lake Michigan)
*Eubosmina maritima**	????	Eurasia (Baltic Sea)	Laurentian Great Lakes

Species marked with an asterisk (*) are cryptic invaders detected by molecular methods. Modified from Havel and Medley (2006) *Biological Invasions* 8: 459–473, with kind permission of Springer Science and Business Media.

invaders are likely cryptic, this list is probably a gross underestimate. Once introduced to a new continent, natural vectors (e.g., wind, flowing water, and birds) would be more important for increasing dispersal opportunities. In addition, recreational boaters likely increase the range expansion of these exotic species, as many boaters travel between lakes and the residual water in their live wells often contains living zooplankton. Such repeated introductions increase the probability that at least one invasion may succeed.

Exotic introductions may have important effects on freshwater communities. Invasion by *Bythotrephes* and subsequent selective predation resulted in reduced densities of some Great Lakes zooplankton species, and has had even more dramatic effects in inland lakes of the Canadian Shield, where cladoceran species richness rapidly declined relative to uninvaded lakes. However, long-term changes in these food webs are difficult to predict, as densities of exotics may become depressed over time while colonization can replace native species.

Glossary*

Amictic – Forming eggs without sexual recombination (see also parthenogenesis); also refers to a lake that never mixes.

Biogeography – Global distribution patterns of plants and animals, seen in both currently-living and extinct forms.

Biramous – Appendages that split into two branches, as seen in crustaceans.

Branchiopod – Class of crustaceans characterized by flattened leaf-like legs.

Chitin – A polysaccharide that is part of the exoskeleton of crustaceans.

Clone – All the genetically-identical asexual descendents from a single female.

Cosmopolitan – Having a world-wide distribution.

Cyclic parthenogenesis – Life cycle of cladocerans and rotifers that alternates between parthenogenetic and sexual phases.

Dimictic – Stratifying twice a year (summer and winter), with mixing in between.

Emigration – Movement of individuals out of a population.

Endemic – Geographic range restricted to a particular region.

Environmental sex determination – Gender is determined by environmental cues, not by genotype.

Ephippium – Thick, modified part of the carapace that forms a case surrounding the resting eggs of many cladocerans.

Epibionts – Organisms, such as protozoans, that live on the surface of other organisms.

Euryhaline – Tolerant of a wide range of salinities.

Exotic species – Species whose distribution has extended to a new continent.

Exuvium – The molted exoskeleton from an arthropod.

Holarctic – Having a distribution extending across the North America, Europe, and northern Asia.

Immigration – Movement of individuals into a population.

Instar – The life stage of arthropods in between molts.

Instinsic growth rate – The rate of exponential growth, measured by changes in numbers over time $[r = (\ln Nt - \ln N_0)/t]$.

Invasibility – The susceptibility of a community to invasion by new species.

Keystone species – A species that, when removed, has a large influence on the structure and function of a community.

Life table – A schedule of age-specific survival and reproduction.

Meiofauna – Community of small invertebrates (0.1–1.0 mm) living in the sediments.

Metapopulation – A group of conspecific populations separated in space and joined by dispersal.

Mictic – Forming eggs through meiosis and thus the eggs require fertilization to develop.

Molt – The process that arthropods undergo to shed their old exoskeleton.

Monophyletic – Derived from the same ancestral lineage.

Parthenogenesis – Asexual reproduction where eggs develop without fertilization.

Pelagic – Open water zone of a lake.

Polyphyletic – Derived from two or more ancestral lineages.

Propagule load – Rate of individuals immigrating to a new locale, which helps explain habitat susceptibility to invasion.

---

* List does not include taxonomic categories (see Table 1).

Resting egg – A stage of arrested development that requires a specific cue to hatch.

Reynolds number – A dimensionless number that represents the ratio of inertial forces to viscous forces.

Setae – Stiff feather-like bristles.

Setose – Bearing setae.

Somite – A body segment during development.

Subitaneous egg – An egg that has no pause in its development (cf. resting egg).

Systematics – Classification based on evolutionary relationships.

See also: Competition and Predation; Copepoda; Cyclomorphosis and Phenotypic Changes; Diel Vertical Migration; Egg Banks; Role of Zooplankton in Aquatic Ecosystems; Rotifera.

Further Reading

Benzie JAH (2005) *Cladocera: The Genus Daphnia (Including Daphniopsis)*. Leiden, Belgium: Backhuys Publishers.

Brooks JL (1959) Cladocera. In: Edmondson WT (ed.) *Freshwater Biology*. New York: John Wiley.

Dodson SI and Frey DG (2001) Cladocera and other Branchiopoda. In: Thorp JH and Covich AP (eds.) *Ecology and Classification of North American Freshwater Invertebrates*, pp. 850–914. San Diego: Academic Press.

Dumont HJ and Negrea SV (2002) *Introduction to the Class Branchiopoda*. Leiden, Belgium: Backhuys Publishers.

Frey DG (1982) Questions concerning cosmopolitanism in Cladocera. *Archiv für Hydrobiologie* 93: 484–502.

Fryer G (1987) Morphology and the classification of the so-called Cladocera. *Hydrobiologia* 145: 19–28.

Havel JE and Shurin JB (2004) Mechanisms, effects, and scales of dispersal in freshwater zooplankton. *Limnology and Oceanography* 49: 1229–1238.

Hebert PDN (1978) The population biology of *Daphnia* (Crustacea, Daphnidae). *Biological Review* 53: 387–426.

Korovchinsky NM (2006) The Cladocera (Crustacea: Branchiopoda) as a relict group. *Zoological Journal of the Linnean Society* 147: 109–124.

Martin JW (1992) Branchiopoda. Harrison FW and Humes AG (eds.) *Crustacea, Microscopic Anatomy of Invertebrates*, vol. 9, pp. 25–224. New York: Wiley-Liss.

Peters RH and de Bernardi R (eds.) (1987) *Daphnia*. Verbania Pallanza: Instituto Italiano di Idrobiologia.

Rivier IK (1998) *The Predatory Cladocera (Onychopoda: Podonidae, Polyphemidae, Cercopagidae) and Leptodorida of the World*. Leiden, Belgium: Backhuys Publishers.

Smirnov NN (1996) *Cladocera. The Chydorinae and Sayciinae (Chydoridae) of the World*. Leiden, Belgium: Backhuys Publishers.

Sommer U (ed.) (1989) *Plankton Ecology: Succession in Plankton Communities*. Berlin: Springer-Verlag.

Vogel S (1996) *Life in Moving Fluids: The Physical Biology of Flow*. Princeton, NJ: Princeton University Press.

Relevant Websites

Zooplankton of the Northeast United States (http://cfb.unh.edu/CFBkey/index.html). Images and keys to cladocerans, copepods, and rotifers of the northeastern United States.

Zooplankton Project (http://www.cnas.missouristate.edu/zooplankton/). Images of cladocerans and copepods from reservoirs of the central United States, plus a listing of references to monographs and taxonomic keys.

Cladocera (http://www.cladocera.uoguelph.ca). Images, taxonomy, and distribution of *Daphnia* in North America.

Freshwater Biology Association. (http://www.fba.org.uk/) Links to information on freshwater biota of Europe.

Copepoda

C E Williamson, Miami University, Oxford, OH, USA
J W Reid, Virginia Museum of Natural History, Martinsville, VA, USA

Introduction

The first known observation of copepods was recorded 23 centuries ago by Aristotle, who in his 'History of Animals' described a 'worm under the fins' of tuna and swordfish. The much smaller free-living copepods were only perceived after the development of modern optical instruments in Holland in the 17th century. In 1688, Stephan Blankaart of Amsterdam described a 'water-louse in a cistern' and clearly illustrated a cyclopoid copepod, and in 1699 Antony van Leeuwenhoek observed and drew copepods in Delft, Holland. Today, there are over 13 500 known species of these small crustaceans in the subclass Copepoda H. Milne Edwards, 1830. The word copepod derives from the Greek words for oar (*kope*) and foot (*podos*).

The copepods of inland waters are generally only 1–2 mm or less in body length, with an exoskeleton surrounding a cylindrical, segmented body and multiple jointed appendages that are used for swimming and feeding (**Figures 1–3**). A few of the larger species found in lakes with few or no visual predators may reach lengths of 3 mm or more (**Figure 4**). Copepods are generally either transparent or pale gray or brown, but may be bright green, blue, yellow, red, purple, or pink. Their abundance, generally omnivorous habits, and their high energy content per unit body mass make copepods an important component in aquatic food webs. In addition to the pivotal role that copepods play in providing ecosystem services through their intermediate role in aquatic food webs, they are important to humans since they serve as vectors of serious diseases such as cholera, and as potential biological control agents for disease vectors such as mosquitoes.

Although copepods are of marine origin and most extant species live in the world's oceans, they are often very abundant in inland waters. They can reach densities of 100s or more rarely 1000s per liter in productive inland waters, and may reach densities of more than $100\,000\,m^{-2}$ in moist soils. In Lake Baikal, Siberia, which contains 20% of the volume of unfrozen, fresh surface water on the Earth, a single species, *Epischura baikalensis*, may account for over 95% of the zooplankton. The zooplankton of many other large lakes such as Lake Tahoe in California-Nevada, USA, and the Laurentian Great Lakes, and many of the large clear lakes in the Southern Hemisphere are also dominated by planktonic copepods. The plankton of the highest-elevation alpine lakes around the world is frequently composed of a single species of large, omnivorous or predatory calanoid copepod containing bright red, photoprotective pigments (**Figure 4**). Copepods can also comprise most of the biomass of the crustacean plankton in smaller lakes. In spite of this dominance of copepods in many lakes, only a small proportion of the species found in inland waters are planktonic. Most copepods are associated with substrates in benthic, littoral, riverine, groundwater, or even terrestrial systems such as moist soils and mosses.

Copepods are included in the Maxillopoda, crustaceans that have their first pair of swimming legs modified to function with the mouthparts. The subclass Copepoda includes two infraclasses: Progymnoplea and Neocopepoda. The Progymnoplea contain only a single order of primarily marine free-living Platycopioida, while the Neocopepoda contain the other eight orders. The members of the order Siphonostomatoida, many Cyclopoida and Poecilostomatoida, and some Harpacticoida are parasitic, whereas the Calanoida, Misophrioida, and Mormonilloida are free-living. Inland waters have been colonized by free-living members of the Calanoida, which are primarily planktonic; the Harpacticoida, which are essentially all benthic, subterranean, or otherwise associated with substrates; and the Cyclopoida, which are mainly benthic with some planktonic species. A few fish parasites of the orders Poecilostomatoida, Siphonostomatoida also occur in fresh water. A tiny group with three known species, the order Gelyelloida, which is closely related to the Cyclopoida, are all subterranean in fresh waters.

External Morphology and Behavior of Copepods

Adult copepods have a segmented, tubular to flattened body with many jointed appendages (**Figures 2** and **3**). A major articulation separates the anterior portion of the body, called the prosome, and the posterior portion or urosome (**Figures 1** and **4**). This articulation is between the fifth and sixth thoracic segments in cyclopoids, harpacticoids, and gelyelloids, but between the sixth and seventh thoracic segment in calanoids. The genital segment is the last of the seven segments of the thorax.

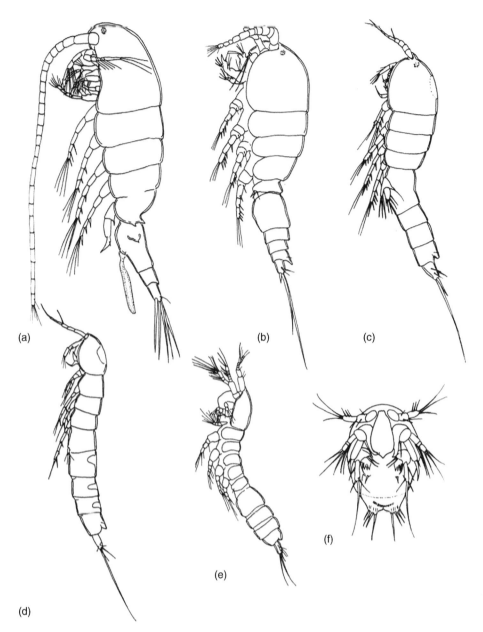

Figure 1   Lateral view of body forms of the major groups of copepods occurring in fresh water: (a) Calanoida (family Diaptomidae); (b) Cyclopoida (family Cyclopidae); (c) Harpacticoida (family Canthocamptidae); (d) Harpacticoida (family Parastenocarididae); (e) Gelyelloida; (f) Ventral view of a nauplius larva. Not to scale.

The prosome is composed of the head and first one or two thoracic segments (collectively referred to as the cephalosome), as well as most of the thoracic segments that bear the swimming and feeding appendages. The urosome includes the last one or two thoracic segments and the abdominal segments. Posterior to the urosome are the caudal rami. The five pairs of appendages on the head include two pairs of antennae, one pair of mandibles, and two pairs of maxillae (**Figures 2** and **3**). The first antennae, or antennules, contain a complex array of chemoreceptors

and mechanoreceptors that are used to detect and separate potential food items, avoid predators, and locate mates (**Figure 3**). The antennules are also used for locomotion, and in males they are modified for use in mating. In male cyclopoids and harpacticoids, both antennules have a single major joint, or geniculation, that allows them to bend sharply to grasp the female during mating. In male calanoids only the right antennule is geniculate. The first maxillae, or maxillules, and the second maxillae are used to capture and handle food before it is macerated by the mandibles and

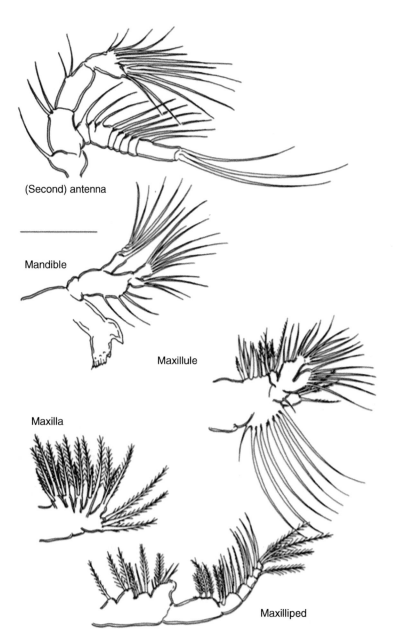

Figure 2 Antenna (second antenna) and mouthparts of a diaptomid calanoid copepod. Together, these five paired appendages act as paddles to create water currents that bring food items to the vicinity of the copepod's mouth and handle them in order to ingest them. Scale bar = 100 μm.

ingested. The maxillipeds, the last pair of mouthparts, function together with the maxillae to manipulate the food. Cyclopoids actively grasp their food after encountering it by swimming through the water or over substrates. Harpacticoids feed largely by foraging on surfaces and grasping their prey as they encounter it, although some species can also create feeding currents by vibrating their first and second maxillae and mandibular palps. Calanoids feed primarily by creating active feeding currents by vibrating their second antennae, mandibular palps, maxillae, and maxillipeds.

These feeding currents serve to increase the effective encounter area of harpacticoids and calanoids by bringing food particles close enough to be grasped and ingested.

The thoracic appendages of copepods include the maxillipeds on the first thoracic segment and four pairs of morphologically similar swimming legs on the next four segments. These swimming legs are used primarily in locomotion. The fifth pair of swimming legs is reduced in cyclopoids and harpacticoids. In female calanoids the fifth pair of swimming legs is

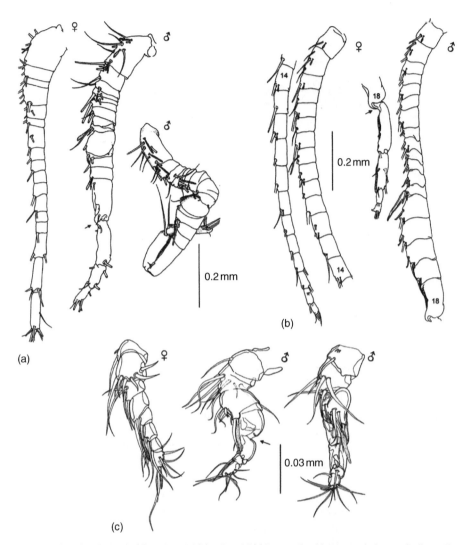

Figure 3 Antennules of a female and male (a) cyclopoid (b) calanoid (c) harpacticoid copepod. Arrows indicate the geniculate joint of the male antennule. (a, b) Adapted from Einsle U (1993) Crustacea: Copepoda: Calanoida und Cyclopoida. *Süßwasserfauna von Mitteleuropa* 8: 4–1. Gustav Fischer Verlag, Stuttgart, with permission from Elsevier. (c) Adapted from Reid JW and Rocha CEF (2003) *Pindamoraria boraceiae*, a new genus and species of freshwater Canthocamptidae (Copepoda, Harpacticoida) from Brazil. *Zoological Journal of the Linnean Society* 139: 81–92, with permission from The Linnean Society of London.

well developed but somewhat smaller than the other four pairs, while in male calanoids the fifth legs are modified for grasping the female during copulation (**Figure 4**). Sixth legs are vestigial or absent in all of these orders. In addition to the swimming legs, copepods also use their urosome for locomotion. In the presence of predators copepods tend to reduce their swimming speeds, but when attacked they exhibit some of the fastest escape responses ever recorded when scaled to body length (>350 body lengths s^{-1} for nauplii). In the presence of prey, on the other hand, the more predatory copepods may increase both vertical and horizontal looping behavior, which serves to increase prey encounter rates.

Internal Anatomy and Sensory Organs in Copepods

The nervous system of copepods includes a simple brain consisting of a supraesophageal ganglion and a ventral nerve cord with several thoracic ganglia. In contrast to some of the marine copepods that have compound eyes with lenses, inland species possess only a simple eye spot that cannot form images. In most species, a frontal organ called the sensory pore X organ is located on the anterior tip of the rostrum and thought to function in chemoreception. Copepods are well equipped with sensory organs. Their bodies are covered with two types of integumental sensilla that

Figure 4 Adult male *Hesperodiaptomus shoshone*. Note that just anterior to the major body articulation is the fifth leg modified for grasping the female during mating. The dorsal testis is visible through the exoskeleton of the prosome. The large size (~ 3 mm body length) and dark red coloration of this copepod is characteristic of copepods in high-elevation alpine lakes that lack visual predators. Photograph by Robert E. Moeller with permission.

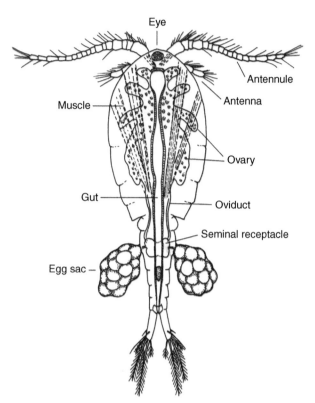

Figure 5 Major reproductive, gut, and some striated muscle structures of a cyclopoid copepod (in dorsal view). This female bears paired lateral egg sacs. The posterior part of the gut contains a fecal pellet.

are thought to function in chemoreception (peglike sensilla basonica) or both chemoreception and mechanoreception (hairlike sensilla trichodea). Small innervated spines located near the joints function to sense the position and movement of the segmented body parts. The antennules have an array of setae used in mechanoreception, while the mouthparts have numerous chemoreceptors.

The circulatory system of calanoid copepods consists of a simple dorsal heart and both lateral and ventral ostia. Cyclopoids and harpacticoids have no heart, and circulation is driven by movements of the body and gut through a hemocoel. The digestive system in copepods includes a mouth with surrounding labral glands that secrete mucus to aid in food passage, a chitinized foregut (esophagus), midgut, chitinized hindgut, and an anus (**Figure 5**). The midgut secretes a peritrophic membrane that surrounds the egested food to form a discrete fecal pellet. Excretion of nitrogen is through antennal glands in the juvenile naupliar stages and maxillary glands in adults. Osmoregulation in freshwater copepods may be aided by a dorsal window in the integument of their cephalothorax.

Female copepods have a single or paired ovary that is located dorsal to the midgut (**Figure 5**). Oviducts extend anteriorly, with diverticuli that extend posteriorly to the gonopore on the genital segment. Female calanoids have an atrium inside the gonopore with paired seminal receptacles, while cyclopoids have a single seminal receptacle connected to the gonopore. Specialized skin glands secrete the egg sac membrane. Male copepods have a single dorsal testis and vas deferens that includes the seminal vesicle, spermatophore

sac, and ejaculatory duct. The spermatophores that the males place on the females range from small, compact, bean-shaped spermatophores in cyclopoids to larger and longer cylindrical spermatophores in harpacticoids and calanoids.

Reproduction and Life History of Copepods

Reproduction in copepods is sexual, with the exception of a very few harpacticoids. Finding a mate in open-water pelagic systems can be difficult if copepod densities are very low. In many marine copepods, the females produce pheromones that attract males to assist in mate location. Densities of copepods in inland waters are generally higher than in the oceans, and their use of sex pheromones is correspondingly rare or absent. Mates are located largely through random encounters. Following an encounter, a male calanoid actively pursues the female, grasps the female with the right geniculate antenna ('precopula 1'), grasps the female by the urosome with the chelate right fifth leg ('precopula 2'), and then copulates by using the left fifth leg to attach a spermatophore to the female's genital segment. The mating process may last for several minutes, during which the urosomes of the male and female are juxtaposed and the prosomes extend off at an angle in opposite directions in a V-like pattern. In cyclopoids and harpacticoids, the males are much more active swimmers than the females, and sexual encounters involve the male grasping the female with both geniculate antennules (**Figure 6**). During mate coupling the male and female combined are more visible to potential predators with a resulting cost in terms of increased vulnerability to visually-hunting fish predators.

Copulation and the union of haploid gametes lead to the formation of a fertilized diploid zygote. Female calanoids in the family Diaptomidae must mate before each clutch is produced, and pass through three or four reproductive phases that last for different lengths of time depending on the availability of food and mates (**Figure 7**). The relative abundance of these different phases can be used as an index of food or mate limitation, as long as abiotic conditions are suitable for reproduction. Cyclopoids and harpacticoids can store sperm and thus produce many clutches after a single mating. Following fertilization, the embryos are extruded into either a single ventral sac (most calanoids and harpacticoids) or two lateral sacs (cyclopoids), or are released individually, directly into the water (a few calanoids such as *Epischura* and *Senecella*, and some cave-dwelling cyclopoids). Clutch sizes can range from 2 to more than 50 eggs per clutch, depending on temperature and food conditions and the size of the adult female. Larger females and larger clutch sizes are often observed at lower temperatures. A single female copepod can produce dozens of clutches and many hundreds of eggs in her lifetime. At warmer temperatures, females may produce a clutch every few days, while at colder temperatures interclutch duration may be several weeks.

The normal mode of reproduction is through subitaneous eggs. Some harpacticoids and calanoids may

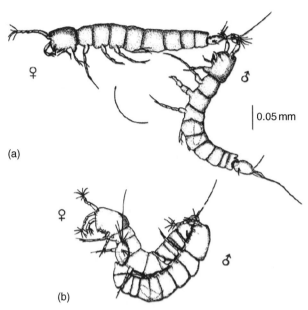

Figure 6 Mating of the harpacticoid *Parastenocaris phyllura* follows a precise sequence. (a) The male clasps the female's caudal setae with his antennule and immediately swings to press his ventral side to hers. (b) The male clasps the female firmly with his 3rd pair of legs held around her genital double-somite and his caudal rami behind her 3rd legs, paddles her caudal rami with his antennae, and strokes her body with his second and fourth legs. After several minutes, the male extrudes the bean-shaped spermatophore, attaching it to the female's gonopore. Adapted from Glatzel T and Schminke HK (1996) Mating behaviour of the groundwater copepod *Parastenocaris phyllura* Kiefer, 1938 (Copepoda: Harpacticoida). *Contributions to Zoology* 66: 103–108, with permission from the University of Amsterdam.

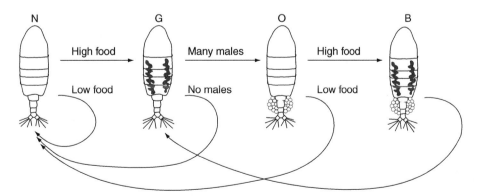

Figure 7 Dorsal view of reproductive phases of calanoid copepods in the family Diaptomidae. Females must mate before the production of each clutch of eggs. If food and mates are abundant, the females become gravid (G phase) and then extrude eggs to become ovigerous (O phase), or if the development time of the eggs is longer than the time it takes the oocytes to develop, females may be both ovigerous and gravid (B phase). When food is limiting, females are neither gravid nor ovigerous (N phase). If food is plentiful but mates are limiting, the female remains in the G phase until she finally extrudes an amorphous mass of unfertilized eggs that disintegrate as the female reverts to the N phase.

also produce resting eggs that enter a period of diapause which may last from several weeks or months to decades or even several hundred years. These prolonged periods of diapause combined with variable hatching rates create 'egg banks' in the sediments of lakes. These egg banks provide a valuable record of genetic variability in changing environments through time and have been used to demonstrate microevolution in copepods. Cyclopoids do not produce resting eggs, but they are able to enter a quiescent state or true diapause as either subadults or adults. Some harpacticoids can encyst and enter diapause as subadults. These diapausing stages are thought to be stimulated by the onset of overcrowding, photoperiod, or age, although some recent evidence suggests that diapause may reach a maximum level under optimal conditions for growth. These resting stages allow copepods to survive adverse conditions ranging from low levels of food and freezing in winter to desiccation or increases in predation pressure during the summer. These resistant resting stages can also aid

UV and zooplankton	Lake age	DOC (mg/L)	Lake name
	10 years	1.0	Little Esker
	35 years	2.6	Plateau 1
	90 years	4.2	Klotz Hills

Figure 8 Copepods are some of the first zooplankton to colonize lakes following deglaciation in remote areas such as in Glacier Bay, Alaska. This ability to disperse and colonize remote young lakes is likely aided by a combination of the desiccation-resistant diapause stages and the relatively high UV tolerance of some copepods. The absence of terrestrial vegetation within the watersheds of the youngest lakes leads to very low dissolved organic carbon concentrations (DOC), making these lakes highly UV-transparent (blue shading in lake basins). The zooplankton symbols represent established (filled symbols, >1 l^{-1}) or sparse (open symbols, <1 l^{-1}) populations of (from left to right in bottom figure) *Asplanchna priodonta*, *Daphnia* spp., *Cyclops scutifer*, *Ceriodaphnia quadrangula* (top) and *Bosmina longirostris* (bottom). Modified with permission from Williamson CE, and Zagarese HE (2003) UVR effects on aquatic ecosystems: A changing climate perspective. In: Helbling EW and Zagarese HE (eds.) *UV Effects in Aquatic Organisms and Ecosystems*. Royal Society of Chemistry, Cambridge, UK, pp. 547–567, Häder DP and Jori G. (Series Editors) *Comprehensive Series in Photochemistry and Photobiology*. See also Williamson CE, Olson OG, Lott SE, Walker ND, Engstrom DR, and Hargreaves BR (2001) Ultraviolet radiation and zooplankton community structure following deglaciation in Glacier Bay, Alaska. *Ecology* 82: 1748–1760.

in the dispersal of copepods across land, and may account for copepods being some of the first colonizers in remote post-glacial lakes (**Figure 8**). Termination of diapause may be stimulated by low oxygen, changes in light or temperature, physical disturbance, or rehydration following desiccation. Emergence from resting eggs or diapausing copepodids may occur within hours or a day or two after the appropriate environmental stimulus.

The larval stages that hatch from eggs are referred to as nauplii (**Figure 1**). There are six naupliar stages (N1–N6), followed by metamorphosis into the subadult or first copepodid stage. The first instar nauplius is rounded and has three pairs of appendages including first and second antennae and mandibles. As the nauplius grows and molts, it becomes elongated and develops additional appendages. Upon metamorphosis into the first subadult or copepodid stage, the immature copepod more closely resembles the adult, with an elongate body clearly divided into a prosome, urosome, and caudal rami with terminal setae. As with the nauplii, there are six copepodid stages (C1–C6) with the last stage being the adult. The normal life span of copepods under optimal conditions is generally on the order of a few months. Males generally mature more rapidly and do not live as long as females. Cold temperatures can slow growth, development, and aging, and extend the life span substantially.

Adult copepods are sexually dimorphic, with the females generally being larger and more voracious feeders than the males. In cyclopoids the female:male body length ratio averages 1.44, and in calanoids it averages 1.13. Harpacticoids and to a lesser extent cyclopoids may exhibit sexual dimorphism of their swimming legs. Males and females also differ in the structure of their antennules, their fifth and sixth legs, and the number of urosomal segments. Seasonal variation in body size may be pronounced, with larger individuals occurring at lower temperatures. In copepods, allometric growth is subtle, without the pronounced seasonal patterns of cyclomorphosis in helmets and spines exhibited by other freshwater invertebrates such as rotifers and cladocerans.

Ecology of Copepods

Copepods have played an important role in our understanding of several fundamental ecological relationships in aquatic invertebrates ranging from basic feeding biology to the ecology and evolution of life history strategies. Due to the dominance of so many inland water bodies and the oceans of the planet by copepods and the role of copepods as primary and secondary consumers a great deal of attention has been paid to their feeding mechanisms. Although calanoids were initially thought to 'filter' fine food particles such as phytoplankton from the water, more recent research has demonstrated their ability to detect larger food particles such as rotifers or nauplii at some distance and actively attack, capture, and ingest these individually-selected food particles. Pelagic copepods generally have a reduced impact on phytoplankton biomass in comparison to large cladocerans such as *Daphnia* due to copepods selecting comparatively larger food particles and their consequent inability to suppress rapidly-growing smaller phytoplankton in addition to their feeding on and thus suppressing micrograzers such as protozoa and rotifers. Copepods also sequester less of the limiting nutrient (phosphorus) in comparison to *Daphnia*, thus increasing P availability for the pico- and nanophytoplankton.

In addition to sequestering less P than *Daphnia*, copepods also require more nitrogen. This suggests that changes in dominance of copepods versus cladocerans may alter nutrient cycling. Atomic ratios of C:N:P for copepods are 212:39:1 while for *Daphnia* they are closer to 85:14:1. While these nutrient ratios would suggest that copepods have a competitive advantage over *Daphnia* under conditions of P limitation, their nauplii tend to have relatively higher P requirements.

A variety of factors influence the distribution, abundance, and species richness of copepods in inland waters. In planktonic systems, the abundance of cyclopoids often increases with increasing lake productivity, while calanoid abundance changes little. This may be related to an increase in the abundance of micrograzers that serve as food for the cyclopoids, or to increases in transparency to ultraviolet radiation (UVR) with decreasing productivity since calanoids tend to be more UVR-tolerant than cyclopoids. Temperature and pH may also influence the dominant genera of copepods in planktonic systems. For example, members of the genus *Mesocyclops* tend to be more abundant in warmer lakes with higher pH or during warmer seasons, while representatives of the genus *Cyclops* tend to be more abundant in cooler waters. A few species, such as *Leptodiaptomus minutus*, tend to be more abundant in lakes with lower pH. In temperate and Antarctic waters in the Southern Hemisphere, calanoids of the genus *Boeckella* tend to dominate the plankton.

Species richness of copepods in benthic habitats may exceed 20 species – much greater than in the pelagic zone where there are generally only two or three dominant species. Abundance is generally greatest near the sediment surface, but copepods occur at depths of 10–20 cm or greater. Most copepods cannot survive low oxygen conditions for prolonged periods of time unless they are in diapause, though a few cyclopoids can survive total anoxia for a few hours to several days.

Copepods play a key intermediate role in food webs as both prey and predators. While highly evasive, copepods also serve as food for both vertebrate and invertebrate predators. Nauplii are a primary food of fish larvae and predatory zooplankton. While most copepods are omnivorous, some are detritivorous, and habits range from herbivory to carnivory; in more carnivorous species there is an ontogenetic shift wherein the size and nature of food changes as nauplii mature into copepodids and adults. Impacts of copepods on prey densities can be substantial. In some larger predatory calanoids, a single individual can clear all of the prey from a liter of water in a single day. Cyclopoid clearance rates can reach 150 ml per predator per day. When harpacticoids are abundant, they can reduce bacteria densities in streams by half in a single day. Many predatory copepods are cannibalistic, though the escape responses of juveniles are highly effective at reducing cannibalism. Food limitation in copepods is common, and may reduce adult body size as well as lower reproductive output by reducing clutch size and prolonging interclutch duration.

Planktonic copepods also exhibit different responses to damaging solar ultraviolet radiation (UVR) than do other zooplankton. In contrast to many rotifers and cladocerans, copepods are able to sequester high levels of photoprotective compounds such as mycosporine-like amino acids (MAAs) and carotenoids from their diet and use them to reduce UVR damage (**Figures 4** and **9**). The photoprotective compounds are more concentrated in juvenile stages, including eggs, nauplii and young. Concentrations of photoprotective compounds may vary widely with temperature and season. Concentrations of these photoprotective compounds and the UVR tolerance of copepods increase with exposure to UVR (**Figure 9**). In contrast, the less UVR-tolerant cladoceran *Daphnia* generally responds to high UVR during the day by migrating downward more strongly than copepods. Both copepods and cladocerans have a substantial ability to repair UVR-damaged DNA with photoenzymatic repair, though this ability in copepods may vary among species even within the same genus.

Dissolved organic matter (DOM) concentrations are increasing in lakes in many parts of the world. DOM selectively absorbs the shorter wavelengths of damaging UVR. Thus, habitats with lower DOM generally have higher UV exposure levels. In recently

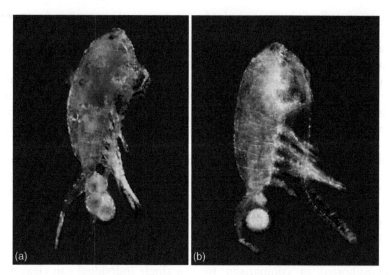

Figure 9 Many copepods contain concentrations of photoprotective compounds such as carotenoids and mycosporine-like amino acids that protect them from damaging UV radiation. The small calanoid *Leptodiaptomus minutus* shown here sequesters these compounds from its diet. When copepods were fed algae containing carotenoids and exposed to UV they accumulated more carotenoids (a) than when they were not exposed to UV (b). Photographs by Robert Moeller with permission (Moeller retains copyright), see Moeller RE, Gilroy S, Williamson CE, Grad G, and Sommaruga R (2005) Dietary acquisition of photoprotective compounds (mycosporine-like amino acids, carotenoids) and acclimation to ultraviolet radiation in a freshwater copepod. *Limnology and Oceanography* 50: 427–439.

deglaciated lakes in Alaska where DOM concentrations are very low and UV transparency is high, copepods are some of the earliest colonizers (**Figure 8**). In an experimental study in NE Pennsylvania, USA, manipulation of UV, DOM, and pH demonstrated that increases in DOM depress populations of the copepod *Leptodiaptomus minutus* because of its sensitivity to reduced pH, while the cladoceran *Daphnia catawba* benefits from the DOM additions because of the reduced UVR exposure levels. Related experiments demonstrated that moderate levels of UVR exposure (20 KJ m^{-2} of 320 nm over a 9 day period) may actually have a positive effect on *L. minutus* survival and reproduction. This beneficial effect of UVR seems to be related to higher food levels in the presence of UVR. Other experiments that manipulated UVR and DOM in a subalpine lake in Wyoming, USA demonstrated that *Leptodiaptomus ashlandi* adults were similarly resistant to UVR damage even at higher exposure levels (30 KJ m^{-2} or 320 nm over a 7 day period), although reproduction and naupliar survival were negatively impacted. Collectively these results suggest that increasing DOM concentrations may alter UV exposure and lead to a shift in zooplankton communities toward fewer copepods and more *Daphnia*.

Relevance of Copepods to Humans and Infectious Diseases

Copepods play important and often contrasting roles in human infectious diseases. Copepods can serve as either vectors of diseases such as cholera, or as biological control agents for mosquitoes that are vectors of human disease. Copepods are important intermediate hosts for a variety of parasites, including flukes, nematodes, and tapeworms that infect organisms ranging from fish and amphibians to birds and mammals. Cyclopoid copepods are an intermediate host for the human nematode parasite *Dracunculus medinensis* (guinea worm) in tropical regions of Africa and India. Experimental studies in the Chesapeake Bay, USA and in Bangladesh have demonstrated the ability of the human pathogenic bacterium *Vibrio cholerae* to adhere to copepods, particularly to the oral region and the eggs. In Bangladesh, outbreaks of cholera are most severe in the fall following blooms of phytoplankton and copepods. Filtration of drinking water through sari cloth to remove the copepods can reduce cholera outbreaks by as much as 48%.

Copepods can also prey on a variety of pathogenic organisms or their vectors, and thus have the potential to be valuable biological control agents. Some of the larger cyclopoid copepods prey on the early instars of mosquito larvae at high rates, and can be cultivated in large quantities to assist in biological control efforts. Both cyclopoids and harpacticoids are also known to prey on nematodes. Interestingly, copepods are less sensitive to pesticides than cladocerans, and may actually increase in abundance following pesticide application. Experiments with the two common insecticides Sevin and Malathion and two common herbicides

2,4-D and glyphosate (Roundup®, Rodeo®) revealed high death rates in two species of the cladoceran genus *Daphnia* in the presence of both insecticides; while the two copepods tested (*Eurytemora* and *Mesocyclops*) showed either no response or a positive effect of the insecticides. The only negative effect on copepods for the four pesticides was the negative effect of glyphosate on *Eurytemora*. This suggests that the combined use of insecticides and copepods as biological control agents to reduce disease vectors may be possible. Copepods may have other effects on humans and human byproducts in that they tend to accumulate toxic trace metals such as cadmium to a lesser extent than do cladocerans. Thus, greater concentrations of copepods may reduce levels of toxic metals in aquatic foodwebs on which humans depend for food.

Glossary

Allometric Growth – The unequal growth of body parts during development of an organism.

Caudal Rami – The bifurcated appendage that extends posteriorly from the urosome of a subadult or adult copepod, and generally bears several setae.

Cephalosome – The anterior portion of the prosome of a copepod composed of the fused head and first two thoracic segments.

Copepodid – The subadult and adult stages of a copepod, which consist of six instars (C1–C6). The first copepodid instar (C1) metamorphoses from the sixth naupliar instar, and the last copepodid instar (C6) is the adult.

Cyclomorphosis – Cyclic changes in morphology in sequential generations of organisms that involve the growth of individuals with exuberant forms, often with spines or helmets, which alternate with generations of individuals with less exuberant forms.

Nauplii – The plural of nauplius.

Nauplius – The larval stage of a copepod, which consists of six instar stages (N1–N6); the last naupliar stage metamorphoses into the first copepodid stage.

Ovigerous – Carrying eggs.

Prosome – One of the two major body sections of a copepod. The prosome is anterior to the urosome and composed of the cephalosome and the thoracic segments that collectively bear the antennules and antennas, feeding appendages, and swimming legs.

Urosome – One of the two major body sections of a copepod. The urosome is posterior to the prosome and composed of the last one or two thoracic segments and the abdominal segments, but does not include the caudal rami.

See also: Competition and Predation; Egg Banks; Role of Zooplankton in Aquatic Ecosystems.

Further Reading

Boxshall GA and Defaye D (2008) Global diversity of copepods (Crustacea:Copepoda) in freshwater. *Hydrobiologia* 595: 195–207.

Boxshall GA and Halsey SH (2004) *An Introduction to Copepod Diversity*, 2 vols. London: The Ray Society. 966 pp.

Damkaer DM (2002) *The Copepodologist's Cabinet: A Biographical and Bibliographical History*. Philadelphia: American Philosophical Society. 300 pp.

Dussart BH and Defaye D (2001) Copepoda: Introduction to the Copepoda. In: Dumont HJF (ed.) *Guides to the Identification of the Microinvertebrates of the Continental Waters of the World 16*, 2nd edn. The Hague: SPB Academic Publishers. 344 pp.

Fryer G (1957) The food of some freshwater cyclopoid copepods and its ecological significance. *Journal of Animal Ecology* 26: 263–286.

Marten GG and Reid JW (2007) Cyclopoid copepods. In: Floore TE and Becnel J (eds.) Biorational Control of Mosquitoes Bulletin No. 7. *American Mosquito Control Association Bulletin* 23 (2): 65–92.

Schminke HK (2007) Entomology for the copepodologist. *Journal of Plankton Research* 29(Suppl. 1): i149–i162.

Smith DG (2001) *Pennak's Freshwater Invertebrates of the United States*, 4th ed. New York: Wiley. 638 pp.

Williams-Howze J (1997) Dormancy in the free-living copepod orders Cyclopoida, Calanoida, and Harpacticoida. *Annual Review of Oceanography and Marine Biology* 35: 257–321.

Williamson CE and Reid W (2009) Copepoda. In: Thorp J and Covich A (eds.) *Ecology and Classification of North American Freshwater Invertebrates*, 3rd edn. In preparation.

Other Zooplankton

L G Rudstam, Cornell University, Ithaca, NY, USA

Other Zooplankton

Several other organisms in addition to copepods, cladocerans, and rotifers can be considered part of the zooplankton. This is a diverse group that includes pericarid crustaceans such as mysid shrimps and amphipods, both larval and adult insects – particularly larvae of the phantom midges in the genus *Chaoborus*, water mites, fairy shrimps, freshwater jellyfish, and planktonic mussel larvae (*Dreissena veligers*). Several of these species are relatively large predators or omnivores (5–30 mm), can be abundant, and do strongly affect the population dynamics of their prey. Because of their large size, they are also sensitive to predation by fish. In contrast, veligers are small filter feeders (0.07–0.2 mm).

Mysid Shrimps

Mysid shrimps, or opossum shrimps, are members of a mostly marine order (Mysidacea). There are some 30 species occurring in freshwater lakes and rivers and the group has a worldwide distribution. Lacustrine species are primarily of the genus *Mysis* and *Neomysis*, but there are several Ponto-Caspian species that are spreading to Europe and North America. For example, *Hemimysis anomala* invaded European lakes from the Ponto-Caspian region in the last century and reached North America in 2006. In marine systems, mysids are primarily found in benthic and nearshore habitats and they are an important component of estuarine food webs.

Mysis relicta (**Figure 1**) is the most studied mysid shrimp. It is a glacial relict with a circumpolar distribution in deep cold lakes of the northern hemisphere. Mortality increases when *Mysis* has been held long-term in temperatures above 13 °C. Their absence from deep cold lakes within a few kilometers of their natural range – areas covered by ice lakes after the last glaciation – is a testimony to their poor natural dispersal ability. The *Mysis relicta* species group consists of four species, one in North America (*M. diluviana*) and three in Eurasia (*M. relicta, M. segerstralei, M. salemaai*). Members of the genus *Mysis* also occur in marine and brackish water and in the Caspian Sea where three morphologically distinct but closely related species coexist. The abundance of mysids can be high and their biomass can exceed that of planktivorous fish. Density and biomass vary seasonally and densities over $500\,m^{-2}$ and a biomass of $10\,g$ ww m^{-2} are not uncommon; peak densities over $1000\,m^{-2}$ have been reported. Mysids do not have gills; oxygen is taken up through the thin carapace. They are rarely found in lakes without oxygenated hypolimnion in the summer.

Mysis relicta can reach 25–30 mm, grow slowly and produce one or two clutches of 10–50 young in a lifetime. Growth rates and therefore generation time depend on lake productivity. Growth rates increase in more productive lakes and range from $0.2\,mm$ month^{-1} in Lake Tahoe to $1.5\,mm$ month^{-1} in mesotrophic lakes. Corresponding generation times are 1 year in productive lakes and 2–3 years in oligotrophic lakes; up to 4 years has been reported from ultra-oligotrophic Lake Tahoe. Some lakes have both a 1- and a 2-year generation time with slow growing individuals taking 2 years to mature. One Swedish population with a 2-year life cycle switched to a 1-year life cycle for several years when transplanted to a more productive lake. The embryos are carried by the female in a brood pouch (marsupium) for several months; hence the name opossum shrimp. Males die after copulation, but females may molt and produce a second clutch. The young ones are released at a length of 2–3 mm and at that size they appear similar to the adults. There are no free swimming naupliar stages. A springtime release is common, but some populations also release young in the fall or throughout the year, and therefore consist of overlapping generations. With such a long generation time and small brood size, mysid populations are sensitive to high mortality rates and would therefore not persist in systems without a refuge from fish predation. A population with a 2-year generation time producing 30 young once per female requires a 25% annual survival rate to persist in the system. This is a very high survival rate compared with, for example, fish, which can have a first year survival of 1% or less.

Mysis performs remarkable diel migrations at dusk and dawn from their daytime refuge in dark, deep water to the meta- or epilimnion, where they feed on zooplankton or algae. These migrations can be over 100 m long (**Figure 2**) and migration speed can exceed $1–2\,m$ min^{-1}. During the day, part of the mysid population may remain in the water column if the lake is deep enough. Some animals stay on the bottom day and night. Their migration is limited by temperatures above 12–16 °C and light levels above 10^{-4} lux, light levels that limit fish visual feeding. *Mysis* have one type of photo-pigment with peak sensitivity between 490–540 nm (adapted to the

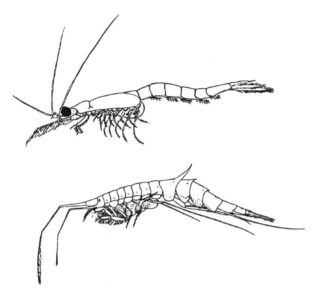

Figure 1 Adult female *Mysis relicta* (top) and adult female *Macrohectopus branickii* (bottom). Both animals are around 20 mm long. Adapted with permission from Rudstam LG, Melnik NG, and Shubenkov SG (1998) Invertebrate predators in pelagic food webs: Similarities between *Macrohectopus branickii* (Crustacea: Amphipoda) in Lake Baikal and *Mysis relicta* (Crustacea: Mysidaceae) in Lake Ontario. *Siberian Journal of Ecology* 5: 429–434.

local light environment). They are not sensitive to long wavelength red light. Moonlight will affect their nighttime distributions and they are tens of meters deeper during full moon nights compared with new moon nights. Larger mysids from Lake Ontario have a strong preference for temperatures between 6 °C and 8 °C; smaller mysids have less pronounced temperature preferences but avoid temperatures above 14 °C. The vertical distribution of mysids is largely predictable from their response to temperature and light. There is a cost to residing at low temperatures of 4 °C during the day, as mysid development time is slower at 4 °C than at 8 °C.

Mysids are omnivores and capable of both filter-feeding and raptorial feeding. A feeding current is created by the thoracic legs, which also provide for regular swimming motion. Diatoms are consumed during spring and to a lesser extent during the fall diatom blooms and algae can then be a major component of the diet. Small mysids are more herbivorous than large mysids. They also feed on benthic prey, detritus, and sediment during the day. By feeding in both the benthic and pelagic layer, mysids can be a vector for redistribution of nutrients and contaminants between these two subsystems. Zooplankton typically dominates the diet. Zooplankton are captured by fast movement towards the prey. The size of prey captured increases with the size of the mysids, and larger mysids can feed on amphipods and smaller conspecifics.

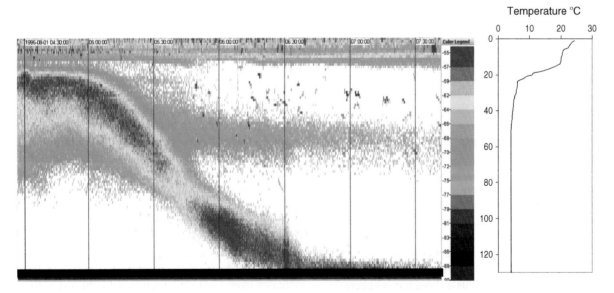

Figure 2 Downwards migration of *Mysis relicta* in Lake Ontario, July 31, 1995, from 02:30 to 05:40 observed with a 420 kHz echosounder. Vertical lines represents 30 min time step. Sunrise at 04:00. Bottom depth 130 m. The temperature profile is also given. The mysid layer moved 100 m in 1.5 hours, for a migration speed of 1 m min^{-1}. Net samples prior to 02:30 showed no mysids in the top 10 m, 5.1 m^{-3} in 10–30 m, 0.9 m^{-3} in 30–70 m, and 0.3 m^{-3} in 70–120 m depth. Some fish and what appear to be a shallow invertebrate layer, possibly small mysids, are also visible in the water column after the main mysid layer has descended.

Small mysids actively avoid larger individuals, potentially to reduce predation risk. Feeding rate of mysids decreases with mysid density in the laboratory, indicating interference among conspecifics. Mysids consume more cladocerans than copepods because copepods are more efficient at avoiding capture. Representative hard parts of crustaceans, rotifers, and diatoms can be identified in the foregut. Mysid feeding rates decline in the presence of fish or fish kairomones. Clearance rates from experiments vary with prey species and range from $0.05 \, l \, hour^{-1}$ when feeding on large copepods to $1 \, l \, hour^{-1}$ when feeding on cladocerans. Short-term specific feeding rates can be as high as 25% of body weight per day, although most reports from field and long-term laboratory experiments indicate values of 5–10% per day. Assimilation efficiency is high (80–90%) when feeding on zooplankton or chironomids, lower when feeding on detritus. Gross growth efficiency for animals feeding on zooplankton is between 7% and 29%. At least one species (*Mysis stenolepis*) is able to assimilate raw sterilized cellulose with an efficiency of 30–50%.

Mysids are an important food resource for fish and several species rely primarily on mysid shrimps. Mysids have a fast avoidance reaction to fish attacks (80 body length s^{-1}) through tail flipping and can often avoid capture. In lakes with native mysid populations, fish have evolved adaptations for migrating with their mysid prey, such as reduced swim bladders, high fat content and large fins. Mysids are rich in essential fatty acids and may be even more important for fish than suggested by their quantitative proportion in the diet. Polyunsaturated fatty acids common in mysids are important for over-winter survival of several fish species.

Mysis relicta shrimps were introduced in 1949 to Kootenay Lake, BC, Canada and are believed to be at least partly responsible for a large increase in growth rate of kokanee salmon (*Oncorhynchus nerka*) and the spectacular fishery that developed after the introduction. After this reported initial success, mysids were introduced to many lakes and reservoirs in western USA, Canada, and Scandinavia to increase fish growth and production. However, results were not often as intended, as mysid predation caused declines in cladocerans, in particular *Daphnia*. Since *Daphnia* is an important prey of juvenile stages of several fish species, these fish species often declined after mysid introductions. Even in Kootenay Lake, kokanee declined in the 1970s although it is uncertain to what degree this was caused by mysids. Although the consequences of mysid introductions were often detrimental to established fisheries, these introductions did show that mysids can affect food web dynamics.

Indeed, mysids can be quantitatively more important zooplanktivores than fish. In Flathead Lake, Montana, the introduction of mysids led to a marked decline in kokanee and subsequent decline in bald eagles and other birds and mammals that fed on the spawning run of the kokanee. There is now a moratorium on mysid introductions.

Members of the *Mysis relicta* species complex are not the only mysids in fresh water. Several species of *Neomysis* are present in lakes and estuaries. These species can be found in warmer waters than *Mysis relicta*, are omnivores, major predators on zooplankton, and perform diel migrations. *Neomysis* sp. are smaller (10–15 mm) than *Mysis* sp., and may have two or more generations per year. *Neomysis mercedis* is a major zooplanktivore in Lake Washington and other lakes in the Pacific North-West of North America. *Neomysis integer* is a European species primarily found in brackish lakes and estuaries as well as in the Baltic Sea. In shallow brackish lakes in Denmark, *Neomysis integer* is abundant only when the fish community is dominated by small fish, such as sticklebacks. When abundant, this species is an important predator on zooplankton and an important prey for fish, and increases in *Neomysis* following fish removal have in some cases negated the intended effect of increases in *Daphnia* grazing. Some of the highest mysid densities ($11\,000 \, m^{-2}$, $40 \, g \, ww \, m^{-2}$ assuming a 15% dry weight) have been reported for *Neomysis intermedia* in hyper-eutrophic Lake Kasumigaura, Japan. *Tenagomysis chiltoni* is abundant in turbid, shallow New Zealand lakes. Littoral mysids, such as *Hemimysis anomala* and *Limnomysis benedeni* can be abundant in both small and large lakes, and will then also affect zooplankton species composition.

Other Crustaceans

In some lakes, a similar life history strategy to mysids have evolved among amphipods. For example, the amphipod *Macrohectopus branickii* is abundant in the open water of Lake Baikal and is similar in appearance to mysids (**Figure 1**). Females can be up to 38 mm long although they mature at 16–20 mm. Brood size increases with female length from 90 to 400 eggs per female. Males mature at smaller size (5–6 mm). The species is an omnivore that migrates over 100 meters each night to feed in the surface water. As in *Mysis relicta*, mortality of *Macrohectopus* increases over 12–13 °C; they avoid light levels above 10^{-4} lux, and become more predatory as they grow larger. Available data suggest that *Macrochectopus* has a generation time similar to *Mysis relicta* making this species similarly sensitive to predation. The biomass of *Macrohectopus* (4–10 g ww m $^{-2}$) is as high or higher than the biomass of *Mysis* in the Great Lakes (2–5 g ww m $^{-2}$ in Lake Ontario). This species is endemic to Lake Baikal and may be the amphipod species with the most extreme ecological divergence from its closest relative. There are no mysids in Lake Baikal. The remarkable similarity in the ecology of *Mysis* and *Macrohectopus* is likely due to the similar selective pressures on large migrating invertebrate predators in deep cold lakes. There is a cost to spending the day in deep, cold water both because there is less food in that area, and because cold water decreases metabolism, increases digestion time, and decreases growth (when food is plentiful). Slower growth leads to longer generation times, and therefore to populations that are even more sensitive to predation than if they resided in the warmer epilimnion both day and night. This will further increase the selective pressure for predator avoidance through migrations. Amphipods are an important component of the limnetic zooplankton in other lakes as well. For example, the omnivorous amphipod *Jesogammarus annandalen* is abundant and performs nocturnal migrations into the water column of Lake Biwa, Japan.

In the East African Great Lakes a major component of the zooplankton are decapod shrimps in the families Atyidae and Palaemonidae. The up to 25 mm-long atyid shrimp *Caridina nilotica* reach densities of 1000 m $^{-2}$ in Lake Victoria. They increased in abundance after the introduction of the Nile perch (*Lates niloticus*), likely an indirect effect of the decline in haplochromine cichlids associated with the increase in Nile perch. *Caridina* feed on detritus and phytoplankton and perform diel migrations to feed at the surface at night. Only small individuals remain in the upper water also during the day.

Fairy shrimps (Anostraca) can be very common in fish-less ponds and salt lakes. Some are predators but most feed by filtering algae and detritus. The brine shrimps (*Artemia* sp.) are found worldwide in high salinity lakes and in ponds that are too saline for fish. This is one of the few zooplankton that are harvested commercially. Large quantities of *Artemia* resting eggs are used in the fish culture industry. These resting eggs can be stored dry for long periods of time, and made to hatch by high temperature, salinity and light.

Chaoborus Larvae – the Phantom Midge

The phantom midge larvae of the genus *Chaoborus* are a major component of the planktonic communities of many lakes. There are about 50 extant species in 6 genera and 2 sub-families in the family Chaoboridae. They are found in lakes worldwide from the tropics to the arctic regions and occur in some of the largest lakes in the world (e.g., Lakes Victoria and Malawi) as well as in small ponds. The adult insects are medium-sized, non-biting, live for about a week and deposit several hundred eggs in rafts on the water surface. Eggs hatch in a few days. There are four larval instars. The first two are limnetic and last a couple of weeks. The third instar occurs both in the open water and burrowing in the sediment. The fourth instar (**Figure 3**) typically represents the largest biomass and has the largest effect on its prey. This stage can last more than one year and is often the stage that overwinters. The pupae stage (**Figure 3**) lasts a few days to 2 weeks depending on temperature. Depending on whether the generation overwinters or not, the life cycle duration is a couple of months to more than half a year. Some species have a 1- or even 2-year life cycle.

When *Chaoborus* larvae burrow into the sediment they assume a vertical anterior-end up S-shaped position and do not maintain a tube connection to the overlying water. Because they burrow into the anoxic layer, they are likely tolerant to hydrogen sulfide present in those layers. They also feed on benthic prey such as oligochaetes and benthic harpacticoid copepods.

Different species vary in their sensitivity to fish predation. Larvae of most species are transparent and move little except during migration. Swimming is a jerky motion produced by flexing the body back

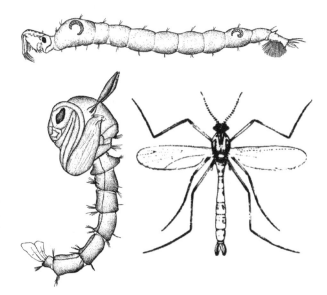

Figure 3 Fourth instar larva (top), pupa (left), and adult (right) of *Chaoborus punctipennis*. Adapted from Johannsen OA (1934) Aquatic Diptera. Part I. Nemocera, exclusive of Chironomidae and Ceratopogonidae. *Memoirs of the Cornell University Agricultural Experimental Station* 164: 1–70.

and forth. The larva has two paired air sacks that are used to maintain buoyancy without the need for conspicuous swimming movements. The eye is the most visible part of the body and larger eyes as well as increased stomach content increase predation risk. The diel migration is a predator avoidance behavior. Predation by fish on *Chaoborus* is more intense on third and fourth instar larvae. These are also the instars that migrate. Most fish will not enter water with oxygen levels below $1-2 \, mg \, l^{-1}$ and the low oxygen levels in the sediment or hypolimnion therefore provide a refuge from fish predation during the day. The migration is triggered by decreasing light levels and the animals move using gravity as a cue – downwards in high light levels and upwards at low light levels. When fish are abundant, *Chaoborus* larvae initiate migration at a lower light level. Not all individuals in a population migrate; the proportion that migrates varies with season and is highest in summer and lowest in winter. *Chaoborus* is adapted to low oxygen conditions but may incur an oxygen debt during the day. Pupae that spend part of the day in anoxic conditions develop slower than pupae that remain in oxic conditions. Pupae also often migrate into the water at night, probably to speed up development time in higher oxygen concentrations. Migration distance is typically tens of meters, but in Lake Malawi, *Chaoborus* fourth instar migrate from daytime depths of 150–200 m to the surface waters each night. Not all *Chaoborus* species have larvae that migrate. There are larger species common in fishless lakes that do not. *Chaoborus* will cease migration after some time if fish are removed, but this change in behavior is slower than the induction of migration when fish are added to a lake.

Chaoborus larvae are ambush predators that can be motionless in the water owing to the buoyancy provided by the air sacks. When a prey swims close enough to be detected, *Chaoborus* strike and capture the prey with the modified prehensile antennae and mandibles. This strategy makes *Chaoborus* equally effective in light and dark and at catching copepods and cladocerans, which is unusual for zooplanktivores. The first and second instars feed on rotifers and larger algae such as dinoflagellates, whereas the third and fourth instars feed on cladocerans and copepods as well as other dipterans and benthic organisms. Most studies show a selection for copepods over cladocerans, likely the result of differences in their ability to handle different prey and on the swimming speed of the prey – more active prey will be encountered more often. Prey size increase with the size of the larvae – the width of the prey is important, not the length. Well-fed *Chaoborus* are more selective as to which prey to attack. Handling of cladocerans is more difficult when neck teeth, elongated tail spines, and helmets are present. Such structures are induced in *Daphnia* in the presence of chemicals (kairomones) released by *Chaoborus*. The animals also feed on smaller *Chaoborus* larvae and other insect larvae. Although several species of *Chaoborus* co-exist in many lakes, smaller species are often depressed by larger species and therefore the smaller species can be more abundant when fish are present than when fish are absent. *Chaoborus* are sensitive to kairomones both from their prey and their predators. Their activity increases in the presence of prey kairomones and migrations increase in the presence of fish kairomones.

Chaoborus can be abundant (over $10\,000 \, m^{-2}$ reported), have high production, and feeding rates as high as 10% of their body weight per day or approximately one crustacean per hour. Although the presence of *Chaoborus* does not always affect zooplankton species composition, there is clear evidence from laboratory enclosures and field experiments showing that this predator can have strong effects on their prey (**Table 1**). But the effect varies with predator and prey size and morphology, and can change over time owing to induction of structural defenses in the prey species. In a biomanipulation experiment in a German reservoir, removal of fish caused an increase in *Chaoborus* followed by a decline in zooplankton, in contrast to the intended effect of increasing zooplankton abundance through fish removal. *Chaoborus* and fish can also interact to produce a stronger effect than when either is present alone because the presence of *Chaoborus* can affect the ability of cladocerans to use low oxygen refuges from fish predation.

Dreissena Veligers

The planktonic larva (veliger) of dreissenid mussels is also considered here. Most marine bivalves have a planktonic larval stage, but most freshwater bivalves do not. Dreissenids release gametes in large quantities (up to a million eggs per female) in synchronized spawning events throughout the season, often with peaks in June and August. Zebra mussels (*Dreissena polymorpha*) have optimum temperatures for spawning and veliger growth between 12 °C and 24 °C, and pH between 7.4 and 9.4. Quagga mussels (*Dreissena bugensis*) occur in large densities in deep, consistently 4 °C cool water, and likely spawn at those temperatures. Following established nomenclature for marine bivalves, Ackerman and colleagues recognized the following stages for dreissenids (**Figure 4**): (1) the egg and embryonic period, (2) the free swimming trochophore, (3) the D-shaped veliger (formed when the velum – a larval organ used for feeding

Table 1 Some case studies of the ecological effects of mysids and *Chaoborus* on zooplankton communities

Group, Location	Peak density	Lake trophic status	Summary of results	Source
Mysis relicta				
Lake Tahoe, NV, CA	400 m^{-2}	Ultra-oligotrophic	*Daphnia* and fisheries declined after introduction of *Mysis*	1
Lake Ontario, USA, Canada	1100 m^{-2}	Oligotrophic	Native, similar feeding rates as planktivorous fish, inferred to be important planktivore	2
Lake Michigan, USA, Canada	1000 m^{-2}	Oligotrophic	Native, omnivore, temperature refuge for zooplankton limits *Mysis* predation in summer months	3
Flathead Lake, Montana, USA	130 m^{-2}	Oligotrophic	Strong food web effects through declines in *Daphnia* and kokanee after introduction of *Mysis*. Also affecting eagles and mammals feeding on kokanee	4
Lake Pend Oreille, Idaho, USA	1250 m^{-2} (average)	Oligotrophic	Introduced, zooplanktivory similar to fish, *Mysis* affect seasonal zooplankton dynamics and spatial patterns	5, 6
Lake Hiidenvesi, Finland	175 m^{-2}	Eutrophic	Native but limited by low oxygen. *Chaoborus* more important predator on zooplankton	7
Lake Selbusjøen, Norway	200 m^{-2}	Oligotrophic	Strong negative effect on cladocerans after the introduction of *Mysis*	8
Lake Jonsvatn, Norway	130 m^{-2}	Oligotrophic	Decine of *Daphnia* and other zooplankton in two embayments of the lake after *Mysis* introduction, less effect in the main lake	9, 10
Canadian Shield Lakes, Ontario	Varied	Oligotrophic	Native, comparisons among similar lakes with and without *Mysis* show negative effects on cladocerans in the meta and hypolimnion	11
Other mysids				
Lake Fering, Denmark– *Neomysis integer*	900 m^{-2}	Eutrophic, brackish	Native, cladoceran seasonal dynamics due to *Neomysis* predation	12
Lake Kasumigaura, Japan *Neomysis intermedia*	17 000 m^{-2}	Eutrophic	Native, strong seasonal variability in mysids related to zooplankton seasonal dynamics	13
Lake Washington, WA, *Neomysis mercedis*	350 m^{-2}	Mesotrophic	Native, major zooplankton predator before increase in fish (longfin smelt)	14
Muriel Lake, BC, Canada *Neomysis mercedis*	85 m^{-3}	Oligotrophic	Native, *Neomysis* abundant in low sockeye salmon years and then more important than fish as zooplankton predator.	15
Beisbosch Reservoir, Netherlands, *Hemimysis anomala*	6000 m^{-2}	Eutrophic	Strong decline of all zooplankton after *Hemimysis* invasion	16
Chaoborus				
Gwendoline Lake, BC, Canada	1000 m^{-2}	Oligotrophic	Fishless lake with two *Chaoborus* species with 1 and 2 year life cycles. *Chaoborus* predation interact with seasonal algal dynamics to produce zooplankton seasonal dynamics.	17
Temporary pond, Santa Cruz, CA	800 m^{-3}	Eutrophic	*Chaoborus* preferred *Ceriodaphnia*, and this species declines during high *Chaoborus* abundance	18
Lake Hiidenvesi, Finland	14 000 m^{-2}	Eutrophic	Large seasonal changes in abundance of *Chaoborus* causes seasonal changes in zooplankton	19
Lake Iso Vakljärvi, Finland	6000 m^{-3}	Eutrophic	Fish kill released *Chaoborus* and depressed zooplankton	20
Small experimental lakes, Germany		Eutrophic	*Chaoborus* increase after fish removal caused decline in *Daphnia*. Moderate fish abundance beneficial to *Daphnia*	21

Source
1. Richards R, Goldman C, Byron E, and Levitan C (1991) The mysids and lake trout of _ake Tahoe: A 25-year history of changes in fertility, plankton, and fishery of an alpine lake. *American Fisheries Society Symposium* 9: 30–38.
2. Johannsson OE, Rudstam LG, Gal G, and Mills EL (2003) *Mysis relicta* in Lake Ontario Population dynamics, trophic linkages and further questions. In Munawar, M. (ed.) *State of Lake Ontario (SOLO) – Past, Present and Future*, pp. 257–287. Leiden, The Netherlands: Backhuys Publishers.

3. Pothoven SA, Fahnenstiel GL, and Vanderploeg HA (2004) Spatial distribution, biomass and population dynamics of *Mysis relicta* in Lake Michigan. *Hydrobiologia* 522: 291–299.

4. Spencer CM, McClelland BR, and Stanford JA (1991) Shrimp stocking, salmon collapse, and bald eagle displacement: Cascading interaction in a food web of a large aquatic ecosystem. *Bioscience* 41: 14–21.

5. Chipps SR and Bennett DH (2000) Zooplanktivory and nutrient regeneration by invertebrate (*Mysis relicta*) and vertebrate (*Oncorhynchus nerka*) planktivores: implications for trophic interactions in oligotrophic lakes. *Transactions of the American Fisheries Society* 129: 569–583.

6. Clarke LR and Bennett DH (2003) Seasonal zooplankton abundance and size fluctuations across spatial scales in Lake Pend Oreille, Idaho. *Journal of Freshwater Ecology* 18: 277–290.

7. Horppila J, Liljendahl-Nurminen A, Malinen T, Salonen, M, Tuomaala, A, Uusitalo L, and Vinni M (2003) *Mysis relicta* in a eutrophic lake: Consequences of obligatory habitat shifts. *Limnology and Oceanography* 48: 1214–1222.

8. Langeland A, Koksvik IJ, and Nydal A (1991) Impact of the introduction of *Mysis relicta* on the zooplankton and fish populations in a Norwegian lake. *American Fisheries Society Symposium* 9: 98–114.

9. Koksvik JI, Reinertsen H, and Langeland A (1991) Changes in plankton biomass and species composition in Lake Jonsvatn, Norway, following the establishment of *Mysis relicta*. *American Fisheries Society Symposium* 9: 115–125.

10. Naesje TF, Saksgard R, Jensen AJ, and Sandlund OT (2003) Life history, habitat utilisation, and biomass of introduced *Mysis relicta*. *Limnologica* 33: 244–257.

11. Nero RW and Sprules WG (1986) Zooplankton species abundance and biomass in relation to occurrence of *Mysis relicta* (Malacostraca; Mysidacea). *Canadian Journal of Fisheries and Aquatic Sciences* 43: 420–434.

12. Søndergaard M, Jeppesen E, and Aaser HF (2000) *Neomysis integer* in a shallow hypertrophic brackish lake: Distribution and predation by three-spined stickleback (*Gasterosteus aculeatus*). *Hydrobiologia* 428: 151–159.

13. Hanazato T (1990) A comparison between predation effects on zooplankton communities by *Neomysis* and *Chaoborus*. *Hydrobiologia* 198: 33–40.

14. Chigbu P (2004) Assessment of the potential impact of the mysid shrimp, *Neomysis mercedis*, on *Daphnia*. *Journal of Plankton Research* 26: 295–306.

15. Hyatt KD, Ramcharan CR, McQueen DJ, and Cooper KL (2005) Trophic triangles and competition among vertebrate (*Oncorhynchus nerka*, *Gasterosteus aculeatus*) and macroinvertebrate (*Neomysis mercedis*) planktivores in Muriel Lake, British Columbia, Canada. *Ecoscience* 12: 11–26.

16. Ketelaars HAM, Clundert FEL–v d, Carpentier CJ, Wagenvoort AJ, and Hoogenboezem W (1999) Ecological effects of the mass occurrence of the Ponto–Caspian invader, *Hemimysis anomala* G.O. Sars, 1907 (Crustacea: Mysidacea), in a freshwater storage reservoir in the Netherlands, with notes on its autecology and new records. *Hydrobiologia* 394: 233–248.

17. Neill WE (1981) Impact of *Chaoborus* predation upon the structure and dynamics of a crustacean zooplankton community. *Oecologia* 48: 164–177.

18. Riessen HP, Sommerville JW, Chiappari C, and Gustafson D (1988) *Chaoborus* predation, prey vulnerability, and their effect in zooplankton communities. *Canadian Journal of Fisheries and Aquatic Sciences* 45: 1912–1920.

19. Liljendahl-Nurminen A, Horppila J, Malinen T, Eloranta P, Vinni M, Alajärvi E, and Valtonen S (2003) The supremacy of invertebrate predators over fish – Factors behind the unconventional seasonal dynamics of cladocerans in Lake Hiidenvesi. *Archiv für Hydrobiologie* 158: 75–96.

20. Rask M, Järvinen M, Kuoppamäki K, and Poysa H (1996) Limnological responses to the collapse of the perch population in a small lake. *Annales Zoologici Fennici* 33: 517–524.

21. Wissel B, Freier K, Müller B, Koop J, and Benndorf J (2000) Moderate planktivorous fish biomass stabilizes biomanipulation by suppressing large invertebrate predators of *Daphnia*. *Archiv für Hydrobiologie* 149: 177–192.

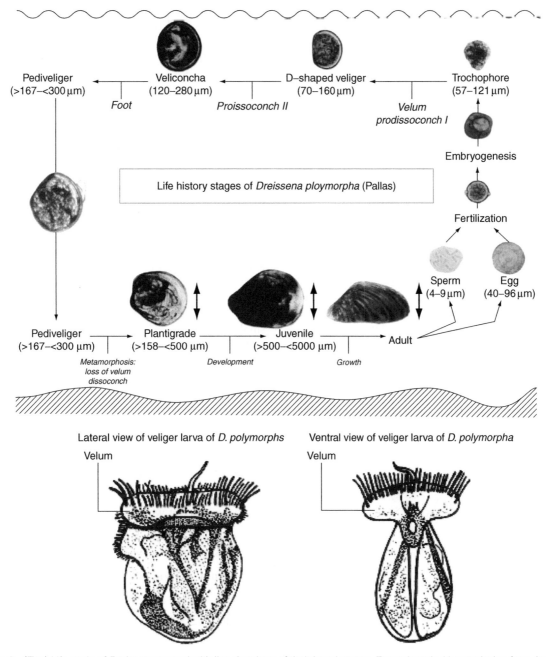

Figure 4 (Top) Life cycle of *Dreissena* mussel with line drawings of their larval stages. Reproduced with permission from Ackerman JD, Sim B, Nichols SJ, and Claudi R (1994) Review of the early-life history of zebra mussels (*Dreissena polymorpha*) – Comparisons with marine bivalves. *Canadian Journal of Zoology* 72: 1169–1179. (Bottom) Free swimming veliger of Dreissena showing the velum. From the zebra mussel information system of the Army Corps of Engineers: http://el.erdc.usace.army.mil/zebra/zmis/

and locomotion – and the first shell has been formed), (4) the velichonca (after a second shell is secreted), (5) the pediveliger (with a foot that can produce byssal treads and is used for both swimming and crawling), (6) the postveliger or plantigrade mussel (formed after the pediveliger has selected a site, attached itself with byssal threads and undergone metamorphosis), and (7) the adult mussel. *Dreissena* veligers can be recognized in plankton samples using

cross-polarized light. The veliger has a row of cilia on the velum that is extended out from the shell. These cilia are in constant motion as the animal filters particles into a feeding groove where the particles are encased in mucus and transported to the stomach. During the larval period the animal increases in size from 70 to over 200 μm.

Veligers can be abundant. They are, at times, the dominant plankton organisms in the St. Lawrence

River estuary and densities of over $500\,000\,m^{-3}$ have been reported from Lake Erie. They feed on $1–4\,\mu m$ particles, including blue-green and small green algae, bacteria, protists, and detritus, and through direct uptake of dissolved organic carbon. Filtration rates are in the order of $10–20\,\mu l\ hour^{-1}$. Although they are consumed by some planktivorous fish, they are not a major component of fish diets. The presence of veligers has not been shown to affect other zooplankton groups, at least not crustacean zooplankton. Filter feeders, such as adult zebra mussels, may be their main predators.

Other Animals In the Zooplankton

There are other animals that should be considered part of the zooplankton of inland waters. These include several species of water mites and jellyfish, and various insect larvae and adults. Water mites and freshwater medusae are predators, and both groups have been shown to affect their zooplankton prey in some lakes. Planktonic water mites occur throughout the world. The most common are in the genus *Piona*. They emerge from the eggs as small, six-legged larvae and metamorphose to the nymphal stage after 8–28 days. These nymphs are predators on zooplankton (**Figure 5**). When encountering a prey, the mite will grab the prey with its legs and tear open the body wall with its chelicerae. The soft body of the prey is predigested and then drawn into the mouth. They select cladocerans over copepods. The larvae do not need to parasitize insect larvae to develop to the nymphal stage (many other water mites require feeding to develop to the nymphs) and it is not known if they are facultative parasites or not.

Freshwater medusae in the family Limnomedusae (e.g., the 20–25 mm *Craspedacusta* sp. in Eurasia and North America and the similar sized *Limnocnida* sp. in Africa and India) are a highly variable component of the plankton, occurring in high numbers for several years followed by years of low abundance. As other jellyfish, they have a complex life cycle with both sessile polyp stages and an open water medusa stage (**Figure 6**). These predators co-occur with high abundance of fish as the fish do not prey on them. Medusae feed on copepod nauplii and copepodites and small cladocerans.

Zooplankton do not only consist of copepods, cladocerans, and rotifers, and some of the main players in planktonic food webs are in the diverse group 'other zooplankton' discussed in this chapter.

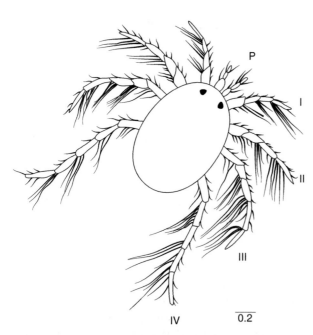

Figure 5 A nymph of the water mite *Piona rufa* (Adapted by Roger Wayman from Wolcott RH (1905) *Water Mites and How to Collect*. Bulletin 11 Roger Williams Park Museum, Providence, R.I. and from Riessen HP (1982) Pelagic water mites: Their life history and seasonal distribution in the zooplankton community of a Canadian lake. *Archiv für Hydrobiologie Suppl* 62: 410–439.

Glossary

Byssal threads – formed by mussels to attach to the substrate.

g ww – gram wet weight.

Kairomone – chemicals used to obtain information on presence of prey or predators.

Marsupium – brood pouch.

Nauplius – larval stage of some crustaceans groups.

Pediveliger – 4th stage of mussel verligers, have a foot.

Planula larvae – free moving larvae of freshwater jellyfish.

Podocyst – resting stage of a freshwater polyp.

Polyp – benthic stage of jellyfish.

Postveliger or plantigrade mussel – 1st attached stage of the mussel.

Trochophore – 1st free swimming stage of mussel veligers.

Velichonca – 3rd stage of mussel veligers– after second shell is formed.

Veliger – free swimming larvae of mussels.

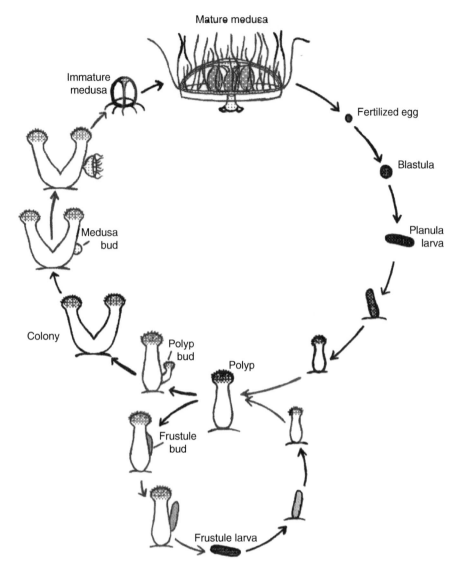

Figure 6 Life cycle of *Craspedacusta sowerbii*. The small stalked polyp lives in colonies attached to a substrate (bottom, docks, plants) and reproduce asexually. Sometimes, medusae are formed that move into the water column and reproduces sexually. Both the medusae and polyps have stinging cells that will fire on contact. Fertilized eggs develop into planula larvae, which will settle and develop into new polyps. The polyp can contract in cold temperature into a resting stage (podocyst) that survives cold temperatures. Podocysts and polyps may be transported by animals or plants to other water bodies. Adapted by M. Thom from Lytle CF (1982) Development of the freshwater medusa *Craspedacusta sowerbii*. In Harrison FW and Cowden RR (Eds.) *Developmental Biology of Freshwater Invertebrates*, pp. 129–150. New York: Alan R. Liss Inc. Reproduced with permission from Terry Peard, www.jellyfish.iup.edu.

See also: Cladocera; Competition and Predation; Copepoda; Cyclomorphosis and Phenotypic Changes; Diel Vertical Migration; Role of Zooplankton in Aquatic Ecosystems; Trophic Dynamics in Aquatic Ecosystems.

Further Reading

Ackerman JD, Sim B, Nichols SJ, and Claudi R (1994) Review of the early-life history of zebra mussels (*Dreissena polymorpha*) – Comparisons with marine bivalves. *Canadian Journal of Zoology* 72: 1169–1179.

Berendonk TU, Barraclough TG, and Barraclough JC (2003) Phylogenetics of pond and lake lifestyles in *Chaoborus* midge larvae. *Evolution* 57: 2173–2178.

Borkent A (1981) The distribution and habitat preferences of the Chaoboridae (Culicomorpha, Diptera) of the Holarctic region. *Canadian Journal of Zoology* 59: 122–133.

Jankowski T, Strauss T, and Ratte HT (2005) Trophic interactions of the freshwater jellyfish *Craspedacusta sowerbii*. *Journal of Plankton Research* 27: 811–823.

Johannsson OE, Rudstam LG, Gal G, and Mills EL (2003) *Mysis relicta* in Lake Ontario: population dynamics, trophic linkages and further questions. In: Munawar M (ed.) *State of Lake Ontario (SOLO) – Past, Present and Future*, pp. 257–287. Leiden, The Netherlands: Backhuys Publishers.

Lasenby DC, Northcote TG, and Fürst M (1986) Theory, practice, and effects of *Mysis relicta* introductions to North American and Scandinavian lakes. *Canadian Journal of Fisheries and Aquatic Sciences* 43: 1227–1284.

Mauchline J (1980) The biology and mysids and euphausiids. In: Blaxter JHS, Russell FS, and Younge M (eds.) *Advances in Marine Biology*. 1–369.

Morgan MD (ed.) (1982) Ecology of Mysidacea. *Hydrobiologia* 93: 1–222.

Nesler TP and Bergersen EP (eds.) (1991) *Mysids in Fisheries: Hard Lessons from Headlong Introductions. American Fisheries Society Symposium 9*. Bethesda, Maryland: American Fisheries Society.

Riessen HP (1982) Pelagic water mites: Their life history and seasonal distribution in the zooplankton community of a Canadian lake. *Archiv für Hydrobiologie Supplement* 62: 410–439.

Sprung M (1992) The other life: An account of present knowledge of the larval phase of *Dreissena polymorpha*. In: Nalepa TF and Schloesser DW (eds.) *Zebra mussels. Biology, Impacts and Control*, pp. 39–54. Boca Raton: Lewis Publishers.

Sweetman JN and Smol JP (2006) Reconstructing fish populations using *Chaoborus* (Diptera: Chaoboridae) remains – A review. *Quaternary Science Reviews* 25: 2013–2023.

Swift MC (1992) Prey capture by the 4 larval instars of *Chaoborus crystallinus*. *Limnology and Oceanography* 37: 14–24.

Wissel B, Yan ND, and Ramcharan CW (2003) Predation and refugia: Implications for *Chaoborus* abundance and species composition. *Freshwater Biology* 48: 1421–1431.

Relevant Websites

www.jellyfish.iup.edu – Freshwater jellyfish. Web site maintained by Dr Terry Peard, Professor of Biology, Indiana University of Pennsylvania, Indiana, PA. This site includes much information, including video clips of freshwater jellyfish.

http://el.erdc.usace.army.mil/zebra/zmis/ – Zebra mussel information. The US Army Corp of Engineers maintain this website 'zebra mussel information system' with excellent information on all aspects of zebra mussel biology.

Cyclomorphosis and Phenotypic Changes

C Laforsch, Ludwig-Maximilians-University Munich, Planegg-Martinsried, Germany
R Tollrian, Ruhr-University, Bochum, Germany

Various freshwater organisms display characteristic phenotypic changes in time, (1) regular seasonal changes in morphology, known as cyclomorphosis, and (2) relatively rapid changes in morphology, life history, and/or behavior in response to predator activity, interpreted as inducible defenses. These two phenomena are not necessarily related but inducible defenses can be cyclomorphic if predators show seasonal changes in abundance.

Cyclomorphosis

Cyclic recurrent polymorphisms in planktonic organisms have been known to scientists since the late nineteenth century. The term cyclomorphosis was coined in 1904 by Lauterborn to describe the seasonal changes in morphology occurring in a variety of taxa including ciliates, rotifers, and cladocerans. The often dramatic morphological changes found in these organisms repeatedly resulted in nomenclatorial and taxonomic confusion.

There has been speculation that the observed seasonal morphological variations could be a succession of different species, a succession of different morphs of a species representing different genotypes or phenotypic plasticity within single genotypes. The latter has been proved in most cases, leading to the current definition of cyclomorphosis: a temporal, cyclic morphological change in successive generations of small aquatic organisms. However, that a seasonal succession of different genotypes might accompany patterns seen in nature in a few species cannot be ruled out (**Figure 1**).

Seasonal variation in morphological traits has been reported for a wide range of organisms including dinoflagellates and rotifers, but most examples have been documented in cladocerans, especially in the genus *Daphnia*.

Conspicuous changes in size have been reported, e.g., a general slenderness and a relative growth of a broad variety of morphological traits, including a marked elongation of the tail spine or the head and the development of spiky or curved helmets or dorsal crests (**Figure 2**). Less obvious temporal changes involve the length of the antennal bristles, the enlargement of the filtering screens on the thoracic limbs, or the size of the compound eye. Other cladocerans such as *Bosmina* have been found to show an elongation of the first antennae in summer months.

Proximate and Ultimate Factors of Cyclomorphosis

Since the discovery of cyclic seasonal polymorphisms in freshwater plankton, there has been plenty of scope for ecologists to unravel the proximate and ultimate factors. Given that natural selection acts on phenotypes, conspicuous cyclomorphic features are unlikely to be neutral. Early research primarily focused on abiotic factors such as temperature, photoperiod, and turbulence as inducing agents. Additionally, food quantity and quality was supposed to determine morphological changes.

While most studies failed to induce responses as distinct as those observed in the field, several hypotheses have been formulated to explain the adaptive value of cyclomorphosis. A first mechanistic interpretation proposed that the elongated morphological traits of the animals produced a 'parachute-effect' to compensate for accelerated sinking rates experienced in summer months, which can be attributed to the decreased viscosity of warmer water. However, experiments failed to prove this hypothesis; on the contrary, it was found that anesthetized animals that possessed longer helmets and tail spines sank even faster than the typical morphs. Other ideas, as to why cyclomorphosis might be adaptive refer to the hydromechanical properties of the plastic traits. Helmets and tail spines were thought to stabilize the movement in *Daphnia* either by working as keels or by increasing swimming strength. Compared to typical morphs, helmeted daphniids were thought to be better able to resist water currents, enabling them to stay in preferred water layers containing higher food concentrations. Further, it was also speculated that elongated antennal muscles, as a result of cyclomorphic changes, improve swimming properties in *Daphnia*.

The pioneering work of de Beauchamp in 1952 and of Gilbert in 1966 switched the view to biotic interactions. They showed that chemical cues released by the predacious rotifer *Asplanchna* induced large divergent spines in *Brachionus calyciflorus* that prevented the predator from ingesting the prey. It has been discovered subsequently that cyclomorphosis in cladocerans is also driven by chemical cues released by a variety of predators, and that induced structures like neckteeth (**Figure 3**) play a defensive role.

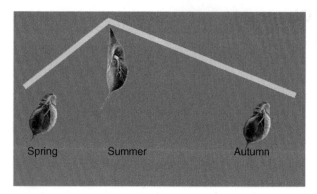

Figure 1 Typical seasonal course of cyclomorphosis in *Daphnia cucullata*, showing the most extreme helmet formation in early summer followed by a reduction in helmet size continuing until autumn (Laforsch and Tollrian, 2004).

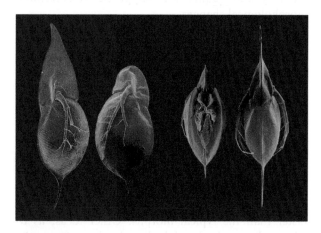

Figure 2 Scanning electron micrograph of genetically identical individuals of *D. cucullata*, cited as the textbook example for cyclomorphosis. (Left to right: lateral view of helmeted morph, also showing an elongated tail spine; lateral view of the typical nonhelmeted morph; ventral view of a nonhelmeted morph; dorsal view of the helmeted morph showing the shape of the blade-like helmet) (© C. Laforsch).

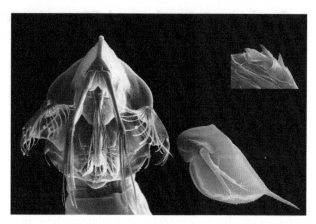

Figure 3 Scanning electron micrograph of a model predator–prey induced defense. Neckteeth formation (upper right) in *Daphnia pulex* (juvenile; lower right) in response to chemical cues released by the predacious dipteran larvae *Chaoborus*. Left: ventral view of *Chaoborus* head region showing an open feeding basket (© C. Laforsch).

The inclusion of predator–prey interactions into the cyclomorphosis framework provided new avenues of research into the ultimate mechanisms controlling the phenomenon. In 1974, Dodson established the antilock-and-key hypothesis, suggesting that the catching apparatus of predatory invertebrates are adapted to the typical morph of the prey and the changed morphology will not fit. Further to this idea, it has been hypothesized that seasonal polymorphisms act as defenses against size-selective predatory invertebrates; predators are less able to ingest individuals possessing enlarged helmets and spines, hence the prey are said to have reached a size-refuge. Observations in natural habitats corroborate these theories because the most conspicuous morphs are expressed

during periods of heavy predation indicating that cyclomorphosis can be driven by inducible defenses.

Inducible Defenses

Phenotypic plasticity in defensive strategies has proved to be a widespread mechanism to cope with predatory selection pressure. In freshwater ecosystems inducible defenses are found in a variety of taxa spanning bacteria, algae, ciliates, rotifers, crustaceans, insect larvae, and even vertebrates.

Inducible defenses are thought to have evolved under specific ecological conditions. Four major factors favoring the evolution of inducible defenses have been identified:

1. The selective pressure of the inducing agent has to be variable but strong.
2. A reliable cue is necessary to indicate the proximity of the threat and activate the defense.
3. The defense must be effective within a relatively short time to avoid lag phases.
4. The inducibility of a defense should incur costs or trade-offs that offset the benefit of a defense during times when the defense is not needed.

The abundance of predators varies temporally because of their ontogeny. Larval fish, for instance, which are probably the most important predators on pelagic crustaceans, reach their highest numbers in early summer but are then drastically reduced in number by both predation and disease until winter Likewise, the selective impact of predatory invertebrates fluctuates seasonally; for example, the ambush predator *Chaoborus* has one or two generations a year in temperate habitats. Larvae pass through the four developmental stages, each of which has a size-dependent prey preference. Phenotypic plasticity allows for adaptations to heterogeneous environments by forming adequate combinations of defensive traits for each specific predation environment.

The seasonally and temporally variable predation impact should be detectable by cues that indicate the actual predation risk for a specific prey species. Infochemicals released by predators have been found to provide reliable signals. These so-called 'kairomones' are exclusively advantageous to the receiver in an interspecific information-transfer context and enable prey organisms to show predator-specific defenses.

Induced defensive traits can occur in several forms. First, organisms can modify life-history factors, including resting-egg induction to temporally avoid predators, or shifts in resource allocation between somatic growth and reproduction. Second, behavioral defenses such as elevated alertness, modified swimming behaviors, the formation of swarms or diel vertical migration may reduce encounters or allow escapes. Third, physiological changes, e.g., toxin production can prevent consumption. Fourth, morphological defenses including crests, helmets, and tail spines have been especially well studied. They mainly interfere with the predator's ability to catch the prey. For example, the development of a large crest in *Daphnia longicephala* is caused by kairomones released by the predatory backswimmer *Notonecta* and results in the predator being unable to obtain a firm grasp on its prey (**Figure 4**). Although conspicuous,

induced traits such as crests, helmets, or spiky protuberances obviously interfere with the ability of a predator to catch or handle the prey – the exact defensive mechanisms still remain unresolved. However, it has been shown that some defenses act on several steps of the predation cycle. Helmets in *Daphnia*, for instance, interfere with the catching apparatus of predatory invertebrates, but in addition they may also disguise the prey individually or result in hydrodynamical alterations leading to the predator incorrectly judging the prey distance. Even tiny cyclomorphic changes such as neckteeth formation in *Daphnia pulex* have been proven to be protective against predatory invertebrates. Recently, it was discovered that the carapace of several *Daphnia* species, including *D. pulex*, possesses a 'hidden' inducible morphological defense, comprising a significantly increased mechanical stability of the armor. A stronger carapace provides a major advantage for the induced morphs because predatory invertebrates need to crush or puncture the cuticle of their prey prior to ingestion. Hence, it is likely that defensive mechanisms cannot be attributed to a single trait (**Figure 5**).

Critically, the overall benefit of inducible defenses depends on the lag phase between when a defense is needed until it is formed. Mechanisms which shorten lag phases have been reported in the form of maternal induction of defenses, where neonates already hatch with preformed defenses. Maternal inductions are good adaptations for situations where the maternal environment is a good predictor for the offspring's environment.

Phenotypic plasticity may additionally have other benefits besides defensive effects. For example, the rate of coevolution on the side of the predator might be reduced if the defenses are exhibited during certain time periods only.

The inducibility of defensive traits implies there is a trade off between the benefits and the costs of a defense. These costs can be saved during times when the defense is not needed. If there were no costs, the defensive traits should become genetically fixed. However, the assessment of costs derived from morphological defenses has led to contradictory results, especially because the simultaneous expression of different defense types hinders the distinction of costs derived from morphological and life-history responses. Nevertheless, costs in the form of a reduced fitness of defended morphs in the absence of predation have been reported in many species. There are many different sources of costs. In ciliates demographic costs due to reduced cell division rates of the induced morphs have been shown. These are likely caused by restructuring processes that affect the

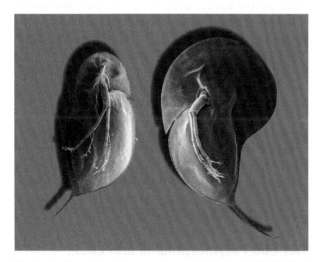

Figure 4 Scanning electron micrograph of genetically identical individuals of *D. longicephala* (a) lateral view of noninduced morph; (b) predator-induced morph showing an elongated tail spine and a defensive crest) (© C. Laforsch).

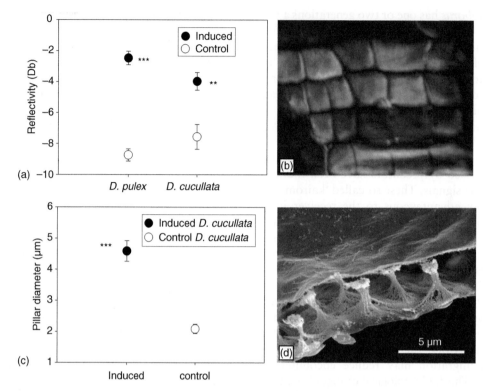

Figure 5 Hidden morphological plasticity in *Daphnia*. (a) Mean reflectivity (measurement of the strength of a material) of the carapace of *D. cucullata* and *D. pulex*. Plastic responses of both *Daphnia* species were induced by water-soluble chemicals released by the predacious phantom-midge larvae. Asterisks indicate significant differences to the control (***$p < 0.001$; **$p < 0.01$). (b) 1.2 GHz PSAM (Phase Scanning Acoustic Microscope) amplitude image of the carapace of *D. pulex* ($140 \times 150 \, \mu m^2$). (c) Mean pillar diameter of predator (induced) and nonpredator exposed (control) *D. cucullata*. Asterisks indicate significant differences to the control (***$p < 0.001$). (d) Scanning electron micrograph showing the lightweight construction of the armour: the inner and outer layer of the carapace is connected by small pillars (Laforsch *et al.*, 2004).

cytoskeletal elements needed for cell division. Allocation costs required for the building and maintenance of the defensive structure might be responsible for conspicuous structures such as crests in *Daphnia*. Operation costs for carrying a defense have been measured in the form of an elevated drag in the deeper bodied morph of the crucian carp (*Carassius carassius*) induced by pike (*Esox lucius*). Recent studies emphasize the importance of operational costs generated by changes in environmental conditions. These environmental costs might be evoked by changes in the predator spectrum or by changes in physical parameters such as higher viscosity of colder water leading to hydromechanical interference for exuberant forms. The consequence of a changing predator spectrum might be that prey animals have to cope with predators showing different size-selectivity or hunting strategies. For instance, visually hunting planktivorous fish prefer larger prey, whereas predatory invertebrates commonly prefer smaller sized prey items. To form the wrong defense would incur tremendous costs, especially for planktonic prey lacking a physical refuge.

Although inducible defenses allow some costs to be saved, they may impose some limitations. There is much speculation about plasticity costs, for example, costs for the infrastructure (e.g., sensors, hormones) to be plastic or from negative pleiotropic effects. So far there are no clear results showing that costs attributable to these sources are significant.

Different inducible traits differ in their effectiveness, their associated costs, and also in the response time between induction and formation. The induction of behavioral traits might be generally faster than the growth of morphological traits, and shifts in life-history traits are likely to be slowest. Thus, a combination of different traits could help to reduce the initial lag phases of each until effective defenses are formed.

Genetic Polymorphisms and Local Adaptations

Different predators, representing different selective forces, can induce a variety of responses including behavioral, life-history, and morphological defenses in the same prey species. Among *Daphnia* species, protective mechanisms differ even against the same predator. For example, *D. pulex* generate neckteeth primarily in the second instar in the presence of *Chaoborus*, whereas *Daphnia cucullata* show helmet formation during all life stages when faced with this predator. The adaptive explanation is, that in the small *D. cucullata* all instars are within the prey size range whereas in the larger *D. pulex* only the juvenile instars are endangered. Likewise, in tadpoles different prey species exhibit several responses to the same predator species, yet vary in their response to alternative predators. Genetic polymorphisms in predator-induced defenses occur within and among populations of several species. Inter- and intrapopulation variations in both character and intensity of defenses have been reported in recent studies comparing several *Daphnia* clones. The extent to which particular clones respond depend on the degree of predation pressure experienced in the habitat from which the prey was derived. Genetic polymorphisms in predator-induced defenses might have evolved in response to environmental constraints such as local variations in the predator spectrum associated with a reduced gene flow between populations. Those local adaptations have been reported for life-history traits in *Daphnia* evoked by fish kairomones and in *Chaoborus*-induced neckteeth formation in *D. pulex*. In both examples, clones originating from ponds inhabited by the predators showed a stronger response. Local adaptations additionally occur between microhabitats, e.g., epilimnion and hypolimnion layers in freshwater lakes. Interestingly, it has been shown that multiple optima exist within lakes, which allow different clones from the same population to reach similar fitness via different combinations of behavioral and life-history defenses.

Multipredator Environments and Uncoupling of Responses

In nature, most prey species are threatened simultaneously by a variety of predators; all of them show different prey preferences and utilize different hunting strategies. As a result, many prey species have evolved both sensitivity to cues derived from different predators and several inducible traits, which allow a flexible response to a changing predator regime. For example, ciliates of the genus *Euplotes* only show the induced bulky morph if it negatively interferes with ingestion by the predator. In contrast, chemical cues released by predatory amoebae, which are still able to feed on the induced morph, are ineffective in

triggering a morphological defense but evoke a behavioral defense instead. Nevertheless, the simultaneous exposure to different predators might cause a dilemma in choosing the adequate defense strategy. Most species are presumed to respond to the most dominant predator, as demonstrated in *Daphnia* where life-history traits induced by fish kairomone have been found to be dominant over those induced by *Chaoborus* kairomone. The impact of specific predators can vary seasonally; e.g., predation on small zooplankters may be dominated by young of the year fish during summertime, but by *Chaoborus* larvae in autumn. These changes in predator abundance will be reflected in the phenotypically plastic responses that are induced in prey organisms.

An adaptive strategy to allow flexible responses to multipredator environments could possibly be the uncoupling of different antipredator responses. This has been found, for instance, in *D. pulex* where the formation of neckteeth in early instars is distinctly separated from life-history shifts induced by *Chaoborus* in later stages. Likewise, in *Daphnia magna* different clones displayed entirely different combinations of independent antipredator responses to fish. These findings lend credence to the hypothesis of uncoupling different adaptive responses.

Furthermore, complex biotic environments such as freshwater habitats account for a variety of phenotypic responses in the same prey species. General defenses may be especially advantageous if they are effective against a variety of attackers. The defenses thus function as a 'multitool.' For instance, in *Daphnia* a combination of an inducible helmet and a tail spine leads to different protective mechanisms against different predators and in different prey size classes. The additive benefits against different attackers may increase the adaptive value of the trait and thus assist a 'diffuse' coevolution of a generalized defensive trait. Multipredator environments are therefore likely to be an important factor for the evolution and persistence of phenotypic defenses.

The Nature of the Cue

The large numbers of organisms that respond to kairomones underpin the adaptive significance of infochemical-driven cyclomorphic changes in freshwater ecosystems. On the other hand, the chemical structures of the majority of kairomones are still undetermined. Kairomones have only been identified in two systems: (1) the kairomones of two important benthic ciliates have been analyzed, and (2) more recently scientists have been able to identify several cues, consisting of aliphatic sulfates, produced by

Daphnia that induce their unicellular green algae food to form grazing resistant colonies. All other attempts to identify kairomones of pelagic organisms have failed at some stage and have only resulted in partial characterizations. The *Asplanchna* kairomone that induces spines in the rotifer *B. calyciflorus* has been shown to have a peptide origin. The active chemical substances released by *Chaoborus* were characterized as low-molecular weight and nonolefinic carboxylic acid carrying a hydroxy group. Likewise, the identification of any fish kairomone has not yet been successful. A recent focus in kairomone research has been on the origin of the kairomones. In some kairomones there is indication that the production is not only related to the predators themselves, but also to other biotic sources. Bacteria, for instance, are thought to be involved in the production of fish kairomones, as fish reared with antibiotics have been found to produce less effective signals compared to untreated fish.

Other factors may also be included, possibly even cues released by conspecifics, and several cues together may form a blend of substances that activate the formation of a defense in nature. Recent studies indicate that cues related to prey are involved in triggering defenses. A stronger reaction in *Daphnia* to kairomones from fish which had been fed *Daphnia* conspecifics was observed compared to cues from fish fed with earthworms. Similarly, in *Daphnia*, cues from crushed conspecifics induced a moderate degree of morphological defenses. Nevertheless, kairomones released by feeding predators had a distinctly stronger effect. These 'alarm cues' are most likely not specifically evolved alarm pheromones but substances within daphniid tissues or hemolymphs, which are released when they are captured or 'latent alarm signals' activated by the metabolism of the predator. 'Alarm cues' in *Daphnia* will provide general information about predation risk but will not provide specific information about the predator. *Daphnia* are prey to visually hunting vertebrates that select larger *Daphnia*, and gape-limited predatory invertebrates that select smaller *Daphnia*. To form adequate defenses *Daphnia* require additional cues which provide information about the actual relevance of different predators. Hence, synergistic effects of both 'alarm cues' and kairomones certainly increase the information quality and the reliability of the predation-risk estimation.

Prey organisms may also utilize cues from the physical environment to predict threat by predators. Photoperiod and temperature cues trigger responses in many organisms. Earlier studies have shown that environmental abiotic cues such as light intensity, photoperiod, and turbulence induced either a minute effect or influenced the response of *Daphnia* to chemical cues released by predators. However, laboratory experiments have not been able to demonstrate responses as distinct as observed in nature suggesting that parameters that are yet to be discovered might be superimposed on cues known to induce cyclomorphic features.

A recent study that combined physical and biotic factors showed that turbulence and *Chaoborus*-kairomones induce the maximum response of morphological plasticity in *D. cucullata*. Fish and predatory invertebrates generate random flow while moving in the water column. It has been suggested that turbulence can act as an additional mechanical cue for the impact of a specific predator. Likewise, a synergistic effect between predator kairomones and a physical contact with the predator itself has been demonstrated in behavioral defenses in ciliates. These results indicate that prey organisms may increase the quality and reliability of environmental information by exploiting a range of cues, and these together provide adequate information about their individual predation risk.

Genetic and Cellular Processes of Phenotypic Changes

Despite the abundance of research on morphological plasticity, knowledge of the underlying genetic and cellular processes remains a matter of speculation. In unicellular organisms, such as the ciliate *Euplotes*, the inducing infochemical is known to bind to specific receptors initiating the cascade responsible for the reorganization of the cytoskeleton to form the typical lateral wings of the induced morph (**Figure 6**). Environmental cues stimulating plastic tissues in multicellular organisms directly seems rather implausible.

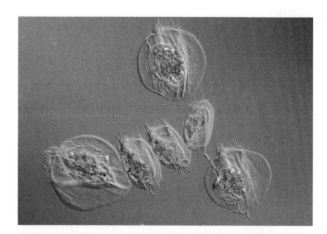

Figure 6 Phenotypic plasticity in protozoa. Chemical cues from predatory ciliates induce individuals of *Euplotes aediculatus* to develop better protected larger winged morphs. Both morphs are shown together (© J. Kusch).

Given that a blend of chemicals, or a variety of biotic and abiotic signals, evokes the same morphological response it is assumed that specific cues are sensed either by tactile or olfactory sensillae. A variety of ontogenetic phases that differ in their sensitivity to kairomones are reported for a variety of species. In rotifers, morphology is determined at an early stage via maternal induction. In *Daphnia*, both embryological and juvenile induction occurs, corresponding to the developmental stage where defense is needed. Even the maternal environment affects cyclomorphic traits in *Daphnia* via maternal activation of genes. Work on *Daphnia* has elucidated that the formation of helmets can also be induced by sublethal concentrations of some pesticides, including carbaryl and endosulfan. These pesticides affect the transmission of nerve impulses in vertebrates and invertebrates. It has been suggested that these chemicals act as acetylcholine esterase inhibitors, resulting in a permanent activation of synaptic activities that stimulate the neurosecretory release the hormones involved in the formation of morphological changes. A concentration gradient of hormones in the hemolymph of the animal might then be responsible for enhanced cell division rates in the target tissue if it exceeds a particular threshold. Moreover, the morphologically plastic tissues in *Daphnia* are characterized by endopolyploid cells that are supposed to act as developmental control centers governing the shape of the defensive traits (**Figure 7**). Facts favoring this hypothesis are the number and distribution of polyploid cells in *Daphnia*, which are strongly correlated to the size and the form of the plastic feature across several species. Additionally, mitotic activity in the cephalic epidermis of a defensive trait has been found to be elevated in regions surrounding these polyploid cells, suggesting the release of a mitogen, which triggers the development of the cyclomorphic trait. However, the cascade responsible for the potential up or down regulation of plasticity genes is still unknown. The implementation of state of the art techniques in molecular biology is an opportunity to unravel mechanisms underlying cyclomorphosis. Functional genomic and proteomic approaches will facilitate work on phenotypic plasticity in many species and the founding of the Daphnia Genomics Consortium (*DGC*) in 2003 highlights the fact that molecular research in some of these areas is well underway.

Ecological Consequences of Inducible Defenses

Although it is obvious that inducible defenses can affect predator–prey relationships, competitive

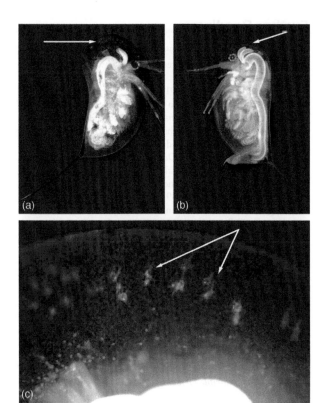

Figure 7 Polyploid cells (white arrows) stained with acridine orange of an induced (a) and noninduced morph (b) of *Daphnia longicephala*. Endopolyploid cells (c) detailed overview) are supposed to act as control centers governing morphological changes in *Daphnia* (© C. Laforsch).

interactions, and potentially ecological processes and ecosystem functions, real studies about population effects are rare. In theoretical models, inducible defenses have been shown to stabilize population fluctuations, reducing the probability that populations will become extinct. These predictions have been verified in empirical tests with inducibly defended rotifers and their predators, and more recently in a study with inducibly defended algae and rotifers in bi- and tritrophic systems.

The reason for the stabilizing function is the density dependence of the defense. It is formed only when the predators reach a density threshold and produce sufficient infochemicals. The defense subsequently reduces the predator population, which in turn leads to a lowered kairomone concentration and a reduced activation of the defenses. The stabilizing function is crucially dependent on short lag times before defenses are initiated and reversed.

See also: Cladocera; Competition and Predation; Diel Vertical Migration.

Further Reading

Chivers DP and Smith RJ (1998) Chemical alarm signalling in aquatic predator–prey systems: A review and prospectus. *Ecoscience* 5: 338–352.

Dicke M and Grostal P (2001) Chemical detection of natural enemies by arthropods: an ecological perspective. *Annual Review of Ecology and Systematics* 32: 1–23.

Dodson SI (1974) Adaptive change in plankton morphology in response to size-selective predation: A new hypothesis of cyclomorphosis. *Limnology and Oceanography* 19: 721–729.

Gabriel W, Luttbeg B, Sih A, *et al.* (2005) Environmental tolerance, heterogeneity and the evolution of reversible plastic responses. *American Naturalist* 166: 339–353.

Gilbert JJ (1999) Kairomone-induced morphological defenses in rotifers. In: Tollrian R and Harvell CD (eds.) *The Ecology and Evolution of Inducible Defenses*, pp. 127–141. Princeton: Princeton University Press.

Havel JE (1987) Predator-induced defenses: A review. In: Kerfoot WC and Sih A (eds.) *Predation: Direct and Indirect Impacts On Aquatic Communities*, pp. 263–278. Hanover, New Hampshire: University Press of New England.

Jacobs J (1987) Cyclomorphosis in *Daphnia*. In: Peters RH and de Bernardi R (eds.) *Daphnia*, pp. 325–352. Verbania, Pallanza: Memorie dell'Istituto Italiano di Idrobiologia.

Kuhlmann HW, *et al.* (1999) Predator induced defenses in ciliated protozoa. In: Tollrian R and Harvell CD (eds.) *The Ecology and Evolution of Inducible Defenses*. Princeton: Princeton University Press.

Laforsch C and Tollrian R (2004) Inducible defenses in multipredator environments: Cyclomorphosis in *Daphnia cucullata*. *Ecology* 85: 2302–2311.

Laforsch C, Ngwa W, Grill W, and Tollrian R (2004) An acoustic microscopy technique reveals hidden morphological defenses in *Daphnia*. Proceedings of the National Academy of Sciences of the United States of America 101(45): 15911–15914.

Lass S and Spaak P (2003) Chemically induced anti-predator defences in plankton: A review. *Hydrobiologia* 491: 221–239.

Miner BG, Sultan SE, Morgan SG, *et al.* (2005) Ecological consequences of phenotypic plasticity. *Trends in Ecology and Evolution* 20: 685–692.

Pohnert G, Steinke M, and Tollrian R (2007) The role of chemical cues and defence metabolites in shaping pelagic interspecific interaction. *Trends in Ecology and Evolution* 22: 198–204.

Relyea RA (2005) Constraints on inducible defenses: Phylogeny, ontogeny, and phenotypic trade-offs. In: Barbarosa P and Castellanos I (eds.) *Ecology of Predator–Prey Interactions*, pp. 189–207. Oxford: Oxford University Press.

Tollrian R and Dodson SI (1999) Predator induced defenses in cladocerans. In: Tollrian R and Harvell CD (eds.) *The Ecology and Evolution of Inducible Defenses*, pp. 177–202. Princeton: Princeton University Press.

Tollrian R and Harvell CD (eds.) (1999) *The Ecology and Evolution of Inducible Defenses*. Princeton: Princeton University Press.

Relevant Website

Daphnia Genomics Consortium, http://daphnia.cgb.indiana.edu.

Diel Vertical Migration

L D Meester, Katholieke Universiteit Leuven, Leuven, Belgium

Introduction

Diel vertical migration (DVM) is a conspicuous and widespread behavior in planktonic organisms in both inland waters and marine environments. The first scientific studies on this behavior date from more than 100 years ago, and DVM has been observed in a wide variety of habitats worldwide. DVM has attracted a lot of attention from researchers because of several reasons: (1) it is a spectacular behavior, (2) its causes have remained enigmatic for a long time, and (3) the behavior has important ecological consequences.

DVM in its Various Forms

DVM is often associated with zooplankton, and zooplankton ecologists have indeed devoted most attention to the phenomenon. However, it should be mentioned that other organisms too have been shown to perform DVM, such as fish, aquatic insects, and phytoplankton. **Figures 1–3** provide illustrations of DVM in zooplankton, chosen because they reveal several key aspects of the behavior and its variation. **Figure 1** illustrates the migration pattern of a calanoid copepod in a subtropical lake. It shows the classical pattern of a 'standard' migration: the animals reside higher in the water column during the night than during the day. During the day, they stay in deep water layers, whereas at night, they distribute themselves more evenly in the water column, and move towards the surface water layers just after sunset and before dawn. The migration is over a long distance. Indeed, the animals move 30–40 m twice every day, which is $>40 \times 10^3$ times their own body length (to human standards, this would translate into a traveling distance of >60 km a day). This illustrates why DVM is considered a spectacular behavior. **Figure 1** also shows that the movement from the upper to the bottom water layers and vice versa is associated with dawn and dusk. Finally, the figure also illustrates another very common feature: there is a clear tendency for more individuals to be caught during the night than during the day. This phenomenon is called the 'daytime deficit,' and is often observed in studies on zooplankton populations showing extensive DVM patterns. In the study system shown in **Figure 1**, it was shown that a fraction of the copepod population actually moves into the sediments during the day. In many systems, the daytime deficit is probably caused by the fact that the zooplankton is either residing so close to the sediments during the day that it cannot be sampled by traditional sampling gear or is even moving into the surface layers of the sediment. This is striking, as the sediment is a harsh environment for a zooplankton individual to reside in during part of the day. In many systems that show a strong oxycline, during the day the zooplankton resides at depths that are characterized by very low oxygen levels. **Figure 2** even shows a population of calanoid copepods that resides part of the day in the anaerobic monimolimnion of a meromictic lake. In this coastal lake, there is no oxygen below 5 m depth. **Figure 2** also shows the pattern of changes in vertical distribution around sunset, and illustrates that different life stages may differ in their daytime distribution and migration pattern. Indeed, it is clear from this figure that the youngest life stages tend to be distributed similarly to their food during the day, whereas most (sub)adult animals reside in the monimolimnion during the night, and rapidly ascend to the mixolimnion after sunset. **Figure 3** illustrates strongly different DVM patterns among two congeneric species in the same habitat. This same study also reported seasonal changes in DVM, with *Daphnia hyalina* migrating only vertically during the summer season. This pattern has been reported for many deep lakes in temperate regions. Summarizing, DVM involves changes in depth distribution over a diel cycle, can take extreme forms and shows tremendous variation through time as well as among life stages, species, and lakes. All the three examples illustrated by the figures represent the most commonly observed pattern of migration, with the animals residing deeper in the water column during the day than during the night ('standard migration'). However, it should be mentioned that many natural zooplankton populations exhibit a clear-cut depth distribution that does not change over a diel cycle, and there are also reports on 'reverse migration,' where populations reside higher in the water column during the day than during the night. This reverse pattern of DVM is less common, but is nevertheless observed in several, often small-bodied species, such as rotifers.

In the face of this overwhelming diversity in DVM patterns among species and populations, it is important to develop a broad perspective that allows structuring the observed variation. DVM can be considered a habitat selection behavior. More specifically, it is a depth selection behavior that consists of several elements that can vary among habitats,

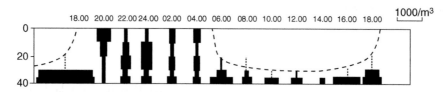

Figure 1 Diel vertical migration of adult females of the calanoid copepod *Pseudodiaptomus hessei* in Lake Sibaya (March 1972). The broken line indicates a light intensity isocline (1 lux). *X*-axis: time of day; *Y*-axis: depth (m); thickness of bars indicates number of animals (see scale bar). Reproduced from Hart RC (1976) The substrate bin – A new sampling device for studying diel vertical migratory movements on to and off lake sediments. *Freshwater Biology* 6: 155–159, with permission from Blackwell Publishing.

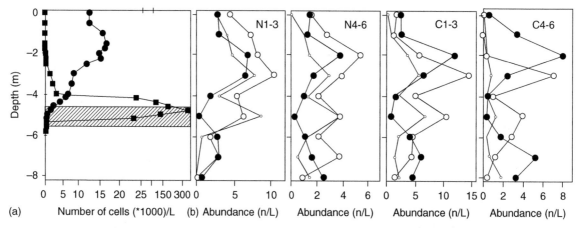

Figure 2 Vertical distribution of calanoid copepods in meromictic coastal Lake Nagada (Papua New Guinea); the lake has a brackish mixolimnion and has no oxygen below 4.5 m (cf. accumulation of hydrogen sulfide by anaerobic metabolism). (a) vertical distribution of algae in a mid-lake station, June 4, 1992. Filled circles: diatoms and flagellates; filled squares: cyanobacteria (*Oscillatoria*); shaded area: bacterial plate. (b) vertical distribution of different ontogenetic stages (small nauplii: N1-3; larger nauplii: N4-6; young copepodites: C1-3; larger copepodites: C4-6) of *Acartia tonsa* 1 h before sunset (17.00; small empty symbols), at sunset (18.00; larger empty symbols) and 1 h after sunset (19.00; filled symbols) at a mid lake station, May 30, 1992. Reproduced from De Meester L and Vyverman W (1997) Diurnal residence of the larger stages of the calanoid copepod *Acartia tonsa* in the anoxic monimolimnion of a tropical meromictic lake in New Guinea. *Journal of Plankton Research* 19: 425–434, with permission from Oxford University Press.

species, and populations: the depth at which the animals reside during the day, the depth at which they reside during the night, and the timing and speed of the migration from one depth to the other. This view has proved productive in structuring the observed variation in DVM patterns. For instance, it allows the viewing of the many observations of strong but constant depth preference during day and night as a DVM phenotype that may be selected for if the optimal depth is the same during the day and night, e.g., because it is only determined by the distribution of food. The ubiquity of standard migration would then reflect the fact that under a wide variety of circumstances it is adaptive to stay deeper in the water column during the day than during the night. It is revealing that DVM, viewed as a habitat selection behavior, is essentially very similar to twilight activity as reported for many small mammals, or the migration of aquatic macroinvertebrates in rivers to the underside of stones during the day. Viewed as a

habitat selection behavior, DVM can also be seen as an alternative to diel horizontal migration (DHM). DHM has been observed in many zooplankton species in shallow lakes with a well-developed littoral zone. In a typical DHM migration, the zooplankton resides between the macrophyte vegetation in the littoral zone during the day, and moves into the open water during the night. DHM shows many parallels to DVM: it shows similar diel dynamics, with the main movements being associated with dawn and dusk, and in both cases, the zooplankton moves to rather marginal habitats during the day.

Causes of DVM

What causes zooplankton to migrate vertically in a diel cycle? It is important to make a distinction between the proximate and ultimate factors leading to DVM in nature.

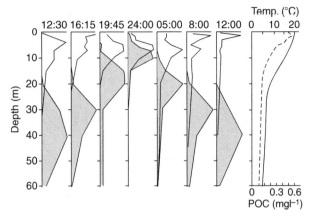

Figure 3 Vertical migration patterns of *Daphnia galeata* (plain) and *D. hyalina* (shaded) during summer in Lake Constance. The right panel shows depth profiles of temperature (solid line) and particulate carbon (broken line) (particles <35 μm, i.e., potential food for *Daphnia*). Reproduced from Lampert W and Sommer U (1997) *Limnoecology: The Ecology of Lakes and Streams.* New York, Oxford University Press; figure based on Stich HB and Lampert W (1981) Predator evasion as an explanation of diurnal vertical migration by zooplankton. *Nature* 293: 396–398, with permission from Nature Publishing.

Proximate Causes

Proximate causes are the stimuli from the immediate environment of the individual organism that trigger the animal to move downwards or upwards. These stimuli–response processes relate to the physiology of the organism, and translate into a specific day- or nighttime depth as well as into a particular timing and speed of migration. It has been convincingly shown that zooplankton DVM is a response to relative changes in light intensity. Elegant laboratory experiments have revealed that a decrease in light intensity that surpasses a specific threshold elicits an upward movement, whereas an increase in light intensity above a specific threshold elicits a downward movement. These responses have been called 'secondary phototaxis' – in contrast to 'primary phototaxis' that involves a response to a constant light intensity. Most detailed experiments have been carried out with the water-flea *Daphnia*, and have shown that changes in light intensity result in eye rotations, which then translate in a change in body orientation and an upward or downward swimming. These simple secondary phototaxis responses may allow predicting the day and night time depth as well as the timing and speed of migration. Moreover, several modifying factors have been identified. For instance, the presence of fish kairomones (which have not been chemically identified yet, but characterized as nonvolatile, low-molecular-weight, lipophylic compounds of medium polarity and high thermal and pH stability) has been

shown to increase the sensitivity to changes in light intensity, which translates into an increase in DVM amplitude in the presence of fish. A strong temperature gradient may reduce the responsiveness to relative changes in light intensity, and the same holds for hunger. The latter results in a less strong DVM behavior in the absence of food.

Ultimate Factors

To the question why the zooplankton engages in DVM, one can also answer with reference to the ultimate factors causing DVM, i.e., its adaptive significance. For a long time, the adaptive significance of DVM has remained enigmatic, as researchers were puzzled by the fact that the zooplankton remained in deep and cold water layers that are characterized by low food concentrations during a large part of the diel cycle. Initially, many authors were convinced that DVM was a side-product of the visual system. Others believed that there were metabolic advantages associated with residing part of the time at a lower temperature. These hypotheses were, however, at odds with model predictions and are unsatisfactory with respect to the synchronized timing of DVM. The currently most widely accepted explanation for DVM in zooplankton is that it acts as a predator-avoidance mechanism. The animals move into the deeper and darker water layers during the day to avoid predation by fish. There are many lines of evidence pointing to the importance of predator-avoidance in DVM, most of them being related to the observation that variation in DVM can often be explained by variation in predator risk:

- First, the hypothesis is logical: fish need light to detect their prey efficiently, and by hiding in the darker water layers, the zooplankton can reduce mortality by visually hunting fishes. During the night, the zooplankton moves to the upper water layers to feed on algae that remain in the epilimnion because they need sufficient light for photosynthesis. Thus, the hypothesis can explain both the daytime and nighttime depth distribution, as well as the timing of the upward and downward migration.
- Overall, there is a tendency for larger zooplankton species to migrate more or with more amplitude than smaller species. For instance, large *Daphnia* species migrate to deeper water layers than smaller crustacean zooplankton, whereas rotifers often do not migrate at all.
- Smaller life stages often migrate with less amplitude or do not migrate at all (e.g., **Figure 2**).
- Individuals that are more conspicuous, such as egg-bearing females, often migrate with more amplitude.

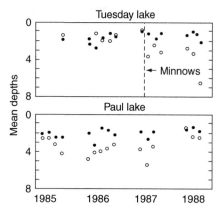

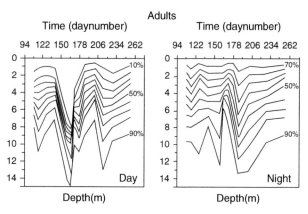

Adults

Figure 4 Impact of fish on DVM: field data from Tuesday and Paul Lake, two kettle lakes at the University of Notre Dame Environmental Research Center. Shown are night (solid symbols) and day (empty symbols) average depth of the *Daphnia* assemblage in the two lakes as determined from vertical profiles taken during several sampling campaigns across four summers. In 1985 and 1986, Tuesday lake was biomanipulated, resulting in a strong reduction in fish predation pressure; in 1987, fish were reintroduced to the lake, resulting in quite high minnow densities. DVM amplitude of the *Daphnia* assemblage is associated with the presence of planktivores. Paul Lake was not biomanipulated, and harbored planktivorous fish during the whole period. Reproduced from Dini ML and Carpenter SR (1991) The effect of whole-lake fish community manipulations on *Daphnia* migratory behavior. *Limnology & Oceanography* 36: 370–377, with permission from American Society of Limnology and Oceanography.

Figure 5 The changes in day (left) and night (right) depth of adult *Daphnia hyalina x galeata* hybrids in Lake Maarsseveen during the course of the growing season (April 18 – September 14 1989). The dip in the bundle of lines indicates the period of strong DVM. This period corresponds with the massive appearance of juvenile perch in the pelagial of the lake. Reproduced from Ringelberg J, Flik BJG, Lindenaar D, and Royackers K (1991) Diel vertical migration of *Daphnia hyalina* (sensu latiori) in Lake Maarsseveen: Part 1. Aspects of seasonal and daily timing. *Archiv für Hydrobiologie* 121: 129–145, with permission from E. Schweizerbart'sche Verlagsbuchhandlung.

- It has been reported repeatedly that also within a population there is a tendency for a correlation between body size and daytime residence depth within a given lake.
- Among-population variation in DVM amplitude can sometimes be related to differences in predation risk by fish (**Figure 4**). This is illustrated by the absence of migration in many fishless lakes.
- Many populations show seasonal variation in DVM amplitude that can be explained by seasonal variation in predation risk by fish (highest in summer when the young-of-the-year are roaming through the pelagial; e.g., **Figure 5**).
- Very direct evidence for the predator-avoidance hypothesis is provided by the many experimental studies that have shown that DVM behavior can be induced by the presence of fish-specific chemicals (kairomones; **Figure 6**).
- Finally, several mathematical models have shown that DVM behavior can indeed be adaptive as a predator-avoidance strategy, i.e., that the benefits outweigh the costs of staying at low food and temperature during part of the diel cycle.

The evidence for the predator-avoidance hypothesis is overwhelming. Yet, predator avoidance need not be the sole adaptive value of DVM. Indeed, some cases of day depth distribution or DVM cannot be satisfactorily explained by the predator avoidance hypothesis. For instance, zooplankton has been shown to migrate vertically in fishless alpine lakes, and in some studies the amplitude of the migration has been shown to be unrelated to fish predation risk. Recently, evidence has accumulated that an important adaptive value of DVM may lie in the avoidance of damage caused by UV-radiation, or more generally, by high light intensity. Especially in transparent high-altitude lakes, the presence of fish may not be the only or even the main reason for DVM behavior to develop. UV-avoidance can also explain the pattern and timing of typical normal DVM behavior. Moreover, there is an intrinsic interaction between UV- and predator-avoidance, as animals have two mechanisms to avoid photodamage: they may accumulate protective pigments or they may migrate vertically. In the absence of predators, protective pigments may be the more beneficial option, but in the presence of predators, pigmentation bears a high cost of increased risk of predation, and DVM may be the better alternative. It is often observed in alpine lakes that copepods are pigmented whereas cladoceran zooplankton are not and migrate vertically. What drives this difference is insufficiently understood, but it might be related to the fact that most copepods are relatively small and better swimmers than water fleas. Alternatively, it may be that copepods have better capacity to accumulate photoprotective pigments from their food.

Abundance (%)

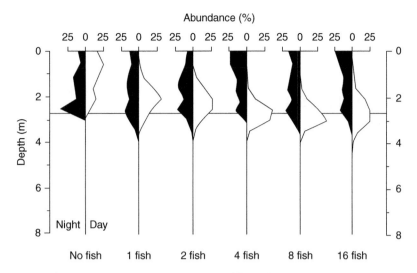

Figure 6   Induction of DVM by fish kairomone in a clonal population of *Daphnia galeata x hyalina* hybrids in an experiment using the Plankton Towers at the Max Planck Institut for Limnology, Plön. The experiment was started without fish. Then, the upper 3 m of the tank received water that was conditioned by one fish (*Leucaspius delineatus*, 5 cm body length). After four days of adaptation, the vertical profile was recorded again at night and during the day. Subsequently, the number of fish was doubled, and the same procedure was continued three more times (cf. up to 16 fish). Reproduced from Loose CJ (1993) *Daphnia* diel vertical migration behavior: Response to vertebrate predator abundance. *Archiv für Hydrobiologie Beihefte Ergebnisse der Limnologie* 39: 29–36, with permission from E. Schweizerbart'sche Verlagsbuchhandlung.

A third ultimate reason to exhibit a typical depth distribution or DVM may be avoidance of competition. This has been shown experimentally in rotifers, and is indeed likely in small zooplankton, which may avoid depths at which larger bodied zooplankton accumulate. It should be noted, however, that this mechanism can only operate in a situation where the competitively superior species is restricted to its depth distribution by additional factors. If competitively dominant large bodied water fleas are forced to hide in the deeper water layers by predation risk or UV radiation, then this may open a window for coexistence of smaller species that occupy the upper water layers during the day. During the night, these smaller species may avoid the upper water layers to reduce interference competition with the large bodied species. The result is a reverse migration. A reverse migration may also be associated with the avoidance of invertebrate predation. Indeed, small bodied species are more vulnerable to invertebrate predators, and as these may be forced to hide in deeper waters during the day because of predation risk by fish, the small bodied species may increase their fitness by residing in the upper water layers during the day and then spread over the water column during the night, when the invertebrate predators move up into the open water to search for prey.

Costs

Viewing DVM as a habitat selection behavior provides a flexible approach to the overwhelming variation in DVM patterns, as the observed day and night time depth distribution as well as timing and speed of migration can be considered to be selected because they provide a high benefit to cost ratio. The benefits of DVM are, as mentioned, often related to reduced mortality by predation, but also reduced damage from high light intensities, and in some cases, reduced damage by competition. The costs are largely cast in terms of reduced food intake, reduced food quality, increased competition (when several competing taxa are migrating to the same refuge), and metabolic costs associated with residing at a lower temperature. In addition, there may be costs associated with residing in truly marginal habitats, such as near-anoxic conditions or being buried in sediments. Residing at or near the sediments can also increase the risk of infection by increasing the likelihood of taking up infective parasite spores, as has been shown in the water flea *Daphnia*. The costs associated with the movement from one depth layer to the other has, however, been shown to be relatively low or even negligible. This is because the animals in general must actively swim to keep their position in the water column anyway, coupled with the observation that the descent is often passive. During the descent, the animals become less active, and they become more active during the ascending phase. In simple wording: if a water flea shows its typical hop–sink–hop–sink swimming behavior to keep its vertical position, descending is accomplished by hop–sink–sink–hop–sink–sink and ascending by hop–hop–hop–sink behavior. One can imagine that the energy lost during the ascending phase matches more or less the energy gained during the descending phase.

Linking Proximate and Ultimate Factors in DVM

The previous paragraph pictures a situation in which the most advantageous DVM behavior may strongly differ from lake to lake. The question then arises as to how this is matched with the response to proximate stimuli. How do the physiological responses of the animals lead to the right vertical distribution during the day and night and the right timing and speed of the ascent and descent? There are two possible mechanisms that are not mutually exclusive, and both are supported by experimental evidence. On the one hand, part of the variation in DVM behavior can be accomplished by phenotypic plasticity acting at the physiological responses. It has, for instance, been shown that predator kairomones and hunger influence the sensitivity of zooplankton individuals to relative changes in light intensity, resulting in a modification of DVM behavior. Populations may thus adjust their DVM behavior by showing the appropriate phenotypic plasticity in response to changes in environmental conditions. At the same time, it has been shown that there is ample genetic variation for DVM behavior, both through field studies as well as through laboratory quantitative genetic analyses of the variation in response to a light gradient (**Figure 7**) and, to a lesser extent, changes in light intensity. Moreover, it has been shown that there is also genetic variation in phenotypic plasticity for DVM, at least in the water flea *Daphnia*. This sketches a picture of very high flexibility: populations can adjust their DVM behavior by phenotypic plasticity of the individuals as well as by changes in genetic composition with respect to DVM. It has been shown that populations show local genetic adaptation for DVM in relation to predation risk. Moreover, reconstruction of microevolution in DVM behavior on laboratory populations hatched from dormant egg banks that were isolated from a dated sediment core has shown that local populations can genetically track changes in fish predation pressure. Finally, it has also been shown that animals isolated from different depths during the day are often genetically different and that seasonal changes in DVM behavior may also have a genetic component.

The picture that emerges is that DVM is an important component of an antipredator strategy in many populations of zooplankton. It should be emphasized that DVM should not be seen in isolation from other antipredator traits. It has, for instance, been shown

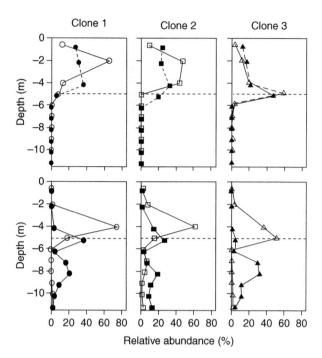

Figure 7 Vertical distribution of three *Daphnia hyalina* x *galeata* hybrid clones isolated from the same population (Schöhsee, Germany) in the Plankton Towers of the Max Planck Institute for Limnology (Plön), showing genotypic differences in DVM behavior. Top: average vertical distribution of adult females during the day (open symbols) and night (solid symbols) in the absence of fish chemicals or fish; bottom: daytime distribution in the presence of fish chemicals (open symbols) and fish (solid symbols). The dotted line indicates the thermocline. The experiment involved mixed populations of the three clones, and individuals were identified using allozyme markers. Clone 3 has a clearly different vertical distribution from clones 1 and 2. Reproduced from De Meester L, Weider LJ, and Tollrian R (1995) Alternative antipredator defences and genetic polymorphism in a pelagic predator–prey system. *Nature* 378: 483–485, with permission from Nature Publishing.

that there is often a relationship between size at maturity and day-depth, with larger animals residing in deeper water layers than smaller ones. Different genotypes and species may thus differ in the antipredator strategy they employ. This is nicely illustrated in the differences in migration pattern of *D. galeata* and *D. hyalina* in Lake Constance shown in **Figure 3**. The strategy of *D. galeata* involves a high predation risk associated with continuous residence at higher temperatures and under good food conditions. This results in high birth and death rates. The strategy of *D. hyalina* is to reduce mortality rates and bear the cost of having lower birth rates. The fact that both species can co-occur in the lake illustrates that both strategies may have similar fitness. Part of this relationship is also observed at the genotypic level, with larger bodied genotypes showing a stronger response to light gradients. Similarly, DVM behavior and pigmentation are both photoprotection strategies, of which the relative importance may be strongly influenced by predation risk. Both pigmentation and DVM have a cost, but the cost of pigmentation is strongly dependent on predator risk.

Consequences of DVM

DVM has important consequences, as it strongly influences the top–down impact in lake food webs. DVM reduces the direct impact of fish predation on zooplankton. A more subtle consequence of DVM is that populations of large bodied zooplankton may survive longer in the lake, which thus extends the time during which this food resource remains available for the fish to feed on. In this sense, DVM can be expected to strongly buffer predator–prey interactions between fish and zooplankton. With respect to the zooplankton–algae interaction, DVM has been shown to strongly affect the grazing impact of zooplankton on algae. Here again, however, this has two aspects to it. On the one hand, grazing impact would indeed be higher in the absence of fish-induced DVM, because DVM reduces the time during which zooplankton resides in the epilimnion to feed on algae. On the other hand, DVM may allow large bodied zooplankton to survive in lakes that harbor fish. Given that large bodied zooplankton are more efficient phytoplankton grazers than small bodied zooplankton, DVM may thus indirectly promote the grazing impact of zooplankton on algae, as it allows large-bodied zooplankton to graze on phytoplankton during at least part of the day. In lakes in which the zooplankton exhibits a strong DVM, the algae are indeed fed on by large and efficiently grazing zooplankton during the night. The impacts of DVM on population dynamics of zooplankton and algae and on predator–prey interactions are thus very pervasive. A nice illustration of the impact of DVM on population dynamics is given by the observation of a lunar cycle in zooplankton densities in Lake Kariba. The population densities of zooplankton species in this lake have been reported to show cyclic changes associated with increased predation success of fish on zooplankton during moonlit nights at full moon. In the tropics, sunset and moonrise occur quite fast, and at full moon, the zooplankton is trapped in the surficial water layers by the suddenly rising full moon, providing a feast for the fish.

An interesting avenue of thought in this context is the impact of predator-induced plasticity in DVM on predator–prey interactions and top–down control in lakes. Indeed, as DVM has strong impacts on top–down control, it should be recognized that part of the impact of predators is actually mediated by induced avoidance behavior rather than by direct predation itself.

Inverse Migrations

We focused on the DVM behavior of herbivorous zooplankton (mainly cladocerans, copepods, and rotifers). Predatory zooplankton may, however, also strongly engage in DVM. The best examples are the extensive migrations carried out by phantom midge larvae (*Chaoborus*) in fish-inhabited lakes. *Chaoborus* species that inhabit fishless habitats do not migrate vertically, but in fish lakes, the animals often migrate to the sediments during the day and appear in the water column at night. This is believed to be a strong structuring force on the zooplankton in lakes. During the day, large-bodied zooplankton, including *Chaoborus* larvae, hide in the deep water layers as a refuge from fish predation. By doing so, they provide enemy-free space to small-bodied zooplankton. The small-bodied zooplankton may, however, move out of the high food upper water layers during the night, as these layers are then invaded by efficiently grazing large-bodied zooplankton and gape-limited invertebrate predators that hunt for the small-bodied zooplankton. The resulting pattern is a strong standard migration for crustacean zooplankton and *Chaoborus* and a reverse migration for small zooplankton such as rotifers.

There is an interesting parallel in large, motile phytoplankton. Algae often perform a 'reverse' DVM, residing deeper in the water column during the night than during the day. There is much less literature on DVM in algae, but here too, it is a profitable approach to consider it a habitat selection behavior. There are two main reasons for DVM in algae. First, it may be a strategy to combine efficient photosynthesis during the day with a reduction of mortality by grazing of zooplankton during the night, by spreading out over the water column. Alternatively, DVM of algae may also be a strategy to increase nutrient uptake. By migrating to deeper and nutrient-rich layers during the night, the phytoplankton may increase nutrient availability for photosynthesis during the day, thus reducing the impact of nutrient depletion in the epilimnion.

DVM in Fish

Fish have also been reported to migrate vertically. They may do so for two reasons. First, they may follow their prey. Even though it is much less efficient to hunt in deeper water layers, it may still be more profitable to follow your food in suboptimal conditions than remaining in a habitat that is devoid of any food. It has been shown that fish may even venture for short dives into the near-anoxic conditions to hunt for zooplankton that use these inhospitable layers as a refuge. This may actually explain why zooplankton often migrate deeper than the border zone of the refuge. A second reason why fish migrate may be to avoid their own predators. By avoiding the surface waters during the day, they may reduce their mortality from fish-eating birds. Many fish populations have been reported to either change their vertical or horizontal distribution. In the latter case, they hide in the littoral zone during the day and invade the pelagial during the night. Given that studies often show that both fish and zooplankton avoid the surficial layers of the pelagial zone during the day, one may wonder why the zooplankton does not simply reinvade the food-rich upper water layers. There are several explanations for this. First, it should be acknowledged that in zooplankton populations that show a very strong DVM behavior, intensive sampling of the superficial water layers often reveals some individuals residing there at very low densities. Secondly, it is easy to imagine that, if densities in the superficial layers would become higher, the fish would also rapidly start exploring this food source. Given that hunting at high light intensities is so efficient, even short excursions of the fish to these water layers would decimate the zooplankton. So one expects densities of larger bodied zooplankton to be very low in the epilimnion in such lakes, which is the pattern that is observed.

See also: Cladocera; Competition and Predation; Copepoda; Phytoplankton Population Dynamics: Concepts and Performance Measurement; Role of Zooplankton in Aquatic Ecosystems; Rotifera.

Further Reading

Bayly IAE (1986) Aspects of diel vertical migration in zooplankton, and its enigma variations. In: De Deckker P and Willams WD (eds.) *Limnology in Australia*. Dordrecht: Junk.

Cousyn C, De Meester L, Colbourne JK, Brendonck L, Verschuren D, and Volckaert F (2001) Rapid local adaptation of zooplankton behavior to changes in predation pressure in absence of neutral genetic changes. *Proceedings of the National Academy of Sciences USA* 98: 6256–6260.

De Meester L, Weider LJ, and Tollrian R (1995) Alternative antipredator defences and genetic polymorphism in a pelagic predator–prey system. *Nature* 378: 483–485.

Gliwicz ZM (1986) A lunar cycle in zooplankton. *Ecology* 67: 882–897.

Haney JF (1988) Diel patterns of zooplankton behavior. *Bulletin of Marine Sciences* 43: 583–603.

Lampert W and Sommer U (2007) *Limnoecology: The Ecology of Lakes and Streams*. 2nd edn. Oxford: Oxford University Press.

Ohman MD (1990) The demographic benefits of diel vertical migration by zooplankton. *Ecological Monographs* 60: 257–281.

Ringelberg J (ed.) (1993) *Diel Vertical Migration of Zooplankton, Archiv für Hydrobiologie – Advances in Limnology*, vol. 39. Stuttgart: E. Schweizerbart'sche Verlagsbuchhandlung.

Ringelberg J (1999) The photobehaviour of *Daphnia* spp. As a model to explain diel vertical migration in zooplankton. *Biological Reviews* 74: 397–423.

Stich HB and Lampert W (1981) Predator evasion as an explanation of diurnal vertical migration by zooplankton. *Nature* 293: 396–398.

Tollrian R and Harvell CD (eds.) (1999) *The Ecology and Evolution of Inducible Defenses*. Princeton: Princeton University Press.

Egg Banks

N G Hairston Jr., Cornell University, Ithaca, NY, USA
J A Fox, Drew University, Madison, NJ, USA

Introduction

'Egg bank' is both a specific term describing the accumulation of diapausing eggs of zooplankton within the sediments at the bottom of a lake or pond, and a more general term describing the sediment accumulation of a diversity of dormant life-history stages made by a variety of aquatic organisms. These stages, known collectively as 'dormant propagules,' include embryos of invertebrates (**Figure 1**), encysted gametes of protists and algae, and spores of bacteria and cyanobacteria. Many of these stages have the ability to remain dormant for long periods of years, decades, and in some cases more than a century while retaining the capacity to emerge from dormancy to become metabolically and ecologically active in lake ecosystems. This 'prolonged dormancy' (i.e., dormancy that lasts longer than from one growing season to the next) leads to the formation of an egg bank as each year more dormant propagules are deposited on the lake bottom than the number that emerge.

Dormant propagules typically possess traits that permit them to survive harsh conditions that would be lethal to individuals in the active stage; these harsh conditions include, but are not limited to, temperatures that are too warm or too cold, low oxygen, desiccation, freezing, low food availability, and high predator densities. Prolonged dormancy permits otherwise short-lived organisms to persist in environments that vary in how hospitable they are from one growing season to the next (i.e., occasional occurrence of a harsh season). In a temporally unreliable environment, a parent organism that produces multiple dormant propagules, some of which emerge the following growing season and some of which remain in dormancy for longer periods, increases the chance that its descendants will persist in the long term. A single season that is so harsh that individuals not in dormancy fail either to reproduce or to survive, would quickly cause population extinction if it were not for the fact that some individuals survive through the harsh season as dormant propagules. A life history in which some of the offspring of an individual emerge to try out one growing season while others wait to try out other growing seasons is called a 'bet-hedging strategy.'

Because dormant propagules are resistant to harsh conditions, they can also provide a means for individuals to disperse from one lake to another. Dormant eggs, cysts, or spores may become attached to the fur, feathers or integument of vertebrates or insects that visit first one lake and then another. Propagules disperse spatially if they attach to the visitors in one lake and then fall off in another. Some dormant propagules, consumed by a predator, can survive gut passage and can disperse between lakes if they are ingested in one lake and defecated in another. Still other dormant propagules may be dispersed by wind. The dispersal of dormant propagules to numerous lakes, each with a different environmental condition can result in spatial bet-hedging, which is similar conceptually to the temporal bet-hedging already described.

Egg banks are important in lake ecosystems in three ways.

1. They are a means by which populations survive through periods of stressful or uninhabitable conditions in a lake, whether those harsh conditions are natural in origin or are caused by human activity. Species or genotypes produced in the past, but absent from the water column in the present, can emerge from dormancy and reinvade rapidly in response to a change in the environment. Reinvasion by hatching from an egg bank provides a mechanism for the long-term maintenance of biological diversity in a lake.
2. Species or genotypes that establish large water-column populations after a period of absence by reinvading from an egg bank can affect how the lake ecosystem functions and how lakes respond to environmental change.
3. The sediments of lakes typically accumulate year by year, with the most recent sediments at the surface and older sediments occurring progressively at greater depths. Buried eggs, cysts, and spores provide a historical record of the species, genotypes, and phenotypes that existed in the past and how they responded qualitatively and quantitatively to changing environments.

Prolonged Dormancy as an Adaptive Life History Strategy

Dormancy and Diapause Defined

Dormancy is any state of reduced metabolic activity; either a temporary loss of the ability of cells to divide (in single-celled organisms such as bacteria, algae,

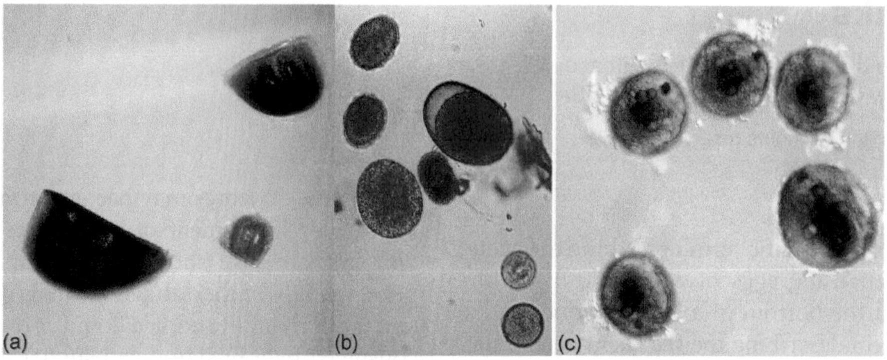

Figure 1 Diapausing eggs of freshwater planktonic crustaceans: (a) ephippial egg cases of three cladoceran species of the genus *Daphnia*, each case contains two eggs; (b) resting eggs of seven rotifer species; (c) the copepod genus *Onychodiaptomus*, one at the gastrula stage, the rest at the eyed-embryo stage. Photos by JA Fox (a) and CM Kearns (b and c).

and protists), or the suppression of growth, development, and reproduction (in complex organisms such as rotifers and crustaceans). Many dormant life-history stages also possess a resistant covering that aids survival in harsh conditions. Nondiapause dormancy is seasonal quiescence – a reversible state of suppressed metabolism directly imposed by exposure to harsh conditions (e.g., low temperature, low dissolved oxygen). Diapause-dormancy is induced by environmental factors that do not directly reduce metabolic rate but which act as a signal to the organism that conditions will soon deteriorate (e.g., day-length, temperature, food concentration, and population density). Life history stages in 'prolonged dormancy' do not all emerge at the first return of favorable conditions in the lake; it is these long-lived dormant stages that accumulate in sediments to produce an egg bank.

Egg and Cyst Viability

Dormant propagules can remain viable for amazingly long time periods. Eggs of cladocerans have been found to remain viable in lake sediments for as long as a century and the eggs of a population of copepods have been hatched after more than 300 years (**Figure 2**). Egg and cyst viability of several decades is common. At least some reduce their metabolic rate in dormancy to undetectable levels. The physiological mechanisms maintaining prolonged embryonic diapause and hence egg banks include accumulation of sugars, carotenoid pigments, and heat-shock proteins that inhibit physical–chemical damage during dormancy.

Very short time scales (2–3 years) seem to have little effect on the hatching of diapausing eggs. Over longer time periods, from a few years to several decades, viability generally decreases with age. Reduced viability is evidenced by decreased hatching rates, and by

both the increased time to hatch of older eggs and the decreased health of the individuals that do hatch (**Figure 3**). There is some evidence that the longevity of an egg or the dormant stage may represent adaptation to local conditions. For example, the eggs of some copepod species remain viable in the dry sediments of temporary ponds but do not show prolonged dormancy when found in large, permanent lakes.

Temporal Migration, Temporal Dispersal, and Temporal Bet-Hedging

Many environments vary seasonally. In the temperate zone and higher latitudes, winter conditions are harsh for organisms. In tropical regions and elsewhere, the dry season can be stressful. Mobile organisms avoid these physiologically challenging periods by migrating to a region where conditions are more benign. Organisms of limited mobility, including the zooplankton and phytoplankton of lakes, lack the ability for long-distance directional movement from one favorable habitat to another. For them dormancy and diapause provide a capacity for 'temporal migration' when spatial migration is not possible. For example, cladoceran crustaceans in a pond that dries every summer can produce diapausing eggs that are able to survive in the pond sediments in the absence of standing water. When the pond refills the following spring the eggs hatch, moving the organisms from one favorable period to another.

One way for organisms to survive in an unpredictable and spatially patchy environment is to disperse their eggs (or their offspring) randomly (similar to spatial seed dispersal in plants). Some individuals will land on unfavorable habitat patches but others will land on good sites and can grow and reproduce to disperse again. The dormant propagules of lake organisms may be dispersed to other lakes by more

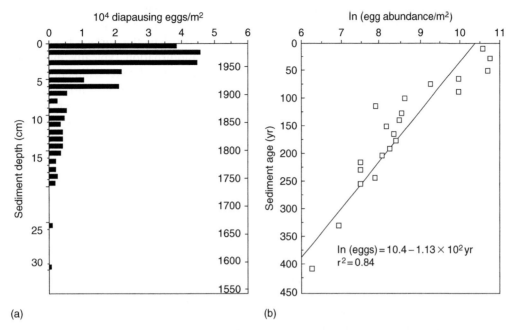

(a) (b)

Figure 2 (a) The density of diapausing eggs of the copepod *Onychodiaptomus sanguineus* in the sediments of Bullhead Pond, RI, as a function of sediment depth and sediment age; (b) The slope of the relationship between the natural logarithm of viable-egg density and sediment age is the mortality rate of the eggs, which in this case is 1.13% per year. From Hairston *et al.* (1995), see references in **Table 1**.

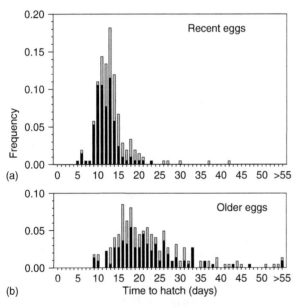

Figure 3 The time to hatch and incidence of inviable hatchlings (grey portion of bars) for diapausing eggs of *D. mendotae* taken from sediments less than 5 years old (a) and 15 to 20 years old (b) in Onondaga Lake, NY. Hatchlings were considered inviable if they had obvious physical deformities or survived less than 7 days in culture. Adapted from Fox JA (2007) Hatching timing of *Daphnia mendotae* diapausing eggs of different ages. *Fundamental and Applied Limnology* 168: 19–26.

mobile organisms, as described earlier. Long-lived dormant propagules, however, also possess the capacity for 'temporal dispersal' in which the propagules stay in one place, but emerge from dormancy in different years. For example, a copepod crustacean may produce a clutch of 20 diapausing eggs in a temporary pond. In some years when the pond refills, predator densities may be so great that any individual hatching has little or no chance of survival. In other years, however, predator density may be low and individuals hatching can mature and produce diapausing eggs of their own. Temporal dispersal occurs when the eggs from a single clutch of diapausing eggs hatch over a range of years, some of which are favorable and some of which are not.

In a temporally varying environment, years differ in their suitability for the survival and reproduction of an organism. The fraction of propagules that should theoretically emerge from the egg bank in any given year depends upon the likelihood of a year being either good or bad. The optimal emergence (or hatching) fraction can be calculated and depends upon the probability of a good year occurring. The greater the probability of any given year being good, the greater the optimal hatching fraction. This optimal hatching fraction is called a 'bet-hedging strategy.' It is probable, however, that year suitability will not simply be either good or bad; mediocre years can also occur in which survival and reproduction is not zero but is less than in good years. Again, an optimal hatching fraction can be calculated depending upon the probability of the occurrence of particular year types. Finally, environments may or may not become uninhabitable (e.g., a pond may or may not dry depending upon the amount of rainfall in a given year). Organisms may make some eggs that hatch immediately and some that enter diapause, and again, an optimal allocation of reproduction to diapausing and nondiapausing eggs can be calculated as a bet-hedging strategy.

Structure of Egg Banks
Egg Bank Formation and Dynamics

An egg bank represents the summation of the allocation to prolonged dormancy by the organisms living in a particular lake or pond. Dormant propagules accumulate in lake sediments both because long-lived propagules settle to the lake bottom when they are released, and because sediment processes cause them to become buried (**Figure 4**). Only those propagules within the top few millimeters of the sediment surface can successfully emerge. Propagules that become buried below this depth may remain viable for decades or more, but they cannot emerge unless brought to the surface by a dramatic event such as burrowing by a vertebrate or sediment disturbance by a boat anchor.

Sediment deposition buries propagules at rates varying between less than a millimeter to greater than a centimeter per year, depending on lake and watershed conditions. Sedimentation rates also vary spatially within a lake. Sediment focusing – the tendency for sediments to accumulate in the deepest portion of a lake basin – makes it likely that eggs deposited in the deepest portion will be buried more quickly, and so will be less likely to hatch, than those near shore. Sediment mixing by insects and worms living in the mud at the bottom of lakes (called 'bioturbation'),

transports propagules near the sediment surface deeper into the mud, while propagules at depth may be transported to the surface from which they can emerge. Bioturbation typically mixes only the top few centimeters of lake sediments. Physical disturbance (e.g., wave action, currents, fluctuations in water depth) is typically greater for shallow near-shore sediments. Macrophytes also influence egg movement by stabilizing sediments and channeling movement of benthic invertebrates.

Lake size influences the relative importance of biological and physical disturbance of sediments. In larger lakes physical mixing, particularly in shallow, near-shore areas where wind action is greatest and near inlets where currents are strongest, may play an important role in mixing sediments. Smaller bodies of water are often more protected from wind action and have a greater proportion of shallow, near-shore regions where bioturbation may be more important.

Egg density is typically greatest in deep-water sediments and lowest in near-shore areas (**Table 1**); however, the probability of receiving a hatching cue is greatest in shallow-water sediments. Thus, different regions of the same lake play different roles in the egg bank dynamics. Hatching from near-shore sediments may act as an important contribution to ongoing population processes while long-term storage in deeper sediments serves as an archive of population history in a lake.

Number Versus Depth

The density of dormant propagules or eggs in sediments varies widely by lake and species, with densities of 10^3 to 10^6 per square meter typical for algae,

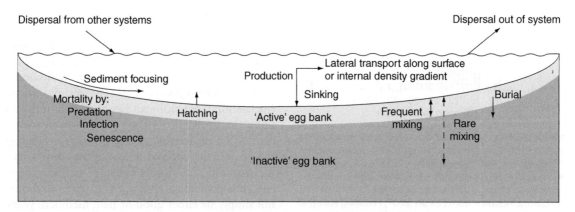

Figure 4 Side-view schematic diagram of a lake and its sediments depicting the distribution of dormant propagules within the egg bank. Because deeply buried propagules are viable but unlikely to hatch, the egg bank consists of 'active' and 'inactive' compartments. The active fraction receives the new propagules from the water column and is defined as the depth to which propagules can receive the cue to exit from dormancy and emerge successfully, most likely less than one centimeter into the sediment. Although propagules in the inactive compartment may remain viable for many years, they cannot emerge from dormancy unless some localized disturbance returns them to the active compartment. Adapted from Cáceres CE and Hairston NG Jr. (1998) Benthic-pelagic coupling in planktonic crustaceans: the role of the benthos. *Archiv für Hydrobiologie Special Issues Advances in Limnology* 52: 163–174.

Table 1 The densities of zooplankton diapausing eggs in central deep sediments compared with shallow near-shore sediments in lakes

Taxon	Lake	Egg densities (no. m^{-2}) and depth (m)			Source
		Deep	Shallow	D/S	
Cladocera					
Holopedium[a]	Windgfällweiher	2.5×10^6 (4.7)	$<1.0 \times 10^5$ (0–2)	25	1
Ceriodaphnia[a]	Piburger See	$>3.0 \times 10^5$ (>20)	$<1.0 \times 10^5$ (<5)	3	2
Bosmina[a]	Piburger See	$1–6 \times 10^5$ (>5)	$<5.0 \times 10^4$ (<5)	2	2
Daphnia[b]	Schöhsee	7.2×10^4 (21–25)	1.1×10^4 (5–10)	6.5	3
Daphnia[b]	Kellersee	1.2×10^5 (21–25)	3.8×10^4 (5–15)	3.2	3
Daphnia[a]	Oneida Lake	$>1.0 \times 10^5$ (12)	$<3.5 \times 10^3$ (3)	28	4
Copepoda					
Diaptomidae[a]	Oneida Lake	$>1.0 \times 10^6$ (10)	$<6.0 \times 10^5$ (3)	1.7	5
Onychodiaptomus[a]	Bullhead Pond	$>2.0 \times 10^5$ (4)	$<2.0 \times 10^4$ (1)	10	6, 7

[a]Total density of diapausing eggs.
[b]Total diapausing egg-case densities.

Sources

1. Lampert W and Krause I (1976) Zur Biologie der Cladocere *Holopedium gibberum* Zaddach im Windgfällweiher (Schwarzwald). *Archiv für Hydrobiologie Supplement* 48: 262–286.
2. Moritz C (1988) Die Verteilung der Ephippien von *Bosmina longirostris* und *Ceriodaphina pulchella* im Sediment des Piburger Sees (Ötztal, Tirol) (Cladocera, Crustacea). *Ber. Nat.-Med. Ver. Innsb.* 75: 91–107.
3. Carvalho GR and Wolf HG (1989). Resting eggs of lake-*Daphnia*. I. Distribution, abundance and hatching of eggs collected from various depths in lake sediments. *Freshwater Biology* 22: 459–470.
4. Cáceres CE (1998) Interspecific variation in the abundance, production and emergence of *Daphnia* diapausing eggs. *Ecology* 79: 1699–1710.
5. Hairston NG Jr. and Van Brundt RA (1994) Diapause dynamics of two diaptomid copepod species in a large lake. *Hydrobiologia* 292/293: 209–218.
6. De Stasio BT Jr. (1989) The seed bank of a freshwater crustacean: copepodology for the plant ecologist. *Ecology* 70: 1377–1389.
7. Hairston NG Jr., Van Brundt RA, Kearns CM, and Engstrom DR (1995) Age and survivorship of diapausing eggs in a sediment egg bank. *Ecology* 76: 1706–1711.
8. Hairston NG Jr. and Kearns CM (2002) Temporal dispersal: Ecological and evolutionary aspects of zooplankton egg banks and the role of sediment mixing. *Integrative and Comparative Biology* 42: 481–491.

rotifers, copepods, and cladocerans (**Table 1**). The vertical distribution of eggs within the sediments is not uniform. The abundance of eggs at a given depth below the sediment surface is a function of the history of the rates of production, mortality, and emergence, as well as sedimentation rate and disturbance. The number of eggs typically decreases with depth in the sediments, often with a peak in abundance just below the actively mixed sediment layer near the sediment–water interface (**Figure 2**). The distribution of eggs in the sediments may show gaps or breaks, corresponding to periods of time when the species was absent from the water column or present in smaller numbers (**Figure 5**).

The vertical profile of eggs in the sediments also varies spatially. Egg density tends to be highest in the central, deep portion of a lake, as a result of greater hatching near-shore combined with sediment focusing. Although egg densities may be lower in shallow, near-shore sediments, these areas have higher hatching rates because greater physical mixing and bioturbation expose eggs to hatching cues more frequently. Thus, eggs hatching from shallow sediments may make a disproportionate contribution to the active population, despite representing a smaller proportion of the total egg bank.

Evidence of Lake Recolonization from the Egg Bank

Resistant diapausing eggs play an important role in facilitating colonization of new environments and recolonization of temporarily unsuitable water bodies. An egg bank allows species to persist years after no adults survived to reproduce. Several experiments monitoring emergence of hatchlings from sediments have shown that copepods and cladocerans hatched from the egg bank years after no adults were present in the water column. A comparative study of two lakes, in which newly stocked fish eliminated water-column populations of copepods, showed that after the fish were removed, copepods returned to the water column of the lake that had a large sediment egg bank but did not return in the lake that had no egg bank (because the eggs had been consumed by a large population of predatory benthic invertebrates).

Studies of the genetic structure of populations provide indirect evidence of recolonization from the resting egg bank. A bryozoan population showed the same genetic composition before and after a significant demographic bottleneck, suggesting recolonization from the dormant propagule bank rather than from neighboring populations that are genetically

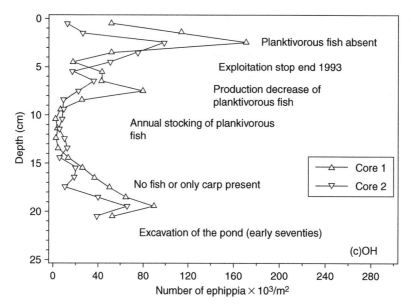

Figure 5 The depth distribution of dormant egg cases (ephippia) of *Daphnia magna* in two sediment cores taken from an artificial pond created in Belgium in the 1970s. Noted on the figure are periods of high and low abundance of fish that consume *Daphnia*. Adapted from Cousyn C and De Meester L (1998) The vertical profile of resting egg banks in natural populations of the pond-dwelling cladoceran *Daphnia magna* Straus. *Archiv für Hydrobiologie Special Issues Advances in Limnology* 52: 127–139.

distinct. Genetic diversity and structure suggest that the recovery of *Daphnia* populations following the neutralization of acidified Canadian lakes may have been due to hatching from the local egg bank.

In most species examined, hatchlings from the egg bank exhibit tremendous genetic diversity, which is often greater than the diversity of the planktonic population at the end of the previous season. However, genetic analyses have also shown that not all individuals in the egg bank are equal. In rotifers, for example, genetic differences between the egg bank and water-column populations suggest that some genotypes are more likely than others to hatch from the egg bank.

The Role of Egg Banks and the Storage Effect in the Maintenance of Lake Biodiversity

Temporal bet-hedging by long-lived dormant propagules present in egg banks can play an important role in maintaining species diversity in lakes. This occurs when the environment in a lake varies from year to year and species differ in the environmental conditions under which they do best. Without an egg bank, two species with similar ecological requirements may not be able to coexist due to competition. Even when the environment varies and species differ in the environmental condition that is favorable to them, on average one species will be better at using

the available resources and over time will eliminate the other. If, however, individuals emerge each year from an egg bank into a varying environment, newly emerging individuals will have high reproductive success in years that are favorable and low success in unfavorable years. In favorable years, the individuals can build up a large population, produce a large number of new dormant propagules, and restock the egg bank before the end of the season. If each species is successful in survival, reproduction, and ultimately restocking its egg bank under environmental conditions when the other does poorly, then conditions exist for the two species to coexist over the long term. This mechanism, by which biological diversity is maintained, is called the 'storage effect' and makes it possible for many species to coexist. In one study, *Daphnia pulicaria* and *Daphnia mendotae* were shown to coexist for several decades even though the former species was the superior competitor in most years. The latter had a more substantial egg bank that it was able to restock in years when high fish predation gave it a fitness advantage.

The egg banks and storage effect also serve to maintain genetic diversity within a species if genotypes differ in the environmental conditions under which they do best. In a population of the copepod *Onychodiaptomus sanguineus*, genetic variation for a trait that influences their ability to survive predation by fish is apparently maintained via the storage effect in the presence of year-to-year variation in predation intensity.

Egg Banks as Historical Archives

Dormant propagules buried in lake sediments provide a paleolimnological record of the changes in a population over years, decades, or centuries. Several different approaches are used, often together, to reconstruct the history of populations using egg banks: (1) determining the distribution and relative abundance of species through time, (2) hatching individuals from eggs of different ages to compare phenotypes, and (3) performing genetic analyses on the dormant propagules. These methods provide insights into the evolutionary history of the species being examined as well as community and ecosystem processes.

The abundance of dormant eggs in the sediments is correlated with population size at the end of the season in which the eggs were produced. Therefore in the absence of significant disturbance of the sediments, changes in egg density throughout the sediments can be used to estimate the size of past populations. Egg banks have been used extensively to document the impact of human-induced changes on zooplankton community structure. Changes in cladoceran community structure over two centuries in Lake Naivasha, Kenya, showed that the introduction of exotic fish played a key role in determining which species were most abundant, with large-bodied species being absent during periods of intense fish predation. In addition, increased erosion from agricultural lands caused increased turbidity and decreased the abundance of submerged aquatic plants, which in turn led to a decline of smaller species that are found in macrophyte beds. Conversely, changes in the abundance of macrophyte-associated species in an egg bank may indicate past vegetation or other ecosystem changes.

The effects of heavy metal pollution on zooplankton populations in lakes in Europe and North America have been recorded in the egg banks of these lakes. For example, analysis of the *Daphnia* egg bank in Canadian lakes recovering from acidification reveals a species shift associated with changes in the lakes' acidity levels and the *Daphnia* egg bank and *Bosmina* microfossils show declines in abundance during periods of peak industrial activity in a small lake near Lake Superior.

Egg banks also store information about phenotypic changes in populations, thereby documenting the evolution of populations. The history of size-selective predation by fish is recorded by changes in the size or density of the cladoceran ephippia in the sediments. Studies of individuals hatched from different time periods demonstrated that traits such as predator-avoidance behavior can evolve rapidly in *Daphnia* populations as a result of changes in fish predation pressures (**Figure 6**).

The rapid evolution of populations in response to environmental change is recorded in the egg bank. Eutrophication of Lake Constance caused an increase

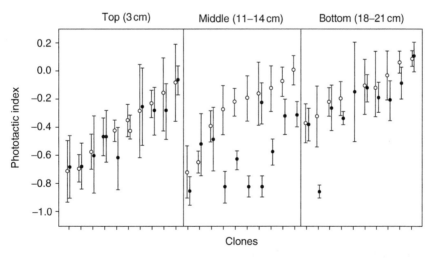

Figure 6 The predator-avoidance behavior of *D. magna* hatched from diapausing eggs collected from three different sediment depths of the egg bank of a Belgian pond. The top and bottom sediments are from time periods when *Daphnia*-eating fish were either rare or absent; the middle sediments are from a time when these fish were abundant (see **Figure 5**). For each time period, the vertical distribution in an illuminated cylinder of water ('phototactic index') of 10 different *Daphnia* clones (each clone descended from a single diapausing egg) in the presence (closed circles) or absence (open circles) of fish. Clones have been sorted according to mean depth without fish. The data show much greater migration away from illuminated surface waters – where vulnerability to fish predation is greatest – by *Daphnia* from the middle time period when fish were abundant. Because the *Daphnia* from different time periods are genetically distinct, the results indicate that the *Daphnia* population evolved rapidly to changes in fish density. Adapted from Cousyn CL, De Meester L, Colbourne JK, *et al.* (2001) Rapid, local adaptation of zooplankton behavior to changes in predation pressure in the absence of neutral genetic changes. *Proceedings of the National Academy of Sciences USA* 98: 5256–6260.

of cyanobacteria in the water column that can be toxic. Experiments on *Daphnia* hatched from sediments corresponding to times after a decade of eutrophication show that they are more resistant to toxic cyanobacteria than are *Daphnia* hatched from sediments deposited before eutrophication, when cyanobacteria levels were much lower. This rapid evolution of phenotypes is associated with a change in molecular genetic frequencies over the same time period.

PCR-based genetic techniques enable genetic analyses to be done directly on dormant propagules, without introducing the bias of examining only those individuals capable of hatching. Although DNA degradation increases with increasing egg age, DNA has been successfully extracted and amplified from *Daphnia* eggs as old as 200 years. As molecular techniques for working with ancient DNA improve, studies of DNA stored in egg banks will become more informative.

Studies of egg banks have led to greater understanding of the processes of colonization and the relative importance of population re-establishment from the egg bank and from external sources. For example, the historical species record in the egg bank revealed that the establishment of a particular *Daphnia* species in lakes recovering from acidification must be from external sources, despite the existence of a large bank of viable eggs of other *Daphnia* species.

Egg banks also register unusual past events, such as the invasion of exotic species. In Onondaga Lake, NY, the presence of two salt-tolerant exotic *Daphnia* species during periods of high salt-loading into the lake were documented by the presence of large numbers of ephippia in the sediments. These species had been undetected or misidentified in samples taken during those years, and were later identified through a combination of morphological, hatching, and genetic analyses of ephippia and eggs stored in the sediments. Genetic analyses revealed that the population of one of these species was established by the introduction of one or two individuals into the lake, a reminder of the tremendous dispersal and establishment capabilities of these dormant stages.

See also: Bacteria, Distribution and Community Structure; Cladocera; Competition and Predation; Copepoda; Other Zooplankton; Phytoplankton Population Dynamics: Concepts and Performance Measurement; Phytoplankton Population Dynamics in Natural Environments; Role of Zooplankton in Aquatic Ecosystems; Rotifera.

Further Reading

Alekseev VR, De Stasio B and Gilbert JJ (eds.) (2007) *Diapause in Aquatic Invertebrates: Theory and Human Use*, pp. 1–257. Dordrecht: Springer.

Brendonck L, DeMeester L, and Hairston NG Jr. (eds.) (1998) Evolutionary and ecological aspects of crustacean diapause. *Archiv für Hydrobiologie Special Issues Advances in Limnology* 52: 1–561.

Cáceres CE (1997) Temporal variation, dormancy, and coexistence: A field test of the storage effect. *Proceedings of the National Academy of Sciences USA* 94: 9171–9175.

Cáceres CE (1997) Dormancy in invertebrates. *Invertebrate Biology* 116: 371–383.

Chesson PL (1994) Multispecies competition in variable environments. *Theoretical Population Biology* 45: 227–276.

Ellner SP and Hairston NG Jr. (1994) Role of overlapping generations in maintaining genetic variation in a fluctuating environment. *American Naturalist* 142: 403–414.

Fryer G and Alekseev VR (eds.) (1996) *Developments in Hydrobiology* 114, 241 pp.

Hairston NG Jr. (1996) Zooplankton egg banks as biotic reservoirs in changing environments. *Limnology and Oceanography* 41: 1087–1092.

Hairston NG Jr. and Cáceres CE (1996) Distribution of crustacean diapause: Micro- and macroevolutionary pattern and process. In: Fryer G and Alekseev V (eds.) *1st International Symposium on Crustacean Diapause. Hydrobiologia* 320: 27–44.

Hairston NG Jr. and DeStasio BT Jr. (1988) Rate of evolution slowed by a dormant propagule pool. *Nature* 336: 239–242.

Hairston NG Jr. and Kearns CM (2002) Temporal dispersal: Ecological and evolutionary aspects of zooplankton egg banks and the role of sediment mixing. *Integrative and Comparative Biology* 42: 481–491.

Hairston NG Jr., Ellner S, and Kearns CM (1996) Overlapping generations: The storage effect and the maintenance of biotic diversity. In: Rhodes OE Jr., Chesser RK, and Smith MH (eds.) *Population dynamics in ecological space and time*, pp. 109–145. Chicago: University of Chicago Press.

Okamura B and Freeland JR (2002) Gene flow and the evolutionary ecology of passively dispersing aquatic invertebrates. In: Bullock JM, Kenward RE, and Hails RS (eds.) *Dispersal Ecology*, pp. 194–216. Oxford: Blackwell Publishing.

Seger J and Brockmann HJ (1987) What is bet-hedging? *Oxford Surveys in Evolutionary Biology* 4: 182–211.

Competition and Predation

Z M Gliwicz, Warsaw University, Warsaw, Poland

Introduction

There is fierce competition between planktonic her-bivores because resources are not as diverse offshore as they are inshore and onshore. Filter feeding on suspensions of small particles does not allow for much partitioning of food resources by selection other than that based on particle size. On the other hand, zooplankton are intensively exploited by planktivorous fish and invertebrate predators, and the impacts of predation are equally severe offshore and inshore in spite of different relative importance of antipredation refuges (deep water and the structural diversity of the habitat).

It has never been resolved whether planktonic her-bivores coexist peacefully or cooccur neutrally. Also, it has not been determined whether a high diversity of zooplankton and the cooccurrence of a multitude of ecologically similar species are possible by avoiding competition through resource partitioning, niche diversification, and character displacement, or rather by low population densities because of predation-induced mortality.

Competition and predation used to be looked upon as mutually exclusive, so the question often asked was: is the factor responsible for zooplankton abun-dance and composition, for the limitation of growth and reproduction caused by a shortage in food supply (bottom-up regulation from the base of the food web), or for the enhancement of mortality by predation (top-down control)? These two divergent approaches to population and community ecology of zooplankton are now replaced by the notion that the forces of competition and predation work side by side, and that an animal's life history and behavior, its popula-tion density, and the structure of the community it belongs to depend on both bottom-up and top-down impacts of competition and predation (**Figure 1**). Both are strongly selective with regard to the animal's species-specific and age-specific body size, with greater size producing superiority in food acquisition ('size–efficiency hypothesis,' **Figure 2**), and smaller size reducing susceptibility to predation by visually oriented fish (reduced 'reaction distance' at which foraging fish sees its prey, **Figure 3**).

Competition by Superiority in Resource Acquisition and by Interference

Animals, plants, and bacteria compete to gain essen-tial resources such as energy, nutrients, water, and territory. Water may be a limiting resource in terres-trial systems, territory may be crucial to benthic algae and invertebrates that need solid substrates to attach to, and light may be essential to planktonic algae and cyanobacteria. In contrast to other organisms, which have to seek and sequester various resources in differ-ent ways and at different locations (consider an alga in the water column with light available close to the surface and nutrients accessible in the dark hypolim-nion), planktonic animals have all the necessary resources (energy and nutrients) combined into parti-cles suspended in the water medium. The 'package character' of the essential resources for zooplankton allows competition for these resources to be quanti-fied by a single niche dimension: food particle size scaled in micrometer. The mean particle diameter is a reasonable approximation, since most food particles (algal and bacterial cells/colonies, organic debris, par-ticles of silt or clay with adsorbed organic com-pounds) are more or less spherical. This makes niche dimensions easier to identify and experimental stud-ies on competition more feasible in zooplankton com-pared with that in other organisms (**Figure 4**).

The abundance of resources shrinks when one or more coexisting species increase in number. When resources are depleted they become harder to seques-ter, and feeding effort may not be compensated by assimilation rate, causing individual growth to be halted. At population densities approaching the 'car-rying capacity of the habitat' (an equilibrium at which mortality is compensated by recruitment, and reduc-tion in food abundance is compensated by food pro-duction), severe food limitation may lead to a halt in reproduction, particularly in species that are infer-ior in the competition for resources. This may result in competitive exclusion of inferior species with a sin-gle competitively superior species monopolizing the resources. Such exclusions are more likely (1) in homo-genous habitats, (2) where disturbance is rare, (3) in the absence of effective predators and parasites, (4) in habitats with low fertility/productivity, and (5) when ecological niches of coexisting species overlap.

In highly dynamic systems, species coexistence stems from frequent change induced by external dis-turbance or density oscillations in interacting species. However, in systems with steady-state equilibria, evo-lutionary change and plasticity may lead to species coexistence as a consequence of improved niche par-titioning and finer niche packing allowed by 'charac-ter displacement' of species (a difference in niche dimensions between similar species accentuated

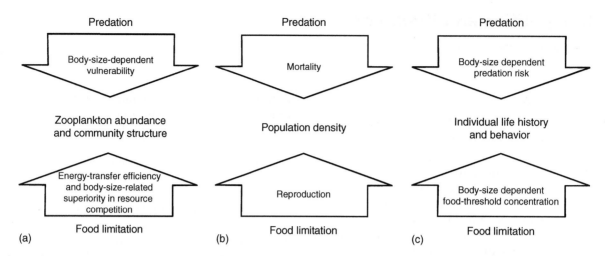

Figure 1 Diagrammatic representation of the equal importance of bottom-up and top-down impacts (food limitation and predation) with regard to zooplankton abundance and community structure (a), population density and age structure (b), and individual behavior and life history in a planktonic animal (c). Reproduced from **Figure 1** in Gliwicz ZM (2002) On the different nature of top-down and bottom-up effects. *Freshwater Biology* 47: 2296–2312, with permission from Blackwell.

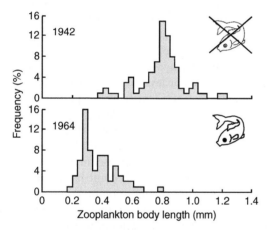

Figure 2 Origin of the 'size–efficiency hypothesis': Large-bodied species, superior in competition for resources, dominate the zooplankton community in a lake in New England sampled in 1942 (top), but are absent in 1964 due to their inferiority in evading predation by a visually oriented planktivorous fish (bottom). The change in size distribution of the zooplankton occurred between 1942 and 1964, following the establishment of a landlocked population of *Alosa pseudoharengus* in the lake. Adapted from **Figure 4** in Brooks JL and Dodson SI (1965) Predation, body size and composition of plankton. *Science* 150: 28–35, with permission from the American Association for the Advancement of Science.

when species coexist). Niche partitioning between coexisting species may be through (1) selective feeding on different resources; (2) exploitation of different resources in separate ways; (3) different uses of space and time, reflecting different vulnerability to predation and parasites; (4) different vulnerability to algae and cyanobacteria that are toxic or interfere with food collection; and (5) distinct requirements with regard to food suspension, food particle shape and dimensions, texture, taste, palatability, handling

time, and food quality, i.e., different proportions of organic carbon, nitrogen and phosphorus, or specific content of particular compounds such as polyunsaturated fatty acids (PUFA).

Both intraspecific (between members of a population) and interspecific (between coexisting species) competition may be classified either as 'exploitation competition' (by indirect effect, via depleting food resources or reducing the quality of resources needed by all coexisting competitors) or 'interference competition' (by directly preventing establishment of competitors in the habitat). Interference competition is probably less important than exploitation competition because it acts between morphologically different taxa rather than closely related species with a large niche overlap. Such interference competition has been observed between suspension feeders from distinct taxa such as tiny ciliates and rotifers, or rotifers and large cladocerans. It is responsible not only for reduced growth and reproduction, but also for high mortality rate in the smaller competitor, thus making competition asymmetrical. The best known example is that of the rotifer *Keratella* injured by filtering machinery when swept into the branchial chamber of *Daphnia* (**Figure 5**).

Exploitation competition should lead either to (1) resource partitioning and niche packing by character displacement, or (2) competitive exclusion of the competitively inferior species as the superior competitor monopolizes resources with its population density at the carrying capacity of the habitat, as predicted by the logistic equation and revealed in Gause's historic experiments with two *Paramecium* species and similar experiments with *Daphnia* (**Figure 6**). Though smaller species have occasionally been reported to outcompete

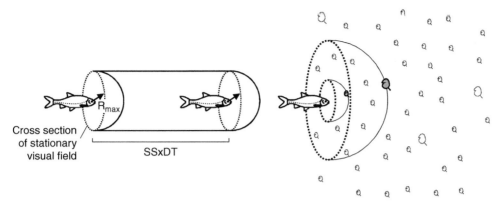

Figure 3 Mechanisms of prey selection by planktivorous fish. Left: 'Search volume' or visual field volume of a foraging fish, where R_{max} is the reaction distance or the maximum distance of perception of a prey, SS is the swimming speed, and DT is the time engaged in the search for a single prey category. Right: Reaction distances for two prey categories of different body sizes. Probability of encounter or assessment of prey density by a foraging planktivore is different for the two prey categories. A foraging fish will perceive the densities of both prey categories as equal, even though their real densities differ by 10-fold, because the individuals depicted as dotted silhouettes are beyond the field of fish vision. Adapted from, respectively, **Figure 1** in Eggers DM (1977) The nature of prey selection by planktivorous fish. *Ecology* 58: 46–59, with permission from the Ecological Society of America, and from **Figure 10** in Gliwicz ZM (2001) Species-specific population-density thresholds in cladocerans? *Hydrobiologia* 442: 291–300, with permission from Kluwer.

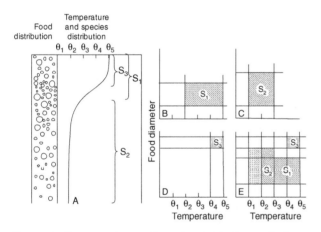

Figure 4 Example of a two-dimensional niche in a vertically stratified lake. The niche dimensions are food particle size (abscissa) and depth-related temperature (ordinate). Niche partitioning and/or possible niche overlap is shown for three suspension-feeding species (S_1, S_2, S_3) in the first step of Hutchinson's fundamental niche definition as a 'multi-dimensional hypervolume of resource axes' based on zooplankton studies (b, c, d). Species S_1, S_2 and S_3 differ in either of the two niche dimensions. They can partition resources by exploiting distinct ranges of grazed particle size at different depths (a). The niche may be exclusive (as between species S_3 and the two others) or inclusive (as between species S_1 and S_2). Diel differences in feeding activity or different vulnerabilities to toxic or interfering algae may be added as a third dimension to construct a 3-dimensional niche. Adapted from **Figure 71** in Hutchinson GE (1967) *A Treatise on Limnology, Volume II, Introduction to Lake Biology and the Limnoplankton.* Wiley, New York, with permission from Wiley.

larger ones in laboratory culture experiments, most field data from habitats free of fish have shown that larger species are the superior competitors. However, the structure of a zooplankton community is often

diverse with many species coexisting in spite of the fact that they feed on the same resources in a similar way. This 'paradox' (Hutchinson, 1961) is explained in two different ways. A nonequilibrium explanation assumes that even the most superior competitor cannot approach the carrying capacity density level. This is so because external disturbance reduces its competitive capacity and intrinsic oscillations preclude equilibrium. The steady-state explanation assumes that high mortalities induced in coexisting populations by predation or parasites may hold population densities much below at which resource competition would cause exclusions of competitively-inferior small-bodied taxa.

The superiority in competition for resources does not stem from the rates of individual growth, reproduction, or population growth (**Figure 6**). Rather, it is due to an individual's ability to grow at a food level lower than that required by its competitors. To grow means to sequester sufficient resources to support the assimilation rate, A, which is greater than the respiration rate, R. This may result from either (1) a wider size spectrum of food particles available to a superior competitor (**Figure 4**), or (2) the species- or instar-specific food threshold concentration. An 'individual's threshold' is the food level at which respiration, R, is just compensated for by the energy of assimilated food, A (**Figure 7**). At a slightly higher food level, $A > R$, and this allows for a net increment in body mass. At a slightly lower food level, $A < R$, and the body mass shrinks. The 'population threshold' is the food level at which population losses due to death or emigration are just compensated by growth and recruitment due to reproduction and immigration, so that the population density or population standing mass remain

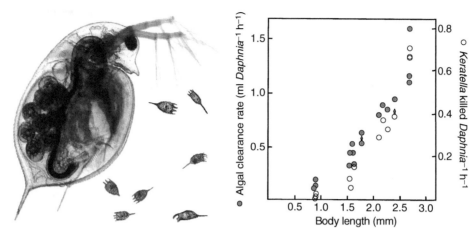

Figure 5 An example of interference competition between two common suspension-feeding zooplankton herbivores. Left: Size relationship between the small (0.1 mm) rotifer, *Keratella cochlearis*, and the large (0.9–2.7 mm body length) cladoceran, *Daphnia pulex*. Right: The lethal effect on *Keratella* swept into the branchial chamber of *Daphnia* increases with *Daphnia* body size in a manner that is proportional to the *Daphnia* clearance rate (i.e., the volume of medium swept clear of algal cells). Adapted from photographs in www. mikroscopia.com, and from **Figure 1** in Burns CW and Gilbert JJ (1986) Effects of daphnid size and density on interference between *Daphnia* and *K. cochlearis*. *Limnology and Oceanography* 31: 848–858, with permission from the American Society of Limnology and Oceanography.

constant. Although the individual threshold, C_0, has no equivalent in Tilman's mechanistic approach to competition in algae, the population threshold corresponds to Tilman's resource level necessary to maintain a stable equilibrium population.

Experimental data on growth rate in *Daphnia* species of different body sizes grown in different food levels have shown food threshold concentrations to be lower in large-bodied species than in small-bodied species, thus strongly supporting the predictions of the size–efficiency hypothesis (**Figure 2**). Further experiments have revealed that food threshold concentrations are lower in older than in younger *Daphnia* instars. This result suggests that (1) adults are superior competitors over juveniles, (2) competitive exclusions may result from the vulnerability of juveniles to starvation, and (3) when facing severe food limitation, adults should either refrain from reproduction or make greater per-offspring investments by allocating resources into fewer eggs per clutch. This may lead to a single-cohort population of long-lived individuals, with younger individuals excluded by severe intraspecific competition (**Figure 8**). The superiority of large body size may be reduced or turned to inferiority in the presence of food particles that interfere with food collection, such as filamentous cyanobacteria. Large body size may be inferior in rotifers.

Predation: Selective and General Predators

Although highly diverse in their modes of foraging for planktonic prey, aquatic predators can be grouped into two distinctly different categories: invertebrate predators and planktivorous fishes. The specificity of each group is related to their taxonomic position, their mode of feeding, and their impact on the planktonic prey, which may include (1) phenotypic adjustments and evolutionary changes in prey morphology, behavior, and life histories; (2) shifts in the density and age structure within the prey population; and (3) changes in the local structure and diversity of the zooplankton community.

Invertebrate predators are part of the zooplankton domain and their effect on planktonic crustaceans, rotifers, and protozoans are comparable with the impact of lions on ungulates of the African savannah. Their encounters with prey are frequent and long-lasting, each being a typical sequence of behaviors a foraging predator has to go through to locate, pursue, attack, handle, and digest its prey. Their assaults are frequently unsuccessful – the molested prey often left alive anticipating a subsequent threat. Their impact on prey through mortality, as well as their indirect effects through reduced rates of prey growth and reproduction, is therefore easier to evaluate than the impact of fish predation. Invertebrate predators are found in all major planktonic taxa. Because of their greater body sizes, they are more vulnerable to fish predation than to herbivore zooplankton, so their importance is amplified in the absence of fish. Where fish are present, antipredator defenses of predacious zooplankton are more apparent than those of herbivorous zooplankton, i.e., they become more transparent or swifter in evasion, and they migrate to deeper strata (e.g., *Chaoborus*).

Planktivorous fishes are members of the 'nekton' – an assemblage of pelagic organisms that groups large and

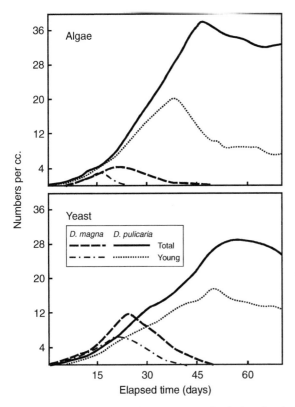

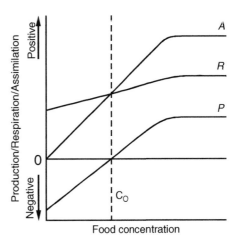

Figure 7 Threshold food concentration: C_0 is the species- or instar-specific food level at which assimilation (A) equals respiration (R). The body mass of an animal is just maintained at a constant level when body growth (net production, P) equals zero. Adapted from **Figure 10.1** in Lampert W (1984) The measurement of respiration. In: Downing JA and Rigler FH (eds.) *A Manual on Methods for the Assessment of Secondary Production in Fresh Water.* IBP Handbook, 2nd ed., vol. 17, Blackwell, Oxford, with permission from Blackwell.

Figure 6 Exploitation competition between *Daphnia pulicaria* and *D. magna* growing on algae or yeast in laboratory cultures. *Daphina pulicaria* wins even when its instantaneous per capita growth rate (*r*, slope of the curve) is lower that that of *D. magna* (yeast as food). This shows that competitive superiority is not only associated with greater initial reproduction potential and maximum possible population growth. The competitor with the lower *r* can win in the long run if it needs lower threshold food concentration to grow (see **Figure 7**). Adapted from **Figure 5** in Frank PW (1957) Coactions in laboratory populations of two species of *Daphnia*. *Ecology* 38: 510–519, with permission from the Ecological Society of America.

highly mobile animals that rapidly propel themselves from one prey aggregation to another, independently of water currents. They are prey harvesters rather than hunters, so the nature of fish predation differs from that of invertebrate predators in a number of ways:

1. They are capable of fast prey collection (up to 1 item per second) and can rapidly eliminate a preferred prey population, particularly in confined habitats.
2. They are long-lived, feeding on tiny prey of short lifespan, thus making numerical responses one-way: only in the prey population (unless the predators are allowed to move in and out), with the functional response operating at its entire prey density spectrum down to its zero level (prey extinction).
3. They are selective, but diet-flexible, readily switching to the most rewarding prey category,

which is consumed until its population density drops to a species-specific threshold level, related to specific reaction distance (**Figure 3**).

4. They are capable of moving quickly into regions in which prey species have just multiplied or aggregated.

Moreover, fish are large, and hence difficult to study alongside their small zooplankton prey in the confined space of experimental chambers, tanks, or in situ enclosures. The enormous disproportions between planktivorous fish and their prey, and consequent disparity in prey–predator interactions, determine that the usual fate of a zooplankton prey is to be eaten by a fish. Fitness does not mean avoiding this fate at all costs, but rather growing and maturing fast enough to produce and release offspring before falling prey. This is another reason why the numerical responses at the zooplankton–fish interface are asymmetrical, and why the nature of antipredation defenses is different for fish and invertebrate predators (**Figure 9**). The impacts of fish predation cascade down to the phytoplankton and are eventually reflected in the properties of the water. Water is clearest in the absence of fish, and when planktivorous fish are kept at bay by piscivores, assuring that the impact on zooplankton is reduced, and that phytoplankton is effectively controlled by abundant and large-bodied planktonic herbivores (**Figure 2**). This can be achieved by top-down 'biomanipulation' (e.g., fishery management promoting piscivores at the expense of planktivorous fishes).

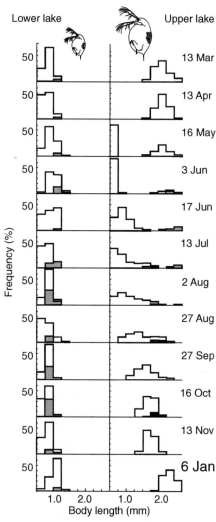

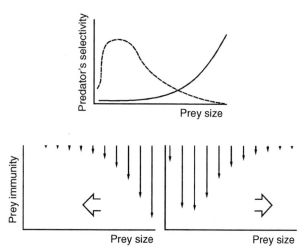

Figure 9 The impact (top) and the effect (bottom panels) of selective predation. Top panel: prey size selectivity by 'size-dependent' planktivorous fish (solid line) and 'gape-limited' invertebrate predators (dashed line) adapted from **Figures 1** and **17** in Zaret TM (1980) *Predation and Freshwater Communities.* Yale University Press, New Haven, with permission from Yale University Press. Bottom panel: The body size distribution of prey immunity to predators shown as the area below the points of the arrows for 'size dependent' predators with a preference for large (better visible and more rewarding) prey resulting in selectivity increase with prey body sizes (left), and for predators with 'gape limitation' with preference for small prey size (right). The change in the arrows' lengths is in both cases equivalent to the respective curve from the top panel turned upside down. Adapted from **Figure 7.2** in Sommer U (ed.) (1989) *Plankton Ecology. Succession in Plankton Communities.* Springer-Verlag, Berlin, with permission from Springer.

Figure 8 Body size distributions (as percent of the total population, examined from 13 March 1996 to 6 January 1997) of *Daphnia* in two neighboring lakes in the Tatra Mountains as a consequence of the presence or absence of fish predation. The lower lake, Morskie Oko (LL, left), is inhabited by natural fish populations and several coexisting cladoceran species. The upper lake, Czarny Staw (UL, right), is free of fish and has *D. pulicaria* as the sole cladoceran species. This large-bodied filter-feeding herbivore monopolizes resources. Its reproduction is synchronized within a narrow time window when food levels increase during the spring overturn, allowing juvenile survival and growth. Grey: fraction of egg-bearing females; black: fraction of ephippia-bearing females. Note that the size at first reproduction is 0.75 mm in LL and 1.5 mm in UL. In contrast to the continuous reproduction in the LL population controlled by fish predation, the reproductive effort in UL is synchronized from 3 June to 2 August, when two cohorts are present: the old generation of adults born in summer 1995, and the 1996 generation hatched from ephippia (May–Jul) or released from the brood chambers of females of the old and the new generations (Jul–Aug). Adapted from **Figure 2** in Gliwicz ZM, Slusarczyk M, and Slusarczyk A (2001) Life-history synchronization in a long-lifespan single-cohort *Daphnia* population of an alpine lake free of fish. *Oecologia* 128: 368–378, with permission from Springer.

Most inland waters are inhabited by fish, and most freshwater fish species readily feed on zooplankton, at least during their adolescence (**Figure 10**). When feeding on zooplankton, fish are highly selective but they are also generalist predators. They are selective in that they preferentially go after prey species whose individuals are most conspicuous and most rewarding. They are also general predators that tend to feed upon the most abundant prey, shifting to the prey category which is most rewarding because of both the properties of an individual prey item and the density of the prey population (**Figure 3**). These two predatory behaviors are not mutually exclusive. On the contrary, these behaviors must be combined and coordinated in every decision concerning prey choice, regardless of whether the subject of the choice is a different prey species or a different instar of the same species. The two modes work together in a simultaneous decision process regarding which individual from which prey category should be taken; for instance, by choosing which is apparently the largest prey. In this situation, a planktivorous fish is assumed to actively select its prey, pursuing the individual prey item that appears largest at the

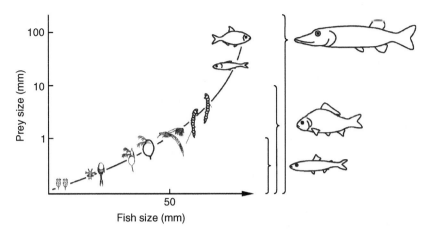

Figure 10 Ontogeny and the feeding habits of European fishes. Both piscivores and benthivores are planktivorous during the first stages of their lives. Adapted from **Figure 4** in Jachner A (1988) Biomanipulation IV. Density and feeding activity of planktivorous fish. *Wiadomosci Ekologiczne* 34:143–163, with permission from the Center of Ecological Studies, Polish Academy of Sciences.

initiation of the thrust. In addition, the two feeding modes must be compromised in a decision to switch from one prey to another; for instance, by choosing to feed on a prey category that has just been found to be more rewarding. This may be (1) the prey category (species or single ontogenetic stage) that is relatively more abundant, (2) that which offers a higher return for the energy or time expended in a successful individual encounter, or (3) that which the fish has found to be easier to capture than an alternative prey category. Each of these reasons should be equally valid justifications for a switch from one prey category to another as soon as the other becomes more rewarding.

Planktonic animals have various strategies to reduce their vulnerability to highly selective visual predation by fish:

1. They can avoid encounters with fish at high light intensity by selecting a depth at which illumination is below a threshold level. This is the reason for diel vertical migrations.
2. They can become inconspicuous to fish vision by minimizing body size and body opaqueness.
3. They can improve their ability of effective evasion while being attacked or swept into a fish's mouth.
4. They may confound the predator by an unusual motion (e.g., somersaulting in *Daphnia*) or by aggregating. Synchronized behavior within a swarm increases predator confusion and within an aggregation the risk is further reduced because of risk spreading.

The risk of visual predation may also be reduced at very low prey density levels at which no single individual of a given prey category is present within the reaction field volume of a foraging fish (**Figure 3**). Such a 'low-density antipredation refuge' may cause the predator's attention to shift to another prey category that has recently become more frequently encountered by a foraging fish.

Trading Growth for Safety: An Individual's Perspective

Until its reproduction, each planktonic animal must trade food for safety, survival, and growth. This is more apparent in animals offshore than elsewhere, because distinct gradients in underwater illumination permit a clear distinction between habitat profitability with respect to the two components which determine an individual's fitness: its ability to survive by avoiding the risks of predation, and its ability to assimilate sufficient resources to enable fast growth to maturity and high reproductive effort. Day-to-day life is conducted, and hour-to-hour behavioral decisions have to be made under the constant threat of predation and the never-ending hazard of starvation. Each decision is made after the magnitude of the two risks has been assessed from the state of satiation and chemical and visual cues in the medium. Both risks are related to the vertical gradient of light intensity, and its shift with the change in incipient radiation over the course of a day. A prey individual's instantaneous decisions concern the depth it selects and the intensity of its feeding activity, both of which are factors vital to safety and food assimilation. Taking a greater risk in better illuminated upper strata of the water column allows increased food ingestion and faster growth enhanced by higher temperatures, but the accumulation of food in the intestine makes an animal more conspicuous to a visually feeding predator.

These instantaneous decisions are influenced by long-term life history, affecting (1) age and body size at maturation, (2) number and size of offspring, and (3) whether or not to enter diapause, represented by egg dormancy (cladocerans and rotifers) or resting stages. Each is a compromise allowing zooplankton prey 'to run faster for its life than the predator runs for its dinner' (Dawkins and Krebs, 1979). Small body size provides an effective escape from the danger of falling prey to a planktivorous fish or another visually oriented predator, but it makes an animal vulnerable to gape-limited invertebrate predators (**Figures 1c** and **9**), and also a less efficient competitor for food resources. A large number of eggs in a clutch enhances reproduction output but makes the egg-carrying female more vulnerable to predation (**Figure 8**), and the eggs possibly too small to assure the survival of juveniles at low food levels (**Figure 12**). The production of dormant eggs and resting stages in response to physical discomforts, increased risks of predation, and/or hazards of severe food limitation diminishes the immediate reproductive success, but increases the chances of success elsewhere and in the future by higher dispersal and building egg banks. In the presence of fish, the most reasonable compromise (assuring the highest individual fitness) between the traits of small body size at first maturation and the highest possible rate of individual growth $(A - R)$ is starting reproduction at an earlier instar rather than reducing growth rate by decreasing A.

Risking life is different from risking hunger. The risk of becoming subject to predation may become fatal within seconds, while the risk of starvation may persist for days or weeks with there always being the possibility of future compensation with an increase in food abundance. This might be the reason why it is not a simple objective to invent a common currency for life-history or behavioral decisions in regard to the risk of predation on the one hand, and the hazards of starvation on the other. The major difference between the two is that in the case of predation a mistake is likely to become terminal, whereas with regard to food limitation many mistakes may be tolerated within an individual's lifespan. This may explain why responses to increased predation risk appear stronger than those to decreased food levels. For an individual, satiation can be at any value within its entire 0–100% range, and long periods with an empty gut (zero satiation) may easily be compensated when food levels increase again in the future. In contrast, individual survival can never be lower than 100%, and death cannot subsequently be compensated. The trading of the hazards of starvation with the risk of predation cannot thus be compared via a common currency; for instance, by *percentage increase*: an increase in feeding rate and an increase in the risk of predation. It is clear that the two are not compatible. This might be the reason

why the different nature of top-down and bottom-up effects has never been ignored in regards to individual behavior and life histories.

Disparity of Competition and Predation: Population Density and Population Growth

The disparity in the nature of top-down and bottom-up effects seems easier to ignore at the population level since mortality and reproduction combine with each other. They do so because there is a common currency for both: the individual. The effect of food limitation would be reflected in birth rate, b, and the effect of predation in death rate, d, and the two would merge into r, the intrinsic rate of population-density increase $(b - d = r)$. For this reason, the 'sandwich' concept of full symmetry between top-down and bottom-up impacts has been approved so readily, with regard to the population as well as to the community level (**Figure 1**). Consequently, it is simple to assume that an observed increase in the overall abundance of herbivores has resulted either from reduced predation or from relaxation of competition and food limitation.

However, full symmetry at the population level must be refuted as it is recognized that the state variables are top-down controlled while the rates of change are bottom-up controlled (**Figure 11**). This fundamental difference is reflected in Holling's

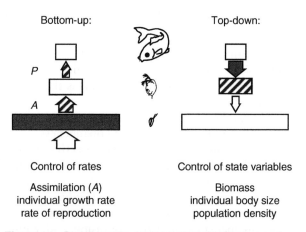

Figure 11 Diagrammatic representation of bottom-up and top-down impacts (food limitation and predation) on zooplankton. Note that the shaded element in each panel controls striped element(s). Abundance (standing crop) is controlled top-down by planktivorous fish (right), while processes (energy/carbon flow) are controlled bottom-up by food levels (left). Reproduced from **Figure 11** in Gliwicz ZM (2002) On the different nature of top-down and bottom-up effects. *Freshwater Biology* 47: 2296–2312, with permission from Blackwell.

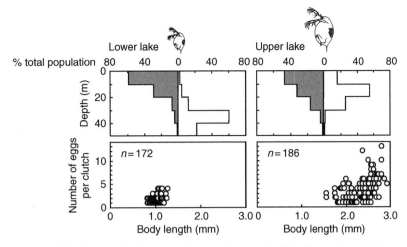

Figure 12 *Daphnia* behavior and life history traits in two neighboring lakes in the Tatra Mountains (cf. **Figure 8**) as a consequence of the presence or absence of fish predation. Depth distribution at midnight (shaded) and noon (unshaded) as percent of the total population on 26 August (LL) and 27 August (UL) 1996 (top). Clutch size of *Daphnia* in relation to body length between 3 June and 27 August 1996 (bottom). Most clutches are smaller than possible, due to selective predation by fishes in LL and a greater per-offspring investment needed for juveniles to survive at low food levels in UL. Reproduced from **Figure 49** in Gliwicz ZM (2003) *Between Hazards of Starvation and Risk of Predation: The Ecology of Offshore Animals*. International Ecology Institute, Oldendorf/Luhe, with permission from the International Ecology Institute.

functional responses, as the predator capture rate is a function of prey density, not prey reproduction and recruitment. It can also explain similar densities of a given zooplankton species in lakes of different trophic state. Higher reproduction rates at increased food levels are counteracted by higher mortality when densities increase beyond a single individual per species-specific reaction field volume (**Figure 3**).

Food Limitation and Predation in Structuring Zooplankton Communities

Coexistence of large- and small-bodied zooplankton in proportions inversely reflecting their body sizes, in habitats of wide productivity spectrum, reflects the interspecific difference in species-specific reaction field volume (**Figure 3**). By switching from one zooplankton prey to another, planktivorous fish hold the density of each zooplankton prey species much below the carrying capacity of the habitat. Each prey-density increase would be followed by a shift in fish diet to the most rewarding prey in the present situation. The most conspicuous prey (the large-bodied and thus competitively superior species) would be held at the lowest density, corresponding to its low 'relative density' resulting from the high vulnerability of individuals at maturation (large body size at first reproduction). The low abundance of superior competitors with high clearance rates would result in higher food levels, and in the coexistence of other

suspension-feeding species, including small-bodied cladocerans, rotifers, and ciliates.

At low fish densities, the average body size of zooplankton increases (**Figure 2**), and the number of coexisting species declines. In the absence of fish, a superior large-bodied competitor may become the sole cladoceran or copepod species that monopolizes resources. Severe intraspecific competition selects for late reproduction at large body size that occurs within a short-lasting time window when food levels are high enough for the juveniles to survive and start a new cohort of long-lifespan individuals (**Figure 8**). Allowed to ignore predation, the members of such populations may be more effective in competition for resources by remaining in the warmer more food-abundant upper strata of the water column for day and night, without costly diel vertical migrations, and maximize their reproductive success by delaying reproduction until attaining advanced age and large body size (**Figure 12**). Contrary to the well-fed but short-lived animals of the time-limited populations in habitats with fish, animals from fishless habitats are hungry but have a longer lifespan, and their populations are typically resource-limited.

See also: Cladocera; Copepoda; Cyclomorphosis and Phenotypic Changes; Diel Vertical Migration; Egg Banks; Phytoplankton Population Dynamics: Concepts and Performance Measurement; Role of Zooplankton in Aquatic Ecosystems; Rotifera.

Further Reading

Chase JM, Abrams PA, Grover JP, *et al.* (2002) The interaction between predation and competition: A review and synthesis. *Ecology Letters* 5: 294–315.

Dawkins R and Krebs JR (1979) Arms races within and between species. *Proceedings of the Royal Society London B* 205: 489–511.

Gilbert JJ (1988) Suppression of rotifers by *Daphnia*: A review of the evidence, the mechanisms, and the effects on zooplankton community structure. *Limnology and Oceanography* 33: 1286–1313.

Gliwicz ZM (2003) *Between Hazards of Starvation and Risk of Predation: The Ecology of Offshore Animals*, 379 pp. Oldendorf/Luhe: International Ecology Institute.

Hutchinson GE (1961) The paradox of the plankton. *American Naturalist* 95: 137–146.

Kerfoot WC and Sih A (eds.) (1987) *Predation: Direct and Indirect Impacts on Aquatic Communities*. Hanover, NH: The University Press of New England.

Lampert W (2006) *Daphnia*: Model herbivore, predator and prey. *Polish Journal of Ecology* 54: 607–620.

Lima SL (1998) Stress and decision making under the risk of predation: Developments from behavioral, reproductive, and ecological perspectives. In: Møller AP, Milinski M, and Slater PJB (eds.) *Stress and Behavior. Advances in the Study of Behavior*, vol. 27, pp. 215–290. San Diego: Academic Press.

Schoener TW (1973) Population growth regulated by intraspecific competition for energy or time: Some simple representations. *Theoretical Population Biology* 4: 56–84.

Sommer U (ed.) (1989) *Plankton Ecology Succession in Plankton Communities*, 369 pp. Berlin: Springer-Verlag.

Stearns SC (1989) Trade-offs in life-history evolution. *Functional Ecology* 3: 259–268.

Tokeshi M (1999) *Species Coexistence Ecological and Evolutionary Perspectives*. Oxford: Blackwell Science.

ECOSYSTEM INTERACTIONS

Contents

Phytoplankton Nutrition and Related Mixotrophy

J A Raven, University of Dundee at SCRI, Dundee, UK
S C Maberly, Centre for Ecology & Hydrology, Lancaster, UK

Introduction

Phytoplankton are algae and cyanobacteria which grow free-floating in a water body and consequently rely on the water column for their resources. Most phytoplankton are photolithotrophs, which means that light provides the energy and inorganic molecules and ions provide the materials, including carbon, required for growth. Some phytoplankton are unable to use external organic carbon as a supplementary or sole source of energy and carbon, a condition termed obligate photolithotrophy. However, some obligate photolithotrophs can take up and metabolize low molecular mass nitrogen-containing organic carbon compounds (e.g., urea, amino acids), thus contributing to the nitrogen budget of the organisms. Others can exploit external reserves of organic nitrogen or phosphorus by producing extracellular enzymes that convert the organic molecule into an inorganic molecule that is subsequently taken up.

Although phytoplankton are the main primary producers in at least larger bodies of inland water, a significant number of these organisms are able to use external organic compounds. These organisms can use external organic compounds as a supplemental source of energy and carbon, or of nitrogen or sulfur, and are termed mixotrophs. Some mixotrophs take up organic compounds into the cytosol on a molecule-by-molecule basis across the plasmalemma; these are sapromixotrophs (e.g., the green alga *Chlamydobotrys*, and some species of *Chlamydomonas*).

Other mixotrophs (e.g., some chrysophytes, cryptophytes and dinoflagellates, and some species of the prasinophyte flagellate *Pyramimonas*) take up organic particles, with digestion of the complex organic molecules in food vacuoles followed by uptake of individual molecules across the food vacuole membrane in the cytosol. These organisms are phagomixotrophs, and can obtain nitrogen, phosphorus, sulfur, iron, and other elements from their particulate food.

Some mixotrophic algae are able to grow in the absence of light. This characteristic is known as chemoorganotrophy, and the algae and cyanobacteria which exhibit it are termed facultative chemoorganotrophs. Some organisms that are very closely related to photosynthetically competent algae have, during evolution, lost the capacity to photosynthesize and are obligate chemoorganotrophs and live as saproorganotrophs, e.g., *Astasia*, or phagoorganotrophs, e.g., *Peranema*. Both of these chemoorganotrophs are close relatives of the photosynthetic *Euglena*.

The precise definitions of these different terms, which are used to categorize the nutritional characteristics of phytoplankton, are given in **Box 1**. It is now possible to suggest how these characteristics are involved in determining the ecological occurrence of the algae. However, almost all the work describing nutritional characteristics of phytoplankton has been obtained from laboratory cultures with inherent limitations of using a small subset of phytoplankton found in inland waters that can be readily cultured,

---

Box 1 Definitions of Terms Related to Trophic Modes of Organisms

Chemoorganotroph: An organism which obtains the energy for growth and maintenance from organic carbon, which also serves as the source of carbon skeletons for growth.

Mixotroph: An organism which combines photolithotrophy (qv) and chemoorganotrophy (qv).

Phagomixotroph: A mixotroph (qv) which obtains its organic carbon as do phagotrophs (qv).

Phagotroph: A chemoorganotroph (qv) which obtains its organic carbon and other resources by ingesting particulate matter.

Photolithotroph: An organism which obtains its energy for growth and maintenance from electromagnetic radiation, and its carbon from inorganic carbon, by photosynthesis. Other elements are taken up in the available chemical form by uptake of individual molecules across the plasmalemma. Photolithotrophy is combined in many phytoplankton cells with a requirement for external supply of one or more of the vitamins Vitamin B$_{12}$, Thiamine, and Biotin, in that decreasing order of the number of species which require them.

Sapromixotroph: A mixotroph (qv) which obtains its organic matter as do saprotrophs (qv).

Saprotroph: A chemoorganotroph (qv) which obtains its organic carbon, and other nutrients, by uptake of individual molecules across the plasmalemma.

---

using vessels with a large solid surface area in relation to the volume of water and non-natural spectral composition of light and levels of ultraviolet radiation.

Photolithotrophy by Phytoplankton in Inland Waters

Background

To describe the process of photolithotrophy in phytoplankton in inland waters, the mechanisms by which electromagnetic radiation as photosynthetically active radiation (PAR; 400–700 nm) is converted to chemical energy is first described. The sources, mechanisms of uptake and uses of the essential elements are then considered in turn in (approximate) order of abundance (by atoms) in the phytoplankton of inland waters, i.e., H, C, O, N, P, K, S, Mg, Ca, Cl, Fe, Mn, Cu, Zn, Mo, Ni, Co, and, for some heterokonts such as diatoms, Si.

Electromagnetic Energy

Photolithotrophy by oxygen-producing organisms, including algae and cyanobacteria, proceeds according to the general equation [1]:

$$2H_2{}^*O + C^\delta O_2 + \geq 8\,photons\,(400-700\,nm) \rightarrow (CH_2{}^\delta O)$$
$$+ H_2{}^\delta O + {}^*O_2 \tag{1}$$

This apparently simple equation carries two significant messages compared to other representations. One important point is the occurrence of the photons on the left-hand (substrate) side of the equation; in some representations light is written above the reaction arrow, which is where catalysts should be placed. The photons are substrates for photosynthesis just as are water and carbon dioxide.

A second important point is the presence of two water molecules on the left-hand side and one water molecule on the right-hand side. The reason for this is shown by the superscripts which indicate how the oxygen from water (*) ends up in the evolved oxygen, while the oxygen from carbon dioxide ($^\delta$) ends up in carbohydrate and water. Although this equation indicates carbohydrate as the product of photosynthesis, and much of the organic products of photosynthesis are stored momentarily, or for a longer time, in carbohydrate, a significant fraction is taken from the photosynthetic carbon reduction cycle (PCRC) as a more oxidized compound, 3-phosphoglycerate, in the synthesis of lipids and of many amino acids and pyrimidines. Notwithstanding the diversity of ultimate end products of photosynthesis, i.e., all organic cell constituents and organic matter lost to the medium, eqn [1], describes the approximate stoichiometry of cell growth in acid-base terms, although different nitrogen sources alter the stoichiometry. Maintaining intracellular acid-base balance even when bicarbonate (see below) is the inorganic species take up by the cells requires excretion of about one hydroxyl ion for each bicarbonate ion taken up.

A third important point is that a minimum of eight photons are needed to reduce one carbon dioxide to carbohydrate, using four electrons from water. The requirement for the energy of two photons to transfer each electron from water to carbon dioxide is that two different photochemical reactions are involved in series, each using one photon to transfer one electron.

Light energy is absorbed by pigments associated with proteins in or on thylakoid membranes which, in eukaryotic algae, are located within organelles known as plastids. Chlorophyll *a*, a ring-shaped molecule with a central magnesium atom, is the main photosynthetic pigment in all oxygen-evolving organisms such as algae and cyanobacteria, and absorbs blue and red light most strongly. The taxonomic diversity of algae and cyanobacteria is reflected in the diversity of the other thylakoid-associated pigments (secondary pigments) which absorb a wide variety of wavelengths and pass the energy on to chlorophyll *a* in reaction centers. In each of the two types of reaction center photochemistry takes place, with the energy of photons converted into oxidation-reduction (chemical) energy. Nonphotochemical reactions, involving among other things a cycling of protons across the thylakoid membranes, result in the energy from the photochemical reactions being converted into the energy used to phosphorylate ADP to produce ATP, and to reduce $NADP^+$ to NADPH. Much of the energy available from reactions involving ATP and NADPH is used to power the PCRC mentioned above in which CO_2 is reduced to carbohydrate. The PCRC enzymes occur in the stroma of the plastids of eukaryotes, and in the cytosol of cyanobacteria.

At low PAR fluxes, the rate of photosynthesis and growth increases proportionately with PAR. As PAR increases further there is a relatively abrupt transition to rates of photosynthesis being independent of the incident PAR (light saturation). Further increases in PAR can cause a decrease in the photosynthetic rate as a result of photoinhibition. This process can occur even at low, rate-limiting, fluxes of PAR but at high light additional photodamage can occur when the rate of absorption of light energy exceeds the capacity for photochemistry to dissipate this excess energy. Algae and cyanobacteria have a number of mechanisms, such as state transitions, which limit the fraction of absorbed energy which reaches the most sensitive site, and nonphotochemical quenching processes such as xanthophyll cycles, which limit photodamage. They also possess repair mechanisms which can cause a net reduction in photodamage; this repair is most obvious after incident PAR flux has decreased.

The relevance of these relationships to the ecology of the phytoplankton of inland waters is that PAR can markedly change with time and depth. The temporal changes result from annual and diel changes in PAR at the water surface made less predictable by variable cloud cover and atmospheric attenuation. Light is lost by reflection at the water surface (particularly at low solar angles) and declines with depth as a result of attenuation by dissolved and particulate material, including the algae themselves, within the water column. Thus, an individual algal cell may, over a day, experience rapid changes in incident PAR from limiting to photoinhibiting as it is moved through different depths by water currents. In inland waters, ultraviolet radiation tends to be attenuated more rapidly with depth than does PAR, particularly where dissolved organic carbon is present. Consequently, exposure to ultraviolet radiation relative to PAR is greater at the surface of the water column than at the depth.

Net primary productivity is the difference between gross production and losses over the full day–night period from respiration (as carbon dioxide) and as soluble organic carbon. Dark respiration is essential for growth and maintenance of photolithotrophs. Loss of dissolved organic matter generally has no obvious function as a possible contributor to evolutionary fitness; an

exception is the secretion of siderophores by cyanobacteria which are involved in iron acquisition. The difference between gross and net productivity means that the maximum depth at which primary productivity can occur (in the absence of vertical water movements) is less than the depth at which significant gross primary productivity can occur, taking into account unavoidable back-reactions in the photosynthetic mechanism. Further processes such as grazing, sinking, and hydraulic flushing can lead to losses of phytoplankton from a system.

Hydrogen

Essentially all of the hydrogen in photolithotrophs comes from water. While water is the most abundant molecular species in inland waters, regulating the net influx of water poses problems both in low-osmolarity inland waters and in those of very high salinity. For the low-osmolarity (fresh) waters the cell contents necessarily have a higher osmolarity than the medium, and cellular volume regulation involves energy input either to pump water out of the cells if there is no effective cell wall (as in flagellates such as chrysophytes and dinoflagellates), or to build the cell wall which resists pressure in turgid cells. For the hypersaline inland waters there is a very low diversity of phytoplankton (mainly *Dunaliella* spp.). The volume regulation problem here is maintaining the internal osmolarity equal to the external osmolarity using organic compounds (compatible solutes) which permit cell function. This is energy-expensive, even using glycerol, the compatible solute with the lowest molecular mass.

Carbon

There is great variability in the concentration, and speciation, of inorganic carbon in inland waters. Inorganic carbon comprises carbon dioxide, bicarbonate and carbonate interlinked by equilibria which are related to pH. When a lake is in equilibrium with the atmosphere, the concentration of inorganic carbon is controlled largely by the geology of the catchment, producing a spatial variation among lakes. Many lakes, however, can be far from equilibrium. When averaged over time, most lakes have excess CO_2 and lose it to the atmosphere because of input from the catchment of CO_2, or organic carbon which is broken down photochemically and biologically, to produce CO_2. In productive lakes there are periods, typically during summer stratification, when rates of photosynthetic inorganic carbon uptake exceed rates of resupply. This removal of what amounts to CO_2 increases the pH, and the near zero CO_2 concentrations can potentially limit rates of photosynthesis. However, net photosynthesis can continue if algae make use of bicarbonate and/or have an effective carbon concentrating mechanism (see below). In productive lakes, pH can vary by 1 or 2 units during 24 h. Spatially, pH can vary from between about 1 in waters influenced by volcanic activity to 10 or 11 in waters of high acid neutralizing capacity and/or that are experiencing severe inorganic carbon depletion. Stratified productive lakes also show strong depth gradients of inorganic carbon and pH with lower concentrations and higher pH in the productive epilimnion.

The core carboxylase of photosynthesis in all oxygen-producing organisms is ribulose-1,5-bisphosphate carboxylase-oxygenase (Rubisco). This enzyme uses carbon dioxide as the inorganic carbon substrate, and reacts not only with carbon dioxide but also, competitively, with oxygen. The phylogenetically variable, but always relatively low, affinity of Rubisco for carbon dioxide means that the carbon dioxide concentration in inland waters can be rate-limiting for gross photosynthesis if the supply of the inorganic carbon substrate to Rubisco is by diffusion of carbon dioxide. The majority of the investigated phytoplankton from inland waters have inorganic carbon concentrating mechanisms (CCMs). These CCMs can be based, depending on the energized transport across a membrane or membranes, on carbon dioxide, bicarbonate or protons. The main inland water taxa known to lack CCMs are the chrysophytes and synurophytes.

Oxygen

Oxygen in the organic matter of photolithotrophs comes mainly from carbon dioxide (eqn [1]); a minor component comes from molecular oxygen via oxygenase reactions.

Nitrogen

The reduced nitrogen in organic matter in photolithotrophic phytoplankton from inland water comes mainly from 'combined' nitrogen, e.g., ammonium, nitrite, nitrate and dissolved low molecular mass organic nitrogen. All phytoplankton can apparently use ammonium as their nitrogen source, most can use nitrate and nitrite, and many can use organic nitrogen. The exceptions are filamentous cyanobacteria (e.g., *Anabaena*) with heterocysts which are specialized in using molecular nitrogen by nitrogen fixation, i.e., diazotrophy. The heterocysts restrict access of molecular oxygen, an irreversible inhibitor of the enzyme nitrogenase, to this key enzyme in the first step of

nitrogen fixation. The extent to which the availability of combined nitrogen limits phytoplankton productivity in inland waters is still a matter of debate. It is widely held that phosphorus is the most frequent limiting element for growth of inland water photolithotrophic phytoplankton. This situation would agree with the geochemical prediction that a shortage of combined nitrogen relative to other nutrients required for photolithotrophic growth would be corrected by increased growth of diazotrophic organisms. However, in some lakes nitrogen is limiting or colimiting with phosphorus, possibly because the environment is unfavorable for nitrogen-fixing organisms.

The present situation is distorted by anthropogenic inputs of combined nitrogen to a greater extent, relative to the quantities needed for phytoplankton growth, than that of phosphates. This combined nitrogen input includes nitrate inputs from agricultural run-off, and oxidized and reduced nitrogen from atmospheric pollution. How this reflects the 'natural,' preindustrial situation is not clear, since at least the atmospheric component of anthropogenic inputs is quantitatively widespread over many remote inland waters.

Phosphorus

The oxidized phosphorus in organic matter in photolithotrophic phytoplankton in inland waters comes from inorganic and organic phosphates, and to a very limited extent from phosphonates, dissolved in the epilimnion. As mentioned in the section Nitrogen, phosphorus is predicted from geochemical considerations to limit phytoplankton productivity in inland waters. However, there are many cases in which combined nitrogen limits phytoplankton growth in inland waters and diazotrophy does not correct this limitation.

Potassium

Potassium has occasionally been suggested as a growth-limiting nutrient in the phytoplankton of inland waters of low salinity, although there are few data which support this contention.

Sulfur

Atmospheric inputs of anthropogenic sulfur dioxide, largely derived from burning of high-sulfur coal in the absence of scrubbing this gas from the flue gases, have increased the sulfate concentration in many inland waters. However, there are still a number of inland water habitats, all of low salinity, in which sulfur availability can be close to limiting the primary productivity of phytoplankton. Sulfate is very abundant in some saline inland waters.

Calcium and Magnesium

These elements do not seem to be a limiting nutrient for phytoplankton primary productivity in inland waters.

Iron

There are some examples of iron limitation of phytoplankton primary production in inland waters with combined nitrogen as the nitrogen source. While diazotrophy has a higher iron requirement than does the use of nitrate, nitrite or, especially, reduced nitrogen sources, there is little evidence of specific iron limitation of nitrogen fixation in inland waters.

Molybdenum

The requirement for molybdenum by inland water phytoplankton mainly relates to nitrogenase in diazotrophs, and to a quantitatively smaller extent to growth with nitrate as the nitrogen source. High sulfate concentrations in some saline inland waters could competitively inhibit the uptake of molybdate, the predominant natural source of molybdenum. For diazotrophs, there are also organisms which can express nitrogenases which use vanadium, or (additional) iron, instead of molybdenum.

Chlorine, Manganese, Copper, Zinc, Nickel, and Cobalt

There seem to be no cases in which these elements limit phytoplankton primary productivity in inland waters.

Silicon

Silicon is only essential for the growth of diatoms and, probably, many chrysophytes, synurophytes, and perhaps prasinophytes with deposits of silica. Silicon is taken up as silicic acid, and the supply of this nutrient can limit the extent to which silicon-requiring organisms contribute to phytoplankton primary productivity in productive habitats at some habitats at some times, often in spring.

The Role of Mixotrophy in the Phytoplankton of Inland Waters

Sapromixotrophy

Dissolved organic matter in inland waters originates in terrestrial primary productivity (see section Hydrogen), from photolithotrophs (see section Electromagnetic Energy) and from downstream processes based on phytoplankton primary productivity. The extent to which sapromixotrophy can contribute to phytoplankton primary productivity in inland waters is unclear. Globally, the conclusion seems inevitable that there is a net efflux of dissolved organic matter from phytoplankton to the epilimnion of inland waters. This conclusion does not preclude the occurrence of locations in which the reverse is true, at least at some times. An example is acidic lakes in disused open-cast lignite mines.

Phagomixotrophy

In a manner in some ways resembling that for sapromixotrophs, the global net flux of particulate organic matter to consumers is from phytoplankton to nonalgal phagotrophs in inland waters, with a relatively small uptake by phagomixotrophs. Of course, the planktonic phagomixotrophs only ingest the smaller particles, and so are best considered as part of the 'microbial loop.' Here, as well as picophytoplankton, the prey for small phagotrophs (including phagomixotrophs) includes bacteria growing on dissolved organic matter, thus giving a link with the dissolved organic carbon pool (see sections Hydrogen and Sapromixotrophy). Locally, e.g., in some lakes in Antarctica, phagomixotrophic algae are predominant (in biomass and in species number) members of the plankton.

While sapromixotrophy deals solely with organic carbon and nitrogen, phagomixotrophy involves the uptake into food vacuoles of elements in the ratio in which they occur in the prey. This means that phagotrophy could do much more than supply organic carbon as a carbon and energy source: it is also a nitrogen, phosphorus and iron source. While some evidence is consistent for a role of phagomixotrophy in the supply of nitrogen, phosphorus and iron for the organisms, the evidence overall is somewhat equivocal. It is also known that, at least in the laboratory, phagomixotrophs can acquire iron from particulate/colloidal iron not readily available to non-phagotrophic phytoplankton.

Conclusions

Phytoplanktonic algae and cyanobacteria face a wide variation in availability of energy and materials among different lakes, and within a given lake at different times and depths. This variation is matched by the different strategies employed in obtaining resources by different taxa, at a phylogenetic or functional group level, and by the flexibility to exploit different types of resources depending on their relative availability. Seen as a 'compound organism,' the phytoplankton are important as 'ecosystem engineers' that alter the availability of the materials within an inland water body.

See also: Algae; Phytoplankton Productivity; Protists.

Further Reading

Falkowski PG and Raven JA (2007) Aquatic Photosynthesis. 2nd edn. Princeton, NJ: Princeton University Press.

Jaworski GHM, Talling JF, and Heaney SI (2003) Potassium dependence and phytoplankton ecology: An experimental study. *Freshwater Biology* 48: 833–840.

Laybourn-Parry J, Marshall WA, and Marchant HJ (2005) Flagellate nutritional versatility as a key to survival in two contrasting Antarctic saline lakes. *Freshwater Biology* 50: 830–838.

Maberly SC (1996) Diel, episodic and seasonal changes in pH and concentrations of inorganic carbon in a productive lake. *Freshwater Biology* 35: 579–598.

Maberly SC, King L, Dent MM, Jones RI, and Gibson CE (2002) Nutrient limitation of phytoplankton and periphyton growth in upland lakes. *Freshwater Biology* 47: 2136–2152.

Raven JA (1997) Phagotrophy in phototrophs. *Limnology and Oceanography* 24: 198–205.

Reynolds CS (2006) The Ecology of Phytoplankton. Cambridge: Cambridge University Press.

Titell J, Bissinger V, Gaedke U, and Kanjuke N (2005) Inorganic carbon limitation and mixotrophic growth in *Chlamydomonas* from an acidic mining lake. *Protist* 156: 63–75.

Chemosynthesis

A Enrich-Prast, University Federal of Rio de Janeiro, Rio de Janeiro, Brazil
D Bastviken and P Crill, Stockholm University, Stockholm, Sweden

Introduction

All known life forms depend on the biosynthesis of large and complex organic compounds. Hence, the capability to produce such biomolecules from abiotic carbon sources, e.g., one-carbon (C-1) compounds such as carbon dioxide (CO_2) and methane (CH_4), was, and still is, crucial for the development of life. The formation of biomolecules from C-1 compounds (often referred to as carbon fixation or autotrophy) requires energy. *Chemosynthesis* can be defined as the biological production of organic compounds from C-1 compounds and nutrients, using the energy generated by the oxidation of inorganic (e.g., hydrogen gas, hydrogen sulfide, ammonium) or C-1 organic (e.g., methane, methanol) molecules. This process contrasts with *photosynthesis*, a process in which organic compounds are synthesized from CO_2, using energy generated from sunlight radiation.

Organisms that use sunlight or chemicals as energy source are called *phototrophs* and *chemotrophs*, respectively (**Table 1**). Chemotrophs can be classified into *chemoorganotrophs* or *chemolithotrophs* depending on their use of organic or inorganic energy sources, respectively (**Table 1**). The term chemolithotroph literally means "rock eaters" and is used to designate organisms that generate energy by the oxidation of inorganic molecules for biosynthesis or energy conservation via aerobic or anaerobic respiration.

Many chemosynthetic organisms are usually referred to as *chemoautotrophs*. The term *autotroph* (self-feeding) denotes organisms that derive the carbon needed for biosynthesis from inorganic molecules (i.e., CO_2) independent of their energy source. *Photoautotrophs* and *chemoautotrophs* obtain their energy from sunlight and the oxidation of inorganic molecules, respectively (**Table 1**). All chemoautotrophs are chemosynthetic organisms.

Some chemosynthetic organisms can also assimilate organic molecules. There are facultative chemolithoautotrophs that can grow autotrophically or heterotrophically depending on conditions such as substrate availability. Chemolithotrophic heterotrophs use energy from the oxidation of reduced inorganic compounds as carbon source to support growth on organic substrates. Chemosynthetic mixotrophs are organisms that use CO_2 and organic carbon compounds as carbon source and organic and inorganic substrates as energy source. These classifications based on the energy and carbon sources and on electron acceptor utilization are reviewed in **Table 1**.

Redox processes associated with chemosynthesis are important drivers of many biogeochemical cycles and have profound roles in the production and cycling of greenhouse gases. In addition, chemosynthetic processes have to be considered in studies regarding biochemical or microbial diversity and the origin of life on earth. The idea that the earliest organism on earth was a heterotroph is no longer universally accepted. G. Wächterhäuser suggested that the earliest self-sustaining metabolism was based on chemosynthesis, an idea that has growing support. This hypothesis is supported by the fact that many of the deepest branching lineages of the phylogenetic tree contain autotrophic organisms.

This chapter primarily focuses on autotrophic chemosynthesis (i.e., inorganic carbon fixation) and aims to present chemosynthesis from an inland water perspective. In aquatic sciences, the term 'dark carbon fixation' is widely used and applies primarily to chemosynthetic reactions where organic matter production occurs without sunlight. Consideration of all existing processes associated with chemosynthesis, their biochemistry, and the organisms involved is outside the scope of this chapter and can be found in more specialized literature cited below.

Carbon Fixation Pathways

The most studied CO_2 fixation pathways for chemosynthesis are the Calvin–Benson–Bassham cycle and the acetyl-CoA pathway. In eukaryotes, autotrophic CO_2 fixation only occurs via the Calvin cycle. The Calvin cycle is found in many chemosynthetic microorganisms, and the first step of CO_2 fixation is the Rubisco reaction combining CO_2 with ribulose-1,5-bisphosphate, a 5-carbon compound, to form a 6-carbon compound, which is then cleaved into two molecules of glycerate-3-phosphate to be used in cellular metabolism. The acetyl-CoA pathway leads to the fixation of CO_2 into acetate and is the main pathway of acetogenic microorganisms. Methylotrophs can metabolize C-1 compounds as energy and carbon sources by three major assimilation pathways (**Figure 1**): (1) the Calvin cycle, (2) the ribulose monophosphate (RuMB) pathway, and (3) the serine pathway. For all aerobic methylotrophs, formaldehyde is a central intermediate that is partly oxidized to CO_2 and is partly assimilated into cell carbon via the serine or RuMB pathways. Other methylotrophs are able to oxidize C-1 compounds to CO_2 with assimilation via the Calvin cycle. For chemosynthetic sulfate reducers, the synthesis of acetyl-CoA from CO_2 was demonstrated to occur via the citric-acid cycle or the CO-dehydrogenase pathway.

Table 1 Classification of microbial metabolism on the basis of energy source, the type of electron donor and the type of carbon source[a]

Energy source		
Chemical	Chemo-	
Light	Photo-	
Electron donor		
Inorganic	-litho-	
Organic	-organo-	
Carbon source		
CO_2		-auto-
Organic		-hetero-
Both		-mixo-
		-troph

[a]Reproduced from Overman, J. (2006). Principles of enrichment, isolation, cultivation and preservation of prokaryotes. In: Dworkin M, Falkow S, Rosenberg E, Schleifer K.-H, and Stackebrandt E (eds.) *The Prokaryotes*, 3rd ed., pp 80–136. New York: Springer.

Energetics of Chemosynthesis

Redox reactions involving the transfer of electrons from reduced compounds (*electron donors*) to more oxidized molecules (*electron acceptors*) are important in the energy metabolism of all organisms. Electron transfer in redox reactions is used both for ATP production and for acquiring reducing power for biosynthesis (e.g., production of NADPH). In this text, both mechanisms are included when considering the energy yield of the reaction. A limited number of inorganic redox reactions are used for chemosynthesis. Examples of such reactions, along with their thermodynamic energy yield ($\Delta G_0'$), are presented in **Table 2**. In general, chemosynthetic microorganisms grow very slowly and their growth rates may be limited by the energy yield of the associated redox reactions. However, energy limitation is not always the case. For example, the doubling time of anaerobic ammonium oxidizers (ANAMMOX) in the laboratory varies from 11 days to 3 weeks, a much higher value than that for aerobic ammonium oxidizers, although the biomass yield and Gibbs free-energy change of both groups are approximately the same. This means that the low growth rate of ANAMMOX bacteria is not caused by inefficient energy conservation but by other constraints such as a low substrate conversion rate.

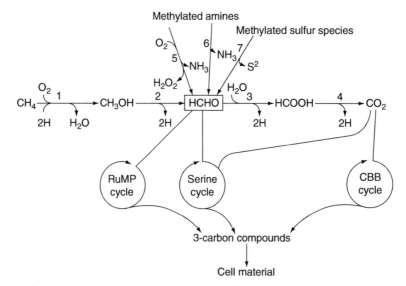

Figure 1 Metabolism of one-carbon compounds in aerobic methylotrophic microorganisms. 1, methane monooxygenase; 2, methanol dehydrogenase; 3, formaldehyde oxidation system; 4, formate dehydrogenase; 5, halomethane oxidation system; 6, methylated amine oxidases; 7, methylated amine dehydrogenase, 8, methylated sulfur dehydrogenase or oxidase; RuMB, ribulose monophosphate; CBB, Calvin–Benson–Bassham cycle. Reproduced from Lindstrom ME (2006) Aerobic methylotrophic prokaryotes. In: Dworkin M, Falkow S, Rosenberg E, Schleifer K-H, and Stackebrandt E (eds.) *The Prokaryotes, vol. 2: Ecophysiological and Biochemical Aspects*, 3rd edn., pp. 618–634. New York: Springer. For more information about RuMP, serine, CBB, and other cabon fixation pathways, refer to Dworkin M, Falkow S, Rosenberg E, Schleifer K-H, and Stackebrandt E (eds.) (2006) *The Prokaryotes, vol. 1: Symbiotic Associations, Biotechnology, Applied Microbiology*, 3rd ed. New York: Springer. Some of the cycles such as CBB are also described in most general microbiology textbooks.

Table 2 Examples of the principal redox reactions associated with chemosynthesis and their Gibbs free-energy yields ($\Delta G^{o\prime}$). The more negative $\Delta G^{o\prime}$, the greater the energy yield

Process	Substrates[a]		Products	$\Delta G^{o\prime}$ (kJ mol substrate^{-1})[b]
Aerobic				
Methane oxidation[c]	$CH_4 + 2\,O_2$	$\rightarrow$	$CO_2 + 2\,H_2O$	−818
Hydrogen oxidation	$H_2 + 0.5\,O_2$	$\rightarrow$	H_2O	−237
Ammonium oxidation	$NH_4^+ + 1.5\,O_2$	$\rightarrow$	$NO_2^- + 2\,H^+ + H_2O$	−275
Nitrite oxidation	$NO_2^- + 0.5\,O_2$	$\rightarrow$	NO_3^-	−76
Sulfur oxidation	$H_2S + 2\,O_2$	$\rightarrow$	$SO_4^{2-} + 2\,H^+$	−796
	$HS^- + H^+ + 0.5\,O_2$	$\rightarrow$	$S^0 + H_2O$	−209
	$S^0 + H_2O + 1.5\,O_2$	$\rightarrow$	$SO_4^{2-} + 2\,H^+$	−587
Thiosulfate oxidation	$S_2O_3^{2-} + H_2O + 2\,O_2$	$\rightarrow$	$2SO_4^{2-} + 2\,H^+$	−823
Iron oxidation[d]	$Fe^{2+} + H^+ + 0.25\,O_2$	$\rightarrow$	$Fe^{3+} + 0.5\,H_2O$	−31
	$FeS_2 + 3.5\,H_2O + 3.75\,O_2$	$\rightarrow$	$Fe(OH)_3 + 4\,H^+ + 0.25\,H_2O + 0.5\,SO_4^{2-}$	−164
Anaerobic				
Sulphate reduction	$4\,H_2 + SO_4^{2-} + H^+$	$\rightarrow$	$HS^- + 4\,H_2O$	−152
Sulfite reduction	$3\,H_2 + SO_3^{2-} + 2\,H^+$	$\rightarrow$	$H_2S + 3\,H_2O$	−173
Sulfur reduction	$H_2 + S^0$	$\rightarrow$	H_2S	−34
Thiosulfate oxidation	$4\,H_2 + H_2S_2O_3$	$\rightarrow$	$SO_4^{2-} + HS^- + H^+$	−22
Hydrogenotrophic methanogenesis	$4\,H_2 + CO_2$	$\rightarrow$	$CH_4 + 2\,H_2O$	−131
Methanogenesis	$H_2 + CH_3OH$	$\rightarrow$	$CH_4 + H_2O$	−113
	$4\,CO + 2\,H_2O$	$\rightarrow$	$CH_4 + 3\,CO_2$	−210
Methane oxidation[c]	$CH_4 + SO_4^{2-} + H^-$	$\rightarrow$	$CO_2 + HS^- + H_2O$	−21
	$CH_4 + 1.6\,NO_3^- + 1.6\,H^+$	$\rightarrow$	$CO_2 + 0.4\,N_2 + 2.8\,H_2O$	−766
	$CH_4 + 2.7\,NO_2^- + 2.7\,H^+$	$\rightarrow$	$CO_2 + 0.7\,N_2 + 3.3\,H_2O$	−928
Acetogenesis	$4\,H_2 + 2\,CO_2$	$\rightarrow$	$CH_3COOH + 2\,H_2O$	−95
	$CH_3OH + 2\,CO_2$	$\rightarrow$	$CH_3COOH + 2\,H_2O$	−93
	$4\,H_2 + 2\,HCO_3^- + H^+$	$\rightarrow$	$CH_3COO^- + 4\,H_2O$	−105
Ammonium oxidation (ANAMMOX)	$NH_4^+ + NO_2^-$	$\rightarrow$	$N_2 + 2\,H_2O$	−238
Iron reduction	$H_2 + 2\,FeOOH + 4\,H^+$	$\rightarrow$	$2\,Fe^{2+} + 4\,H_2O$	−110
Manganese reduction	$3\,S^0 + 4\,H_2O + MnO_2$	$\rightarrow$	$2\,H_2S + SO_4^{2-} + 2\,H^+ + Mn^{2+}$	−297

[a] The first substance is the electron donor and the last the electron acceptor.

[b] Standard Gibbs free-energy ($\Delta G^{o\prime}$) in kJ mol substrate^{-1}, at pH 7 and 25 °C, calculated after Thauer RK, Jungermann J, and Decker H (1977). Energy conservation in anaerobic chemotrophic bacteria. *Bacteriological Reviews* 41: 100–180.

[c] The main carbon sources for all processes is CO_2. The main carbon sources for methane oxidizers are CH_4 or C-1 compounds.

[d] Calculations for Fe^{2+} at pH 2.

Chemosynthetic Organisms and Associated Redox Processes

Chemosynthetic organisms are highly diverse in terms of phylogeny, substrates used, morphology, habitats, and metabolism. They can be strictly aerobic, strictly anaerobic, or facultative anaerobes. Similarly, some are strictly autotrophic, while others can perform heterotrophic or mixotrophic metabolism depending on the environmental conditions and substrate availability. Major known groups of chemoautotrophs and associated key processes are presented briefly below. It should be remembered that much of the information referred to is based on studies of specific strains grown in the laboratory, and many aspects of the organisms and their metabolism *in situ* are still unknown.

Ammonium and Nitrite Oxidizers

Nitrification is an aerobic two-step chemosynthetic process in which ammonium is oxidized to nitrite, which is then oxidized to nitrate (**Table 2**). Nitrifiers fix CO_2 via the Calvin cycle, but to a minor extent, and also via the phosphoenolpyruvate carboxylase pathway. Some nitrifiers use the reverse tricarboxylic acid cycle (TCA cycle), where CO_2 is eventually fixed into acetyl-CoA and pyruvate. By convention, the ammonium oxidizers are characterized by the prefix nitroso- and the nitrite oxidizers by the prefix nitro- (i.e., yielding strain names like nitrosobacter and nitrobacter, respectively). The two groups of Gram-negative bacteria are not phylogenetically related. Some heterotrophic bacteria, fungi, and algae can also oxidize ammonia to nitrate in a process known as heterotrophic nitrification. This process depends on the oxidation of organic compounds and, in opposition to chemosynthetic nitrification, is not related to energy generation. The relative importance of heterotrophic nitrification in inland waters has not been clarified yet.

Ammonium and nitrite oxidizers are slow-growing bacteria with doubling times that can vary from

7 to 10 h in the laboratory and up to weeks in natural environments depending on conditions. Aerobic ammonium oxidation is regulated by oxygen and substrate availability, temperature, and pH. Under low oxygen tension, aerobic ammonium-oxidizing bacteria can act mixotrophically using carbon dioxide and organic compounds as carbon sources and can also reduce nitrite (NO_2^-) to nitric (NO) and nitrous oxide (N_2O). Ammonium oxidizers develop very well at low ammonium concentrations and they can survive ammonium starvation for up to a year. Under anoxic conditions, some ammonium and nitrite oxidizers can also perform denitrification.

A previously unknown redox process yielding energy for chemosynthesis, anaerobic ammonium oxidation (ANAMMOX), has been described during the last decade. This process is performed by the members of the Planctomycetales order, which oxidize ammonium with nitrite as electron acceptor (**Table 2**). Bacteria that perform this process are very slow-growing organisms using different metabolic pathways compared with aerobic ammonium oxidizers. ANAMMOX is inhibited by low oxygen (<1 μM) and high nitrite (>10 mM) concentrations.

Sulfur Reducers

Sulfur reducers reduce oxidized sulfur compounds (e.g., sulfate) in association with a respiratory type of energy conservation (**Table 2**). Sulfate reduction usually occurs via heterotrophic anaerobic respiration, i.e., oxidation of organic matter. However, sulfate-reducing bacteria are metabolically very versatile and many can use H_2 as sole electron donor to synthesize cell material from both acetate and CO_2 (mixotrophs) or only from CO_2 (chemolithoautotrophs). Some sulfate-reducing bacteria can use nitrate and nitrite as electron acceptors, even in the presence of sulfate. The end product of this reaction is ammonium and not N_2 as it is for denitrifiers. In the absence of sulfate or other inorganic electron acceptors, some sulfate reducers can grow by fermentation of several organic substrates.

Sulfur Oxidizers

Reduced sulfur compounds such as sulfide (H_2S), thiosulfate ($S_2O_3^{2-}$), tetrathionate ($S_4O_6^{2-}$), and elemental sulfur (S^0) are used as the electron donors by sulfur oxidizers (**Table 2**). Such organisms are capable of performing extensive chemosynthesis in inland waters having high concentrations of, e.g., hydrogen sulfide, sulfite (HS^-), and sulfur (**Table 3**). Reduced sulfur compounds are usually formed as a product of anaerobic heterotrophic respiration with sulfate, but some waters receive large inputs of sulfide via ground water. Oxygen is the most common electron acceptor, but other potential electron acceptors such as nitrate have been proposed. Sulfur reducers and oxidizers can be active under extreme pH, temperature, and saline conditions.

Iron and Manganese Oxidizers

In aquatic environments, iron and manganese are found both in oxidized (Fe^{3+} and Mn^{4+}) and reduced (Fe^{2+} and Mn^{2+}) forms depending on pH and O_2 concentrations (**Table 2**). Dissimilatory Fe^{3+} and Mn^{4+} reductions are processes by which microorganisms transfer electrons from the oxidized to the reduced form of the metal. Most Fe^{3+} reducers seem able to reduce Mn^{4+} as well. Fe^{2+} and Mn^{2+} are more soluble than Fe^{3+}. The oxidized forms (Mn^{4+} and Fe^{3+}) are highly insoluble at neutral pH, but highly soluble at low pH. Recent studies indicate that sediment microorganisms may be able to access particle-associated Fe^{3+} in spite of its low solubility.

Methanogens

Methane-producing microorganisms belong to the Archaea domain and are called methanogens and more recently methanoarchaea. Methanogens can be found in a variety of environments across extreme pH, temperature, and salinity conditions. Methanogens are limited in their capacity to use different substrates, with acetate and CO_2/H_2 being the most common. The use of acetate is a heterotrophic process, whereas hydrogenotrophic methanogens rely on H_2 as the electron donor and CO_2 as the carbon source and hence carry out a chemosynthetic process (**Table 2**). About half of the described species of methanogens are hydrogenotrophic. Methanogens are abundant in habitats with very low availability of oxygen, sulfate, nitrate, Fe^{3+}, and Mn^{4+}. In the presence of these electron acceptors, methanogens are thermodynamically out-competed (**Table 2**) for reduced substrates (e.g., H_2 and acetate). Hydrogenotrophic methanogenesis contribution to total methane production varies from 0 to 100% in lake sediments, and temperature is one of its main regulating factors, as higher temperatures usually favor hydrogenotrophic methanogenesis.

Methylotrophs and Methanotrophs

Methylotrophs are chemosynthetic microorganisms that can use C-1 organic molecules as their sole source of carbon and energy. Methylotrophs are phylogenetically diverse and they obligately require an

Table 3 Examples of in situ chemosynthesis rates in various aquatic systems

System	Primary electron donor[a]	Habitat[b]	Max $S_2^-/NH_4^+/CH_4$ $(\mu M)^c$	Volumetric rate $(mmol\ m^{-3}\ d^{-1})$	Areal rate $(mmol\ m^{-2}\ d^{-1})$	% of total C fixation	Area integration[a,d]	Notes[e]	Source
L Kinneret, Israel, Monomictic lake	S^{2-}	w	382/79/nd	0.8–80	26–44	8–92	ns	Chemos 20–25 % of C fixation annually	8
Solar Lake, Sinai, Saline lake, seepage from sea into lake	S^{2-}	w	1000/nd/nd	0–35		Up to 33	WL		10
L Belovod, Russia, Meromictic lake	S^{2-}	w	4471/nd/nd	0–2.3					17
L Gek-Gel, Russia, Meromictic lake	S^{2-}	w	79/111/nd	0–1.6	18	99	ns		17
L Maral-Gel, Russia, Meromictic lake	S^{2-}	w	7/nd/nd	0–4.2					17
Ilersjön, Sweden, Eutrophic lake	CH_4	w	nd/nd/138	0–1.0	0.6–1.4	1–6	WL	MBP corresponded to 4–120% of HBP	2
Mårn, Sweden, Eutrofic, humic lake	CH_4	w	nd/nd/56	0–1.0	0.1–0.6	2–7	WL	MBP corresponded to 2–39% of HBP	2
Lillsjön, Sweden, Humic lake	CH_4	w	nd/nd/68	0–0.3	0.01	0.3–5	WL	MBP corresponded to 0.5–14% of HBP	2
Cadagno, Switzerland, Meromictic lake	ns	w	0.3/60/nd	20–188	260	ca. 50	WL		3
Big Soda Lake, USA, Saline lake	ns	w	12 059/2500/nd		2.5–57	1–72	1 m²	Chemos 30% of C fixation annually	5
Lake Cisó, Spain, Monomictic lake, gypsum rich drainage area	ns	w	5000/nd/nd		5.8–141	11–75	1 m²		7
L Mekkojärvi, Finland, Humic, acidic lake	ns	w	nd/29/nd	0–7.5		>50 during stratification	ns		12

Continued

Table 3 Continued

System	Primary electron donor[a]	Habitat[b]	Max S_2^-/NH_4^+/CH_4 (μM)[c]	Volumetric rate (mmol m⁻³ d⁻¹)	Areal rate (mmol m⁻² d⁻¹)	% of total C fixation	Area integration[a,d]	Notes[e]	Source
Lake Baikal, Russia, Large, deep, permanently stratified	ns	sed	nd/20/nd		0.03–0.1	<2	1 m²		13
Rybinsk reservoir, Russia	ns	w		0.03–1.8				Chemos up to 50% of microbial production in bottom layers and 1–2% in nonstratified water column	16
Kuibishev reervoir, Russia	ns	w	132/nd/nd	0–2.3		5	1 m²		16
Scheldt estuary, Netherlands/Belgium, Tidal estuary	NH_4^+	w	nd/150/nd	0.01–3.6				Chemos limited by O_2 and CH_4 or H_2 concentrations Higher rates in freshwater part	1
Rhône river estuary, France, Nontidal estuary	NH_4^+	w	nd/10/nd	0.02–0.15				Higher rates in freshwater part	6
Ebro River salt wedge, Spain, Estuary	ns	w	20/74/nd	0.5–10	2.4–5	37–61	ox/an		4
Saanich inlet (fjord), Canada, Stratified fjord	ns	w	15/nd/nd	0–2					11
Aburatsubo inlet, Japan, Coastal marine	ns	w		0–8.3		2–90	Individual, depths compared		15
Aburatsubo inlet, Japan, Coastal marine	ns	s			7–250	5–80	1 m²		15
Black sea, Marine	S^{2-}	w	80/8/nd	0–0.32	2–5.3	Up to 15	ox/an		9
Cariaco Trench, Venezuela, Marine basin	ns	w	30/nd/nd	0.02–0.26	7–18	17–38	1 m²		19

| Cariaco Trench, Venezuela, Marine basin | S^{2-}/ NH_4^+/ Mn^{2+}/ Fe^{2+} | w | 76/0.2/nd | 0–2.52 | 26–157 | 10–333 | 1 m² | Chemos on average 66% of C fixation | 18 |
| Surface ocean water | ns | w | | 0.2–86 | | | Individual, samples compared | Chemos <10 % when total C fixation >1 µg C L^{-1} h^{-1} | 14 |

[a] ns denote not specified.

[b] w indicates water column and sed denotes sediment.

[c] nd denote no data.

[d] WL denotes that area integration was made on a whole lake basis taking the volumes of different water layer into account. 1 m² denotes that a 1 m² water column was used, and ox/an indicates that depth integration was made over the oxic/anoxic interface only.

[e] Chemos, MBP, and HBP denote chemosynthesis, methanotrophic bacterial production, and heterotrophic bacterial production, respectively.

Sources

1. Andersson MGI, Brion N, and Middelburg JJ (2006) Comparison of nitrifier activity versus growth in the Scheldt estuary – a turbid, tidal estuary in northern Europe. Aquatic Microbial Ecology 42: 149–158

2. Bastviken D, Ejlertsson J, Sundh I, and Tranvik L (2003) Methane as a source of carbon and energy for lake pelagic food webs. Ecology 84: 969–981.

3. Camacho A, Erez J, Chicote A, et al. (2001) Microbial microstratification, inorganic carbon photoassimilation and dark carbon fixation at the chemocline of the meromictic Lake Cadagno (Switzerland) and its relevance to the food web. Aquatic Sciences 63: 91–106.

4. Casamayor EO, Garcia-Cantizano J, Mas J, and Pedros-Alio C (2001) Primary production in estuarine oxic/anoxic interfaces: contribution of microbial dark CO_2 fixation in the Ebro River Salt Wedge Estuary. Marine Ecology Progress Series 215: 49–56.

5. Cloern JE, Cole BE, and Oremland RS (1983) Autotrophic processes in meromictic Big-Soda Lake, Nevada. Limnology and Oceanography 28: 1049–1061.

6. Feliatra F and Bianchi M (1993) Rates of nitrification and carbon uptake in the Rhone River plume (northwestern Mediterranean Sea Microbial Ecology 26: 21–28.

7. Garcia-Cantizano J, Casamayor EO, Gasol JM, Guerrero R, and Pedros-Alio C (2005) Partitioning of CO2 incorporation among planktonic microbial guilds and estimation of in situ specific growth rates. Microbial Ecology 50: 230–241.

8. Hadas O, Pinkas R, and Erez J (2001) High chemoautotrophic primary production in Lake Kinneret, Israel: a neglected link in the carbon cycle of the lake. Limnology and Oceanography 46: 1968–1976.

9. Jorgensen BB, Fossing H, Wirsen C, and Jannasch H (1991) Sulfide oxidation in the anoxic Black Sea chemocline. Deep-Sea Research 38: (Suppl. 2), S1083–S1103.

10. Jorgensen BB, Kuenen JG, and Cohen Y (1979) Microbial transformations of sulfur-compounds in a Stratified Lake (Solar Lake, Sinai). Limnology and Oceanography 24: 799–822.

11. Juniper SK and Brinkhurst RO (1986) Water-column dark CO_2 fixation and bacterial-mat growth in intermittently anoxic Saanich Inlet, British Columbia. Marine Ecology Progress Series 33: 41–50.

12. Kuuppo-Leinikki P and Salonen K (1992) Bacterioplankton in a small polyhumic lake with an anoxic hypolimnion. Hydrobiologia 229: 159–168.

13. Maerki M, Muller B, and Wehrli B (2006) Microscale mineralization pathways in surface sediments: a chemical sensor study in Lake Baikal. Limnology and Oceanography 51: 1342–1354.

14. Prakash A, Sheldon R, and Sutcliffe WHJ (1991) Oceanic variation of ^{14}C dark uptake. Limnology and Oceanography 36: 30–39.

15. Seki H (1968) Relation between production and mineralization of organic matter in Aburatsubo inlet Japan. Journal of the Fisheries Research Board of Canada 25: 625–687.

16. Sorokin JI (1964) On the trophic role of chemosynthesis in water bodies. Internationale Revue der gesamten Hydrobiologie 49: 307–324.

17. Sorokin YI (1970) Interrelations between sulphur and carbon turnover in meromictic lakes. Archive fur Hydrobiologie 66: 391–446.

18. Taylor GT, Iabichella M, Ho TY, et al. (2001) Chemoautotrophy in the redox transition zone of the Cariaco Basin: a significant midwater source of organic carbon production. Limnology and Oceanography 46: 148–163.

19. Tuttle JH and Jannasch HW (1979) Microbial dark assimilation of CO_2 in the Cariaco trench. Limnology and Oceanography 24: 746–753.

available methylated compound (methylated amines, halogenated methanes, and methylated sulfur species). Many methylotrophs are capable of N_2 fixation and can use N_2 as a nitrogen source. Those methylotrophs capable of growing on methane are distinguished as methanotrophs. Methane is the main electron donor for methanotrophs, but other C-1 compounds like methanol, methylamine or carbon monoxide, and formate can also be used.

Methane oxidizers have been divided into two groups (Type I and Type II) depending on their internal cell structure and carbon assimilation pathway. Type I assimilate C-1 carbon via the RuMB cycle and are α-Proteobacteria, while Type II assimilate C-1 compounds via the serine pathway and are γ-Proteobacteria. Both types use the enzyme methane monooxygenase (MMO) to oxidize methane to methanol. Methanotrophs are found in most environments with both methane and O_2. Methanotrophs generally seem to oxidize ammonium to nitrite, while some ammonium oxidizers oxidize methane and incorporate some methane as cell material; however, methanotrophs and ammonium oxidizers are taxonomically different.

Anaerobic methane oxidation can occur with sulfate or nitrate as electron acceptors (**Table 2**). Anaerobic methane oxidation has primarily been studied in marine systems, but probably also occurs in inland water sediments with high nitrate or sulfate concentrations. This process appears to be mediated by microbial consortia consisting of sulfate or nitrate reducers and methanogens. Isotopic studies suggested that methane is not directly used by sulfate- or nitrate-reducing bacteria, but by the methanogens.

Hydrogen Oxidizers

Molecular hydrogen (H_2) is a powerful electron donor in cellular metabolism in both prokaryotes and eukaryotes. The capability of using H_2 is spread across many taxonomic groups of different and independent phylogenetic origin. Hydrogen oxidation occurs in restricted habitats and is catalyzed by hydrogenase enzymes. Aerobic hydrogen oxidizers, sulfate reducers, methanogens, acetogens, and cyanobacteria are some groups that metabolize H_2. The main biogenic source of H_2 is anaerobic metabolism and the fermentation of organic matter in sediments. However, the H_2 turnover time in sediments is estimated to be on the order of 2 min and therefore almost all H_2 that is produced in the sediment is rapidly consumed. H_2 concentrations in lake sediments vary from 1 to 150 nM. The main sedimentary H_2 consumers that have been studied are methanogens and sulfate-reducing bacteria, and both groups

compete with each other for H_2. Sulfate reducers outcompete the methanogens in the presence of sulfate because they have a higher affinity for H_2 and there is a higher thermodynamic yield for the reaction and thus a higher growth yield (**Table 2**). The occurrence of one or the other process seemingly depends on sulfate availability, which is often low in freshwater sediments. The importance of, and competition for, H_2 in anoxic microbial metabolism is illustrated by the development of consortia of fermenters and H_2 oxidizers. Such consortia facilitate 'interspecies hydrogen transfer,' a process whereby H_2 that is produced by the fermenter is rapidly transported to the H_2 consumer. This favors the H_2 consumer and also keeps H_2 concentrations low, which increases the energy yield of the fermentation.

Acetogens

Acetogens are organisms that use the acetyl-CoA pathway for the synthesis of acetate as cell carbon from CO_2. Homoacetogen is a more specific term and is often used to describe organisms that only produce acetate. However, most acetogens that use the acetyl-CoA pathway often form products other than acetate, such as butyrate, succinate, and ethanol. Acetogens are anaerobes but they can tolerate small amounts of O_2. Acetogens can grow chemoautotrophically on H_2, and also heterotrophically depending on environmental conditions (**Table 2**). Because CO_2 is the most common terminal electron acceptor for acetogens, CO_2 availability may regulate their growth and determine what substrate will be metabolized. However, other electron acceptors can also be used, allowing these microbes to adapt to a wide range of redox conditions and optimize their growth efficiency. Their competitive ability for H_2 increases when alternate electron acceptors (e.g., nitrate) are used.

All acetogens are members of the bacteria domain. Different species of sulfate-reducing bacteria have some acetogenic capabilities. Although theoretical considerations suggest that acetogens are less competitive than methanogens and sulfate-reducing bacteria, they occur in diverse habitats and the capacity of acetogens to utilize various substrates simultaneously might contribute to their competitiveness in lake sediments.

Spatial and Temporal Distribution of Chemosynthesis in Aquatic Environments

Chemosynthesis depends on the presence of both reduced and oxidized compounds to be used as electron donors and acceptors, respectively. At oxic–anoxic interfaces, the simultaneous access to, e.g., S^{2-}, NH_4^+ or CH_4, and O_2 can support chemosynthesis. However, electron acceptors other than O_2, such as NO_3^-, SO_4^{2-}, and Fe^{3+}, can also be used and thereby other interfaces (e.g., the sulfide–nitrate interface) may also be important (**Figures 2** and **3**). Therefore, in general terms, chemosynthesis is expected to be most extensive at sites with steep redox gradients. Such redox interfaces are common in sediments and stratified water columns and can also exist over small scales within microbial mats and biofilms (**Figure 2**). Sites with seepage of reduced compounds of geochemical origin can also be important for chemosynthesis. Examples include hot springs, lakes fed by sulfide-rich ground water, and marine vent environments.

Direct studies of in situ chemosynthesis in inland waters (usually measured as dark fixation of $H^{14}CO_3$) have so far primarily addressed oxic–anoxic interfaces in sulfide-rich water columns (**Table 3**). These studies confirm a highly stratified distribution of the chemosynthesis with highest rates of radiocarbon incorporation at the oxic–anoxic interface. Some cases also

suggest considerable chemosynthetic activity in the anoxic water below the interface (**Figures 3** and **4**). The most likely explanation for anoxic chemosynthesis is that electron acceptors other than O_2 are being used deeper in the anaerobic water column. The stratified location of the associated redox processes implies that chemosynthesis is generally most extensive in narrow zones at redox interfaces under field conditions (**Figure 4**).

Because of the limited number of systems in which chemosynthetic carbon fixation has been measured, the spatial distribution of chemosynthesis among different types of inland waters is poorly known at present. Very high rates of chemosynthesis have been measured in saline sulfide-rich water columns (**Table 3**), but comparable methods to estimate chemosynthesis have rarely been reported so far from sulfide-poor water columns, where chemosynthesis associated with other redox reactions could dominate. Increasing ammonium oxidation-associated chemosynthesis with decreasing salinity along the Scheldt estuary (see **Table 3**) indicates that sulfide

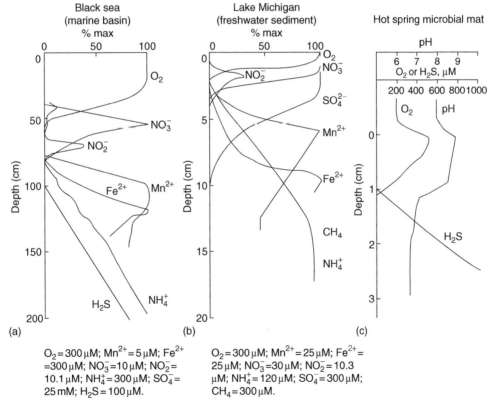

$O_2 = 300\,\mu M$; $Mn^{2+} = 5\,\mu M$; $Fe^{2+} = 300\,\mu M$; $NO_3^- = 10\,\mu M$; $NO_2^- = 10.1\,\mu M$; $NH_4^+ = 300\,\mu M$; $SO_4^- = 25\,mM$; $H_2S = 100\,\mu M$.

$O_2 = 300\,\mu M$; $Mn^{2+} = 25\,\mu M$; $Fe^{2+} = 25\,\mu M$; $NO_3^- = 30\,\mu M$; $NO_2^- = 10.3\,\mu M$; $NH_4^+ = 120\,\mu M$; $SO_4^- = 300\,\mu M$; $CH_4 = 300\,\mu M$.

Figure 2 Distribution of oxidized and reduced compounds related with chemosynthesis at (a) the water column from the Black Sea, (b) the sediment from Lake Michigan, and (c) a hot spring microbial mat. The scales vary from hundreds of meters in the water column to centimeters and millimeters at the sediment and microbial mat, respectively. Note the different scales at the y-axis. Reproduced from Stahl DA, Hullar M, and Davidson S (2006) Prokaryotes and their habitats. In: Dworkin M, Falkow S, Rosenberg E, Schleifer K-H, and Stackebrandt E (eds.) *The Prokaryotes, vol. 1: Symbiotic Associations, Biotechnology, Applied Microbiology*, 3rd ed., pp. 299–327. New York: Springer.

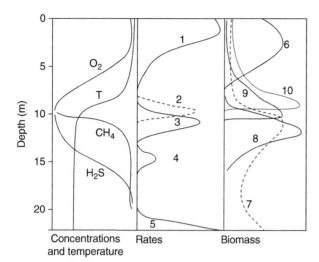

Figure 3 Idealized vertical profiles in a temperate lake during summer stratification, according to Schlegel HG and Jannasch HW (2006) Prokaryotes and their habitats. In: Dworkin M, Falkow S, Rosenberg E, Schleifer K-H, and Stackebrandt E (eds.) *The Prokaryotes, vol. 1: Symbiotic Associations, Biotechnology, Applied Microbiology*, 3rd ed., pp. 137–184. New York: Springer. Note that this illustration is conceptual only and that absolute concentrations, rates, or biomasses are not given. Profiles: 1, CO_2 fixation in the light (oxygenic photosynthesis); 2, dark CO_2 fixation; 3, CO_2 fixation in the light (anoxygenic photosynthesis); 4 and 5, sulfate reduction; 6, biomass of algae and cyanobacteria; 7, total bacterial biomass; 8, biomass of phototrophic bacteria; 9, biomass of protozoa; 10, biomass of Copepoda and Cladocera. Chemosynthesis is directly related with profiles 2, 4, 5, 6, and 7 and indirectly with profiles 9 and 10 as microbial biomass formed via chemosynthesis is consumed by predators.

and ammonium may be important electron donors for chemosynthesis in different environments. It has also been proposed that sulfide oxidizers out-compete ammonium oxidizers in sulfide-rich environments. Hence, substantial chemosynthesis associated with electron donors other than sufide is possible.

Intensive chemosynthesis can occur inside macro-fauna. This is well known from marine hydrothermal vents, where endosymbiotic sulfide reducers perform chemosynthesis supporting hosting animals such as giant tubeworms (e.g., *Riftia pachyptila* and *Lamellibrachia* cf. *luymesi*) with organic material. Likewise, some bivalves (e.g., *Calyptogena magnifica* and *Bathymodiolus thermophilus*) at marine cold seeps rely on endosymbiotic methane oxidizers for access to organic material. At least one freshwater mussel of the genera *Solenaia* sp. has been suggested to rely on chemosynthesis associated with oxidation of reduced sulfur compounds. This mussel burrows into muddy anoxic sediments and has a sulfur yellow foot suggested to function in sulfur storage. By venting down O_2 into environments rich in reduced sulfur

compounds, the mussel can greatly enhance the transport across the redox gradient, and thereby also chemosynthesis rates.

The temporal distribution of chemosynthesis is related to the development of redox gradients. Water column studies indicate that chemosynthesis is most extensive during stratification. In sediments overlain with oxic water, redox gradients are likely to be permanently present, although the oxygen penetration and consequently the depth of the redox interfaces may vary over time.

Implications of Chemosynthesis in Aquatic Ecosystems

Chemosynthesis and Carbon Cycling

The highly stratified distribution of free-living chemosynthetic microorganisms implies that intensive rates of chemosynthesis can be spatially restricted. In some systems, chemosynthetic processes appear to account for a major share of the total carbon fixation and may exceed 90% of the total carbon fixation under certain conditions (**Table 3**). The highest rates have been reported for saline sulfide-rich environments, such as mono- or meromictic lakes with haloclines. Lakes fed by sulfide-rich ground water also have high rates of chemosynthesis. Saline inland waters with haloclines may dominate globally in terms of water volume if the Caspian Sea, Lake Baikal, the African rift lakes, and large estuaries such as the Baltic are included in the accounting. From a global perspective, inland waters without haloclines and with moderate-to-low concentrations of sulfide are more abundant in number. The limited data from such nonsaline inland waters make it unclear to what extent chemosynthesis contributes to the total CO_2 fixation.

The contribution of chemosynthesis to overall carbon fixation in aquatic ecosystems is limited by the availability of electron donors (e.g., sulfide, ammonium, etc). In most inland waters, the electron donors that are the energy source for chemosynthesis typically originate from the degradation of organic matter. The organic matter was typically produced by photosynthesis, either in the aquatic system itself or in the terrestrial surroundings from which it was transported to the aquatic environment. Hence, from an energetic point of view, most chemosynthesis in inland waters is indirectly driven by sunlight energy, given the dependence of chemosynthesis on photosynthesis supplying of organic material (**Figure 5**). An exception would be systems with extensive supply of electron donors from hydrothermal or geochemical

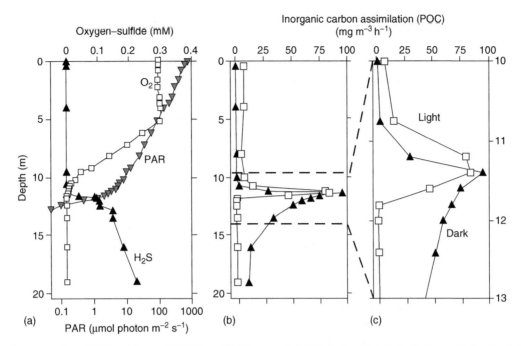

Figure 4 Concentrations of H_2S and O_2, and PAR (a), and light versus dark CO_2 fixation (b) in Lake Cadagno, Switzerland, September 13, 1999, at noon. (c) represent detailed rates from (b) between 10 and 13 m. Reproduced from Camacho A, Erez J, Chicote A, Florín M, Squires MM, Lehmann C, and Bachofen R (2001) Microbial microstratification, inorganic carbon photoassimilation and dark carbon fixation at the chemocline of the meromictic Lake Cadagno (Switzerland) and its relevance to the food web. *Aquatic Sciences* 63: 91–106.

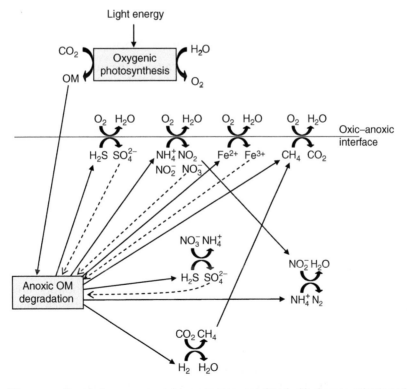

Figure 5 Illustration of the connections between some redox processes associated with chemosynthesis and anoxic organic matter (OM) degradation. Solid arrows indicate contribution of reduced compounds produced by anoxic OM degradation processes. These reduced compounds function as electron donors in the redox processes associated with chemosynthesis. The contribution of OM (functioning as electron donor in anoxic OM degradation) from oxygenic photosynthesis is also indicated. Note that OM can be both produced in the aquatic system (aquatic photosynthesis) and transported from the terrestrial surroundings (i.e., depend on terrestrial photosynthesis). Dashed arrows indicate resupply of oxidized compounds that can function as electron acceptors in anoxic OM degradation.

external sources. Therefore, chemosynthesis should only be able to dominate carbon fixation relative to aquatic photosynthesis in (1) systems receiving substantial input of organic matter from the surroundings or (2) systems with large inputs of reduced compounds directly suitable as electron donors (e.g., waters receiving reduced compounds of geochemical origin, such as hot springs or some saline sulfide-rich lakes).

Chemosynthesis and Greenhouse Gases

Globally, it is now clear that climate changes can be caused by changes in atmospheric content of greenhouse gases including carbon dioxide (CO_2), methane (CH_4), and nitrous oxide (N_2O) in the atmosphere. Chemosynthetic processes are directly related with the natural production and consumption of these gases. Inland waters can be simultaneously both carbon sinks for organic matter accumulating in sediments and major sources of greenhouse gases such as carbon dioxide and methane. Chemosynthetic processes are involved in both production and consumption of methane and nitrous oxide. The relative balance of these biological reactions along with transport regulates emissions from inland water to the atmosphere. While the role of inland waters in nitrous oxide emissions is unclear at the moment, wetlands and lakes together contribute more than 75% of the natural methane emissions to the atmosphere. However, these methane emissions would be much higher without methane oxidation, since aquatic methane oxidation may consume >50% of the produced methane and thereby substantially limit emissions.

Chemosynthesis and Food Webs

Oxic–anoxic interfaces are known to be sites of high biological activity with high abundance and biomass of microorganisms. In fact, several studies have indicated that chemosynthetic organisms such as methane oxidizers account for up to 41% of the total bacterial population at certain depths. Hence, chemosynthetic microorganisms can provide an abundant and concentrated food resource for higher trophic levels. Chemosynthesis can contribute carbon to food webs both through endosymbiosis (see above) and by direct or indirect predation on chemosynthetic microorganisms. Stable isotope studies confirm that carbon being fixed by chemosynthesis enters food webs (**Table 4**). Chemosynthetic biomass is depleted in ^{13}C relative to terrestrial organic carbon in inland waters. Likewise, ^{34}S signatures can be used to trace the contribution of sulfide-dependent chemosynthesis. Although the studies presented in **Table 4** show a contribution of chemosynthesis to aquatic food webs, the magnitude and importance of this contribution is still unknown.

Chemosynthesis and Element Cycling

Redox processes associated with chemosynthesis are crucial for the biogeochemical cycling of many common elements. For example, nitrogen cycling depends upon nitrification (i.e., ammonium oxidation to nitrite). Nitrification appears to be primarily a chemoautotrophic process and chemosynthesis is thereby vital for remineralization of nitrogen and indirectly for denitrification. Another example is anaerobic methane and ammonium oxidation, which have recently been estimated to account for more than 75% and 30–50% of all marine methane oxidation and ammonium oxidation, respectively. The overall importance of these processes in inland water ecosystems is still unknown, but the number of lakes and rivers where these processes have been measured is increasing. There are many other examples where chemosynthetic processes are critical in biogeochemistry, such as the cycling of sulfur, iron, and manganese (**Table 2**). From a general element cycling point of view, chemoautotrophic microorganisms function as catalysts for a large number of redox reactions, greatly increasing reaction rates.

Knowledge Needs

Because of the small number of studies on chemosynthesis in inland waters, several aspects require future attention. One fundamental shortcoming of current research is related to methodological limitations. Typically, chemosynthesis is estimated from dark ^{14}C fixation, which is a method with many advantages such as rapid incubations, high sensitivity, and straight forward interpretation of results. However, the method does not tell what organisms are fixing the carbon, and because heterotrophic microorganisms and plants are capable of limited dark carbon fixation, using this approach for assessing chemosynthesis is problematic. Some studies have used selective inhibitors to allow separation of C fixed by different microbial guilds. Inhibitors, however, are not always effective, but at least such approaches provide some help in partitioning chemoautotrophic from other types of carbon fixation. Further method development would clearly be valuable and perhaps the use of multiple isotopes in conjunction with inhibition studies could provide a way forward.

Another obvious need is to increase the number of studies on chemosynthesis in many common aquatic environments. More measurements in nonsaline waters, sediments, and common microenvironments

Table 4 Examples of studies suggesting contribution of chemosynthesis to different parts of the food web

Food web compartment	System studied	Methods (indicator, and type of study)	Sources
S^{2-} oxidizers			
Protozoans and zooplankton	Lake Kinneret, Israel	^{13}C	8
Snail (*Semisulcospira libertina*)	Headwater stream, Japan	^{13}C, ^{15}N, ^{34}S	4
Chironomids	Eutrophic lakes, UK	^{13}C, ^{34}S	6
Bivalve (*Solenaia oleivore*)	Muddy lake sediments, China	Field and morphological observations	16
CH_4 oxidizers			
Zooplankton	Billabongs, Australia	^{13}C, field observations	2
	Humic lakes	^{13}C, field observations	11
	Eutrophic and humic lakes, Sweden	^{13}C, carbon mass balance, field observation	1
	Polyhumic lake, Finland	^{13}C, carbon mass balance, laboratory and field experiments	12
Chironomids	Large mesotrophic Lake Biwa, Japan	^{13}C, field observations	14
	Eutrophic lakes, Germany	^{13}C, field observations	7
	Forest lakes, Finland	^{13}C, field observations	10
	Large mesotrophic Lake Biwa, Japan	^{13}C, fatty acid compostion, field observations	13
	Eutrophic lake, Germany	^{13}C, microbial community analyses, field observations	5
Chironomids and several other benthic primary consumers	Pantanal, Brazil	^{13}C, field observations	3
Offshore chironomids and oligochetes	Small oligotrophic arctic lakes, Canada	^{13}C, field observations	9
Several macroinvertebrates	Small stream, Japan	^{13}C, ^{15}N	15

Sources

1. Bastviken D, Ejlertsson J, Sundh I, and Tranvik L (2003) Methane as a source of carbon and energy for lake pelagic food webs. *Ecology* 84: 969–981.
2. Bunn SE and Boon PI (1993) What sources of organic carbon drive food webs in billabongs? A study based on stable isotope analysis. *Oecologia* 96: 85–94.
3. Calheiros DF (2003) *Influência do pulso de inundação na composição isotópica $\delta^{13}C$ e $\delta^{15}N$ das fontes primárias de energia na planície de inundação do rio Paraguai (Pantanal-MS)*. Piracicaba-SP: Universidade de São Paulo.
4. Doi H, Takagi A, Mizota C, et al. (2006) Contribution of chemoautotrophic production to freshwater macroinvertebrates in a headwater stream using multiple stable isotopes. *International Review of Hydrobiology* 91: 501–508.
5. Eller G, Deines P Grey J, Richnow HH, and Kruger M (2005) Methane cycling in lake sediments and its influence on chironomid larval partial derivative C-13. *Fems Microbiology Ecology* 54: 339–350.
6. Grey J and Deines P (2005) Differential assimilation of methanotrophic and chemoautotrophic bacteria by lake chironomid larvae. *Aquatic Microbial Ecology* 40: 61–66.
7. Grey J, Kelly A, Ward S, Sommerwerk N, and Jones RI (2004) Seasonal changes in the stable isotope values of lake-dwelling chironomid larvae in relation to feeding and life cycle variability. *Freshwater Biology* 49: 681–689.
8. Hadas O, Pinkas R, and Erez J (2001) High chemoautotrophic primary production in Lake Kinneret, Israel: a neglected link in the carbon cycle of the lake. *Limnology and Oceanography* 46: 1968–1976.
9. Hershey AE, Beaty S, Fortino K, et al. (2006) Stable isotope signatures of benthic invertebrates in arctic lakes indicate limited coupling to pelagic production. *Limnology and Oceanography* 51: 177–188.
10. Jones RI and Grey J (2004) Stable isotope analysis of chironomid larvae from some Finnish forest lakes indicates dietary contribution from biogenic methane. *Boreal Environment Research* 9: 17–23.
11. Jones RI, Grey J, Sleep D, and Arvola L (1999) Stable isotope analysis of zooplankton carbon nutrition in humic lakes. *Oikos* 86: 97–104.
12. Kankaala P, Taipale S, Grey J, et al. (2006) Experimental delta C-13 evidence for a contribution of methane to pelagic food webs in lakes. *Limnology and Oceanography* 51: 2821–2827.
13. Kiyashko SI, Imbs AB, Narita T, Svetashev VI, and Wada E (2004) Fatty acid composition of aquatic insect larvae *Stictochironomus pictulus* (Diptera: Chironomidae): evidence of feeding upon methanotrophic bacteria. *Comparative Biochemistry and Physiology B Biochemistry and Molecular Biology* 139: 705–711.
14. Kiyashko SI, Narita T, and Wada E (2001) Contribution of methanotrophs to freshwater macroinvertebrates: evidence from stable isotopes. *Aquatic Microbial Ecology* 24: 203–207.
15. Kohzu A, Kato C, Iwata T, et al. (2004) Stream food web fueled by methane-derived carbon. *Aquatic Microbial Ecology* 36: 189–194.
16. Savazzi E and Yao PY (1992) Some morphological adaptations in fresh-water bivalves. *Lethaia* 25: 195–209.

such as biofilms would be valuable. A focus on non-sulfide-dependent chemosynthesis would also supplement current data greatly, e.g., redox processes associated with iron or manganese could potentially be important in many soft water ecosystems. It would also be interesting to further investigate the anaerobic (non-O_2-dependent) chemosynthesis, which is shown to be extensive in some anoxic hypolimnia where it is unclear what electron acceptors are used (**Figure 3**). In general, there is a risk that important processes are ignored because of the limited number of systematic studies in many common types of inland waters.

The importance of chemosynthesis as a source of organic matter to macrofauna, with or without endosymbiosis, is another important area of study. Redox interfaces with high microbial activity and high biomass abundance could represent hot spots in terms of microbial food web interactions. In addition, chemosynthesis could be very important for specific species having certain feeding strategies or depending on endosymbiotic microorganisms. An apparent extension of the studies confirming a carbon transfer from chemosynthetic organisms to higher trophic levels would be to quantify the magnitude of this contribution.

Glossary

Acetogenesis – Synthesis of acetate as cell carbon from CO_2 via the acetyl–CoA pathway.

Anoxic – Denotes the absence of molecular oxygen (O_2).

Autotroph – Organism that derives the carbon needed for biosynthesis from inorganic molecules (carbon dioxide) independent of their energy source. *Photoautotrophs* and *chemoautotrophs* obtain their energy from sunlight and the oxidation of inorganic molecules, respectively.

Bacterial growth efficiency – The fraction of the total carbon in the metabolized substrates that is used for biomass production (the remainder is used for energy generation). Bacterial growth efficiency is usually calculated as the bacterial production divided by the sum of bacterial production and bacterial respiration.

Bacterial production – Production of bacterial biomass C per time unit.

Carbon fixation – The formation of organic biomolecules from inorganic carbon (e.g., CO_2).

Chemolithotrophic heterotrophs – Organisms that use energy from the oxidation of reduced inorganic compounds to support growth on organic substrates as carbon source.

Chemosynthesis – Can be defined as the biological production of organic compounds from inorganic C-1 molecules (e.g., carbon dioxide and methane) and nutrients using the energy generated by the oxidation of inorganic (e.g., hydrogen gas, hydrogen sulfide, ammonium) or organic C-1 (e.g., methane, methanol).

Chemosynthetic mixotrophs – Organisms that use CO_2 and organic carbon compounds as carbon source and organic and inorganic substrates as energy source.

Chemotrophs – Organisms that use chemical reactions as energy source.

Chemoorganotrophs or chemolithotrophs – Obtain their energy for biosynthesis or energy conservation via aerobic or anaerobic respiration from the oxidation for organic or inorganic molecules, respectively.

Chironomids – A family of nonbiting midges. Larval stages can be found in surface sediments in most freshwater habitats. The larvae of some species are tolerant to low O_2 conditions.

Dark carbon fixation – Term widely used in aquatic sciences and applies primarily to chemosynthetic reactions where organic matter production occurs without sunlight.

Electron acceptor – A chemical entity that accepts electrons transferred to it from another compound. It is an oxidizing agent that, by accepting electrons, is itself reduced in the process.

Electron donor – A chemical entity that donates electrons to another compound. It is a reducing agent that, by donating electrons, is itself oxidized in the process.

Endosymbiosis – Term for a symbiotic relationship when one organism live inside another organism and when both organisms benefit from this relationship.

Epilimnion – The uppermost surface wind mixed water layer of a thermally stratified lake.

Fermentation – The incomplete anoxic degradation of organic matter resulting in, e.g., small organic acids, alcohols, H_2, or CO_2.

Heterotrophs – Organisms relying on organic matter as their source of carbon.

Hypolimnion – The bottom water layer in thermally stratified lakes positioned below the strong temperature gradient.

Mineralization – The complete degradation of organic matter to CO_2 under oxic conditions, or CO_2 and CH_4 under anoxic conditions.

One-carbon (C-1) compounds – Carbon dioxide (CO_2) and carbon monoxide (CO) are examples of inorganic molecules, and methane (CH_4), methanol (CH_3OH), methylamine (CH_3NH_2), and methane thiol (CH_3SH) are examples or organic molecules.

Oxycline – The steep O_2 concentration gradient that represent the border between oxic and anoxic environments.

Oxidation – Describes the loss of electrons by a molecule, atom, or ion.

Phototrophs – Organisms that use sunlight as energy source.

Phylogeny – The study of evolutionary relatedness among various groups of organisms.

Primary production – The production of organic matter from carbon dioxide (CO_2).

Redox – (Short term for reduction/oxidation reaction) describes all chemical reactions in which atoms have their oxidation number (oxidation state) changed.

Reduction – Describes the gain of electrons by a molecule, atom, or ion.

Syntrophic consortia – Term for the aggregates of different microbial groups depending on each other for optimizing their metabolic energy yield. An example is H_2-producing fermenting bacteria living together with H_2-using methanogenic or sulfate-reducing bacteria. The organization into a consortium results in an efficient and rapid H_2 transfer to the H_2-using bacteria and the rapid removal of H_2 increases the thermodynamic energy yield of the fermentation process.

Thermodynamic energy yield ($\Delta G_0'$) – The amount of Gibbs free energy yielded in a chemical reaction.

See also: Archaea; Bacteria, Distribution and Community Structure; Comparative Primary Production; Microbial Food Webs; Sulfur Bacteria.

Further Reading

Andersson MGI, Brion N, and Middelburg JJ (2006) Comparison of nitrifier activity versus growth in the Scheldt estuary – A turbid, tidal estuary in northern Europe. *Aquatic Microbial Ecology* 42: 149–158.

Camacho A, Erez J, Chicote A, *et al.* (2001) Microbial microstratification, inorganic carbon photoassimilation and dark carbon fixation at the chemocline of the meromictic Lake Cadagno (Switzerland) and its relevance to the food web. *Aquatic Sciences* 63: 91–106.

Dworkin M, Falkow S, Rosenberg E, Schleifer KH, and Stackebrandt E (eds.) (2006) *The Prokaryotes, vol. 1: Symbiotic Associations, Biotechnology, Applied Microbiology,* 3rd edn., 959p. New York: Springer.

Dworkin M, Falkow S, Rosenberg E, Schleifer KH, and Stackebrandt E (eds.) (2006) *The Prokaryotes, vol. 2: Ecophysiological and Biochemical Aspects,* 3rd edn., 1107p. New York: Springer.

Garcia-Cantizano J, Casamayor EO, Gasol JM, Guerrero R, and Pedros-Alio C (2005) Partitioning of CO_2 incorporation among planktonic microbial guilds and estimation of in situ specific growth rates. *Microbial Ecology* 50: 230–241.

Madigan MT, Martinko J, and Parker J (2005) *Brock Biology of Microorganisms,* 11th edn. Englewood Cliffs, NJ: Prentice Hall.

Prakash A, Sheldon R, and Sutcliffe WHJ (1991) Oceanic variation of ^{14}C dark uptake. *Limnology and Oceanography* 36: 30–39.

Sorokin JI (1964) On the trophic role of chemosynthesis in water bodies. *Internationale Revue der gesamten Hydrobiologie* 49: 307–324.

Sorokin YI (1970) Interrelations between sulphur and carbon turnover in meromictic lakes. *Archive fur Hydrobiologie* 66: 391–446.

Strous M and Jetten MSM (2004) Anaerobic oxidation of methane and ammonium. *Annual Review of Microbiology* 58: 99–117.

Taylor GT, Iabichella M, Ho TY, *et al.* (2001) Chemoautotrophy in the redox transition zone of the Cariaco Basin: A significant midwater source of organic carbon production. *Limnology and Oceanography* 46: 148–163.

Phytoplankton Population Dynamics: Concepts and Performance Measurement

C S Reynolds, Centre of Ecology and Hydrology and Freshwater Biological Association, Cumbria, UK

Introduction

The phytoplankton of inland waters – streams, rivers, lakes, ponds and wetlands – is collectively comprised by a phylogenetically diverse array of essentially photoautotrophic microorganisms.

However, the particular constraints placed upon the evolution and functional adaptations to plant life in the open water of pelagic environments invoke several common morphological and behavioral traits. Crucially, small organismic size (generally <1 mm) permits the mechanical support and entrainment opportunities to be exploited, while the shear stresses imposed by even the smallest turbulent scales are normally avoided. In turn, the embedding of microscopic chlorophyll-containing phytoplankton in the near-surface convection benefits the photon-harvesting opportunities, which is typically a variable and often suboptimal in the underwater light field. The normal habit of planktic autotrophs is either unicellular (maximal dimensions ranging over three orders of magnitude, from 0.2 to 200 μm), or uniseriately filamentous (up to about 0.3 mm in length), or they occur as coenobia of varying complexity (which might measure between 20 and 200 μm, exceptionally, up to 5 mm across).

As with the numbers (concentrations) of other microorganisms, phytoplankton populations are known to fluctuate conspicuously in nature, increasing exponentially under favorable environmental conditions and declining rapidly in the presence of consumers or pathogens, or following the onset of hostile conditions. Besides being scientifically intriguing, the dynamics of phytoplankton populations influence the direction and efficiency of energy transfer through aquatic food webs. They also affect the balance of dissolved gases and other solutes and so have a bearing upon water quality. Moreover, because not all phytoplankton is regarded as benign, the dynamics of 'harmful algal blooms' remain a relevant topic for research. There are good reasons to seek an understanding of the basis of phytoplankton dynamics in nature.

In spite of the apparent simplicity of accounting for algal cells recruited by a process of binary division, the study of phytoplankton dynamics has suffered historically from two serious difficulties. One owes to the complexity of the interacting factors known to influence the attainable rates of cell division in the phytoplankton (temperature, deficiencies in the supply of light, carbon and other nutrients); separating their effects and those of resource-limitation thresholds has proved daunting. The second is that it has been extremely difficult to infer true recruitment rates in the field because, supposing that even a discrete population can be serially sampled with appropriate frequency, the observable rate of population increase is always net of simultaneous loss processes (mortalities, consumption by grazing animals, sedimentation, whose rates inherently require independent determination). Unfortunately, the direct measurement of anabolic processes, such as photosynthetic carbon fixation or nutrient uptake, are rarely analogues of true growth; they provide little more than the capacity to be able to sustain growth. Most past studies have had to confine themselves to estimates of net rates of population change in the field, recognizing these to be, to varying extents, underestimates of dynamic performance.

A third obstruction to understanding is entirely conceptual and self-imposed: it concerns the difference between limitations on growth rates and on growth yields, and it should be addressed at once. Every living process (taking up resource, transporting it within the cell, synthesizing proteins, and assembling these into relevant structures and organelles) occupies a finite period of time. For as long as the supply of resource meets the demand of assembly, the assembly cannot be 'resource-limited.' Should the resource be depleted to an extent where the uptake rate falls below the optimal deployment rate, however, then its uptake can be said to be constraining the rate of deployment, and cell assembly rate may be properly claimed to be 'resource-limited.' Plainly, if the resource is effectively exhausted, no further assembly is possible. Thus, the capacity of a habitat to support growth is determined by the resources available, relative to the demand. The least available resource sets a capacity limitation. Before that point is approached, much less reached, a predictably *capacity-limiting* factor need not be *rate-limiting*.

In the following subsections, an attempt is made to illustrate the processes that govern algal growth rates, very much from the standpoint of the individual cell and its clone. Quantitative measurements are drawn from the literature to show how the wax and wane of phytoplankton populations may be estimated and modeled.

Cell Growth

Phytoplankton growth and cell recruitment progress through a cycle in which the cell firsts accumulates the resources required to bring about an approximate doubling of its mass and then embarks upon the mitotic division of its nucleus, before the remainder of the protoplasm is allocated to the separating new daughter cells. This cell growth cycle was characterized by Murray and Kirschner (1991). After separation, the next generation of daughters resumes the photosynthetic fixation (reduction) of carbon, the sequestration of phosphorus, nitrogen, and mineral nutrients (up to a score), and the synthesis of the proteins that will contribute to the generational doubling of biomass. Structures such as flagella, scales, and exoskeleta also have to be reallocated or copied. These processes are mediated by the ribosomal nucleic acids, responding to regulatory proteins that signal the stage of assembly reached. These issue critical cues: ultimately the trigger for the replication of the nuclear DNA, which irrevocably commits the nucleus (and the cell) to mitotic division, depends upon the activation of a maturation-promoting factor. In turn, this is, in most instances, demonstrably dependent upon the satisfaction of intracellular resource required to sustain the replication of the cell material. Correspondingly, resource failure suppresses the division trigger; this is, just one of the ways that cells defend their integrity in suboptimal environments. Moreover, an important corollary of such biochemical inventorizing is that cells of successive generations are capable of achieving the size of the parent cells and that there is a general tendency against radical intergenerational diminution of cells.

Quantifying and Defining Growth Rates

Given that daughter cells are, ultimately, sufficiently similar to the parent, the increase in biomass of a population is conveniently analogized to the increase in the number, or volume-specific concentration (N) of cells. Of course, the potential doubling of cells at each generation determines that populations typically increase exponentially, so that it is usual to express the increase as a logarithmic exponent; including a time

dimension, provides the rate of increase. From the observed change in population during a time period, t, we infer

$$\delta N / \delta t = N_t / N_0 = N_0 e^{rt} \qquad [1]$$

where N_0 and N_t are the populations at two points in time separated by an interval, t, and e is the natural logarithmic base. Thus, r, is the 'exponential increase constant'; its numerical value differs, of course, if logarithms to base10 or base2 are used, as was the case in many early studies. More importantly, it should be stressed that r is not at all a constant: the rate of increase that is thus observed, in the field as in the laboratory, is very much the average of what is happening to all the cells present and is net of simultaneous failures and mortalities that may be occurring. Frequently, the observable rate of increase in the natural population falls far short of what most students understand to be its growth rate. Thus, plankton biologists properly distinguish between 'true growth rates' and 'net growth rates.' The terminology that I have advocated distinguishes 'growth rate' (represented by r) as applying to the rates of intracellular processes leading to the completion of the cell cycle, net or otherwise of respiratory costs as specified. The observable 'net increase rate' (represented as r_n) refers to the rate of accumulation of species-specific biomass, though frequently as detected from change in cell concentration (r_n is substituted for r in eqn [1]). However, the potential increase in the biomass provided by the frequency of cell division and the intermediate growth of the recruited cells of each generation (and which, indeed, may be considerably greater than r_n) is referred to as the biomass-specific population *replication rate*, signified by r'. In this way, the replication rate is equivalent to the rate of cell production before any rate of loss of finished cells to all mortalities (r_L), is computed. Thus,

$$r' = r_n + r_L \qquad [2]$$

It is emphasized that, according to this scheme, replication rate is net of all metabolic losses, not least those due to respiration rate (R). Growth rates are not necessarily well-predicted from indirect measurements of photosynthesis, carbon fixation, or nutrient uptake. In laboratory cultures, when the best estimates of phytoplankton growth have been derived, species-specific replication rates are frequently found to be saturated with respect to the contributory metabolic capacities.

Measuring Growth Rates

The difficulties of estimating growth rate in natural populations are considerably magnified with respect to those of laboratory strains, yet both activities invoke common problems. Serial estimates of population size (N_0, N_t,... and so on) rely upon the application of rigorous, representative, and (so far as possible) nondestructive sampling techniques to ensure that all errors due to selectivity or patchiness of distribution are minimized. Estimating the numbers or the biomass of each species in each of the samples is usually challenging too. The attraction of measuring such analogues as chlorophyll, fluorescence, light absorption, or scatter must be balanced against the compounding of errors of sampling through misplaced assumptions about the biomass equivalence, as well as a lot of species-specific information. There have thus been few substitutes for direct counting, using a good microscope and a prevalidated subsampling method, and subject to known statistical confidence. The original iodine-sedimentation/inverted microscope technique has largely given way to the convenience of flat, hemocytometer-type cells. With the advent and improvement of computer-assisted image analysis and recognition, algal counting is now much less of a chore. Properly used, the computerized aids yield results that can be as accurate as those of any human operator.

Results accumulated over a period of algal increase can be processed to determine the mean rate of population change over that period. Often it is convenient to find the least-squares regression of the individual counts for each species $\ln(N_1)$, $\ln(N_2)$, ..., $\ln(N_i)$, on the corresponding occasions (t_1, t_2, ..., t_i). The slope of $\ln(N)$ on t is, manifestly, equivalent to r_n. Modifications to the counting method are necessary with respect to colony-forming algae, such as *Microcystis* and *Volvox* (see Further Reading).

Other recently pioneered approaches following the *in situ* replication rates of planktic microorganisms include techniques for the detection of the frequency of dividing cells (FDC) in serial samples collected from the water column; this works especially well for populations of relatively large organisms, such as dinoflagellates and desmids. More recently, in situ growth rates have been estimated from FDC among cell suspensions of *Microcystis aeruginosa* and *M. wesenbergii* from Biwa-ko (prepared by ultrasonication of field-sampled colonies). In both species, the FDC varied between 10% and 15% in offshore stations and between 15% and 40% at inshore stations, with the average duration of cytokinesis varying from 25 to 3–6 h. Growth rates of $0.34\,\text{day}^{-1}$ thus appear sustainable in the near-shore harbor areas of Biwa-ko, whence they are liable to become more widely distributed in the circulation of the lake. For phytoplankton species which are less amenable to the tracking of cell division, the principle has been extended to monitoring the frequency of nuclear division. The success of this method relies on the good fixation of field samples followed by careful staining with the DNA-specific fluorchrome, 4,6-diamidino-2-phenylindole (DAPI). There is now interest in the possibility of sensing the DNA replication itself. Since the groundbreaking study of Dortch *et al.* (1983), microbial ecologists have been debating the validity of the ratio of DNA to cell carbon as an index of the rate of DNA replication. The ratio of RNA:DNA is already in use as an indicator of the capacity for protein synthesis and a barometer of the cell growth cycle in marine flagellates.

Occasionally, *in situ* growth-rates of diatom species have been accurately estimated from the rates of depletion of soluble reactive silicon from the medium.

Growth Performance in Culture

The optimum performances of species-specific strains are attained in carefully prepared standard media, designed to saturate resource requirements, under constant, continuous light of an intensity sufficient to saturate photosynthesis, and at a steady, optimal temperature. Even then, maximal replication rates are not established instantaneously. There is usually a significant 'lag phase' during which the inoculated cells acclimatize to the ideal world into which they have been introduced. Within a day or two, however, the isolated population will be increasing rapidly and, generally, will be doubling its mass at approximately regular intervals. It is early in this 'exponential phase' that the maximal rate of replication is achieved, when r' is supposed to be equal to the observed net rate of increase, r_n, solved by eqn [1]. Later, as the resources in the medium become depleted or the density of cells in suspension begins to be self-shading, the rate of increase will slow down considerably (eventually, to the 'stationary phase'). Care is taken to discount the biomass increase in this phase from the computation of the exponent of the maximum value of the specific replication rate, r.

From the relatively small number of published data, it is nevertheless quite striking that the various disparate observations in quite separate, independent studies on, ostensibly, the same species also encounter quite similar maximal light- and nutrient-saturated temperature-dependent replication rates. Various compendia are available (see Further Reading), from which certain graphical and statistical patterns have been deduced. Besides confirming that, even under the idealized conditions of laboratory culture, there is a finite and consistent species-specific, temperature-dependent, time requirement for the completion of the cell cycle and the attendant biomass doubling, these analyses reveal that consistent interspecific differences in replication rates are related to morphological differences among the individual species compared. Thus, the replication rates determined for a selection of species of a variety of habits and phylogenetic affinities at $20 \pm 1\,^\circ\text{C}$ were found to conform best with the regression equation,

$$r'_{20} = 1.142(s/v)^{0.325}\,\text{day}^{-1} \qquad [3]$$

where v is the average unit volume of the alga (in μm^3) and s is its surface area (in μm^2); the coefficient of correlation is 0.72 and the equation accounts for 52% of the total variance. The finding confirms the anticipation that the replication rates of planktic algae, as of all other poiklothermic organisms, conforms to an allometric relationship with organismic size but it is interesting that the slope of the regression is closer to one-third than to one-quarter, which earlier physiologists predicted. Note also the use of *unit* volumes and areas: these are necessarily invoked to include algae that are usually coenobial in habit or whose cells are typically embedded in mucilaginous sheaths; only in the case of unsheathed coenobial species are the values comparable to those of individual cells of similar dimensions. Subject to this qualification, the relationship adequately and simultaneously describes population increase rates in cultures of a wide range of planktic species, from colonial *Microcystis* (r'_{20}, 0.4–0.5 day^{-1}, biomass doubling time, ≥ 1.5 days) to unicellular *Chlorella* (r'_{20}, 1.8–2.0 day^{-1}, biomass doubling time, ≤ 0.4 days) (see **Figure 1**).

The species-specific replication rates of many freshwater phytoplankters, some acknowledged cold-water stenotherms and known thermophiles apart, generally accelerate from around freezing to a maximum, usually in the temperature range 25–35 $^\circ$C. A few, like *Synechococcus*, continue to increase beyond 40 $^\circ$C but, exposed to their respective supraoptimal temperatures, the replication rates of most of the species considered here first stabilize and, then, fall away abruptly. From 0 $^\circ$C to just below the temperature of the species-specific optimum, the replication rates of most plankters in culture appear, as expected, to increase exponentially as a function of temperature. However, the degree of temperature-sensitivity of the division rate is evidently dissimilar among plankters. The maximum temperature-dependent replication rates of some species roughly double for each 10 $^\circ$C step in temperature, whereas, for others, the response of r' to temperature is steeper. Regressing daily specific replication rates against temperatures, normalized on an Arrhenius scale, wherein temperatures ($\theta\,^\circ$C) are rendered as reciprocals of absolute temperature (in Kelvins; purely for manipulative convenience, the units actually used were calculated in terms of

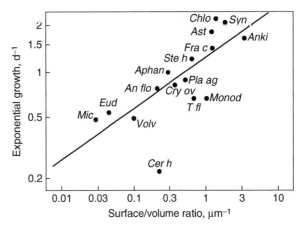

Figure 1 Maximum resource- and light-saturated replication rates of phytoplankton in culture at 20 $^\circ$C, plotted against the surface-to-volume ratios of the algal units (cells, or, where appropriate, colonial aggregates). The species represented are *An flo = Anabaena flos-aquae; Aphan = Aphanizomenon flos-aquae, Ast = Asterionella formosa, Cer h = Ceratium hirundinella, Chlo = Chlorella, Cry ov = Crptomonas ovata, Eud = Eudorina unicocca, Fra c = Fragilaria crotonensis, Mic = Microcystis aeruginosa, Monor = Monoraphidium sp., Pla ag = Planktothrix agardhii, Ste h = Stephanodiscus hantzschii, Syn = Synechococcus sp., Tab = Tabellaria flocculosa, Volv = Volvox aureus.* From data compiled and presented in Reynolds (1997).

$A = 1000[1/(\theta\,\text{K})]$), slopes of between $\beta = -3.50A^{-1}$, for the small unicellular *Synechococcus*, and $\beta = -8.15A^{-1}$ for the colonial *Microcystis*, have been demonstrated. Moreover, the implicit allometric relationship with morphology has been described by

$$\log\,(r'_\theta) = \log\,(r'_{20}) + \beta[1000/(273 + 20) \\ - 1000/(273 + \theta)]\,\text{day}^{-1} \qquad [4]$$

where $\beta = 3.378 - 2.505\,\log(s/v)$; coefficient of correlation 0.84. In essence, the relationship quantifies the observation that the replication rates of small algae or of larger ones having a high surface-to-volume ratio (i.e., most unicellular species, pennate diatoms and filamentous coenobia) are less sensitive to low temperatures (say <10 $^\circ$C) than are large or colonial algae having a low surface-volume relationship (including the bloom-forming Cyanobacteria). Major growth periods of the latter generally coincide with ambient temperatures of >15 $^\circ$C.

Growth Performance in Natural Environments

Away from the carefully regulated environments of laboratory culture, the dynamic performances of ostensibly the same algal species in nature seem considerably inferior, especially in terms of daily recruitment rate. As already recognized, natural habitats are intrinsically more variable, in relation to temperature, photoperiod, and entrainment through the underwater light gradient. One or more nutrients may be in short supply, so that the rates of organismic resource gathering may be sufficiently impeded to impose the principal constraint on cell replication rate, i.e., to limit the rate of growth below that of the phytoplankter's potential. Even at these commuted rates of cell replication, natural population recruitment rates may be further truncated by simultaneous cell losses to mortality, consumers, parasites, and physical removal.

However, experience with simulation models suggests that the maximum performances of algae, corrected for temperature, proportioned to day-length and species-specific dynamic sensitivities to threshold resource concentrations do offer good approximations of the recruitment rates of phytoplankton populations in their natural locations. These effects are considered in a companion article.

See also: Phytoplankton Population Dynamics in Natural Environments.

Further Reading

Alvarez Cobelas M, Velasco JL, Rubio A, and Brook AJ (1988) Phased cell division in a field population of *Staurastrum longiradiatum* (Conjugatophyceae: Desmidaceae). *Archiv für Hydrobiologie* 112: 1–20.

Banse K (1976) Rates of growth, respiration and photosynthesis of unicellular algae as related to cell size: a review. *Journal of Phycology* 12: 135–140.

Braunwarth C and Sommer U (1985) Analyses of the *in situ* growth rates of Cryptophyceae by use of the mitotic index. *Limnology and Oceanography* 30: 89–897.

Carpenter EJ and Chang J (1988) Species-specific phytoplankton growth rates via diel DNA-synthesis cycles. *Marine Ecology–Progress Series* 43: 105–111.

Chang J and Carpenter EJ (1990) Species-specific phytoplankton growth rates via diel and DNA synthesis cycles. IV. Evaluation of the magnitude of error with computer-simulated cell populations. *Marine Ecology–Progress Series* 65: 293–304.

Coleman A (1980) Enhanced detection of bacteria in natural environments by fluorochrome staining of DNA. *Limnology and Oceanography* 25: 948–951.

Dortch Q, Roberts TL, Clayton JR, and Ahmed SI (1983) RNA/DNA ratios and DNA concentrations as indicators of growth rate and biomass in planktonic marine organisms. *Marine Ecology–Progress Series* 13: 61–71.

Elliott JA, Irish AE, Reynolds CS, and Tett P (1999) Sensitivity analysis of PROTECH, a new approach in phytoplankton modelling. *Hydrobiologia* 414: 37–43.

Elliott JA, Irish AE, Reynolds CS, and Tett P (2000) Modelling freshwater phytoplankton communities: An exercise in validation. *Ecological Modelling* 128: 9–16.

Fogg GE and Thake B (1987) *Algal Cultures and Phytoplankton Ecology*, 3rd edn. London: University of London Press.

Frempong E (1984) A seasonal sequence of diel distribution patterns for the planktonic dinoflagellate, *Ceratium hirundinella* in a eutrophic lake. *Freshwater Biology* 14: 301–321.

Heller M (1977) The phased division of the freshwater dinoflagellate *Ceratium hirundinella* and its use as a method for assessing growth in a natural population. *Freshwater Biology* 5: 527–533.

Hoogenhout H and Amesz J (1965) Growth rates of photosynthetic microrganisms in laboratory cultures. *Archiv für Mikrobiologie* 50: 10–25.

Ishikawa K, Kumagai M, Vincent WF, Tsujimura S, and Nakahara H (2002) Transport and accumulation of blooming Cyanobacteria in a large, mid-latitude lake; the gyre-. *Microcystis hypothesis*. *Limnology* 3: 87–96.

Kemp PF, Lee S, and LaRoche J (1993) Estimating the growth rate of slowly growing marine bacteria from RNA content. *Applied and environmental Microbiology* 59: 2594–2601.

Kerkhof L and Ward BB (1993) Comparison of nucleic-acid hybridization and fluorometry for measurement of the RNA/DNA ratio and growth rate in a marine bacterium. *Applied and Environmental Microbiology* 59: 1303–1309.

Kratz WA and Myers J (1955) Nutrition and growth of several blue-green algae. *American Journal of Botany* 42: 282–287.

Lund JWG, Kipling C, and LeCren ED (1958) The inverted-microscope method of estimating algal numbers and the statistical basis for estimation by counting. *Hydrobiologia* 11: 143–170.

Mann NH (1995) How do cells express limitation at the molecular level? In: Joint I (ed.) *Molecular Ecology of Aquatic Microbes*, pp. 171–190. Berlin: Springer-Verlag.

Murray A and Hunt T (1993) *The Cell Cycle: An Introduction.* New York: W.H. Freeman.

Murray AW and Kirschner MW (1991) What controls the cell cycle? *Scientific American* 264: 34–41.

Nielsen SL (2006) Size-dependent growth rates in eukaryotic and prokaryotic algae exemplified by green algae and cyanobacteria: Comparisons between unicells and colonial growth forms. *Journal of Plankton Research* 28: 489–498.

Nielsen SL and Sand-Jensen K (1990) Allometric scaling of maximal photosynthetic growth rate to surface-to-volume ratio. *Limnology and Oceanography* 35: 177–181.

Ojala A and Jones RI (1993) Spring development and mitotic division pattern of a *Cryptomonas* sp. in an acidified lake. *European Journal of Phycology* 28: 17–24.

Padisák J (2003) Phytoplankton. In: O'Sullivan P and Reynolds CS (eds.) *The Lakes Handbook*, vol. 1, pp. 251–308. Oxford: Blackwell Science.

Pollingher U and Serruya C (1976) Phased division in *Peridinium cinctum* f. *westii* (Dinophyceae) and development of the Lake Kinneret (Israel) bloom. *Journal of Phycology* 12: 163–170.

Porter KG and Feig YS (1980) The use of DAPI for identifying and counting aquatic microflora. *Limnology and Oceanography* 25: 943–948.

Raven JA (1982) The energetics of freshwater algae: Energy requirements for biosynthesis and volume regulation. *New Phytologist* 92: 1–20.

Reynolds CS (1983) Growth-rate responses of *Volvox aureus* Ehrenb. (Chlorophyta, Volvocales) to variability in the physical environment. *British Phycological Journal* 18: 433–442.

Reynolds CS (1988) Functional morphology and the adaptive strategies of freshwater phytoplankton. In: Sandgren CD (ed.) *Growth and Reproductive Strategies of Freshwater Phytoplankton*, pp. 338–433. Cambridge: Cambridge University Press.

Reynolds CS (1989) Physical determinants of phytoplankton succession. In: Sommer U (ed.) *Plankton Ecology: Succession in Plankton Communities*, pp. 9–56. Madison: Brock-Springer.

Reynolds CS (1997) *Vegetation Processes in the Pelagic: A Model for Ecosystem Theory.* Oldendorf: Ecology Institute.

Reynolds CS (1998) Plants in motion: Physical-biological interaction in the plankton. *Coastal and Estuarine Studies* 54: 535–550.

Reynolds CS (2006) *Ecology of Phytoplankton.* Cambridge: Cambridge University Press.

Reynolds CS and Jaworski GHM (1978) Enumeration of natural *Microcystis* populations. *British Phycological Journal* 13: 269–277.

Reynolds CS and Wiseman SW (1982) Sinking losses of phytoplankton in closed limnetic systems. *Journal of Plankton Research* 4: 489–522.

Tsujimura S (2003) Application of the frequency-of-dividing-cells technique to estimate the *in-situ* growth rate of *Microcystis* (Cyanobacteria). *Freshwater Biology* 48: 2009–2024.

Utermöhl H (1931) Neue Wege in der quantitativen Erfassung des Planktons. *Verhandlungen der internationalen Vereinigung für theoretische und angewandte Limnologie* 5: 567–596.

Vaulot D (1995) The cell cycle of phytoplankton: Coupling cell growth to population growth. In: Joint I (ed.) *Molecular Ecology of Aquatic Microbes*, pp. 303–332. Berlin: Springer-Verlag.

Youngman RE (1971) *Algal monitoring of water supply reservoirs and rivers.* Technical Memorandum No.TM63. Medmenham: Water Research Association.

Phytoplankton Population Dynamics in Natural Environments

C S Reynolds, Centre of Ecology and Hydrology and Freshwater Biological Association, Cumbria, UK

Introduction

Not surprisingly, phytoplankters perform less well in natural environments than they do in the controlled medium of a laboratory culture. Part of this is due to dynamic loss processes (the cropping of cells by herbivores, settling out of cells to depth, or succumbing to attacks by pathogenic viruses, bacteria, and fungi). These processes are quantifiable in the same exponential units used for expressing replication rates, which may be added to and subtracted from actual growth rate to yield a *net population recruitment rate* or *net increase rate*. This may well have a negative value: even growing populations can decline in numbers. Another difference between nature and cultures relates to the source inoculum. Although many freshwater species of phytoplankton are ubiquitous and cosmopolitan, the flora at any given point in space and time is heavily weighted in favor of species that have been abundant in the recent past or whose propagules have been conspicuously recruited from a nearby source. Even so, to grow successfully, the conditions encountered must be perceived by the alga to satisfy its minimal environmental requirements. It is a fundamental tenet of phytoplankton ecology that algae will grow wherever and whenever they have the opportunity to do so.

That said, it is apparent both from careful analysis of intensive campaigns of sampling and counting and from simulation models invoking eqns [3] and [4] of **Phytoplankton Population Dynamics: Concepts and Performance Measurement** in that *in situ* cell replication rates are as rapid as they can be, comparing favorably with performances in culture. Thus, adjusting for temperature and the length of the photoperiod, we may recognize observed replication rates in the field of over $1.0 \, day^{-1}$ for unicellular chlorococcals (such as *Ankyra*), $0.6 \, day^{-1}$ for the diatom (*Fragilaria*), $0.5 \, day^{-1}$ for motile colonies of *Eudorina*, and for colonies of *Microcystis* ($0.2–0.3 \, day^{-1}$) to be in proportion to the fastest cultured clones. Even observed vernal growth rates of *Asterionella* between 0.15 and $0.20 \, day^{-1}$ demand that the simulation is generously temperature- and photoperiod-compensated.

Suboptimal Growth: The Role of 'Limiting' Constraints

Set against these robust performances, however, are many instances of the growth rates being depressed through the intervention of limiting factors. In all such instances, the rate of supply of one or other of the essential resources contributing to the production of new cells is itself so modest or so slow that it becomes the controlling determinant of the rate of cell replication. 'Limitation' is an intensively studied area of the physiological ecology of phytoplankton and there is a bewilderingly large literature on the subject. The term 'limitation' is used freely and frequently without qualification, so it is as well to set a working definition. Periods of growth may yield a biomass of cell matter that is determined by the effective exhaustion of one or other of the preexisting stocks of bioavailable resources. Among fresh waters, nitrogen or phosphorus usually turns out to be the critical resource. Given the minimal internal quota of each element that is required to sustain the healthy, vegetative state of the biomass (q_0), the potential *yield* of the available stock of resource (B_{max}) may be approximated from the least resource-specific quotient, K_i/q_0, where K_i is the concentration of the *i*th element in the externally available resource pool. Before that point is reached, however, there is phase of resource diminution, during which the sequestration and assimilation of a rarefied resource can proceed at a rate that is so suboptimal with respect to the resource-replete rate of replication that the latter is regulated by the rate of limiting resource acquisition. Thus, it is the *rate* of biomass accumulation that is subject to limitation.

Light Limitation of Growth Rate

Distinction may also be made among rate-limiting constraints on processing (those relating exclusively to the fluxes of reductant) and those that relate also to the progressive depletion of an ultimately yield-limiting resource. The limitation of replication rate through a weakened rate of photon delivery or harvesting provides the clearest example of the first of these alternatives. Photosynthesis not only provides most of the reduced, organic carbon species that will constitute some 50% of the dry weight of the new-formed vegetative biomass but controlled oxidation of photosynthate provides the power that drives the electromotive forces of cell synthesis. It is fairly straightforward to calculate from measurements of light-saturated photosynthetic rates the time required to accumulate the carbon needed to sustain a doubling of the cell content. In the case of *Chlorella*, growing at 20 °C, the alga was found to be capable of fixing the carbon needed to double the cell complement within 5.4 h, and well within the minimum observed biomass doubling time of 9.05 h. For growth rate to exceed photosynthetic rate is, of course, unsustainable over the life time of the cell. Following the logic developed earlier, however, the rate of photosynthesis cannot be said to limit the rate of growth until it fails to meet the growth demand in $\leq$9.05 h.

The intensity of light (or photon flux density) is plainly important to satisfying the growth requirement for photosynthetically fixed carbon. From a variety of field and laboratory studies, it may be recognized that mean specific growth rates (r'_{20}) of up to 1.6 day^{-1} [requiring the fixation of some 70 mg C (g cell C) h^{-1}] are typically saturable at photon flux densities of about 70 μmol photons m^{-2} s^{-1} (or 6 mol photons m^{-2} day^{-1}); this is well within the limits of the conventionally- (if arbitrarily-) defined euphotic zone (being that above a depth to which 1% of the surface reached by incoming light, or some 2–20 μmol photons m^{-2} s^{-1}). Entrainment of phytoplankton in natural mixed layers that extend beyond the euphotic depth, especially in turbid waters, may well take cells for a time into water wherein the instantaneous flux rate fails to satisfy the maximum growth rate, and even to depths below the compensation depth, where the rate of photosynthesis fails to meet even the respiratory demand, and effective darkness. This may be analogized to photoperiod and to a lengthening of the time taken to harvest the photons required to sustain a doubling of the cell carbon (equivalent to some 1–1.4 mol photon per g cell carbon). Again, if the relevant light harvesting is achieved within the doubling time of the cell (i.e., in <9.05 h, in the case of the *Chlorella*), then an energy constraint on growth continues to be avoided. Else, the rate of cell growth must be considered to be 'light limited.'

Healthy cells are generally able to offset the constraining impacts of truncated light harvesting on cell growth over quite a wide band of limiting light doses, through varying degrees of low light adaptation. In essence, this involves increasing biomass-specific rate of light harvesting. There are at least two ways that cells do this, though they are invoked under differing circumstances and to markedly differing species-specific extents. In the face of diminishing photoperiods, chiefly associated with deep entrainment and short day lengths, the cell is required to maximize its photon interception during the truncated windows of near optimal light exposure, which it does through building more light-harvesting centers. This is detected as a biomass-specific increase in the complement

of photosynthetic pigment, notably chlorophyll *a*. As the chlorophyll-specific photosynthetic yield is progressively impaired, the yield specific to cell carbon suffers much less readily, so, plotted against low daily irradiances, the responses of species-specific growth rates appear to markedly nonlinear (**Figure 1**). There is a limit to this adaptation, set partly by just how much chlorophyll can be deployed and maintained and partly by the shape of the alga and its cell-specific capacity to present light harvesting centers to the light field. In this way, the efficiency of the low light adaptation is reflected in the steepness of the slope (α_r) of light-limited growth on low photon doses. These slopes also have a statistically valid allometric correlation with the dimensionless product of maximum unit dimension (m) and its surface-to-volume ratio (s/v):

$$\alpha_r = 0.257(ms/v)^{0.236} \qquad [1]$$

The coefficient of correlation is 0.57; the equation explains 33% of the total variance (Elliott *et al.*, 1999).

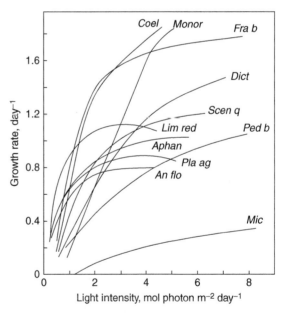

Figure 1 Light-limited, steady-state growth rates of selected species of phytoplankton at 20 °C, plotted as a function of daily light doses. *An flo = Anabaena flos-aquae, Aphan = Aphanizomenon flos-aquae, Coel = Coelastrum microporum, Dict = Dictyosphaerium pulchellum, Fra b = Fragilaria bidens, Lim red = Limnothrix redekei, Mic = Microcystis aeruginosa, Monor = Monoraphidium sp., Ped b = Pediastrum boryanum, Pla ag = Planktothrix agardhii, Scen quad = Scenedesmus quadricauda.* From data compiled and presented in Reynolds CS (1997) *Vegetation Processes in the Pelagic: A Model for Ecosystem Theory.* Oldendorf: Ecology Institute.

The second mechanism allows the organism to adapt to persistently low levels of light (or to fluctuations subject to a low maximum photon flux rate) as might be encountered by plankters able to stratify in relatively nutrient rich, deep metalimnetic layers of clear, generally oligotrophic lakes. Here, the emphasis is on being able to intercept light of all wavelengths, by increasing the complement not just of chlorophyll but also of accessory photosynthetic pigments, such as the phycobiliproteins (phycocyanins, phycoerythrins) present in Cyanobacteria and Cryptophyta. These help to plug the gaps in the activity spectrum of chlorophyll *a* by transferring the energy they intercept to the (longer) wavelengths to which the chlorophyll reacts. The additional pigmentation masks the typical green color of photosynthetic organisms with blues, browns, and purples, which is acknowledged in the term, *chromatic adaptation*. Some of the best known cases involve Cyanobacteria of the genera *Planktothrix*, *Limnothrix*, a.o., in the deep alpine lakes of central Europe, and *Cryptomonas* spp. in karstic dolines in Spain.

Nutrient Limitation of Growth Rate

The conceptual difference between yield- and rate-limiting capacities of available nutrients is probably more readily appreciable than it is in the case of light income. However, even within the yield-limiting capacity of a given resource, the expectation that the rate of cell replication is 'limited' by the rate of its supply is often accepted uncritically. This is particularly so in the context of simultaneous evaluations of several nutrients and of their comparative availabilities in relation to the growth demand for each, and where many students of phytoplankton dynamics look for a controlling influence of resource ratios. Part of the difficulty owes to the relative abundance of data on the characteristics of nutrient uptake among selected planktic algae and of the sparsity of unambiguous experimental evidence of growth rate regulation by the supply flux. What can be estimated fairly easily is the amount of each resource that is required to bring about a doubling of cell biomass, while maintaining a steady, optimal cell stoichiometry through successive generations. For instance, if, during one generation time, the self-replicating cell requires to accumulate 1 mol carbon per mol of cell carbon at the start of the period, then we may suppose it to assimilate roughly $1/106$ mol P and $16/106$ mol N per mol cell C incorporated during the same period if the optimal C:N:P ratio is to be maintained in the next generation [i.e., 9.43×10^{-3} mol P and 0.151 mol N (mol cell C)$^{-1}$]. Taking data for *Chlorella*, it may be argued that, in order to support a

doubling of the cell carbon content of 0.63×10^{-12} mol, the cell requires 5.9×10^{-15} mol P per cell. Yet, at the maximal rate of phosphorus uptake by P-starved cells (13.5×10^{-18} mol P cell^{-1} s^{-1}), the cell is theoretically capable of absorbing this amount in only 0.44×10^3 s, or just 7.3 min! Moreover, from the curve of maximal phosphorus uptake rate as a function of external concentration, it was argued that 6.3×10^{-9} mol l^{-1} ($\sim 0.2\,\mu g\,P\,l^{-1}$) is a sufficient concentration of molybdate-reactive phosphorus to supply the entire phosphorus requirement over the full generation time of 9.05 h. The analogous figures for nitrogen show that a concentration of 7 µmol DIN l^{-1} (100 µg N l^{-1}) delivers to the same cell of *Chlorella* its maximum uptake rate of 175×10^{-18} mol N s^{-1}, and so fulfil its doubling requirement of 0.095×10^{-12} mol N in 540 s (9 min). A concentration of DIN of 0.12 µmol N l^{-1} ($\sim 2\,\mu g\,N\,l^{-1}$) is, in theory, enough to supply the nitrogen requirement of doubling without prolonging the doubling beyond 9.05 h.

These calculations apply strictly to *Chlorella* but less detailed data for other common species do not suggest that the nutrient demands of their specific growth rates are met by concentrations of a dissimilar order. To be able to justify the description of growth rates as being 'nutrient limited' by the availability of phosphorus, nitrogen, or any other element, it has to be first demonstrated that the generation time is significantly longer than it is at growth-saturating levels. As it is, this seems not to be the case until these nutrients fall close to the limits of their detection by conventional chemical analyses.

Of course, there are many instances in particular fresh waters where the ambient concentrations of nutrients are, at times or for long periods, drawn down to low levels consistent with nutrient limitation of growth rate. Indeed, the habitats of phytoplankton may be broadly subdivided among (1) those that are so relatively nutrient-replete that net phytoplankton growth and biomass fluctuations are regulated by processing constraints or through the ('top-down') effects of herbivorous consumers, (2) those in which the potential increase of phytoplankton mass sooner or later becomes severely constrained by the effective non-availability of one or other of the essential nutrients, and (3) those in which chronically low nutrient availability exerts a severe and near-continuous bar against any net accumulation of phytoplankton biomass at all.

The first type includes many eutrophic and anthropogenically enriched water bodies, which are often rather small and well-flushed as well. The phytoplankton flora that they may support is broad in species representation but, at any given point in time, it is supposed to be dominated by a small number of species, conveying an impression of low diversity. Episodes of rapid increase in response to improved physical conditions may typically be dominated by fast-growing, high-s/v forms (including unicellular chroococcals, small flagellates, and diatoms). Longer periods of high biomass may feature grazing-resistant large or colonial algae and Cyanobacteria, or proceed to dominance by filamentous Cyanobacteria, diatoms (such as *Aulacoseira*), or xanthophytes (such as *Tribonema*). In each case, their ascendency can be seen to operate through recruitment dynamics and the relative tolerance of low light and nonselective grazing.

In those water bodies where (say) phosphorus is yield-limiting and where soluble, assimilable inorganic sources become severely depleted, phytoplankton growth rates may well enter a phase of phosphorus limitation. Certainly, the rates of nutrient uptake are constrained. Mathematical description of uptake at subsaturating concentrations, based on Michaelis–Menten kinetics and the Monod equation, enjoys wide acceptance, albeit subject to a recognition that the uptake requirement must take account of the cell's internal store. In a later adaptation of the Monod equation, this dependence is explicit:

$$r' = r'_{max}(q - q_0)/(K_r + q - q_0) \qquad [2]$$

where r'_{max} is the fastest sustainable growth rate, $(q - q_0)$ is the difference between the actual cell quota and its minimum for cell sustainability; K_r is the external nutrient concentration that will sustain half the maximum rate of growth. As shown here, the phosphorus concentration required to saturate fully the instantaneous growth is generally within a range of 0.1–$0.2 \, \mu mol \, P \, l^{-1}$, while uptake rates will always help to maintain an internal reserve. Few dynamic estimates of q are available but it is known that the maximum storage capacities (q_{max}) are such that they are capable of sustaining 2–3 generations, or up to eight, possibly more, times as many cells with a content of q_0. It may be deduced that the dynamic advantage at low phosphorus concentrations should fall to those species that have a high affinity for phosphorus (those that have low K_r values) or high storage capacities (those with a high q_{max}). Against a declining resource base, however, it is difficult to make a case for a competitive advantage accruing to low-resource tolerant species, when the difference between concentrations that will support the growth of most species and those that represent resource famine to all is so absolutely narrow; the biomass will already be dominated by generalists while the window of opportunity provided to specialists is narrow. The true benefit to good resource competitors in frequently resource-deficient waters is probably cumulative: their ability to thrive, then survive, for longer than inferior competitors lends a bias to the provision of inocula in future growth episodes, so that the advantage of better recruitment is magnified over a series of such episodes. Thus, the plankton flora of lakes in which phytoplankton battle the constraint of diminishing phosphorus concentrations through much or most of the vegetative period tends to feature species of acknowledged resource affinity (various diatoms, especially species of the *Cyclotella glomerata-comensis* group, many desmids (especially species of *Staurodesmus* and *Cosmarium*), or species investing stores in perennating propagules (some chrysophytes and dinoflagellates). Other adaptations that benefit certain specialist species in phosphorus-poor lakes include the biochemical flexibility to exploit other sources of phosphorus (including organic sources, through the production of phosphatases, and the phagotrophic ingestion of bacteria) and the extreme motility permitting algae to migrate between depths in the quest for otherwise unexploited nutrient reserves.

Similar reasoning applies to the dynamics of phytoplankton in water bodies where the sources of dissolved inorganic nitrogen are ostensibly yield-limiting and the available sources fall much below $0.2 \, \mu mol \, N \, l^{-1}$. Two important caveats to this relate to the question of dinitrogen fixation in Cyanobacteria, which has often been assumed to free nitrogen fixers from the constraints of deficiencies of dissolved sources of combined nitrogen and, accordingly, to remove the general possibility of N-limitation of the biomass capacity, which may thus be raised to the limit set by the next least available resource. In fact, nitrogen fixation is an energetically expensive activity, and subject to thresholds of carbon fixation rate, phosphorus concentration, and of trace element (iron, molybdenum) availability. The presence of well-known N_2-fixers (Cyanobacteria of the order Nostocales) is not always consequential upon nitrogen depletion. Moreover, the presence of these organisms is not dependent upon a low nitrogen concentration *per se*. Neither is their nitrogen-fixing activity continuous (judging from the frequency of heterocysts, the main fixation sites) but, rather, it is apparently stimulated by diminishing levels of ammonium. Thus, it is not unusual to find heterocystous populations of *Anabaena* or *Aphanizomenon* growing alongside other algae in water having ambient nitrogen concentrations of 6–20 $\mu mol \, NO_3 \cdot N \, l^{-1}$ (Reynolds, 2006). Nevertheless, progressive exclusion of nonnitrogen fixers by growing populations of *Anabaena* or *Aphanizomenon* spp is a familiar outcome in small- to medium-sized, phosphorus enriched lakes, where it is attributable to the unique, dynamic success of Cyanobacteria imparted by the ability to meet their nitrogen requirements from atmospheric sources.

Chronic Nutrient Limitation

The third kind of limitation is the persistent and severe nutrient deficiency, usually associated with the open oceans but it has its analogues among the large lake systems of the world, notably Baykal, Superior, and Tanganyika. Here the capacity limitation set by the bioavailable concentrations of MRP or DIN is so low that phytoplankton concentrations are also continuously small. Their metabolic exchanges are closely enmeshed with those of the heterotrophic bacterioplankton, to the extent that the cap on biomass forces a mutualism in which a significant proportion of the carbon fixed in near-surface photosynthesis is taken up by bacteria and cycled through a web of microbial feeders and their ciliate consumers. The concentration of particles is too sparse to support filter-feeding herbivores, so these ultraoligotropic systems acquire a characteristic functional structure, culminating in calanoid copepods and specialized pelagic fish. The phytoplankton is also generally characterized by a relative abundance (sometimes dominance) of unicells of picoplanktic size (0.2–2 μm in diameter). Such small sizes are beneficial to the sequestration of nutrients from the water; larger cells requiring absolutely greater resources are more reliant than small ones on turbulence to deliver scarce nutrients to their surfaces, whereas there is little prospect of light limitation (at least near the water surface) or of being seriously depleted by herbivores. Net growth rates are low or weakly episodic, in the wake of physical disturbances, but there is a productive yield which drives the higher levels of the pelagic food web. There seems little difference in the outcome whether the oligotrophy is attributable to the severe scarcity of phosphorus or nitrogen, or, as is true of much of the ocean, of iron.

Carbon Limitation

This separate consideration of carbon as an influence on phytoplankton dynamics owes to its role as an essential resource, providing 50% of the ash-free biomass yield of cell growth, the immediate source to phytoplankton (mainly dissolved carbon dioxide in solution) is persistently replenished, albeit at finite rates. There is scarcely a supportive capacity associated with a given carbon content, but phytoplankton growth rates may be potentially or actually severely constrained by such flux terms, if the situation arises when the carbon demand of growing populations depletes the immediate sources of inorganic carbon in the water at a faster rate than they can be replenished, resulting in a dearth of assimilable carbon available to growing algae. The provenance of a part of the carbon dioxide is, self evidently, by invasion of atmospheric gas across the water surface, at a rate that relates to solubility, the difference in partial pressure of the carbon dioxide between the water and air. Migration rates across the surface are accelerated as a function of wind and surface roughness. The air-equilibrated carbon dioxide content of water, $\sim$23 μM at 0 °C, falling to $\sim$13 μM at 0 °C (say 0.15–0.3 mg C l^{-1}), signifies no net gas exchanges but, given a strong partial pressure gradient, wind-dependent invasion rates may achieve the order of magnitude 3–30 $\times$ 10^{-8} mol m^{-2} s^{-1}, or 31–310 mg C m^{-2} day^{-1} (Watson et al.). These are probably rarely realized for, in large lakes, as in the sea, where surface invasion may provide the principal source of carbon, the modest rates of photosynthetic depletion can be balanced, while in smaller lakes, other sources of carbon dioxide are relatively more important. Much of their CO$_2$ reserves are delivered in solution by inflowing streams, or they are derived from the slow oxidation of the large amounts of organic material carbon brought in from the catchment, as terrestrial biomass, particulate detritus, and in solution. Far from being scarce, the concentration of carbon dioxide in many lakes is considerably supersaturated with respect to the air–water equilibrium and the net flux of carbon dioxide across the water surface is actually outward. Perplexingly, most lakes act as organic digesters for their catchments; they are vents rather than sinks for atmospheric carbon dioxide.

In spite of all this, there are plenty of documented instances, especially among fertile, unbuffered soft water lakes during periods of calm, warm weather, where net photosynthetic CO$_2$ consumption depletes lake concentrations to very low levels and raises pH by several points (to >pH 9 or, more rarely to >pH 10). Although the biomass doubling requirement of *Chlorella* can be shown, under conditions of neutrality and saturating carbon dioxide levels (offering $\sim$0.3 mg C l^{-1}), to be satisfied under 40 min, collective carbon demands that outstrip the rate of replenishment and force the upward drift of pH begin to interfere, not just with the capacity of cells to absorb carbon dioxide, but also with its ability to maintain the activity of the RUBISCO enzyme catalyzing photosynthetic carboxylation. Up to pH $\sim$ 8.3, the dissociation of bicarbonate may maintain a proximal source of assimilable carbon but, when this too is exhausted, the rate of cell growth of algae is likely to be already limited by the carbon supply. Interestingly, many algal species possess the ability to mitigate (or, at least, delay) the impact of carbon withdrawal, by invoking ATP-mediated transport mechanisms and the production of carbonic anhydrase, and other enzymes. These mechanisms concentrate and focus dissolved inorganic carbon at the sites of assimilation, so increasing the affinity of the relevant phytoplankters for the modest sources. Expenditure of energy is required and the dependence of dynamic performance on the carbon supply is only partially overcome. Nevertheless, it had been apparent for many years before the various contributory mechanisms were elucidated that phytoplankton algae differ in their abilities to sequester their carbon requirements and, hence, in their sensitivity to diminished CO_2. Among the most tolerant in this respect are the bloom-forming Cyanobacteria, especially *Microcystis*, certain dinoflagellates (e.g., *Ceratium*) and, unusually among desmids (certain species of *Staurastrum*). At the other extreme are the Chrysophyceae, and Synurophyceae, some species of which have been shown to be incapable of using bicarbonate and whose growth is dependent upon a supply of carbon dioxide. In between, there is a scale of species of varying intermediate sensitivities of growth rate to carbon supply: according to Talling's experiments, the sequence, *Aulacoseira subarctica* → *Asterionella formosa* → *Fragilaria crotonensis* → *Microcystis aeruginosa*, represents the increasing pH and diminishing carbon flux at which species-specific growth can be maintained.

Concluding Remarks

On the basis of even this cursory overview, it is possible to assert that the structure of phytoplankton assemblages is, everywhere, the outcome of several simultaneous processes of recruitment and attrition; though they take place at finite rates, the latter are influenced by a wide variety of environmental factors. The extent to which these controls may impinge on the dynamics of natural populations means that there are numerous selective outcomes on the performances of individual species. The frequency and severity with which the dynamic controls are expressed play a large part in determining the qualitative composition of the phytoplankton, at least to the functional, if not the species level.

In much the same way as a plant ecologist can view the vegetation of a given locality and infer so much about the factors controlling its development, so phytoplankton ecology is gradually matching floristics to the dominant moulding processes. Frequent or continuous dynamic dependence upon the supply of certain essential nutrients restricts the biomass-supportive capacity and biases composition in favor of species with high affinity or sourcing versatility for the limiting nutrient. Competitive exclusion is slow but selection of species with relatively better growth performances during resource deficiency is cumulative over many generations. In this way, habitually phosphorus-deficient lakes come to support recognizably similar, defining phytoplankton assemblages.

Relief of phosphorus deficiency allows larger biomasses to be attained and at a faster initial rate. However, if either first encounters limitation by another resource, then the bias of outcome switches away from the poorly adapted species and towards tolerant species. Depletion of nitrogen, silicon, and carbon fluxes each have dynamic consequences, which, in turn, may modify assemblage structure and function in factor-specific ways (respectively, by favoring nitrogen fixers, excluding diatoms, promoting 'eutrophic' species). Sustaining high levels of biomass in water has consequences on light penetration and energy harvesting, and an analogous shift in bias towards low-light-adapted species. More phytoplankton biomass represents a more abundant, directly harvestable food source to animal consumers, with proportionately more varied trophic interactions and compositional outcomes in space and time.

At the basis of all these eventualities are the complex and variable dynamics of the participating organismic populations. Individually, these may be hard to monitor directly and to evaluate relatively but research is gradually revealing the mechanistic rules of their patterns and the empirical sensitivity of their dynamics.

See also: Phytoplankton Population Dynamics: Concepts and Performance Measurement.

Further Reading

Badger MR, Andrews TJ, Whitney SM, *et al.* (1998) The diversity and co-evolution of Rubisco, pyrenoids and chloroplast-based CO_2-concentrating mechanisms in algae. *Canadian Journal of Botany* 76: 1052–1071.

Cole JJ, Caraco NF, Kling GW, and Kratz TW (1994) Carbon dioxide supersaturation in the surface waters of lakes. *Science* 265: 1568–1570.

Droop MR (1973) Some thoughts on nutrient limitation in algae. *Journal of Phycology* 9: 264–272.

Droop MR (1974) The nutrient status of algae cells in continuous culture. *Journal of the Marine Biological Association of the United Kingdom* 54: 825–855.

Dugdale RC (1967) Nutrient limitation in the sea: Dynamics, identification and significance. *Limnology and Oceanography* 12: 685–695.

Elliott JA, Irish AE, Reynolds CS, and Tett P (1999) Sensitivity analysis of PROTECH, a new approach in phytoplankton modelling. *Hydrobiologia* 414: 37–43.

Elliott JA, Irish AE, Reynolds CS, and Tett P (2000) Modelling freshwater phytoplankton communities: An exercise in validation. *Ecological Modelling* 128: 9–16.

Gasol JM, Guerrero R, and Pedrós-Alió C (1992) Spatial and temporal dynamics of a metalimnetic *Cryptomonas* peak. *Journal of Plankton Research* 14: 1565–1579.

Moss B (1973) The influence of environmental factors on the distribution of freshwater algae; an experimental study. II. The role of pH and the carbon dioxide–bicarbonate system. *Journal of Ecology* 61: 157–177.

Omata T, Takahashi Y, Yamaguchi O, and Nishimura T (2002) Structure, function and regulation of the cyanobacterial high-affinity bicarbonate transporter, BCT-1. *Functional Plant Biology* 29: 151–159.

Qiu B and Gao K (2002) Effects of CO_2 enrichment on the bloom-forming Cyanobacterium *Microcystis aeruginosa* (Cyanophyceae): Physiological responses and relationships with the availability of dissolved inorganic carbon. *Journal of Phycology* 38: 721–729.

Raven JA (1997) Inorganic carbon acquisition by marine autotrophs. *Advances in Botanical Research* 27: 85–209.

Raven JA, Ball LA, Beardall J, Giordano M, and Maberly SC (2005) Algae lacking carbon-concentrating mechanisms. *Canadian Journal of Botany* 83: 879–890.

Reynolds CS (1983) Growth-rate responses of *Volvox aureus* Ehrenb (Chlorophyta, Volvocales) to variability in the physical environment. *British Phycological Journal* 18: 433–442.

Reynolds CS (1986) Experimental manipulations of the phytoplankton periodicity in large limnetic enclosures in Blelham Tarn, English Lake District. *Hydrobiologia* 138: 43–64.

Reynolds CS (1990) Temporal scales of environmental variability and the responses of phytoplankton. *Freshwater Biology* 23: 25–53.

Reynolds CS (1997) *Vegetation Processes in the Pelagic: A Model for Ecosystem Theory.* Oldendorf: Ecology Institute.

Reynolds CS, Thompson JM, Ferguson AJD, and Wiseman SW (1982) Loss processes in the population dynamics of phytoplankton maintained in closed systems. *Journal of Plankton Research* 4: 561–600.

Saxby-Rouen KJ, Leadbeater BSC, and Reynolds CS (1998) The relationship between the growth of *Synura petersenii* (Synurophyceae) and components of the dissolved inorganic carbon system. *Phycologia* 37: 467–477.

Shapiro J (1990) Current beliefs regarding dominance by blue greens: the case for the importance of CO_2. *Verhandlungen der internationalen Vereinigung für theoretische und angewandte Limnologie* 24: 38–54.

Talling JF (1976) The depletion of carbon dioxide from lake water by phytoplankton. *Journal of Ecology* 64: 79–121.

Tandeau de Marsac N (1977) Occurrence and nature of chromatic adaptation in Cyanobacteria. *Journal of Bacteriology* 130: 82–91.

Watson AJ, Upstill-Goddard RC, and Liss PS (1991) Air–sea exchange in rough and stormy seas measured by a dual-tracer technique. *Nature* 349: 145–147.

Wolf-Gladrow DA and Riebesell U (1997) Diffusion and reactions in the vicinity of plankton: a refined model for inorganic carbon transport. *Marine Chemistry* 59: 17–34.

Phytoplankton Productivity

M T Dokulil and C Kaiblinger, Austrian Academy of Sciences, Mondsee, Austria

Defining Productivity

Ecologists prefer the term phytoplankton productivity although what is usually measured are photosynthetic rates. Moreover, the term productivity is frequently confused with production. It has therefore been argued to abandon one of the two terms or to replace it by activity. Strictly speaking, there is a clear difference. Productivity is a time dependent process; it is a rate with dimensions of mass per unit area/volume per time. Production is a quantity, with dimensions of mass.

Biological processes perceive and adjust to changes in the external environment. Many regulatory processes occur rapidly to allow photosynthesis to track instantaneously changes in the surrounding medium (**Figure 1**). In nature, photosynthetic processes are therefore constantly modified. Regulation describes adjustments of catalytic or energetic efficiency, which involve slight structural modifications. Physical, biochemical, or physiological response on a short time scale within the life span of individual organisms or assemblages are collectively called acclimations, which are nonlinearly related to temporal variations in irradiance, temperature, and nutrients. Ecological or evolutionary adaptations by assemblages through selection of phenotypic traits occur on a longer time scale. Associated changes in community structure affect photosynthetic rates.

Terminology

Conceptually, photosynthesis or productivity is defined as an instantaneous rate according to dB/dt over infinite time intervals, which cannot be measured. Normally, integral rates over time are obtained. Estimation by fluorescence techniques however come closest to instantaneous rates.

Gross photosynthesis (P_G) is defined as the light-dependent rate of electron flow from water to the terminal electron acceptors (e.g., CO_2) in the absence of any respiration losses. This definition is directly proportional to linear photosynthetic electron transport, and, hence, gross oxygen evolution. Therefore, gross photosynthesis should be defined on the basis of oxygen evolution rather than carbon fixation. The difference is critical, especially when photorespiration is high. Respiratory losses (R_P) in photosynthetic organisms are rates of electron flow from organic carbon to O_2 with the concomitant production of CO_2, including photorespiration. The difference of these two processes is called net photosynthesis (P_N):

$$P_N = P_G - R_P \qquad [1]$$

Gross primary production is identical to gross photosynthesis. Net primary production, however denotes organic carbon produced by photosynthetic processes within a specified time period available to other trophic levels.

In natural aquatic environments, direct measurements of primary production are virtually impossible to obtain because it is difficult to determine the contribution of algal respiration to total respiratory losses. Measurements of respiration include the metabolic contribution of heterotrophs, and therefore reflect community respiration (R_c). In fact, net primary production values, when reported, are usually confused with net photosynthesis.

Estimating Photosynthesis in Aquatic Ecosystems

The simple basic reaction equation of photosynthesis in green plants is as follows:

$$CO_2 + H_2O + 8\ photons \rightarrow \frac{1}{6}C_6H_{12}O_6 + H_2O \qquad [2]$$

Measurements of photosynthesis for autotrophic plankton organisms in aquatic ecosystems are therefore usually based on rates of either oxygen evolution or inorganic carbon uptake. Various modifications and alternatives of the basic techniques exist. In addition, fluorescence techniques recently become increasingly important. All methods are used either in situ or in laboratory simulations. A variety of mathematical models describe photosynthetic rates in space and time and their acclimation to environmental conditions.

The O₂-Technique

The rate of oxygen evolution is usually quantified from chemical, electrochemical or, most recently, from optochemical methods. Because absolute changes in oxygen concentration are usually small and the background concentration of the gas in most situations is very large, precise data are relatively difficult to obtain. Moreover, even if done precisely, evolution of oxygen in the light represents net community photosynthesis, and fluxes of oxygen in the light and dark include total

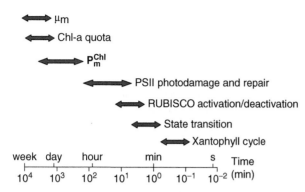

Figure 1 First-order time constants for acclimation of algal photosynthesis to step changes in photon flux density. Modified from Raven and Geider (2003) in Larkum AWD, Douglas SE, and Raven JA (eds.) (2003) Photosynthesis in algae. In: *Advances in Photosynthesis and Respiration*, Vol. 14, p. 407. Dordrecht: Kluwer.

community respiration. Adding dark respiratory losses to light-dependent oxygen evolution gives a measure of gross photosynthesis.

Estimates of net photosynthesis in natural aquatic ecosystems based on measurements of oxygen fluxes are more uncertain than measurements of gross photosynthesis.

The ^{14}C Method

By far the most commonly used method is based on the rate of incorporation of radioactive ^{14}C in the form of inorganic carbon into acid-stable (usually particulate) organic carbon. This method was introduced first to oceanography and is, in principle, relatively straightforward. If we assume that over some finite time period, the change in concentration of total dissolved inorganic carbon in the bulk water is small relative to the photosynthetic rate of the cells, and the addition of a small amount of radioactively labeled inorganic carbon does not perturb the concentration of the dissolved inorganic carbon, then the rate of incorporation of radioactivity into organic material obeys the rules of tracer analysis. These conditions are met in many environments, but care must be taken in fresh waters having low total concentrations of inorganic carbon.

The rationale for the ^{14}C-based tracer method is that the light-dependent rate of incorporation of the radioactively labeled carbon into organic material is quantitatively proportional to the rate of incorporation of nonradioactive inorganic carbon.

$$^{12}C_{uptake} : {}^{12}C_{available} = {}^{14}C_{uptake} : {}^{14}C_{available} \qquad [3]$$

From eqn [3] follows that carbon uptake is

$$^{12}C_{uptake} = \left({}^{14}C_{uptake} / {}^{14}C_{available} \right) * {}^{12}C_{available} \qquad [4]$$

Because ^{14}C is heavier than the stable natural isotope ^{12}C, there is an isotopic discrimination against the radioactive isotope during carbon fixation. The commonly accepted discrimination factor is taken as about 5% and this factor is taken into account in calculations by multiplication with 1.05.

Over short periods of time, before a significant fraction of organic carbon becomes labeled, the method gives a reasonable estimate of gross photosynthetic rates. As the exposure time continues, the organic carbon pool becomes increasingly labeled, ultimately reaching equilibrium with the isotope in the bulk water. A fraction of the labeled carbon is then respired, and the incorporation of the tracer begins to approximate the rate of net photosynthesis.

The rate at which equilibrium labeling is approached is dependent on the growth rate of the organism(s). The faster the growth rate is, the faster the equilibrium will occur. Thus the interpretation of radioactive carbon incorporation as gross or net photosynthesis is generally somewhat ambiguous, and is complicated by the duration of the experiment in relation to the growth rate. In nature, the latter parameter is usually unknown and numerous discussions have emerged concerning the validity, accuracy, and interpretation of the method.

Fluorescence Techniques

The major problem of the methods mentioned previously is the timescale. Photosynthetic rates obtained by in situ incubation techniques are average values over a given time. To overcome theses problems, bio-optical models have been developed in oceanography to extrapolate photosynthesis vs. irradiance curves (*P/E*). In bio-optical models, it is assumed that the *P/E* relationship does not depend on the incubation time, which is not true. Most of the short term adjustments of phytoplankton are expressed in changes of the functional absorption cross-section σPSII. To keep the timescale short, rapid measurement techniques based on in vivo fluorescence have been developed mainly for marine use. Recent active fluorescence techniques opened the possibility for quantifying small-scale, rapid changes in physiology by providing estimates of primary productivity with a temporal resolution of seconds.

Fluorescence is a phenomenon based on the absorption of light by photosynthetic pigments and its reemittance at a longer wave length. When light energy (photons) reaches the light trapping antennapigments of a photosynthetic unit (PSU), the chlorophyll-a (Chl-a) molecule is excited. To return to the ground state, the Chl-a molecule needs to release energy, most of which is lost as heat. What is left is used to 'close' the reaction center or is reemitted as fluorescence. Chlorophyll fluorescence is very much dependent on the wavelength and intensity of the incoming light. Either photosynthesis or thermal dissipation influence fluorescence resulting in photochemical quenching (qP) or nonphotochemical quenching (NPQ), respectively. While high light conditions can cause NPQ, weak light that is continuously applied is correlated to in vivo chlorophyll. When phytoplankton is in a dark adapted state (e.g., during night or when the light is very weak), all reaction centers are in an active, open state and the fluorescence yield is minimal. In this state, fluorescence is at a baseline value F_0. When all the traps are closed the maximum fluorescence level (F_m) is reached. Variable fluorescence is obtained by subtracting F_0 from F_m ($F_v = F_m - F_0$). The potential yield of the photochemical reaction is therefore given by F_v/F_m, which is quantitatively related to the photochemical efficiency of PSII. On the basis of these parameters fluorometric methods have been developed since the early 1970s:

Pump-and-Probe fluorometry Pump-and-Probe (P & P) Fluorometry uses an intense 'pump' flash to close the reaction centers of natural phytoplankton assemblages and reach F_m. After a decay of 80–100 μs, a weak 'probe' flash is induced to reopen the traps and yield F_0. By controlling the intensity of the pump flash and the time-delay between the pump and probe flashes, more physiological information can be retrieved.

Pulse-Amplitude-Modulation fluorometry Pulse-amplitude-modulation (PAM) fluorometry is a variant of the P & P technique and was introduced to distinguish between photochemical and nonphotochemical quenching. Repeated intense flashes are used against a background of low light to obtain F_m and F_0. PAM as well as P & P fluorometry are methods that are based on multiple turnover rates. However, both these two methods have limitations. PAM fluorometry provides only variable fluorescence and some features of the PSU such as the functional absorption cross-section (σPSII) cannot be described, while P&P fluorometry is too slow to follow the dynamic changes in PSII.

Fast repetition rate fluorometry This fluorometry has been developed to overcome the deficiencies of its predecessors. The instrument is based on high frequency emission of blue light flashes at subsaturating energy, which gradually saturates the PSII reaction centers within a single turnover. The rate required to saturate the PSII reaction centre is proportional to the functional absorption cross-section of

PSII (σPSII). After reaching the saturation level (F_m), relaxation flashes with longer interflash delay are emitted to reopen the reaction centre. The rate of relaxation is related to the turnover rate of PSII (τ). The saturation protocol is set to provide 50–100 saturation flashes per sequence, followed by 20 relaxation flashes with a longer time-delay between the flashes to allow reoxidation.

Primary productivity estimates can be derived from variable fluorescence. By introducing the photosynthetic quotient (PQ) in the equation, the rate of carbon fixation can be obtained to qualitatively predict photosynthetic rates from changes in the quantum yield of fluorescence.

Physiology of Photosynthesis

Photosynthesis involves a series of reactions that start with light absorption, which is followed by synthesis of intermediate energy-conserving compounds, finally leading to CO_2 fixation in the Calvin cycle. Although eqn [3] captures the essence of photosynthesis, it is an incomplete model because of the Mehler reaction, photorespiration, allocation of photosynthate to protein synthesis, and incorporation of carbon by nonphotosynthetic reactions. Hence, interpretation of productivity measurements in fresh waters is difficult, dependent on the technique used, and further complicated by the complex and variable composition of natural algal assemblages.

Light Harvesting

Photosynthesis begins with photoabsorption by pigments located in the thylakoid membranes. The only pigment present in all oxygen producing phytoplankton is Chl-a and it is the essential pigment required for photosynthesis. Most light absorption, however, is due to accessory pigments such as other chlorophyll species, carotenoids, or phycobilins. Major differences in pigment composition among higher taxonomic groups can significantly affect the ability of phytoplankton to absorb light. Chl-a-specific light absorption is not only affected by the pigment complement, but also by cell size and intracellular pigment concentration.

Photosynthetic electron transfer (PET), linking oxygen evolution to reductant production, is located in large membrane located complexes harboring Photosystem I (PSI) and Photosystem II (PSII). The structural complexity of the PET chain allows considerable versatility in electron flow to match varying requirements of nutrient transport, photo-, and biosynthesis. The ratio of PSII:PSI however, remains ambiguous in nature because of variability among algal divisions and environmental conditions. In Heterokontophyta the ratio may exceed 2, whereas in Cyanobacteria it is typically 0.25–0.5. The ratio may also change with growth, irradiance, or nitrogen starvation.

Carbon Reduction

Net carbon fixation in photosynthesis involves a cycle of reductions referred to as the Calvin cycle. At the heart of this cycle is the carboxylation of ribulose 1,5-biphosphate (RuBP) catalyzed by the enzyme RUBISCO, a monophyletic enzyme existing in two forms. The Form I enzyme is found in cyanobacteria and in all eukaryotes, with the exception of some peridinin-containing Dinophyta, which contain the Form II enzyme. The activity of the enzyme is low compared with other carboxylases. RUBISCO accounts for a sizeable fraction of the cell (1–10% of cell carbon or about 2–10% of total protein), with the amount dependent on ambient light conditions. Under high irradiance, RUBISCO can account for up to five times as much cell mass as Chl-a.

Understanding the role of RUBISCO in phytoplankton ecophysiology is essential to determine whether RUBISCO or PET limits light-saturated photosynthesis and to estimate the concentration of CO_2 needed to support a specific photosynthetic rate. This has implications for assessing the inhibition of carbon fixation by oxygen and the role of a CO_2-concentrating mechanism. Activity of the enzyme is regulated in vivo by a multitude of mechanisms and external factors best studied in Chlorophytes and Cyanobacteria. Under low light or low CO_2, RuBP is saturating and the rate of CO_2 fixation is limited by RUBISCO activity. However, under conditions of high light or high CO_2, carbon fixation is limited by the regeneration of RuBP, a process known as sink limitation.

Photorespiration

Photorespiration is the oxygenation of RuBP by RUBISCO followed by photorespiratory glycolate metabolism. Competition between O_2 and CO_2 reduces the rate of carbon assimilation, energetic efficiency of photosynthesis, and may reduce the photosynthetic quotient (PQ = O_2 evolved/CO_2 assimilated). Values of PQ of 1.2–1.8 are representative for protein and lipid synthesis. Values of PQ as low as 0.75 can be obtained from photorespiration and glycolate excretion at the CO_2 compensation point, where CO_2 uptake equals CO_2 evolution. PQ values below 0.75 can only be explained if unbalanced growth conditions where respiration contributes to net gas exchange are considered.

Evidence that photorespiration influences net photosynthesis and is linked to photoinhibition stems from the effect of oxygen concentration on the rate of ^{14}C assimilation. The observed reduction in particulate ^{14}C is accompanied by an increase of excreted dissolved organic carbon.

Carbon-Concentrating Mechanisms

Most of the understanding of the biophysical carbon-concentrating mechanisms (CCM) is based on studies of Cyanobacteria and Chlorophytes. Bacillariophyceae and peridinin-containing Dinophytes have less efficient CCM. Direct evidence for a CCM comes from intracellular CO_2 levels exceeding those in the suspending medium. In Cynobacteria, the CCM is associated with the carboxysomes in Chlorophytes and other eukaryotes and it is associated with pyrenoids containing highly varying amounts of RUBISCO.

An alternative to biophysical CCM is C4 photosynthesis, still a controversial issue in algae. The confusion over C4 metabolism arises because the enzymes present in C4- photosynthesis are present in algae as well. These enzymes are present in different algal groups in different amounts and are not sensitive to oxygen like the Calvin cycle enzyme RuBP. The best cases so far have been reported for two marine organisms.

Other Pathways for C-Acquisition

Besides the Calvin cycle, carbon may also be gained via two other pathways: β-carboxylation and mixotrophy. Anapleurotic carboxylation occurs in the dark and does not generally result in a net increase in carbon fixation. Alternatively, carbon can be acquired through mixotrophy. Several algal groups now have been shown to be facultatively mixotrophic, heterotrophic, or may use dissolved organic compounds osmotrophically. The substitution of autotrophy and the regulation of mixotrophy are not well understood.

Response of Phytoplankton Productivity to Environmental Variability

Instantaneous rates of photosynthesis are controlled by external factors. Understanding how the environment influences Chl-a-specific light absorption (α^*), the ratio Chl-a:C, and the parameters of the PE curve are essential to evaluate, model, and predict primary productivity. The rate of photosynthesis is controlled by the efficiency of light utilization so as to drive the ensemble of photosynthetic reactions. As a first approach, the light dependency can be described as:

$$P_E = E_a \Phi_E \qquad [5]$$

where P_E is the photosynthetic rate at incident irradiance E, E_a is the light absorbed by the organism, and Φ_E is the quantum yield at irradiance E. The absorbed light is calculated from incident spectral irradiance $E_{0(\lambda)}$ and α^*:

$$E_a = E_{0(\lambda)}\alpha^* \quad [6]$$

The quantum yield must be specifically related to a product of the photochemical reaction. For oxygenic photoautotrophs, this is usually oxygen evolved or carbon fixed as discussed earlier; hence units for quantum yield are normally moles of O_2 evolved or CO_2 fixed per moles of photons absorbed.

The photosynthesis–irradiance response (PE curves) is commonly obtained from oxygen or carbon-based rates per hour normalized to Chl-a, versus irradiance E. Typically three distinct regions, namely a light-limited region, a light-saturated region, and a region of photoinhibition, can be distinguished (**Figure 2**). Photosynthetic rates are linearly proportional to irradiance at low light levels because the rate of photon absorption determines the rate of steady-state electron transport from water to CO_2. The initial slope of this light limited region is commonly denoted by the symbol α or α^B when normalized to Chl-a. This initial slope of the PE curve is not a photosynthetic rate; rather it is a time independent

measure of efficiency indirectly related to the maximum quantum yield Φ_m.

As irradiance further rises, photosynthetic rates become increasingly nonlinear and approach a saturation level P_m at which photon absorption exceeds steady state electron transport. Light-saturated photosynthesis is explicitly time dependent with dimensions of O_2 evolved or CO_2 fixed per unit chlorophyll per unit time. The curvature can be described by the light-saturation parameter E_k which is never actually directly measured; rather it is derived from the intersection of α and P_m (**Figure 2**):

$$E_k = P_m/\alpha \quad [7]$$

At supraoptimal irradiance beyond light saturation, a reduction in photosynthetic rate may occur. This reduction is dependent on intensity of the light and the duration of exposure and is called photoinhibition. This modification of P_m is brought about by either a reduction in the number of photosynthetic units or by an increase in the maximum turnover time. The degree of inhibition can be estimated from the parameter β, which is derived similar to α from the PE curve (**Figure 2**). Corresponding to E_K, a light intensity E_I can be inferred describing the onset of inhibition. A number of further light intensities can be deduced from PE curves as indicated in

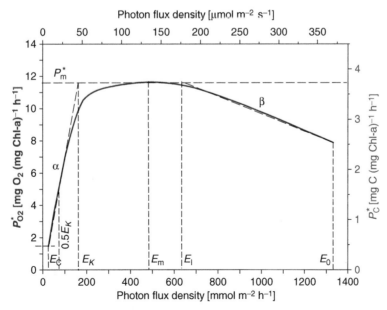

Figure 2 Example of a photosynthesis–irradiance response curve (PE) expressed both as chlorophyll-specific oxygen evolution ($P^*_{O_2}$, left Y-axis) and carbon uptake (P^*_C, right Y-axis) versus photon flux density. The two axes are shifted by $PQ = 1.16$. P^*_m, Maximum chlorophyll-specific rate of photosynthesis; α, slope of the light limited region; β, slope of the region of photoinhibition; E_0 subsurface photosynthetic active radiation (PAR); E_I, photon flux density at onset of inhibition; E_m, photon flux at P^*_m, E_K, photon flux density at the onset of saturation; $0.5E_K$, 50% of E_K used in certain types of column integration calculation; E_C, Photon flux density at the compensation point.

the caption to **Figure 2**. Most notable is the compensation irradiance (E_C) when net gas exchange becomes zero because R equals GPP. The compensation irradiance can be related to α^{Chl} and chlorophyll normalized respiration:

$$E_C = R^{Chl}/\alpha^{Chl} \qquad [8]$$

The irradiance at the compensation point is not fixed but varies primarily with respiration and to a lesser extent with α^{Chl}, both of which depend on environmental conditions.

Physiological acclimation of the PE parameter α^{Chl} appears to follow several simple rules under conditions of balanced growth (**Figure 3**). As a first approximation α^{Chl} can be considered constant, with changes to the carbon-specific light-saturated rate (P_m^c) inducing changes in the ratio Chl-a:C. At constant temperature and when cells are nutrient-repleted, maximum carbon-specific rate is largely independent of irradiance, but is highly correlated with maximum specific growth rate, μ_m and temperature because temperature affects enzymatic reactions. The light-saturated rate often is also determined by CO_2 availability.

Several empirical mathematical descriptions of the PE curve exist. The various formulations mainly differ in the abruptness of the transition from light-limited to light-saturated photosynthesis. Some models include photoinhibition.

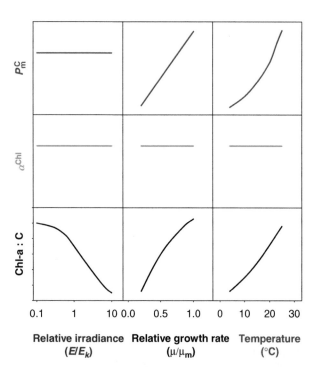

Figure 3 Schematic relationships of photoacclimation of P_m^c, α^{Chl}, and Chl:C for light (*left*), nutrient (*center*), or temperature (*right*) limitations.

One simple mechanistic interpretation using the terms and notations mentioned already is the exponential function:

$$P^{Chl} = P_m^{Chl}[1 - \exp(-\alpha^{Chl}E/P_m^{Chl})] \qquad [9]$$

Productivity in the 'Real' World

Let us now consider photosynthesis in a planktonic system in a water column, a planar representation of surface area projected to a depth z. Light (E) incident on the water surface is attenuated with depth, approximately following an exponential function (**Figure 4**). We define a depth z_{eu} equal to the 1% light intensity as the base of the euphotic zone. This is the portion of the water column supporting primary production in most instances. In some cases, extremely light-adapted assemblages can survive at much lower photon flux densities. The euphotic depth is often confused with the compensation depth (z_c), defined as the depth at which daily gross productivity balances respiratory losses over a day. Above z_c, net photosynthesis is possible; below z_c the balance of oxygen production and consumption is negative. The average compensation depth is dependent on the surface light intensity E_0 and can be highly variable; it may however, frequently correspond to the euphotic depth. If the portion of the water column mixed by wind (z_{mix}) extends beyond the euphotic zone, a critical depth (z_{cr}) can be defined, where integral net photosynthesis balances integral respiratory losses (**Figure 4**).

Also in **Figure 4**, an idealized daily integrated profile of photosynthetic rates is presented with algal biomass assumed to be homogeneous, and therefore unimportant to the shape of the curve. At the surface, photosynthetic rates are typically depressed as a manifestation of photoinhibition, which, at least partly, is an artifact of in situ methodology. Common methods keep algae constantly at high irradiance over prolonged periods of time, which otherwise would float freely through the light gradient. Rates rise to P_m lower in the column and then decline monotonically as light becomes limiting, representing the low irradiance-dependent region of the PE curve. Respiration losses (R_P) are assumed to be constant throughout the water column. Traditionally, respiration of the photoautotrophs has been assumed to be 10% of the maximum photosynthetic rate, and independent of irradiance. Many freshwater studies however, indicate that respiratory losses relative to P_m are variable. Additionally, the euphotic zone is influenced by temperature, supraoptimal light levels, nutrient limitations, nonhomogenous distribution of

The labels on the figure axes (bottom): **Relative irradiance** (E/E_k)   **Relative growth rate** (μ/μ_m)   **Temperature** (°C), with values 0.1, 1, 10, 0.0, 0.5, 1.0, 0, 10, 20, 30. The vertical axes labels are P_m^c, α^{Chl}, and Chl-a : C.

phytoplankton biomass, or even compositional changes of the assemblage within the water column. Plankton respiration on average increases with Chl-a, total phosphorus (TP), and dissolved organic carbon (DOC) concentrations. Accurate measurements of R_P are one of the biggest problem in estimations of primary production, and will also affect the determination of the compensation depth as well as the critical depth.

Depending on the composition and vertical distribution of phytoplankton biomass, vertical profiles of phytoplankton productivity can substantially deviate from the schematic profile shown in **Figure 4**. Three

examples are depicted in **Figure 5**. Deep chlorophyll maxima of various types of algae, such as picoplankton, Cyanobacteria etc., can produce secondary peaks of productivity at a greater depth. Turbid shallow lakes often do not show photoinhibition at the surface and have steep, compressed vertical profiles. High mountain lakes may have profiles reaching P_m near the sediment because motile phytoplankton species dominate, which tend to avoid large UV doses associated to the high photon flux at such elevations (**Figure 5**). Such 'irregular' types of vertical profiles cannot be transferred to PE curves; hence, physiologically meaningful parameters are extremely

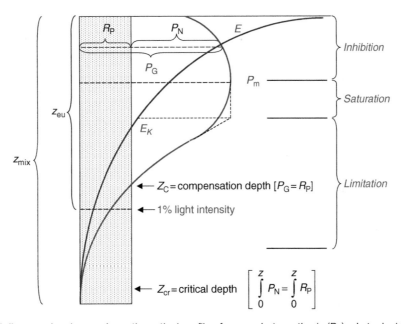

Figure 4 Conceptual diagram showing a schematic vertical profile of gross-photosynthesis (P_G), phytoplankton respiration (R_P), and attenuation of irradiance (E) in a hypothetical lake. Maximum photosynthetic (P_m) and net-photosynthetic rates (P_N) are indicated. The regions of photoinhibition, light saturation, and light limitation are indicated on the right-hand side. The depth of the euphotic zone (z_{eu}), equal to 1% light intensity, the compensation depth (z_c), and the critical depth (z_{cr}) as well as the mixing zone (z_{mix}) are given.

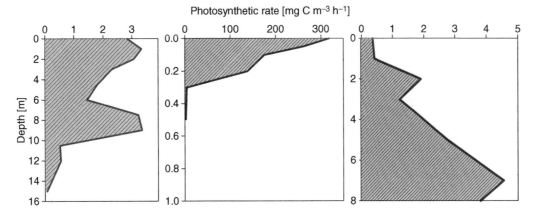

Figure 5 Examples of vertical productivity profiles.

difficult to deduce. Moreover, these profiles cannot be described in simple mathematical terms, and modeling therefore is complicated if not impossible.

Daily, Seasonal, and Interannual Variability of Production

One of the major determinants of variability in photosynthetic response is the daily variation in irradiance. Both the light-saturated and the light-limited chlorophyll-specific photosynthetic rates are generally elevated during the photoperiod. The correlation between these two parameters suggests an increase either in the number of active reaction centers or in their turnover rate. The dial irradiance cycle is the natural synchronizing agent for algal populations in the environment constraining growth rates. Algal assemblages must cope with short term changes in irradiance during a day, and acclimate photosynthetic rates to variations from day to day.

Superimposed is the annual cycle of solar radiation, which depends on latitude and altitude. Depending on the geographical position, annual variation of irradiance ranges from moderate (near the equator) to very pronounced (at high latitudes). Accordingly, the magnitude of primary production strongly varies with climatic regions (**Table 1**).

Assessment of Primary Production at the Global Scale

Several thousand vertical profiles of daily integrated photosynthetic rates have been obtained worldwide from many different freshwater ecosystems. Vertical profiles of photosynthetic rates from a large number of lakes using different measurement techniques are summarized and expressed as volumetric carbon uptake per day in **Figure 6(a)**. The profiles display a high degree of variability from as low as 0.1 mg C m^{-3} day^{-1} to way over 1000 mg C m^{-3} day^{-1}. The range in fresh waters, therefore, is at least four orders of magnitude while it is only three in the sea.

This is not surprising, considering the wide variety of lake types from extremely nutrient poor (ultraoligotrophic) to nutrient rich (hypertrophic), from very shallow lakes (< 2 m) to deep lakes (>200 m). Similarly, average daily column productivity ($\Sigma\Sigma P$) ranges from 50 mg C m^{-2} day^{-1} to greater than 2500 mg C m^{-2} day^{-1} (**Table 2**).

The major source of variation in photosynthetic rates in aquatic systems is related to the amount and distribution of photoautotrophic biomass. To put it simple, under any irradiance condition photosynthetic energy flow is dependent on population density of

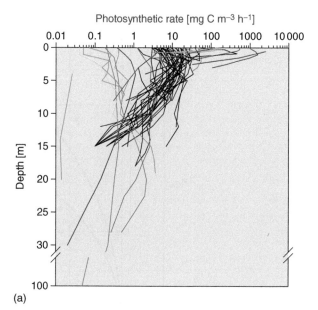

(a)

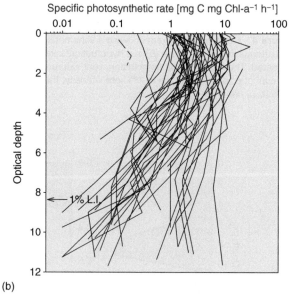

(b)

Figure 6 One hundred vertical profiles of carbon fixation as a function of physical depth (a) and normalized to chlorophyll-a, and optical depth (b). One unit of optical depth is equal to a halving of light intensity.

Table 1 Ranges of average daily and annual net-productivity in ecosystems of the world

Ecosystem	Mean daily production (mg C m^{-2} day^{-1})	Average annual production (g C m^{-2} year^{-1})
Tropical lakes	100–7600	30–2500
Temperate lakes	5–3600	2–950
Arctic lakes	1–170	<1–35
Antarctic lakes	1–35	1–10
Alpine lakes	1–900	<1–250
Temperate rivers	<1–3000	<1–650
Tropical rivers	<1–24 000	1–5000

Table 2 Primary production, biomass, chlorophyll-a (Chl-a), and dominant algal groups for the trophic categories observed in fresh waters

Trophic level	Primary production $(mg\ C\ m^{-3}\ day^{-1})$	Biomass $(mg\ C\ m^{-3})$	Chl-a $(mg\ Chl\text{-}a\ m^{-3})$	Dominant algal group(s)
Ultraoligotrophic	<50	<50	0.01–0.5	Chrysophyceae
Oligotrophic	50–300	10–100	0.3–3.0	Cryptophyceae
				Bacillariophyceae
Mesotrophic	250–1000	100–300	2–15	Dinophyta
Eutrophic	600–2500	250–600	15–30	Cyanobacteria
Hypertrophic	>2500	>600	>30	Chlorophyta
				Euglenophyta
Dystrophic	<50–500	<50–200	0.1–10	

the photosynthetic machinery. Since photosynthetic reaction centers are causally related to the concentration of photosynthetic pigments, normalization to Chl-a, which is universally contained in all algal classes should lead to a reduction in variance. When carbon uptake rates from **Figure 6(a)** are normalized to Chl-a and plotted versus optical depth (which is independent of physical depth), the magnitude is largely reduced (**Figure 6(b)**). Chlorophyll-specific carbon uptake rates range from less than 1 to over $10\ mg\ C\ mg\ Chl\text{-}a^{-1}\ h^{-1}$, similar to observations in the sea.

See also: Phytoplankton Nutrition and Related Mixotrophy.

Further Reading

Blankenship RE (2002) *Molecular Mechanisms of Photosynthesis.* Oxford: Blackwell.

del Giorgio PA and Williams PJ le B (eds.) (2005) *Respiration in Aquatic Ecosystems.* Oxford: Oxford University Press.

Dokulil MT, Teubner K, and Kaiblinger C (2005) Produktivität Aquatischer Systeme. Primärproduktion (autotrophe Produktion). In: Steinberg C, Calmano W, Klapper H, and Wilken R-D (eds.) Handbuch Angewandte Limnologie, IV-9.2, 1–30, 21, Erg. Lfg, 4/05, Landsberg: Ecomed.

Falkowski PG and Raven JA (1997) *Aquatic Photosynthesis.* Malden: Blackwell.

Geider RJ and Osborne BA (1992) *Algal Photosynthesis.* New York: Chapman & Hall.

Goldmann CR (ed.) (1965) Primary productivity in aquatic environments. *Memorie Dell Instituto Italiano Di Idrobiologia* 18(Suppl.).

Larkum AWD, Douglas SE, and Raven JA (eds.) (2003) Photosynthesis in algae. *Advances in Photosynthesis and Respiration,* Vol. 14. Dordrecht: Kluwer.

LeCren ED and Lowe-McConnell RH (eds.) (1980) *The Functioning of Freshwater Ecosystems.* Cambridge: Cambridge University Press.

Li WKW and Maestrini SY (eds.) (1993) Measurement of primary production from the molecular to the global scale. *ICES Marine Science Symposia* 197: 1–287.

Morris I (ed.) (1980) *The Physiological Ecology of the Phytoplankton.* Oxford: Blackwell.

Platt T (ed.) (1981) Physiological bases of phytoplankton ecology. *Canadian Bulletin of Fishery and Aquatic Sciences* 210: 1–346.

Steemann Nielsen E (1975) *Marine Photosynthesis with Special Emphasis on the Ecological Aspects.* Amsterdam: Elsevier.

Williams PJ le B, Thomas DN, and Reynolds CS (eds.) (2002) Phytoplankton Productivity. *Carbon Assimilation in Marine and Freshwater Ecosystems.* Oxford: Blackwell.

Comparative Primary Production

M T Dokulil, Austrian Academy of Sciences, Mondsee, Austria

Comparative Productivity

Inland water bodies are an infinite variety of freshwater and saline lakes, ponds, human-made reservoirs, rivers streams, swamps, and marshes. Total production of these lentic and lotic ecosystems comprises a large number of photoautotrophs that inhabit a variety of habitats. Aquatic systems are conceptually divided into pelagic (free-water), benthic (bottom), and littoral (shore) biotopes, as well as aquatic–terrestrial ecotones. Organisms range from microscopic forms, such as planktonic and benthic algae, via macroscopic visible life forms, such as filamentous algae and charophytes, to submersed and emergent macrophytes.

Primary production can be compared

- between different components within the same freshwater ecosystem,
- between types of freshwater ecosystems,
- between climatic regions, and
- in relation to other ecosystem types.

Estimation of Productivity and Production

In principle, the same or a similar terminology as outlined elsewhere applies to the organisms considered here. Corresponding to the variety of habitats and life forms, a large number of different techniques and methods exist.

Modifications of conventional techniques are used to assess photosynthetic rates of microscopic and macroscopic algal assemblages. Productivity on substrates, however, is often estimated from algal accumulation as either area-specific (dN/dt, e.g., cells cm^{-2} day^{-1}) or biomass-specific rates (dN/dt, e.g., cells $cell^{-1}$ cm^{-2} day^{-1}). The term accrual rate is also used for area-specific accumulation rates. Gross accrual and net accrual rates might be appropriate for biomass changes (biomass cm^{-2} day^{-1}) that do or do not respectively take gains and losses into account. Many measurements are used to estimate benthic algal biomass. Besides cell-counting methods, several inexpensive techniques are used to obtain estimates of biomass, such as determination of chlorophyll a, algal group-specific pigments (marker pigments), dry mass, ash-free dry mass, C, N, or P content of the cells, all expressed usually per square centimeter. These different measurements have their advantages and disadvantages. Intercomparison of data is often difficult because of the multitude of units used.

Production of macroscopic algae and, especially, macrophytes is often deduced from standing crop censuses at times of maximum biomass development. Again, as outlined above, a number of different estimators of biomass are in use, making comparison of results difficult. Production of higher plants (macrophytes) is usually expressed as dry mass per unit area. It must be emphasized, however, that standing crop estimates in macrophytes commonly refer to above-ground biomass only, and do not take losses into account or assume them negligible.

River assessments often use an integrative approach. Total area-specific net production may be calculated for a river segment from diurnal changes in O_2 or CO_2 concentration of the open water.

Components of Freshwater Ecosystems

Benthic Algae

Benthos refers to organisms living on the bottom or associated with substrata. Terms synonymous with benthic algae are periphyton and Aufwuchs. The nature of the habitat in which these organisms are found depends on the habitats present and on the size of the organisms. Macroalgae on substrata are in very different habitats than microalgae on the same substratum type, because they extend further into the surrounding water. However, one common criterion for habitats is the substratum type:

- Epilithic algae grow on hard, relatively inert substrata (gravel, pebble, cobble, stone surfaces).
- Epiphytic algae grow on plants and larger algae.
- Epipsammic algae grow on hard inert sand that is smaller than the smallest diatoms.
- Epipelic algae grow on inorganic or organic sediment. Characteristically, these are large motile diatoms, motile filamentous Cyanobacteria, or large motile flagellates.
- Metaphyton refers to algae of the photic zone not directly attached to substrata. Usually these are clouds of filamentous green algae.

As primary producers, algae are fundamental for wetland ecosystems and their food web. Production estimations are summarized in **Table 1** for a variety of algal and wetland types. Evidently, highest biomass and carbon fixation rates are attained by all components in freshwater marshlands but vary considerably.

Table 1 Summary of biomass and carbon fixation rates for algal components in wetlands

Algal type	Wetland	Chl a (mg m^{-2})	Dry weight (mg cm^{-2})	C-fixation rate (g m^{-2} yr^{-1})
Epipelon	Freshwater marsh	0.4–435		<0.1–29
	Salt marsh	20–231		28–341
	Tundra ponds			4–10
Epiphyton	Freshwater marsh	1–650		2–548
	Peat bog	2–238		
Metaphyton	Freshwater marsh		0.5–80	12–1120
	Shallow fish ponds		1–49	
	Peat bog		0–6	
	Cypress swamp		1–3	
Phytoplankton	Freshwater marsh	3–104		1–381
	Cypress swamp		0.34	
	Tundra ponds			0.6–0.9
	Salt marsh			100

Table 2 Ranges of parameters derived from photosynthesis-irradiance curves (PE curves) for wetland algae

Algal type	α (µg C µg Chl a^{-1} h^{-1} (µE m^{-2} s^{-1})$^{-1}$)	E$_K$ (µE m^{-2} s^{-1})	P$_{max'}$ (µg C µg Chl a^{-1} h^{-1})
Epipelon	0.002–0.006	237–834	0.8–4
Epiphyton	0.008–0.016	198–659	0.5–21
0.003–0.007	36–400	0.2–2.0	Megaphton
Phytoplankton	0.025–0.092	235–446	2–79

In Tundra ponds, annual carbon uptake rates of the epipelon are 10 times higher than those of the phytoplankton.

It is difficult to generalize the quantitative contribution of the types of algal assemblages to gross primary production in wetlands. Few studies have been sufficiently inclusive to measure all potential producers. In constructed wetlands in Illinois, USA, epiphytes contributed 1–65% to total primary production. In an Alaskan tundra pond, annual primary production by epipelon was estimated to be 4–10 g C m^{-2}, as compared to 1 g C m^{-2} for phytoplankton and >15 g C m^{-2} for the emergent macrophyte *Carex aquatilis*. In a dystrophic wetland of South Carolina, contribution by littoral algae, including all possible types, was estimated to over one-third of net primary production, whereas macrophytes made up the remainder. In a shallow clear-water lake, carbon uptake of microphytobenthos contributed more than 70% (5.4 g C m^{-2}) to total production on an areal basis while phytoplankton uptake was less than 30% (2.2 g C m^{-2}).

There is considerable variation in the photosynthesis-irradiance relationship measured for benthic algae from wetlands (**Table 2**). Differences reflect the physiological state of organisms collected from diverse habitats but may also reflect differences in analytical methodology. Part of the variability in epiphyte assemblages with discrete three-dimensional physiognomy can be attributed to accumulated biomass

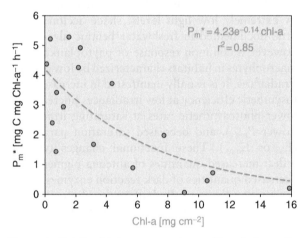

Figure 1 Relationship between maximum chlorophyll-specific photosynthetic carbon uptake and biomass as chlorophyll *a* of epiphytic algae.

because photosynthetic efficiency and biomass are inversely related (**Figure 1**). Based on the few studies conducted, generalizations on the photosynthetic behavior of various algal assemblages are difficult to make since ranges of all three parameters are wide. Maximum chlorophyll-specific carbon uptake of the Metaphyton, however, is comparatively small and light saturation (E$_K$) occurs at low light intensities. In turbid waters and under a macrophyte canopy, irradiance can fall well below E$_K$ (e.g. <5 µmol photons m^{-2} s^{-1} below a dense mat of *Lemna minor*). Evidence

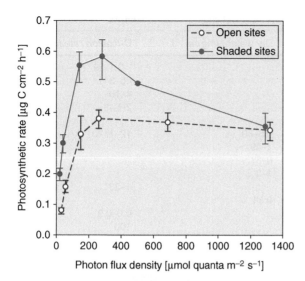

Figure 2 Photosynthesis–irradiance (PE) response by stream periphyton at open sites and at sites shaded by vegetation.

Table 3 Annual net production of wetlands compared to other ecosystems

Ecosystem	Range of annual production (Mt dry wt ha^{-1} yr^{-1})
Wetlands with helophytes	
Temperate	50–70
Tropical	60–90
Wetlands with pleustohelophytes	
Mostly tropical	40–60
Aquatic, fresh water	
Submerged macrophytes, temperate	5–10
Phytoplankton	15–30
Cultivated plants	25–85
Forests	20–60

of photoinhibition in surface waters has been reported from only a few investigations.

Although many benthic habitats are characterized by extremely low light levels, shade acclimation is little documented in freshwater benthic algae. It is, however, a common response of phytoplankton and macrophytes in habitats characterized by low ambient irradiances. It is usually manifested in increased photosynthetic efficiency at low irradiances (increased α), lower photosynthetic rates at saturating irradiances (lower P_{max}), and decreased saturation parameters (E_K or E_{max}). These functional changes typically reflect increased quantities of antenna pigments and decreased quantities of dark reaction enzymes. Shade acclimation theoretically buffers the photosynthesis-limiting effects of low irradiances. If respiratory losses are not too high, it will have a positive effect for production. The best documented shade acclimation by benthic algae originates from a study on a stream (explained later and **Figure 2**).

Macrophyte (Hydrophyte) Communities

Investigations on macrophyte biomass and production are numerous. The most productive communities in the world are generally emergent macrophytes, which are most common in lentic ecotones. In temperate zones, they attain an annual production of 30–45 Mt ha^{-1}, and in tropical regions 65–85 Mt ha^{-1}. Submerged plants are less productive (**Table 3**). Maximum standing crop data for different species and climatic regions are summarized in **Table 4**.

Freshwater Ecosystems

Rivers

Organic carbon in rivers is either autochthonous or allochthonous. Autochtonous carbon derives from primary production within the river itself, while allochthonous carbon is based on decomposed products of terrestrial origin.

An interbiome comparison of first- to third-order streams in North America using open-water diurnal oxygen changes revealed three general patterns of whole ecosystem production:

1. high gross primary production (GPP) with positive net ecosystem production (NEP);
2. moderate GPP during daylight but negative NEP at all times;
3. little or no GPP during daylight and negative NEP over the entire day.

In these streams, GPP ranged from <0.1 to 15 g O$_2$ m^{-2} day^{-1}. Variation of respiration was substantially lower, ranging from 2.4 to 11 g O$_2$ m^{-2} day^{-1}. The NEP was negative for all streams except one. Most of the streams were strongly heterotrophic with P:R ratios <0.25.

Photosynthetic active radiation (PAR) and soluble reactive phosphorus (SRP) explained 90% of the variation in GPP while NEP was correlated only with PAR ($r^2 = 0.53$). The rate of ecosystem respiration was poorly correlated with GPP, but SRP and the size of the transient storage zone explained 73% of the variation.

Rivers and streams often provide a mosaic of open sites exposed to full irradiance and sites shaded by patches of, e.g. trees and bushes. Stream benthic algae have to cope and acclimate to such situations. An example of shade acclimation in stream benthic algae is shown as PE curves in **Figure 2**. At low light intensities,

Table 4 Seasonal maximum standing crop for species of above-ground vegetation of wetlands in different climatic regions

Climate	Species	Standing crop (kg dry wt m^{-2})	Country
Tropical rain, no long dry season (rain forest)	Cyperus papyrus	4.30	Uganda
	Eichhornia crassipes	2.40	Florida, USA
	Paspalum repens	1.14	Brazil
	Cladium jamaicense	1.13	Florida, USA
	Pistia stratiotes	0.46	Florida, USA
	Lepironia articulata	0.29	Malaysia
Tropical rain, long dry season (savanna)	Arundo tonax	10.00	Thailand
	Typha domingensis	2.54	Malawi
	Salvinia molesta	0.60	Zambia
Arid (steppe and warm desert)	C. papyrus	3.60	Chad
	Phragmites australis	3.60	Ukraine
	P. australis	3.20	Chad
	Typha sp.	2.00	Chad, Uzbekistan
	P. australis	0.60	
Temperate rain (broad-leaved, evergreen or deciduous forest)	Typha angustata	3.86	Kashmir
	P. australis	3.00	Austria
	C. papyrus	2.90	Kenya
	E. crassipes	2.13	Alabama, USA
	Schoenoplectus lacustris	2.00	Germany
	Nasturtium officinale	1.26	England
	Carex lacustris	1.15	New York, USA
	Glyceria maxima	1.00	England
	Alternanthera philoxeroides	0.84	Alabama, USA
	Trapa natans	0.52	Japan
Boreal (deciduous or needle-leaved, evergreen forest)	S. lacustris	4.20	Czech Rep.
	Typha latifolia	3.60	Czech Rep.
	P. australis	3.30	Czech Rep.
	P. australis	3.00	Austria
	G. maxima	2.70	Czech Rep.
	Sparganium erectum	1.90	Czech Rep.
	Typha glauca	1.50	Iowa, USA
	Stratiotes aloides	1.20	Belo-Russia
	Carex rostrata	1.10	Minnesota, USA
	Sagittaria sagittifolia	0.47	Belo-Russia

photosynthesis from the shaded sites (average intensity 50 µmol m^{-2} s^{-1}) was twice that from the open sites (average intensity 1100 µmol m^{-2} s^{-1}). In these laboratory experiments, efficiency α is two times higher in the shaded assemblages. Shade acclimated communities saturate at lower intensities, have the highest P_{max}, and show clear evidence of photoinhibition not indicated in the results from the open site.

Temperature appears to be of secondary importance in determining seasonal changes in primary productivity of epilithon in streams. Little changes in primary production are documented when temperature increases, suggesting that respiration increases at the same rate. Temperature however, may interact with changes of other factors. The degree of photoinhibition can for instance be inversely correlated with water temperature. Temperature certainly sets an upper limit to areal primary productivity. As can be seen from **Figure 3**, maximum areal net primary productivity not corrected for biomass increases exponentially for temperatures $<30\,°C$.

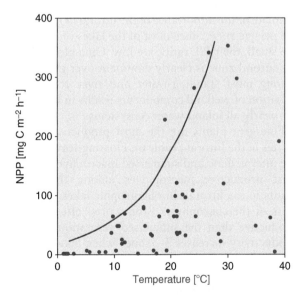

Figure 3 Relation of areal net primary production in lotic epilithon assemblages with temperature. Curve determined by regression of maximum rates versus temperature, $r = 0.96$.

Table 5 Algal biomass range as indicators of primary production in rivers of different qualities

River type	$NO_3 - N$ (mg l^{-1})	$PO_3 - P$ (mg l^{-1})	Biomass range minimum (g m^{-2})	Chl a maximum (mg m^{-2})
Soft water pH 4–5	0.06–0.35	<1	26	92
Soft water pH 6–7	0.5	1–2	40	178
Headwater chalk stream	4–5	10–30	50	200
Chalk stream	4–5	30–50	15–25	150–300
Hyper-eutrophic stream	5–15	1000–3000	25–50	150–250

The Q_{10} value for the regression in **Figure 3** is 2.9, similar to relations obtained for phytoplankton growth rates.

Algal biomass is used as an indicator of primary productivity (see **Table 5**) to characterize different stream quality. Chalk streams and hypertrophic streams attain highest biomass and chlorophyll *a* concentrations whereas acidic soft water streams are among the most unproductive.

In general, attached algal and cyanobacterial productivity and biomass development is moderate in mid-order streams (3–5) and in backwater areas but low in light-restricted, canopied, or high-velocity portions of low-order streams and in more turbid large rivers. In the highly dynamic large rivers of the tropics, such as the Amazon and Paraná River ecosystems, productivity of different vegetation components becomes much more complicated.

Littoral of Lakes

Although the productivity in the littoral of lakes is relatively high, their contribution to the overall lake production largely depends on the morphometry of the system. The pelagic/littoral ratio (P/L) permits an estimate of the importance of the littoral in relation to the pelagic zone. Since most of the lakes of the world are small, their P/L ratios are low. On a global basis, the littoral zone is clearly dominant over the pelagic among most standing-water and river ecosystems. Addition of wetland components results in low ratios for nearly all inland water ecosystems.

Emergent plants are the most productive macrophytes in the littoral (**Table 6**). Floating-leaved plants are intermediate and submersed macrophytes are the least productive macrophytes among the higher plants in the littoral. In oligotrophic lakes, the algal class of the charophytes (stoneworts) often is more productive than the submersed vegetation. Littoral productivity increases 3–5 times when lakes become eutrophic. Since macrophyte biomass in the littoral is high and turnover is low (P/B < 1), nutrients are less readily available in the littoral region. Macrophyte beds buffer the pelagic zone. In general, the littoral is more defined by biomass than by turnover.

Table 6 Annual net primary productivity in the littoral of lakes

Type	Climatic region or trophic level	NPP (g C $m^{-2}yr^{-1}$)
P. australis	Temperate	80–800
	Tropical	>1000
Typha sp.	Temperate	1500–2500
	Tropical	3000–4000
Floating leave plants		50–1400
Submersed plants	Temperate	50–400
	Tropical	~2000
Littoral periphyton	Oligotrophic	~40
	Eutrophic	160 to >400
Littoral phytoplankton	Oligotrophic	~100
	Eutrophic	~1000

Effects of temperature on the productivity of lentic periphyton indicate an increase with temperature mainly due to increased biomass since assimilation efficiency is usually lower at heated sites.

Estimation of Total Lake Production

Estimations of all components of lake primary production allowing calculation of production for the whole lake are not very common. Two examples are given in **Table 7**. Average total carbon uptake amounts to 3310 mg C m^{-2} day^{-1} in the shallow Neusiedlersee, Austria bordering Hungary. With reed being the largest contributor (83%), followed by *Utricularia* and phytoplankton (7% each). All other components are of minor importance.

Estimations of annual production for lakes in Poland (**Table 7**) indicate that over half of the 19 373 kJ fixed under a square meter per year is due to macrophytes while the other half is almost equally divided between periphyton and phytoplankton.

In the littoral zone of Lake Mikolajskie, Poland, attached algae contributed 41% to annual NPP per square meter, submersed macrophytes 28%, metaphyton 21%, and phytoplankton 10%. In a littoral zone overgrown with emergent vegetation, the contribution was almost twice as high from macrophytes (57.2%) and phytoplankton (19.6%), the share from attached algae decreased to 23.1% and metaphyton dropped to a negligible amount of only 0.1%.

Table 7 Two examples for the estimation of total lake production from measurements of individual components

Neusiedlersee component	Annual average (mg C m^{-2} d^{-1})	%	Comment to Neusiedlersee
Phytoplankton (pelagial, open lake)	235	7.10	1.3 m mean depth, area 320 km^2
Phytoplankton (littoral, within reed belt)	20	0.60	(about 60% covered by a Phragmites
Epiphytic algae (on Phragmites)	11.6	0.35	reed belt)
Epipelic algae (on the sediment surface)	50	1.52	
Submersed macrophytes	12	0.36	
Pleustophytes (Utricularia spp.)	242	7.32	
Helophytes (P. australis)	2740	82.75	
Total	3310.6		
Polish lakes component	Annual production (kJ m^{-2} yr^{-1})	%	
Littoral phytoplankton	3871	20.00	
Periphyton	4566	23.00	
Macrophytes	11 300	57.00	
Total	19 373		

Besides other environmental constraints, water-level changes can severely affect total lake production through effects on the littoral vegetation as exemplified in **Table 8** using data from a wetland on the shores of Lake Manitoba. Epiphytic and particularly metaphytic algae totally dominated the littoral algal productivity. Production rates declined markedly as mean water level raised by half a meter.

Production of attached algae is usually higher in eutrophic shallow lakes when compared with oligotrophic deep lakes. Average annual production of epiphytic algae in Priddy Pool, a shallow lake in England, was estimated to be 63.9 mg C m^{-2} h^{-1}, which is 95.2% of the total amount (67.16 mg C m^{-2} h^{-1}). The remaining 3.26 mg C m^{-2} h^{-1} was almost equally shared by epipelic algae and phytoplankton. In contrast, attached algae contributed 28.5%, epilpelic algae 29.1%, and phytoplankton 42.4% to the average total production of 12.63 mg C m^{-2} h^{-1} in the deep oligotrophic Lake Pääjärvi in Finland. Total production in eutrophic lakes is much higher (in this case 5 times) and shifted towards attached algae.

Reservoirs

Reservoirs range from small ponds to huge impoundments. Classification is mainly based on morphological and hydrological features. One important criterion is retention time reaching from through-flow (e.g. river-run dams) to very long residence times (e.g. storage reservoirs).

On a large scale, lakes and reservoirs are distinctly different in behavior and operation. At the small spatial and temporal scales at which algae operate, they are generally similar in both habitats. Broadly, the same type of algae occurs. Responses to fluctuations in resource availability and the severity of processing constraints are similar in both types of waters. The development of biomass and the rate

Table 8 Effect of different water level regimes on primary productivity of attached and planktonic components in a wetland of Manitoba, Canada

	Mean algal biomass (mg Chl a m^{-2})	Mean daily productivity (mg C m^{-2} d^{-1})		
		Water level		
		Low	Medium	High
Phytoplankton	7	169	209	225
Epipelic	4	32	47	15
Epiphytic	67	1003	813	353
Metaphyton	530	2863	2372	1338
Total		4067	3441	1931

Table 9 Mean volumetric and photic zone integrated primary production at three regions of reservoir varying in nutrient concentration and clay turbidity

Region	P$_{avg}$ (mg C m^{-3} d^{-1})	$\Sigma\Sigma$P (mg C m^{-2} d^{-1})	Nutrients/ turbidity
South transition	319	578	High/high
North transition	583	1103	Adequate/ moderate
Lacustrine	269	890	Low/low

of productivity are therefore similar to those commonly observed in lakes. Elongated or highly dendritic reservoirs, however, develop longitudinal gradients as a result of variable input from the tributaries, retention time, and sedimentation of particles. Moreover, clay and other suspended solids can reduce underwater light availability to an extent that primary production is suppressed through light limitation even in reservoirs with high nutrient load (**Table 9**). The high average volume-based production of the south arm is constrained to a shallow photic zone by the high clay load. The water column in the north arm with adequate nutrients and less

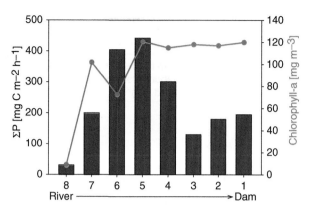

Figure 4 Longitudinal profile of integrated water column productivity rate (areal production) and chlorophyll a from the river to the dam in a Canyon-type reservoir.

Table 10 Annual primary production of aquatic communities

Type of ecosystem	Average net organic dry production (Mt ha^{-1} yr^{-1})	Range of production (Mt ha^{-1} yr^{-1})
Lake phytoplankton	2	1–9
Submersed macrophytes		
Temperate	6	1–7
Tropical	17	12–20
Emergent macrophytes		
Temperate	38	30–45
Tropical	75	65–85

clay-turbidity produces almost twice as much. Even the lacustrine region with less nutrients and low clay-turbidity produces more than the nutrient rich but clay-loaded south arm.

Canyon-shaped reservoirs develop distinct longitudinal gradients (**Figure 4**). Chlorophyll a concentration, low near the inflow, increases rapidly further down the reservoir. Plankton production is also low near the river mouth, reaching maximum values in the middle section of the reservoir declining thereafter towards the dam region.

The high and variable turbidity and the water-level fluctuations in many reservoirs preclude the development of moderately stable and diverse substrate habitat. Limited habitat, substrata, and light therefore often restrict development and productivity of attached algae. Turbidity, basin shape, and retention time are the key variables controlling phytoplankton primary production in reservoirs.

Climatic Regions

There is no obvious correlation between standing crop data for species of wetland vegetation and climate (**Table 4**). High standing crops of around 4 kg m^{-2} are found from tropical to boreal regions. Values below 0.2 kg m^{-2} are equally widespread. Obviously, biomass development of emergent vegetation in wetlands is generally affected by local factors such as species composition, microclimate, soil, and grazing more than by climate.

Summarized across larger climatic regions, however, large differences in the production of organic matter appear (**Table 10**). Production of submersed macrophytes triples on average in the tropics while production from emergent vegetation doubles than in temperate regions.

In an early analysis based on data collected during the worldwide International Biological Program (IBP), latitude and altitude alone explained 49% of the variance of phytoplankton production; chlorophyll, as a substitute for biomass, and latitude explained 78.6%. Thus, it appears that the geographical position on the globe and biomass are largely responsible for the productivity of freshwater ecosystems. Moreover, biomass and productivity are a reflection of the nutritional status of freshwater systems. Both are therefore dependent on the trophic level.

Comparison to Other Ecosystems

The best current estimate of global net primary production (NPP) is 104.9×10^{15} g C year^{-1} (104.9 pg C year^{-1}), which is about 200×10^9 tons dry weight per year, assuming 50% carbon in dry weight. The oceans contribute 46.2%, while production on land, including fresh waters, contributes 53.8%. A more detailed breakdown of NPP for various ecosystem types is presented in **Table 11**. All freshwater systems, lakes, streams, swamps, and marshes, contribute 3.9% to total continental net primary production, and about 2.5% to total global NPP. Almost half of NPP on land is produced by tropical forests, while tundra, alpine, and desert ecosystems together contribute only 2.4%. Cultivated land produces about 8% of NPP on land. The range of wetland production, however, is equal or higher than the production of cultivated plants or forests (see **Table 3**).

A clear latitudinal gradient of increasing productivity from boreal through temperate to tropical conditions has been established for lakes. Similar trends are observed in grassland, tundra, and cultivated crops. Global NPP peaks on either side of the equator with a wide shoulder of higher productivity between 30° and 60° N, and a smaller peak between 30° and 45° S (**Figure 5**).

Average values of productivity are broadly related to biomass for the ecosystem types in **Table 11**. For a

Table 11 Net annual primary productivity estimates for plant associations of the world

Ecosystem type	Net primary productivity (NPP) per unit area dry weight (g m^{-2} or t km^{-2})		World NPP (10^9 t)
	Range	Mean	
Lakes and streams	100–1500	250	0.5
Swamps and marshes	800–3500	2000	4.0
Cultivated land	100–3500	650	9.1
Tropical forests	1000–3500	1900	49.4
Temperate forests	600–2500	1250	14.9
Boreal forests	400–2000	800	9.6
Wood and shrubland	200–1200	700	6.0
Savanna	200–2000	900	13.5
Temperate grassland	200–1500	600	5.4
Tundra and alpine	10–400	140	1.1
Desert	10–250	90	1.6
Extreme desert	0–10	3	0.07
Total continental		773	115.0
Open ocean	2–400	125	41.5
Upwelling zones	400–1000	500	0.2
Continental shelf	200–600	360	9.6
Algal beds and reefs	500–4000	2500	1.6
Estuaries	200–3500	1500	2.1
Total marine		152	55.0
Average globe		333	170.0

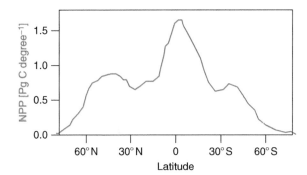

Figure 5 Latitudinal distribution of the global net primary production. Modified from Field CB, Behrenfeld MJ, Randerson JT, and Falkowski P (1998) Primary production of the biosphere: Integrating terrestrial and oceanic components. *Science* 281: 237–240.

given NPP value, however, biomass varies widely. Thus, production–biomass (P/B) ratios (e.g., kg produced per year per kg biomass) are 17 for aquatic communities, 1–2 for helophytes, 0.29 for terrestrial systems, and 0.042 for forests. Differences are mainly due to turnover rates, which are much higher in smaller organisms such as algae, which therefore produce more per unit biomass. In contrast, forests are slow growing and accumulate large portions of biomass in supporting tissue or dead material.

See also: Phytoplankton Population Dynamics: Concepts and Performance Measurement; Phytoplankton Productivity.

Further Reading

Bott TL (2006) Primary productivity and community respiration. In: Hauer FR and Lamberti GA (eds.) *Methods in Stream Ecology*. Amsterdam: Elsevier.

Cooper JP (ed.) (1975) *Photosynthesis and Productivity in Different Environments*. Cambridge: Cambridge University Press.

Dokulil MT (2003) Algae as ecological bio-indicators. In: Markert BA, Breure AM, and Zechmeister HG (eds.) *Bioindicators and Biomonitors*, pp. 285–327. Amsterdam: Elsevier.

Field CB, Behrenfeld MJ, Randerson JT, and Falkowski P (1998) Primary production of the biosphere: Integrating terrestrial and oceanic components. *Science* 281: 237–240.

Janauer G and Dokulil M (2006) Macrophytes and Algae in running waters. In: Ziglio G, Siligardi M, and Flaim G (eds.) *Biological Monitoring of Rivers. Application and perspectives*, pp. 89–109. Chichester: Wiley & Sons.

Likens GE (1975) Primary production of inland aquatic systems. In: Lieth H and Whittaker RH (eds.) *Primary Productivity of the Biosphere*. New York: Springer.

Mulholland PJ, Fellows CS, Tank JL, *et al.* (2001) Inter-biome comparison of factors controlling stream metabolism. *Freshwater Biology* 46: 1503–1517.

Sculthorpe CD (1967) *The Biology of Aquatic Vascular Plants*. London: Edward Arnold.

Steinmann AD, Lamberti GA, and Leavitt PR (2006) Biomass and pigments of benthic algae. In: Hauer FR and Lamberti GA (eds.) *Methods in Stream Ecology*. Amsterdam: Elsevier.

Stevenson RJ, Bothwell ML, and Lowe RL (eds.) (1996) *Algal Ecology. Freshwater Benthic Ecosystems*. San Diego: Academic Press.

Tundisi JG and Straškraba M (eds.) (1999) *Theoretical Reservoir Ecology and Its Application*. Leiden: Backhuys.

Westlake DF, Květ J, and Szczepański A (eds.) (1998) *The Production Ecology of Wetlands*. Cambridge: Cambridge University Press.

Wetzel RG (2001) *Limnology. Lake and River Ecosystems*, 3rd Edn. San Diego: Academic Press.

Eutrophication of Lakes and Reservoirs

V Istvánovics, Budapest University of Technology and Economics, Budapest, Hungary

Definition

Eutrophication is the process of enrichment of waters with excess plant nutrients, primarily phosphorus and nitrogen, which leads to enhanced growth of algae, periphyton, or macrophytes. Abundant plant growth produces an undesirable disturbance to the balance of organisms (structural and functional changes, decrease in biodiversity, higher chance for invasions, fish kills, etc.) and to the quality of water (cyanobacterial blooms, depletion of oxygen, liberation of corrosive gases, and toxins, etc.).

Introduction

Trophic status of a water body can roughly be assessed by using information about the concentration of the limiting nutrient (phosphorus), chlorophyll (an indicator of phytoplankton biomass), and transparency (dependent on both algal biomass and sediment resuspension, expressed as Secchi depth). The most widely accepted limits are those suggested by the Organization for Economic Cooperation and Development (OECD):

Trophic category	Mean total, P ($\mu g\ l^{-1}$)	Mean ($\mu g\ chl$-$a\ l^{-1}$)	Max. ($\mu g\ chl$-$a\ l^{-1}$)	Mean Secchi depth (m)
Oligotrophic	<10	<2.5	<8	>6
Mesotrophic	10–35	2.5–8	8–25	6–3
Eutrophic	>35	>8	>25	<3

During the last four decades, eutrophication has undoubtedly been the most challenging global threat to the quality of our freshwater resources. Survey of the International Lake Environmental Committee has indicated in the early 1990s that some 40–50% of lakes and reservoirs are eutrophicated. Many of these water bodies are extremely important for drinking water supply, recreation, fishery, and other economic purposes. Developing countries in hot and dry climate face particularly serious and rapidly increasing eutrophication-related ecological and economic challenges. Unlike this paper, eutrophication is not restricted to lakes and reservoirs. Large rivers, estuaries, coastal zones, and inland seas are also subject to this undesirable process.

Although nutrient cycling is far from being closed even in the most developed societies, eutrophication has been successfully reversed in several lakes by managing human nutrient emission (low-P detergents, P precipitation at sewage treatment plants, decreased fertilizer application, erosion control, etc.) and by cutting off nutrient loads to recipients (sewage diversion, buffer zones, etc.). Thus, eutrophication seems to be reversible. Nutrient management, however, does not result in an immediate oligotrophication (i.e., reversal of eutrophication) because of the resilience of the aquatic ecosystem. Various methods of in-lake physical, chemical, and biological interventions have been developed to facilitate the efficiency of load reduction, shorten the delay in recovery, and accelerate the rate of reversal. Early recognition of eutrophication, understanding the nutrient load–trophic response relationship in individual lakes, and planning the most suitable combination of management measures require an in-depth knowledge of functioning of aquatic ecosystems.

Basics of Eutrophication

Water bodies are open systems, in which autotrophic organisms, mainly algae and macrophytes, convert inorganic carbon into organic matter using the energy of solar radiation. In addition to C, other nutrients (oxygen, hydrogen, nitrogen, phosphorus, sulfur, silica, trace metals) are universally or taxon-specifically needed for primary production. The German chemist, von Liebig recognized in the nineteenth century that the yield of a plant is proportional to the amount of nutrient, which is available in the lowest supply relative to the plant's demand. This observation is known as the Law of the Minimum. Studying the elemental composition of phytoplankton in the English Channel, Redfield found that healthy algae contained C, N, and P atoms in an average ratio of 106:16:1 (40:7:1 by weight). This means that 1 g of P available in the water allows the algae to assimilate 40 g of C and thus increase fresh weight by about 400 g.

Algae, like any other plants, require nutrients in ratios, which are radically different from the elemental composition of their environment. In fresh water, the mid-summer average demand to supply ratio of P is up to 800 000; that of N is 300 000; that of C is 6000; and that of all other elements is below 1000. Consequently, P and N are the nutrients, which most often determine the carrying capacity of lakes and reservoirs. This has been verified experimentally during the 1970s by Schindler and his team. In the

316

Canadian Experimental Lake Area, untreated lakes were compared with nutrient enriched ones. Standing crop of algae increased with the supply of P. Nitrogen addition alone failed to similarly trigger phytoplankton growth.

Another large-scale study of the OECD (1982) also confirmed the biomass-limiting role of P by the means of statistical modeling. This study considered over 120, mostly temperate lakes and reservoirs. Mean in-lake concentration of total nutrients was calculated from the annual load corrected for flushing. Annual mean and maximum concentration of chlorophyll was regressed against the estimated nutrient concentration. The relations were highly significant for P. In most lakes, biomass was determined by the amount of P, while N was the controlling factor in only a few cases.

Chemical nature of major N and P ions explain why P determines the carrying capacity of lakes more often than N. Both nitrate and ammonium are much more mobile in soils and sediments than orthophosphate, since the latter is chemisorbed by clay minerals, iron(III) oxy-hydroxides, carbonates, etc. As a consequence, nitrogen compounds are preferentially washed out from the soils and phytoplankton in pristine temperate lakes tends to be P limited. For the same reason, terrestrial plants are most often N limited. In productive lowland wetlands dominated by dense stands of emergent plants that release O_2 to the atmosphere, anaerobic conditions may develop in the water and denitrification speeds up. In such systems N deficiency prevails as frequently indicated by the presence of insectivorous plants. Carrying capacity of many subtropical and tropical lakes are also N determined because of the intense denitrification characteristic under warm climates.

However, widespread nutrient limitation is not a universal phenomenon in fresh waters. Besides nutrients, algal growth requires sufficient time and light. In several water bodies, biomass is determined by the availability of these latter factors. Such conditions prevail in running waters in which inorganic nutrients are usually in excess. To support abundant plankton, a river must either be sufficiently long or flow sufficiently slowly to permit seven to eight cell divisions required to raise an inoculum of several cells per milliliter to the order of a few thousand cells per milliliter. When the hydraulic residence time is increased by reservoir construction, the time available for 'undisturbed' algal growth and thus, the biomass may increase without any further nutrient enrichment. In addition to this, enhanced sedimentation within mainstream reservoirs improves light availability both in and downstream of the reservoir. Construction of a series of reservoirs on the upper Danube was recognized as a key factor in downstream eutrophication of this large river.

Eutrophication is not merely an increase in the biomass of various organisms. Structural and functional changes accompany, and – as careful long term observations repeatedly testify – even precede quantitative changes at each trophic level.

Changes in Primary Producers

One of the most conspicuous and the best-known changes associated with eutrophication is the mass development of cyanobacteria, be it N_2-fixing (*Anabaena*, *Aphanizomenon*, *Cylindrospermopsis*, etc.) or nonfixing (*Planktothrix*, *Microcystis*, etc.). Many species may develop toxic strains (e.g., *M. aeruginosa*, *Cylindrospermopsis raciborskii*) and thus, large blooms may directly harm both other aquatic organisms and humans. An enormous scientific literature discusses the ecological traits leading to mass development of Cyanobacteria as well as their manifold influence on the functioning of the aquatic ecosystem. Without going into details, one can recognize three basic lines of adaptations in the background of cyanobacterial success. Each bloom-forming species possesses a certain combination of the following traits:

1. *Nutrients*. Bloom-forming Cyanobacteria are capable of exploiting nutrient reserves that are unavailable, or not readily available, for most other algae. This is the reason why annual mean concentration of chlorophyll may show an abrupt rise upon the establishment of Cyanobacteria at high but 'steady' external nutrient load. Since the inorganic carbon concentrating mechanisms are highly efficient in the case of both CO_2 and HCO_3^-, Cyanobacteria continue to assimilate at high pH. Many species have high affinity for NH_4^+ uptake besides the ability to fix N_2. Either fast maximum rate of P uptake, or exploitation of vertical nutrient gradients, facilitates P acquisition, thanks to buoyancy regulation. The relatively large size allows the storage of considerable amounts of excess C, N, and P beyond the actual needs of growth, thereby providing independence from the fluctuating supply.

2. *Light*. The lowest light saturation values of both photosynthesis and growth have been observed among bloom-forming cyanobacteria (e.g., *Planktothrix rubescens*, *Cylindrospermopsis raciborskii*). In general, their light requirement tends to be lower than that of the eukaryotic algae. Under calm conditions, buoyancy regulation allows optimal positioning of Cyanobacteria in the light gradient. Simple prokaryotic structure results in relatively low maintenance costs that both decreases the light demand and leaves more energy to acquire the limiting nutrient.

3. *Low biomass loss.* Buoyancy regulation prevents sinking loss of healthy Cyanobacteria even under calm conditions. Large size and morphology (large colonies, filaments) reduce zooplankton grazing to negligible levels. The decreased loss of Cyanobacteria is equivalent to the slowdown of nutrient regeneration and diminished internal supply of the limiting nutrient.

Summer blooms of Cyanobacteria cause a major shift in the seasonal pattern of phytoplankton biomass in temperate lakes. In oligo- and mesotrophic lakes, the biomass maximum occurs during the spring when temperature and light increase rapidly, and relatively large amounts of nutrients are delivered into the water by spring floods as well as the spring overturn in deep lakes. In comparison, the summer biomass of phytoplankton is lower as a result of the diminishing external and internal nutrient supply. In productive lakes, increased nutrient availability differentially enhances summer production and leads to a virtually monomodal temporal distribution of biomass with a summer maximum.

The most important functional changes associated with the dominance of bloom-forming Cyanobacteria are (1) the involvement of formerly unavailable resources into aquatic production, (2) the decrease in the rate of turnover of nutrients, most importantly in that of the limiting ones, and (3) the decrease in the efficiency of energy transfer from primary producers to higher trophic levels. Because of these self-stabilizing functional changes, the shift from the noncyanobacteria dominated to Cyanobacteria dominated late summer phytoplankton assemblages can be seen as alternative stable states of the aquatic ecosystem. An important manifestation of the alternative stable states is the hysteresis that occurs when the system is forced from the one state to the other.

Establishment of bloom-forming Cyanobacteria is the last step in the course of restructuring of phytoplankton during eutrophication. Case Study 1 summarizes the history of compositional changes observed in a large, shallow temperate lake.

Submerged macrophytes may cover a significant portion of the area of shallow lakes that are protected from strong wave action or the small surface area of which restricts wind fetch. During eutrophication, such lakes may abruptly turn from a macrophyte-dominated clear water state into a phytoplankton-dominated turbid state. Similar to the case of Cyanobacteria, both states may prevail in a broad range of external nutrient loads since a number of feedback mechanisms stabilize the actual state. Thus, in the clear water state abundant macrophyte stands prevent sediment resuspension by dampening wind work and provide shelter for zooplankton. Visual predatory fish keep efficient control on planktivorous fish, thereby promoting the growth of zooplankton. In turn, grazing may significantly influence growth, biomass, and succession of phytoplankton. In the turbid state, intense resuspension and shading by phytoplankton inhibits macrophyte growth and suppresses foraging of predatory fish. In the lack of sufficient refuge, zooplankton fall victim to planktivorous fish and phytoplankton are released from the top-down control.

Changes in Consumers

Increased primary production supports an elevated production and biomass of consumers. Structural changes in the phytoplankton, first of all those in the size distribution, exert a strong influence on pelagic herbivores by differentially favoring or suppressing one or the other group of zooplankton. Since detritus and associated bacteria make up the main food of most benthic invertebrates, the zoobenthos is less sensitive to eutrophication-related changes in algal composition. At the same time, presence of aquatic macrophytes is beneficial for both pelagic and benthic consumers because of either the mere maintenance of habitat patchiness or to more specific biotic interactions. The retreat of macrophytes during eutrophication may result in a drastic reduction in the species diversity of consumers.

The four main groups of freshwater zooplankton – protists, rotifers, cladocerans and copepods – partition food primarily on the basis of size. Most aquatic grazers consume any particles in the appropriate size range, be it algae, bacteria, another grazer, detritus, or inorganic particle. Size-selective predation by planktivorous fish inserts a top-down control on grazer populations, the larger zooplankton being more vulnerable to predation. In addition to the direct biological interactions, deterrent environmental effects of eutrophication, including elevated turbidity, magnified daily and seasonal fluctuations in the oxygen concentration and pH may adversely affect the sensitive groups of zooplankton. Intense grazing alters species composition of phytoplankton by both selective removal of edible algae and nutrient regeneration. Because of this intricate net of interactions, eutrophication-related changes in the biomass, composition, size distribution, and seasonal pattern of various groups of zooplankton are highly lake-specific. The forthcoming discussion is restricted to a few general trends.

The most abundant freshwater zooplankters are crustaceans. Most cladocerans, such as *Daphnia*, are filter feeders whereas most cyclopoid and calanoid copepods are selective, raptorial grazers. In oligotrophic lakes, the maximum community grazing rate is below 15% while in eutrophic lakes the entire volume of water can be filtered up to 4–5 times in a day (400–500%). *Daphnia* and other cladocerans usually account for up to 80% of the community grazing rate.

Filter feeders collect particulate matter from the water. Anatomy of feeding appendages prevents collecting particles smaller than 0.8 µm. The upper limit to ingestible particle size increase with the size of the animal up to about 45 µm in the case of spherical particles and larger in the case of nonspherical ones. Edible algae are small, naked green algae, nanoflagellates, cryptomonads, and certain diatoms. Algae that cannot be ingested are large, have biologically resistant cell walls and spines, or form colonies. Large, unicellular desmids and dinoflagellates, chain-forming diatoms, and colonial Cyanobacteria respresent this group. The copepods are inefficient at retaining small particles but can process much larger algae than cladocerans. Some copepods, for example, can break up chains of diatoms.

In oligo- and mesotrophic lakes, the spring bloom of phytoplankton is made up by small, fast growing, edible algae such as small diatoms, nanoflagellates, and small greens. This favors the growth of nonselective cladocerans that – similar to their prey – are specialists at fast reproduction by parthenogenesis. Seasonal succession proceeds towards summer and autumn associations of large, slowly growing species such as dinoflagellates, gelatinous greens, colonial or filamentous Cyanobacteria. With the increasing abundance of large or noxious algae, slower growing but more selective copepods become more common. Collapse of algal blooms is associated with an increased availability of detritus and bacteria that constitute an appropriate food for cladocerans. Smaller blooms of diatoms and nanoplankton (<60 µm) during the autumn also promote the growth of filter feeders. As a general trend, the aquatic food chain gradually shifts during the succession from one based on living algae to one based on bacteria and detritus.

Compared with that in unfertile lakes, the shift from the dominance of nanoplankton to that of larger 'net' plankton takes place early in the season in eutrophic lakes. Moreover, filamentous Cyanobacteria that take over the dominance in the summer assemblages inhibit filtration of nonselective grazers by clogging up the feeding appendages. Thus, the share of grazing-resistant algae increases during eutrophication, accelerating the shift toward a bacteria- and detritus-based pelagic food chain that may result in an increasing dominance of cladocerans.

The tendency of an eutrophication-related increase in the cladoceran to copepod ratio is best seen in deep, stratified lakes. In naturally turbid reservoirs and shallow lakes zooplankton inherently rely to a greater extent on detritus and associated bacteria than in deep ones of the same trophic state. With increasing productivity, however, enhanced detritus availability disrupts the natural balance between pelagic and benthic food webs and the share of the latter increases in the total energy budget of both shallow and deep lakes.

The zoobenthos is an extremely diverse group comprising nearly all phyla from protists through large macroinvertebrates to vertebrates. Similar to the zooplankton, eutrophication-related changes in the composition, biomass, and seasonal dynamics of benthic assemblages are determined by food availability, predation, and indirect environmental effects, particularly oxygen concentration. Oxygen conditions in the hypolimnion or in the uppermost sediment layer strongly depend on the downward flux of detritus, and the tolerance to low oxygen is extremely variable among benthic organisms. One of the pioneering discoveries of limnology was in the early twentieth century that the profundal benthic fauna is an excellent indicator of nutrient richness in deep lakes. Naumann postulated a direct relationship between phytoplankton and nutrient conditions in lakes and contrasted the extreme ends as 'eutrophic' (well-nourished) and 'oligotrophic' (poorly nourished). Thienemann recognized two lake types based on hypolimnetic oxygen concentrations and on corrclated differences in the benthic chironomid fauna. The oligotrophic–eutrophic paradigm emerged originally from the crossing of these two lines of research. The oligotrophic water was deep with low nutrient supply, low algal production in the epilimnion, and low flux of detritus to the hypolimnion. Because of the small oxygen consumption, the hypolimnion had an orthograde oxygen profile and the corresponding stenoxybiont benthic fauna exploited mainly by whitefish (*Coregonus*). The eutrophic type was relatively shallow and nutrient-rich, water blooms appeared in the summer. The high flux of detritus to the shallow hypolimnion resulted in a fast depletion of oxygen. The oxygen profile was clinograde, and a euryoxybiont *Chironomus* fauna dominated the benthos. This paradigm had been rephrased during the 1950s and 1960s when eutrophication became a recognized environmental threat in the developed countries.

The two dominant groups of zoobenthos are the oligochaete worms and the dipteran chironomids. Although some oligochaetes are restricted to oligotrophic waters, the tubificids can be extremely abundant in highly eutrophicated as well as in organically polluted lakes if some oxygen is available from time to time and toxic products of anaerobic metabolism do not accumulate in large quantities. In such systems, they benefit from both the rich nutrient supply and the lack of competition with other benthic animals that cannot tolerate the poor oxygen conditions. In contrast to the tubificids, relative abundance of the more oxygen demanding chironomid larvae decreases during eutrophication.

In deep, mesotrophic lakes the biomass of zoobenthos exhibits two maxima: (1) a diverse fauna with high oxygen demand inhabits the littoral sediments and (2) less rich assemblages of species that tolerate low oxygen characterize the lower profundal zone. With increasing productivity, the littoral zone may loose its heterogeneity that results in a single profundal biomass maximum of the zoobenthos. A further increase in fertility and the associated hypolimnetic oxygen deficit may then lead to the decline in the biomass of the benthic fauna in the profundal zone, too.

Restructuring of the zooplankton and zoobenthos during eutrophication should not be perceived as a chain of smooth transitions. On the contrary, abrupt compositional shifts may be associated with stepping over the threshold values of critical environmental variables, including habitat patchiness, food availability, or simple physical and chemical factors. Case Study 2 demonstrates the threshold effect of food availability in the compositional change of the chironomid fauna in a large, shallow lake.

The shift from the pelagic to the benthic food web implies basic alterations in the functioning of the freshwater ecosystem. On the one hand, the consumer control of phytoplankton diminishes for two reasons. First, the diet of zooplankton relies to a much greater extent on living algae than that of benthic animals. Second, a major change occurs in the seasonal pattern and magnitude of nutrient regeneration. The zooplankton regenerate nutrients directly into the trophogenic zone, even though the sedimentation of fecal pellets represents a net outward flux of nutrients from the epilimnion to the sediments. In contrast to this, benthic animals release nutrients to the sediments from where the flux to the water is primarily regulated by abiotic factors. When, however, large oligochaetes or chironomids are present in high densities, their burrowing activity considerably enhances the sediment-water exchange rates of various nutrients (O, N, P, etc.). In shallow lakes and in the littoral zone of deep ones, bioturbation may result in a significant increase in the internal load of nutrients. On the other hand, the annual production to biomass ratios are higher among planktivorous fishes than among bentivorous ones. Therefore, the overall efficiency of energy utilization decreases in the lake with the increasing proportion of the benthic food web.

Although the total biomass of fish tends to increase during eutrophication, the species composition shifts in undesirable directions. Most conspicuously, the relative abundance of visual predators drops drastically with increasing turbidity. This change was repeatedly shown to cascade down along the food chain to the phytoplankton. Unbalanced food availability may lead to enhanced mortality of the fry. Widely fluctuating oxygen conditions may result in mass killing of fish. Although intense fish production requires highly productive lakes, maintenance of a diverse native fish fauna conflicts with eutrophication.

Oligotrophication

Comprehensive studies of lake recovery from eutrophication have shown that the trophic status of lakes is not a linear function of the external nutrient load. The same external load supports a higher biomass during oligotrophication than during eutrophication, that is a hysteresis can be observed (**Figure 1**). Sas and his team recognized four stages of recovery by examining 18 eutrophicated deep and shallow Western European lakes during the 1990s (**Figure 1**).

During Stage 1, the excess of available P is flushed out from the lake without any reduction in the standing crop of algae. This stage can be observed in lakes that received sufficiently high P loads for sufficiently long periods, and thus, phytoplankton biomass has no more been P-dependent. In such lakes, a considerable time may pass before P regains its biomass-limiting role. In deep lakes, the delay is essentially a function of the hydraulic residence time. In shallow lakes, where the

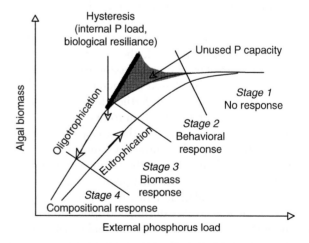

Figure 1 Schematic representation of phytoplankton biomass as a function of the external nutrient load during eutrophication and oligotrophication with the four stages of lake recovery.

excess of available P accumulates in the sediments, diagenetic processes and burial of P in deeper sediments are as, or even more important than flushing. A net annual internal P load is common among shallow lakes during the first years following the reduction of external load, whereas it is rare in the prerestoration period and does not occur in deep lakes.

During Stage 2, increasingly P limited algae disperse to greater depths in order to exploit vertical nutrient gradients characteristic of stratified lakes. As a result, transparency increases rapidly in spite of a negligible decrease in algal biomass. Shallow lakes do not provide such refuge options and therefore Stage 2 is lacking.

Stage 3 is the period of significant biomass decrease that occurs in proportion to load reduction. This phase is usually faster and more pronounced in stratified than in shallow lakes. The reason is the elevated internal load from the surface sediments of shallow lakes keep the 'memory' of past loading conditions.

In Stage 4, species composition of phytoplankton changes and a new steady state is established. Both absolute and relative abundance of Cyanobacteria decreases. Blooms disappear; species diversity and stability of the system increase. The delay in the retreat of Cyanobacteria may depend on the life history of species, gaining dominance during eutrophication in various types of lakes.

The data of Sas (1989) allowed investigating the reaction of perennial Cyanobacteria, like *Planktothrix*. Suppression of periodically planktonic species may take a longer time than that of the perennial ones. Akinetes and colonies of many common bloom-forming genera (e.g., *Aphanizomenon*, *Anabaena*, *Microcystis*, *Gloeotrichia*) overwinter in the sediments. In most species, a small portion (up to 5–6%) of the benthic forms is required to initiate the growth of the planktonic population next year. Thus, the benthic reserves of Cyanobacteria that might increase orders of magnitude during eutrophication may inoculate the water for many years after substantial reduction in the external load. Case Study 1 also describes the behavior of periodically planktonic Cyanobacteria in a recovering shallow lake.

One can complete the earlier-mentioned list by adding Stage 5, which covers compositional changes at higher trophic levels. Similar to the restructuring of phytoplankton, delayed responses and hysteretic effects are characteristic of this process, too (cf. Case Study 2). For this reason, reduction of the external nutrient loads is only the first and most crucial step during eutrophication management in lakes, particularly in shallow ones. A series of measures have been developed that aim at speeding up recovery after load reduction by manipulating biotic interactions within the lake. If, however, load reduction is analogous to stopping a tooth, biomanipulation is certainly related to brain surgery.

Case Study 1

Three circumstances make large (596 km^2), shallow ($z_{mean} = 3.1$ m) Lake Balaton an excellent case study when studying eutrophication-related changes in phytoplankton. First, floristic studies date back to the end of the nineteenth century, while Sebestyén initiated quantitative phytoplankton research in the 1930s. Second, due to the specific morphological features (elongated shape, the main tributary enters at the southwestern end, the only outflow starts at the opposite end, relatively closed large-scale circulation patterns develop in the four basins under the influence of the dominant winds; **Figure 2**), the western areas become hypertrophic during the 1970s while the eastern ones remained mesotrophic. Third, after reducing the external P load by about 50% from 1.3 to 0.7 mg P m^{-2} day^{-1} during the late 1980s and early 1990s, the lake recovered surprisingly fast.

Although the development in the methods of phytoplankton counting somewhat biases the comparability of long-term data, neither eutrophication nor oligotrophication has substantially affected the eukaryotic algal flora. In the same time, appearance, disappearance, and reoccurrences of Cyanobacteria, especially Nostocales, have been detected during rapid changes in trophic conditions. Prior to eutrophication, three *Aphanizomenon* spp. (*A. flos-aquae*, *A. klebahnii*, *A. gracile*) and four or less *Anabaena* spp. represented the order in Lake Balaton. During the period of rapid eutrophication in the 1970s, many new species appeared abruptly in the flora, including *A. aphanizomenoides*, *A. issatschenkoi*, *Anabaenopsis elenkinii*, *Cylindrospermopsis raciborskii*, and *Rhaphidiopsis mediterranea*. Most cyanobacteria present in the pre-eutrophication period maintained at least modest populations but *A. gracile* became virtually extinct. In the 1980s, when the ecosystem 'stabilized' at a high trophic level, the only newcomer was *Anabaena contorta*. Fast oligotrophication from the mid-1990s was coupled with frequent appearance of new species (*Anabaena compacta*, *Anabena circinalis*, *Aphanizomenon hungaricus*, *Anabaenopsis cunningtonii*, *Komvophoron constrictum*), disappearance of some 'eutrophic' ones (*Anabaenopsis elenkinii*, *Rhaphidiopsis mediterranea*), and reappearance of *Aphanizomenon gracile*.

A definite increase had already been observed in the biomass of the dominant summer alga, *Ceratium hirundinella* between the 1930s and the 1950s. The

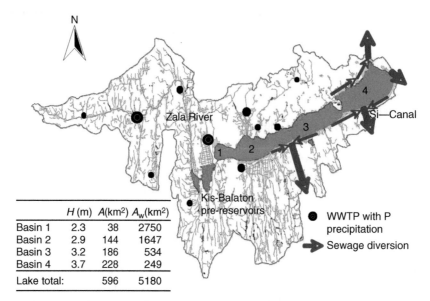

	H (m)	A(km²)	A_w(km²)
Basin 1	2.3	38	2750
Basin 2	2.9	144	1647
Basin 3	3.2	186	534
Basin 4	3.7	228	249
Lake total:		596	5180

Figure 2 Lake Balaton and its catchment. The most important management measures are also indicated. *H*, average water depth; *A*, surface area; A_w, corresponding subwatershed area; WWTP, waste water treatment plant.

annual mean biomass of phytoplankton increased from about 0.3 mg fresh weight per liter by a factor of 4. Up to the mid 1970, *Cyclotella bodanica* and *C. ocellata* dominated the spring plankton. Thereafter the abundance of pennate diatoms (*Synedra acus* and *Nitzschia acicularis*) increased conspicuously.

In the eastern areas of the lake, the biomass of summer phytoplankton showed a moderate further increase during the period of eutrophication compared with the 1950s, but annual maxima did not exceed 5 mg l^{-1}. In the same time, maxima reached 40–60 mg l^{-1} in the western areas during the early 1980. Interannual differences in biomass among successive years increased substantially in both areas. Prior to eutrophication, late summer assemblages were dominated by *Ceratium hirundinella, Aphanizomenon klebahnii,* and *Snowella lacustris*. Depending on the wind regime, various meroplanctonic diatoms (*Aulacoseira granulata, Cyclotella radiosa, C. ocellata,* small *Navicula* and *Nitzschia* spp.) contributed significantly; in very windy years they could dominate. These species, however, have been gradually replaced by Cyanobacteria during eutrophication. The key cyanobacterium, *Cylindrospermopsis raciborskii* is a subtropical species the akinetes of which require exceptionally high and narrow temperature range (22–24 °C) for germination when compared with other Cyanobacteria present in Lake Balaton. Warm conditions are incidental in this temperate lake, and this is certainly one of the main factors leading to the irregularity of *C. raciborskii* blooms.

C. raciborskii was first detected in Lake Balaton in 1978, and it produced its first large bloom in 1982.

This bloom was exceptional compared with cyanobacterial blooms during the 1970s in two respects: (1) the maximum biomass exceeded those of the previous blooms by a factor of about 2, and (2) with a delay of about 3 weeks, the blooms spread to the mesotrophic areas. The external nutrient load was known in the westernmost basin from daily measurements since 1975. These data evidenced that *C. raciborskii* achieved a much higher biomass than other N$_2$-fixing Cyanobacteria under relatively constant external load conditions. This is clearly indicative of a superior exploitation of the available resources by *C. raciborskii*. Simple mass balance models indicated that in summers of *C. raciborskii* dominance, the internal P load was much higher than in other years. Indirect evidence suggests that it is not the enhanced internal P load that induces the blooms, but presence of *C. raciborskii* leads to an elevated internal load. Eastward extension of *C. raciborskii* blooms has repeatedly been observed in subsequent years including 1992 and 1994, when the external loads have already been reduced close to the present levels. It has been shown that dispersion along the longitudinal axis may result in an inoculation of the mesotrophic eastern areas from the eutrophic western ones.

Case Study 2

The water of shallow, wind-exposed and highly calcareous Lake Balaton is always turbid, the mean vertical light attenuation coefficient varies between 2 and 4 m^{-1}. Gut content analysis of zooplankton revealed that grazers collect huge amounts of inorganic ballast and they starve even if the table is set in a near-optimal way. Because of this geomorphological drawback, community grazing rate of zooplankton is low (usually <5%), and phytoplankton production is channeled to fish primarily by zoobenthos. Investigation of the benthic macrofauna was rather sporadic in Lake Balaton before the mid-1990s. This sporadic information had to be supplemented with paleolimnological data, as well as with comparative data obtained along the west–east trophic gradient in order to reconstruct composition and biomass of the benthic macrofauna during the preeutrophication period.

Gastropods (*Potamopyrgus jenkinsi*, *Lithoglyphus naticoides*) comprised a quantitatively important group of the macrobenthos until about the middle of the twentieth century. In the late 1990, however, no living gastropods could be collected during a detailed, long-term zoobenthos survey. Disappearance of gastropods started shortly after the invasion of *Dreissena polymorpha* in the early 1940s, most probably independent of eutrophication. At the present time, chironomids and oligochaetes are the dominant groups of the benthos. Tubificidae, however, make up less than 20% of the total benthic biomass.

Of the more than 50 chironomid taxa known from Lake Balaton, only seven can be found in the profundal sediments. Three of the latter species, *Procladius* cf. *choreus*, *Tanypus punctipennis*, and *Chironomus balatonicus* give 80–90% of the biomass and annual production. These three species responded specifically to eutrophication and oligotrophication, and the reaction could sufficiently be explained by differences in their feeding habits.

Procladius choreus is a predator. It is able to utilize a wide range of food, including macrobenthic animals, zooplankton, microphytobenthos, and detritus, while the most important food is meiozoobenthos. As a consequence, the biomass of *P. choreus* was more or less independent of the changes in trophic conditions both in time (eutrophication and oligotrophication) and in space (the west-east trophic gradient).

Tanypus punctipennis preys other chironomid larvae, but it may also ingests a great deal of plant material and detritus. Its biomass increases towards the mesotrophic areas of Lake Balaton. Prior to eutrophication and in the mesotrophic eastern areas, a *Procladius–Tanypus* chironomid community was characteristic of Lake Balaton. The biomass of this community showed only a weak positive correlation with the biomass of phytoplankton.

Chironomus larvae are filter- and surface deposit feeders that take the advantage of the large amount of fresh detritus sedimenting after major algal blooms. In this way, their biomass and production strongly depends on the biomass of phytoplankton. In Lake Balaton, there was a strong positive correlation between the spring biomass of *Chironomus*-dominated benthic community and the late summer phytoplankton biomass in the previous year. The relationship, however, was not linear. *Chironomus* larvae showed a presence–absence type response. When the late summer algal biomass exceeds 20–30 μg chl-a l^{-1}, that is the annual primary production is higher than 220–250 g C m^{-2} year^{-1}, *Chironomus* dominates the zoobenthos in the next year and the chironomid biomass attains high values (0.6–3.4 g m^{-2}). Below this level of primary production, *Chironomus* is absent, Tanypodinae dominate the zoobenthos, and the total benthic biomass varies from 0.3 to 0.5 g m^{-2}.

One of the reasons why *Chironomus* depends so strongly on the last year's primary production in Lake Balaton is the low organic content (2–4%) of the sediments. This is due to the frequent resuspension that keeps the water and surface sediments permanently aerobic, thus enhancing bacterial decomposition. It was estimated that only 1.1–3.2% of the primary production was assimilated by benthic chironomids. Fast aerobic decomposition may also explain that chironomid densities are rather low in Lake Balaton when compared with other European lakes of similar trophic state.

The threshold-like shift in the dominance of *Tanypus* to *Chironomus* with increasing trophy is accompanied by a 2.5-fold increase in energy transfer efficiency from the phytoplankton to chironomid fauna. The reason is that phytodetritiphagous *Chironomus* directly harvests primary producers, while a trophic loop of one or two steps length connects predatory *Tanypus* to algae. Variability of the energy transfer efficiency is of vital importance for benthivorous fish. For example, stock density and growth rate of the dominant fish species, the common bream (*Abramis brama*) is basically influenced by chironomid production.

See also: Egg Banks.

Further Reading

Istvánovics V and Somlyódy L (2001) Factors influencing lake recovery from eutrophication—The case of basin 1 of Lake Balaton. *Water Research* 35: 729–735.

Istvánovics V, Clement A, Somlyódy L, Specziár A, G-Tóth L, and Padisák J (2007) Updating water quality targets for shallow Lake Balaton (Hungary), recovering from eutrophication. *Hydrobiologia* 581: 305–318.

Padisák J and Reynolds CS (1998) Selection of phytoplankton associations in Lake Balaton, Hungary, in response to eutrophication and restoration measures, with special reference to the Cyanoprokaryotes. *Hydrobiologia* 384: 41–53.

Porter KG (1977) The plant–animal interface in freshwater ecosystems. *American Scientist* 65: 159–170.

Reynolds CS, Dokulil M, and Padisák J (eds.) (2000) *The Trophic Spectrum Revisited*. Springer.

Sas H (1989) *Lake Restoration and Reduction of Nutrient Loading: Expectations, Experiences, Extrapolations*. St Augustin: Academia Verlag Richarz.

Scheffer M, Hosper SH, Meijer M-L, Moss B, and Jeppesen E (1993) Alternative equilibria in shallow lakes. *Trends in Ecology and Evolution* 8: 275–279.

Schindler DW (1974) Eutrophication and recovery in experimental lakes: Implications for lake management. *Science* 184: 897–899.

Specziár A and Vörös L (2001) Long-term dynamics of Lake Balaton's chironomid fauna and its dependence on the phytoplankton production. *Archive für Hydrobiologie* 152: 119–142.

Vollenweider RA and Kerekes JJ (1982) Background and summary results of the OECD cooperative programme on eutrophication. OECD report, Paris.

Wetzel RG (2001) *Limnology. Lake and River Ecosystems*. San Diego: Academic Press.

Relevant Websites

http://www.ceep-phosphates.org (European Union Eutrophication Guidance Document information. (2006). SCOPE Newsletter No. 64).

http://www.ilec.or.jp/eg/

http://www.umanitoba.ca/institutes/fisheries/

Role of Zooplankton in Aquatic Ecosystems

R W Sterner, University of Minnesota, St. Paul, MN, USA

Introduction

The freshwater zooplankton include representatives from the Protozoa, the Rotifera, and the Crustacea, as well as some less common but still widespread and often important members from such groups as the Insecta. Zooplankton are herbivorous, carnivorous, or perhaps most frequently, omnivorous. They make up one to several trophic levels in lake ecosystems. Their role as herbivores has been especially well studied. Herbivory in planktonic ecosystems has a number of characteristics that distinguishes it from herbivory in terrestrial systems:

- Small planktonic primary producers possess relatively little structural support; hence, planktonic primary producers overall are often high-quality forage.
- By virtue of its small size relative to its consumer and its short life span, the individual primary producer in planktonic systems cannot easily defend itself chemically from herbivores.
- In contrast to terrestrial systems, planktonic herbivores almost invariably are as large as or larger than the prey items they consume.

Much of what follows in the rest of this article flows from these three facts. We will examine the aspects of the basic biology of suspension feeders relevant to ecosystem dynamics. We will also look at some of the effects of zooplankton grazing on reducing algal abundance. We will examine zooplankton relative to both the so-called 'grazing chain' and the 'microbial loop.' We will see that zooplankton actively participate in nutrient cycles and simultaneously stimulate algae and microbes via nutrient remineralization while they are reducing populations of these same organisms by directly consuming them. Finally, we will briefly consider how zooplankton fit into ecosystems as a function of the fish species that are present.

Taxonomy and Biology

We will consider four principal groups.

Protozoa

Protozoans are unicellular or colonial organisms with the capacity for phagotrophy. It is often difficult, if not impossible, to cleanly separate the algae from the protozoans because many possess both autotrophic and heterotrophic nutrition; these are referred to as 'mixotrophs.' Protozoans are the smallest of the zooplankton; most are 2–200 µm long. Only a few studies have estimated the contribution of protozoa to total zooplankton abundance and production in fresh waters. Studies found that they averaged 15% of the zooplankton biomass in Lake Constance and 21% in Neumühler Lake. Sixteen phyla are represented in fresh waters, making this by far the most biodiverse group of zooplankton in water columns. Flagellated protozoans are very abundant in lake water columns while ciliated forms are widespread but dominant mainly in sediments. Certain protozoans are very active consumers of bacteria and the smallest size classes of algae (<10 µm) and protozoans can at times control the abundance of the bacteria. Protozoans are fed upon by larger zooplankton. Protozoans therefore form a bridge between the microbes and the so-called 'grazing' food chain of algae, crustacean zooplankton, and fish.

In contrast to the marine realm where the zooplankton include representatives from numerous phyla, the nonprotozoan zooplankton of fresh waters are not particularly taxonomically diverse.

Rotifera

The rotifers are the smallest multicellular zooplankton. Most are 200–500 µm long. Their name comes from Latin and means 'wheel bearer,' which refers to their food gathering apparatus. This structure, the 'corona,' consists of cilia that direct particles into the body cavity where they are ground and absorbed. Rotifers ingest mainly larger bacteria, heterotrophic flagellates, and small ciliates as well as small algae. They are fed upon mainly by larger zooplankton. Like the protozoa, therefore, they form a bridge between the microbial community and larger organisms. They are not often dominant members of the zooplankton in terms of biomass. Therefore, although one estimate indicated that rotifers accounted for 80% of the annual grazing pressure on algae in a small, eutrophic lake in Vermont, their effects may seldom be great at the ecosystem level. This, plus a general lack of published information on the topic mean they do not play a large role in this review. Perhaps their importance has been underestimated.

Copepoda

The copepods are one of the two major types of crustacean zooplankton in fresh waters. Copepods have complex life cycles. From a fertilized egg, copepods develop through 10 stages ('instars') until they are sexually mature. The long and narrow adults are generally 0.5–2 mm long. The youngest instars, the nauplii, have similar size and feeding niches as rotifers. By the time they reach adult size, however, copepods exhibit complex feeding behaviors, with strong selection for particular food types, based on chemical detection. They combine this selective mode of feeding with a much less selective mode used to capture smaller algae and microbes. Adults may be exclusively carnivorous, herbivorous, or omnivorous. As adults, they may even feed upon juveniles of their own species. Given this wide range of possibilities, it is difficult to characterize even the trophic level of copepods, much less their ecosystem-level impact, but most studies agree that many copepods will select algal particles larger than ~10 μm when available. There is some evidence that copepods actively select protozoans in preference to algae. Copepods package unassimilated food into pellets before egestion; these fecal pellets affect the dynamics of settling of material out of surface layers. Copepods are often numerically abundant in lakes, frequently accounting for >50% of the total zooplankton biomass. They are fed upon by the largest size classes of zooplankton, by aquatic insects, and by fish.

Cladocera

The cladocerans are the second major group of crustaceans. They fall within the taxonomic order Branchiopoda. The term 'Cladocera' is not a monophyletic taxonomic unit and it probably should be discarded from active use, but it is still widely used and it is retained here. Adults of herbivorous zooplankton generally range from 0.5 to 5 mm, making these some of the largest of the zooplankton in lakes, especially among the herbivores and omnivores (some carnivorous juvenile insects can be much larger). They exhibit simple life histories with an absence of metamorphosis and a prevalence of asexual, clonal reproduction. Their feeding and growth rates can be very high, and they may represent large fractions of total zooplankton biomass. Members of this group are widely recognized to have very large ecosystem-level impacts and hence they are a focal group here. There are carnivorous cladocerans too, but they will not be covered. The herbivorous species collect food using eyelash-shaped thoracic limbs, which have an appearance of a fine filtration screen. The actual hydrodynamics of food collection on these screens has been controversial, in part because of the special hydrodynamic properties of water at these small spatial scales. Nevertheless, the particles collected on the

filtration screens are generally those that are larger than the mesh openings, indicating a filter-like functioning. Given this mode of feeding, cladocerans as a rule are not greatly selective, consuming a broad size range of the smaller microbes and algae. They are often thought of as very abundant, very active, generalist feeders with a colonist, high growth rate life history. They are consumed mainly by invertebrate predators and by fish.

Body Size Relations

Aquatic ecologists base a great deal of their understanding of ecosystems on information about the size of individuals. In part, this is because aquatic organisms range enormously in size, from ~10^{-7} m for small bacteria (viruses, which are still poorly studied, are even smaller) to 10^{-3} m for predatory invertebrates up to 1 m for very large fish. A focus on size comes about also because aquatic predator–prey interactions are strongly size structured. In aquatic food chains, predators are invariably at least as large as their prey and usually are larger (**Figure 1**). **Figure 1** also shows how zooplankton occupy a very broad size range and an intermediate trophic position. Moreover, (a) primary producers are among the smallest members of the aquatic ecosystem, (b) larger food packets are generally more energetically advantageous to consume than smaller ones, and (c) the probability of detection of prey by a predator is size-dependent. In addition, larger passive particles sink much faster than small passive particles. When we consider these factors and other strongly size-based rules, it is easy to see how important size is as a factor in aquatic ecosystems.

Aquatic ecologists also must consider some aspects that complicate size-based analysis. For example, there are numerous examples of individual species with ontogenetic niche shifts due to a development from one size class to another. Juvenile fish may inhabit the nearshore and consume zooplankton when young but move offshore and consume fish when mature. Copepods may consume bacteria and small algae as juveniles but prey upon individual rotifers as adults. In general, though, the strong size-based rules provide aquatic ecologists with an important tool to further their understanding of aquatic ecosystems.

Mechanics of the Grazing Processes

Modes of Feeding

There are three principal mechanisms of feeding within the freshwater zooplankton:

1. *Phagocytosis* is a form of endocytosis wherein particles are enveloped by the cell membrane of a

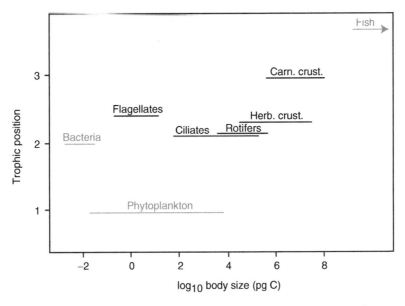

Figure 1 Body size and trophic position of different functional groups within Lake Constance. Data for one annual cycle was averaged. Zooplankton, the subject of this article, are shown in black while other organisms are in gray. Vertical position of each group indicates their trophic position (primary producers defined as position one and bacteria defined as position two, the rest calculated by analysis of diets, see original article for details). The horizontal bars represent the size range of the individuals in the group. Size and trophic position are positively related, particularly in the grazing food chain (phytoplankton, various herbivores, fish). Members of the microbial chain (bacteria, flagellates, ciliates) are generally smaller than members of the grazing chain. Zooplankton occupy most of the size range in the ecosystem. Adapted from **Figure 1** in Gaedke U, Straile D, and Pahl-Wostl C (1996) Trophic structure and carbon flow dynamics in the pelagic community of a large lake. In: Polis GA and Winemiller KO (eds) *Food webs: Integration of pattern and dynamics*, pp. 60–71. New York: Chapman and Hall, with permission.

cell and internalized to form a food vacuole. Organisms performing phagocytosis will normally be larger, but need not be greatly larger than their prey particles. This mode is associated with protozoa.

2. *Filter feeding* occurs when large numbers of small particles are collected at once using specialized feeding appendages that resemble filters. Though the hydrodynamics is complex and varies with the species, these appendages may function just like simple filters. Filter feeders cannot easily discriminate among different types of particles and each 'mouthful' is likely to contain multiple species of algae and microbes, each of which differs to a certain extent in their quality as food. Organisms performing filter feeding will normally be collecting particles much smaller than themselves. This mode is associated with cladocerans.

3. Finally, *individual particle selection* or *raptorial predation* both are terms that refer to purposeful grasping of individual food items that may be detected via chemosensory mechanisms, up to several body lengths away from the consumer. This last mode of feeding is only energetically advantageous when the individual food particle is large and of high quality. Organisms selecting

individual particles by grasping will normally be feeding on particles smaller than themselves, but still large enough to provide a good nutritional payoff. This mode is associated with copepods.

The ratio between predator length and prey length increases with predator size and mechanism of feeding: for small zooplankton such as protozoans exhibiting phagocytosis, this ratio is in the range of 1–3 whereas for large cladocerans it is on the order of 50.

Functional Response

No matter what mode of feeding is employed, the rate of food consumption will increase quickly with food concentration at low food concentration but will plateau at high food concentration as the consumer approaches a maximal feeding rate. Ecologists refer to the relationship between feeding rate and food abundance as a 'functional response' and they have defined three principal types (I, II, and III). A type I functional response has the shape of two line segments (**Figure 2(a)**) whereas types II and III are smooth curves (not shown). In practice it is often difficult to discriminate among the potentially subtle differences among the different kinds of curves using real data.

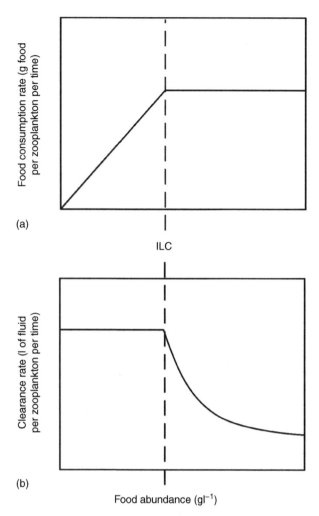

(a)

ILC

(b)

Food abundance (gl⁻¹)

Figure 2 How food abundance affects feeding rate in a type I functional response. (a) Food consumption rate rises linearly with food abundance in the low part of the range but when feeding rate hits a maximum at the Incipient Limiting Concentration (ILC), it plateaus. Panel (a) shows the 'perspective of the consumer' because the primary important variable for the consumer is the rate of food intake. (b) From (a), one can calculate the volume of the environment swept clear of prey items per unit time, or the 'clearance rate.' Note that clearance is maximal at food abundance below the ILC where the consumer is scanning and processing a maximum amount of environmental volume but it declines with food abundance below the ILC because the consumer needs to process smaller and smaller environmental volumes to harvest the same amount of food. Panel (b) shows the 'perspective of the prey' because per capita prey death rate is proportional to clearance rate.

From the functional response, one can calculate a second important quantity, the clearance rate (**Figure 2(b)**). The clearance rate has units of environmental volume per consumer (or consumer biomass) per time. Note from the figure that type I would be an appropriate model for filter feeders processing a maximum volume of the environment in an effort to

extract their food particles up until a food concentration where their ability to ingest food is maximal.

Like many physiological rates, mass-specific clearance rates decline to approximately a quarter power scale of body mass, meaning that an equivalent total mass of small individuals should have a higher community clearance rate than the same total mass of large individuals. We will see, however, that other factors contribute and that large-bodied zooplankton often have a profound influence on algal mortality rates and ecosystem properties.

Ecosystem Impacts of Herbivory

Lakes are Grazing Dominated Ecosystems

Close to one-half of the biomass of living things is carbon; therefore, understanding carbon flow goes a long way toward understanding the ecosystem processing of all matter. Similarly, the flow of carbon in an ecosystem traces the flow of chemical energy, because carbon-containing molecules, whether they are carbohydrates, proteins, lipids, or others, are used as metabolic fuel. Last but not least, organic carbon cycles in lakes, oceans, and on land affect atmospheric carbon dioxide because of rapid exchange of CO_2 between air and the living organisms in ecosystems. A major question for us here is how zooplankton influence organic carbon cycles in lakes.

Most carbon fixed on land enters into detrital pathways, for example via leaves becoming litter at the end of a temperate zone growing season. Aquatic systems dominated by macrophytes are similar; most plant production enters food chains only after that plant matter is no longer part of a living plant. Herbivores in those systems harvest a relatively small fraction of primary production. In contrast, the percent of primary production consumed by herbivores in ecosystems based on aquatic algae is high (**Figure 3**). Note the much higher median for algal (A) based systems than for macrophyte (M) or terrestrial (T) systems. The figure also shows a lot of variability in this parameter even for algal-dominated ecosystems, with an upper range >100% of annual net primary productivity removed by herbivores. This seemingly impossible value may result from error associated with incomplete sampling of a spatially and temporally variable process. Or, it may be explained by subsidies of terrestrial carbon. Given the generally high levels of grazer cropping of primary production in algal-dominated systems, herbivorous zooplankton must often be key players in ecosystem processes in lakes.

Though researchers agree about the relatively high impact of grazers in algal-dominated ecosystems, the reasons for this distinction are not completely sorted

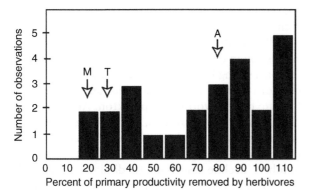

Figure 3 Distribution of the percent of annual net primary productivity removed by herbivores for ecosystems with primary producers consisting of aquatic algae (freshwater and marine planktonic plus marine periphyton). The range extends from ~20% to ~100% with a median of 81% (indicated by the arrow marked 'A' for algae). In contrast, aquatic ecosystems with primary producers dominated by macrophytes (arrow marked 'M') or terrestrial systems in general (arrow marked 'T') exhibit much lower median percentages (data for M and T systems not shown here). Adapted from **Figure 1** of Cyr H and Pace ML (1993) Magnitude and patterns of herbivory in aquatic and terrestrial ecosystems. *Nature* 361: 148–150, with permission.

out and probably several causes are at work. Microalgae have relatively high growth rates and high nutrient content. In general, therefore, algae may be better quality food for herbivores than are larger plants containing a lot of nutrient-poor structural biomass. In addition, by virtue of being so small and numerous, microalgae are less able to defend themselves via antiherbivore chemical compounds in comparison to larger plants. High grazing rates in microalgal systems can potentially come about due to higher overall biomass of grazers in those systems or higher activity per unit biomass. Researchers do not agree about the extent to which these two components contribute.

Inedible, Poor Quality, and Toxic Algae

The discussion here is not meant to lend the impression that all algae are good-quality food for all herbivores all the time. There are multiple factors that influence the quality of algae as food. One important one is the size itself. Because of the size restrictions of the grazing process, not all algae can be harvested by all planktonic herbivores. Some may be too small to be efficiently retained on filters, or they may offer too small a chemical signal or too small an energetic payoff to provoke individual particle capture behaviors. Chemical content too plays a role. Different major groups of algae differ in significant ways in their biochemical content. One of these is the complement of fatty acids. Some algae have high concentration of unsaturated fatty acids, such as the

'fish oils,' and these may become limiting to higher trophic levels. Nutrients themselves also play a role, and low-growth-rate algae can have low enough nutrient content that nutrients may limit herbivores feeding on them. Finally, consuming some algae has deleterious consequences such that zooplankton are worse off eating them even than eating nothing. In fresh waters such toxic algae come mainly from the cyanobacteria. When considering the generally high levels of herbivory by zooplankton and the different factors that influence edibility and food quality, it would seem as though quite often food webs would be steered toward dominance by inedible algae. Surprisingly, little agreement exists today as to how grazing steers algal communities in this way. Perhaps this is because of the importance of other, indirect effects of herbivores (discussed later).

Clear Water Phases

Plankton populations fluctuate greatly on subseasonal time scales. One common temporal pattern in temperate lakes is for an early-spring algal bloom to be followed by an increase in herbivorous zooplanktons, presumably fostered by the combination of high food abundance and moderating temperature, supplemented in some cases by recruitment from resting stages from the sediments. This late-spring buildup in herbivore abundance can produce a feature referred to as a 'clear water phase.' In a clear water phase, algal abundance rapidly declines and water transparency increases. Changes in transparency on the order of one to several meters can occur over a time frame of days to weeks during such clear water phases. An example is shown in **Figure 4**. Careful studies confirm that these clear water phases occur during periods of very heavy grazing, and that without the heavy grazing pressure, they would not occur. In productive, shallow lakes, strong peaks of herbivorous zooplankton are not limited to spring conditions. These too can result in clear water phases. An example from a shallow, hypertrophic Danish lake is given in **Figure 5**. This lake showed strong depressions of algal abundance associated with peaks in zooplankton abundance several times over a 3-year period. Indeed one such period occurred throughout most of the summertime in 1986, a very prolonged phase of apparently grazing-induced clear water.

Grazer-caused increased water transparency can have other effects on lake ecosystems. The higher availability of light may have significant effects on algal photosynthesis or on the match of carbon and phosphorus in algae and microbes. Also, when transparency increases, heat penetrates into the water column to greater depth (water is heated by the absorption of

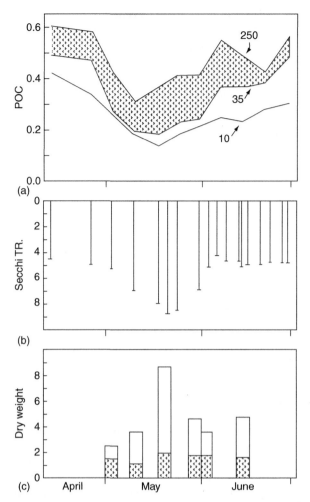

Figure 4 A very well-documented example of a clear water phase. In this small lake in Germany, a several-week-long depression of algal abundance (a) beginning in late April and continuing into May resulted in a dramatic clearing of the lake waters, with secchi transparency approximately doubling from 4 to 8 m (b). The clear water in mid-May corresponded to the peak abundance of herbivorous zooplankton in this lake (c). Reproduced from **Figure 2** of Lampert W, Fleckner W, Rai H, and Taylor BE (1986) Phytoplankton control by grazing zooplankton: A study on the spring clear water phase. *Limnology and Oceanography* 31: 478–490, with permission.

light, and so the deeper that light penetrates, the deeper the heat appears in the lake's waters). This alteration in lake physics can even cause a noticeable change in the depth of the lake's thermocline. Hence, high grazing systems with lower algal biomass will have thicker epilimnia and deeper thermoclines.

An effort to ascertain the common patterns of seasonal successions in temperate lakes was made some years ago by the Plankton Ecology Group (PEG), a group of mainly European limnologists. Based on their experiences with a diversity of lakes, they put together the PEG model of seasonal succession of planktonic events in fresh waters. The model is a

consensus view and consists of 24 sequential statements which describe seasonal events step by step. The PEG model codifies the role of grazers in seasonal succession in the following way. Early in the growing season a spring bloom of small, rapidly growing algae develops but is then eliminated due to the later buildup of herbivorous zooplankton. An overgrazed, clear water phase results, characterized by high grazing pressure and high nutrient recycling. At this point, the herbivores become food limited and their numbers decline due both to diminished fecundity and to increasing predation pressure from fish. As the season progresses, a complex set of processes take over in summertime, which include grazing on edible species and limited algal growth by one or more chemical elements. Larger zooplankton are replaced by smaller ones. The PEG model does not indicate a strong role of grazing in late-season dynamics as summer gives way to fall and winter.

The PEG model is a very useful 'ideal' consensus view that summarizes some of the more commonly reoccurring steps of seasonal succession in the freshwater plankton. However, the timing and interrelationships of these different seasonal features in temperate lakes is still only partially worked out. They may be sensitive to climate change because not all of the different seasonal features respond to the same signals. In Lake Washington, Seattle, it has been suggested that spring algal blooms are largely driven by warming and thus should occur earlier and earlier as climate warms. In contrast, late-spring zooplankton blooms in this lake are probably triggered by the hatching of resting eggs from the sediments. Resting egg hatching is related at least in part to photoperiod. Because day–night cycles are not affected by climate, a consequence of climate warming in this lake could be to alter the relative timing of the algal bloom and the bloom of herbivores that can help control summer algae. It already appears that zooplankton are hatching from resting eggs in this lake too late to make use of the now too early algal blooms. These kinds of shifts in seasonal timing events may have profound consequences that we are just beginning to understand.

Complex Interactions: Nutrient Recycling by Zooplankton

No organism is 100% efficient at retaining the matter in the food it ingests. Part of the ingested food is assimilated and the rest is egested. In addition, some assimilated matter is subsequently metabolized and rereleased to the environment. From the food they eat, zooplankton recycle the two important nutrient elements N and P in soluble forms that are readily

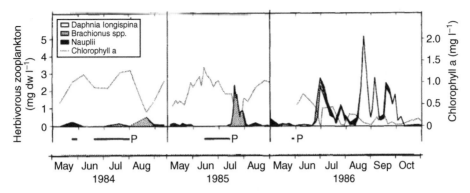

Figure 5 Seasonal dynamics of chlorophyll *a* and several forms of grazing zooplankton over three growing seasons in a shallow, hypertrophic Danish lake. Phytoplankton populations were markedly reduced during a year of high zooplankton abundance. Adapted from **Figure 2** of Jeppesen E, Sondergaard M, Sortkjaer O, Mortensen E, and Kristensen P (1990) Interactions between phytoplankton, zooplankton and fish in a shallow, hypertrophic lake: a study of phytoplankton collapses in Lake Sobygard, Denmark. *Hydrobiologia* 191: 149–164, with permission.

taken up by algae and microbes. The role of zooplankton in ecosystems is not limited to direct effects of herbivory per se, but it also includes impacts that derive from recycled nutrients. In the open ocean, one can estimate the fraction of primary production driven by recycled N vs. newly input N by looking at rates of uptake of ammonium (recycled) vs. nitrate (newly input). Because water columns in lakes are not isolated from lake edges and bottoms, we cannot assume all ammonium used by freshwater plankton derives from regeneration in the water column; thus, we cannot use exactly the same approach to estimate the importance of consumers in regenerating nutrients in inland ecosystems. The question though is still very relevant – how important is in situ nutrient cycling to the functioning of aquatic ecosystems?

An example showing these separate roles of zooplankton in simultaneously consuming and stimulating the growth of algae is shown in **Figure 6**. In this study, a gradient of *Daphnia* densities was established and porous-walled chambers were used to isolate a subset of the algae from direct contact with the grazers. The algae in the chambers, however, were subjected to any nutrients remineralized by the *Daphnia*. As a consequence, algal growth rates were positively affected by the presence of *Daphnia*. This indirect, positive effect was almost perfectly counterbalanced by the negative, direct grazing effect. Thus, the net effect of grazers on algae was to accelerate the population turnover rate (higher growth and loss rates).

With large amounts of matter flowing in and out of zooplankton biomass, a further set of interactions comes into play involving the stoichiometric balancing of elements. In ecological interactions, just as in chemical reactions, mass must balance and there are rules of definite proportions affecting the relationship between the chemical composition of products and

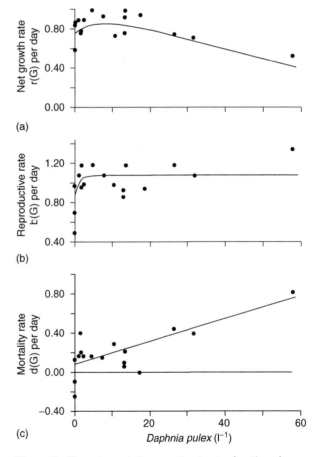

Figure 6 The net population growth rate as a function of *Daphnia* density in a eutrophic pond community. Phytoplankton grew at a relatively high rate in all *Daphnia* densities (a), with only a slight overall herbivore effect. However, this net effect could be decomposed into two stronger but opposing effects, one a stimulation of production as a result of nutrient regeneration (b) and the other a direct, consumption effect (c). Reprinted from **Figure 2** in Sterner RW (1986) Herbivores' direct and indirect effects on algal populations. *Science* 231: 605–607, with permission.

reactants. Consider the conversion of algal biomass (reactant) to zooplankton biomass plus wastes (products). Algal biomass may have a very different chemical composition than that of zooplankton biomass, which in turn means that the composition of waste products must conform to this difference for mass in the overall reaction to balance.

Much has been learned about the relative proportions of C, N, and P in different plankton species within different trophic levels in aquatic ecosystems. Phytoplankton exhibit a wide range of C:N:P ratios, depending on taxonomy and growth conditions. Rapidly growing algae exposed to high levels of resources have relatively high concentration of N and P within algal biomass. As N or P become limiting and growth slows, either the C:P or C:N or both increase because algae can continue to photosynthesize and add substances such as lipids or carbohydrates to their biomass even if they are not taking up nutrients at comparable rates. Zooplankton too exhibit variation in their elemental composition; however, variation in zooplankton is much less than in algae. Some zooplankton species have highly characteristic elemental content while other species show ranges of C:P and C:N on the order of about twofold variation. For our purposes, the most significant aspect of zooplankton chemical content is that some species are characterized by high levels of phosphorus ($\sim$1.5% of total dry biomass) and have low N:P ratios while other species are characterized by low levels of P ($\sim$0.5% of total dry biomass) and have high N:P ratios. The basis of this interspecific difference in elemental content is thought to relate primarily to life history differences – species that exhibit high growth rates do so by having high complements of P-rich RNA. Some species of *Daphnia* are particularly rapid growers with high P contents. Low P species include other cladocerans such as *Bosmina* and adult calanoid copepods.

From the stoichiometric relations described above, it is easy to see that shifts in zooplankton community composition can have a large influence on the partitioning of N and P within aquatic ecosystems. For example, shifts between *Bosmina* or copepods and certain *Daphnia* can result in large swings in the storage and flux of N and P in zooplankton. These shifts in turn can feed back to influence the N or P limitation in algae and microbes. One of the first such documented cases is presented in **Figure 7**. In this study of three small lakes in Michigan, USA, when the zooplankton was dominated by large- bodied *Daphnia*, two different measures of N vs. P limitation in algae yielded low values, indicating strong algal P limitation. The pattern of nutrients limiting primary production was influenced strongly by the species composition of the consumers in the lake ecosystem.

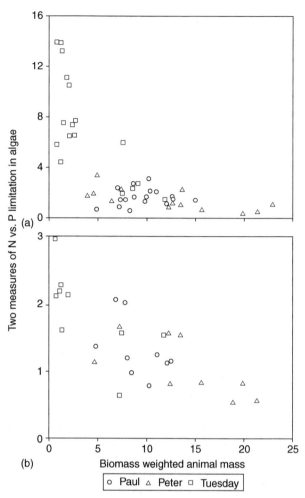

Figure 7 In this study of three lakes in northern Michigan, when large *Daphnia* dominated the crustacean zooplankton community, two different measures of N vs. P limitation in algae were low, indicating strong P limitation. Small-bodied zooplankton communities on the other hand were associated with high values of both measures, indicating N limitation. The zooplankton community therefore affected not just grazing rates but also the relative availability of N and P to the algae in the lake waters. These trends are consistent with what is known about the stoichiometry of large vs. small herbivores in this lake. Reprinted from **Figure 5** in Elser JJ, Elser MM, MacKay NA, and Carpenter SR (1988) Zooplankton-mediated transitions between N and P limited algal growth. *Limnology and Oceanography* 33: 1–14, with permission.

Zooplankton Sustaining Higher Trophic Levels

Traditionally, aquatic ecosystems have often been represented as a single linear chain:

Algae → Zooplankton → Planktivorous small fish

→ Piscivorous large fish

This classical view suggests that all fish production is supported by zooplankton. However, today it is widely recognized that the above is too simplistic a view and concepts of the role of zooplankton in sustaining higher trophic levels, including fish, are under revision.

One modification to the classical food chain involves the role of heterotrophic bacteria and their mainly protozoan consumers (recall **Figure 1**). Planktonic bacteria are more numerous than once believed and they are metabolically active. They obtain carbon from dissolved organic pools and they generally compete with algae for dissolved nutrients. The rate of carbon processing in this heterotrophic bacterial assemblage generally rivals the total amount of carbon processing in the entire rest of the aquatic ecosystem, and often exceeds it. To what extent does this very active microbial community contribute to fish production? Because fish do not eat microbes directly, bacterial production will make its way into higher trophic levels mainly by supplementing production of zooplankton over what can be sustained by algae alone. Certain crustacean zooplankton with fine filter screens can graze directly upon bacteria but many do not. Thus, bacterial production can follow a complex, multistep pathway to make it into zooplankton. In most cases, the primary bacterivores are protozoans (flagellates and sometimes ciliates). These, roughly speaking, are algal-sized particles and are grazed upon by 'herbivorous' zooplankton, which should really be considered omnivores. When multiple feeding steps are involved in getting bacterial production into higher trophic levels, the inevitable thermodynamic loss associated with each trophic level means that the magnitude of bacterial production ending up in fish might often be quite small. The same cannot be said however for nutrients such as N or P. Though that is not as well studied as C or energy, nutrients may be passed from trophic level to trophic level with closer to full efficiency, and the microbial pathway of moving nutrients into fish might be an important one.

Another way that the classical food chain is being revised is in an increased appreciation for the importance of boundaries and edges of aquatic ecosystems and their connections to water column processes. Fish populations use the nearshore and benthos to different degrees for feeding, spawning, and other needs. These regions are often highly productive and spatially heterogeneous. Invertebrates in these zones do include some of the same species seen in the open waters but they also include many others that are adapted to utilization of the great amount of surface area present. In some fish species, use of the nearshore and bottom is a function of the size or stage structure of the fish population, and use can vary diurnally and seasonally. Given these many variables it is difficult to characterize the importance of

the nearshore. However, nearshore or benthic production may often be a very large subsidy to classical, open-water food chains.

A tool that has been used very effectively to address these issues of pathways of carbon, energy, and nutrients in food chains is the use of stable isotopes. Many naturally occurring elements exist in multiple isotopic forms. Carbon is found most abundantly in its form with mass equal to 12, but rarer, heavier, isotopes also are found, especially the form with mass equal to 13. Different isotopes of a given element have nearly the same chemical reactivity, but some different ecosystem processes fractionate the isotopes to different degrees. This fractionation can create a distinct isotopic signature in biomass, which can then be used to trace the flow of organic matter throughout an ecosystem. This tool has been used to address the role of pelagic zooplankton in sustaining higher trophic levels. Production by planktonic algae creates a very different carbon isotopic signature than does production by benthic algae. The reason for this difference relates to the different nature of boundary layers surrounding the primary producers in the two kinds of habitats. One study of carbon stable isotopes in lake food webs found that top predators in temperate and arctic lakes had an isotopic signature indicating nearly equal dependence on planktonic and benthic pathways, in spite of the fact that in at least some of these same lakes, total benthic production is thought to be very minimal in comparison to planktonic production. In another recent study, carbon stable isotopes were used to separate allochthonous from autochthonous production in supporting higher trophic levels in lakes. The former (allochthonous) refers to photosynthesis that occurred outside the boundaries of the lake, and was transported, for example, in the form of dissolved organic carbon. The latter (autochthonous) refers to photosynthesis that occurs inside the lake; in this case in this study, mostly in the water column. In looking at the share of these two pathways in supporting higher trophic levels, this study concluded that 22–50% of zooplankton carbon was derived from terrestrial sources, showing that there is significant subsidy of these ecosystems by organic carbon produced outside their boundaries. The most likely pathway accounting for this subsidy is the microbial community.

A food web channels much more than just carbon and energy. Essentially every essential element and many biochemicals follow these pathways through ecosystems. Among these biochemicals are hydrophobic organic compounds such as polychlorinated biphenyl (PCB) contaminants. In general, through a process referred to as biomagnification, PCBs tend to increase in concentration with every trophic level in a single food chain. Accordingly, the concentrations of PCBs in top predators in lakes are highly dependent on food web structure. This is a topic of human health concern because of the potential toxicity of these pollutants. Zooplankton are involved in patterns of biomagnification. For example, in Canadian lakes differing in the number of trophic levels between algae and the top fish trophic level, a very clear pattern of higher levels of PCBs where there are multiple trophic levels of zooplankton. In this study, short food chains occurred when lake trout fed directly upon herbivorous zooplankton. Longer food chains included a second zooplankton trophic level, an invertebrate predator (*Mysis*). The insertion of another zooplankton trophic level yielded longer food chains and greater tissue PCB concentration in lake trout. Later studies indicated that this effect was due both to the structure of the food web and to differing patterns of lipid accumulation in the fish in different types of ecosystems.

In summary, although a great deal of study has moved scientists beyond thinking about the simple, linear classic food chain, zooplankton are still key intermediate trophic levels where matter and energy funnel through between primary producers and top predators.

Top-Down Control

The trophic position occupied by zooplankton – between fish and algae – is well illustrated by the voluminous scientific literature on the effects of fish on aquatic ecosystems. More than 50 years ago, limnologists had shown that the identity of the fish species present in lakes and ponds had many implications for the structure and functioning of aquatic ecosystems. Bottom-feeding fish such as carp resuspend great volumes of sediments, decreasing transparency and redistributing nutrients. More relevant to the present discussion, however, the predation pressure on zooplankton is highly variable with fish community structure.

Just like the zooplankton, fish in lakes can occupy multiple trophic levels. Small fish such as juvenile bluegills or different species of minnows tend to be visually oriented zooplanktivores, consuming the largest zooplankton available. Hence, zooplankton size structure often relates strongly to the abundance of these small fish. Small fish in many lakes will be prey for larger, piscivorous fish. Fish populations fluctuate over subseasonal and interannual time scales. Thus, predation pressure on zooplankton can be highly variable.

The theory of top-down control in lakes puts many of these relationships together. Consider the implications of the introduction of a piscivore species to a lake, adding a new trophic level to the ecosystem. As the piscivore reduces the populations of minnows, predation on the largest size classes of zooplankton is relaxed. The mean size of zooplankton in the community and the total biomass of zooplankton can thereby increase. A second phenomenon can also come into play because some of the largest zooplankton are predators on the small ones. By releasing these invertebrate predators from fish predation, a mortality source to small zooplankton is increased, again with the net result being a gain in the mean size of the zooplankton population. Further size-related shifts can come about because crustacean grazers are predators on smaller, mainly protozoans; hence, a buildup of crustaceans can greatly deplete forms like flagellates that graze upon the smallest algae and the bacteria. Concomitant with such shifts in fish and increase in the mean size of zooplankton, it is often observed that algal populations overall are diminished. Thus, in spite of known diverse side branches and 'webby' trophic connections in aquatic ecosystems, a simple chain of positive and negative effects across multiple trophic levels has been often observed to result from alterations to fish stocks. One study that examined a large set of such experiments concluded that in about one-third of all experiments, alterations in fish results in large changes in algal biomass. Effects were weaker in the other experiments. Purposeful lake management using these principles falls under the concept of 'biomanipulation' and the general ecological theory behind it is called a 'trophic cascade.'

Trophic cascades have been described in many ecosystems. One metaanalysis said that the trophic cascades were strongest in marine benthos, intermediate in lake plankton, lake benthos, and running water benthos, and weakest in marine plankton and terrestrial systems. Another study found that a combination of herbivore and predator metabolic factors and taxonomy explained most of the strength of trophic cascades. In particular, a combination of invertebrate herbivores and endothermic vertebrate predators produced the strongest cascades.

One interesting aspect of trophic cascades concerns the reality of the herbivore trophic level in ecosystems. It is well understood that aquatic ecosystems have a high degree of omnivory (feeding on multiple trophic levels) and that the organisms at intermediate trophic levels are connected in complex ways involving a microbial and a grazing chain and interactions between them. In other words, the known trophic relationships at the bottom of the aquatic food chain are very 'webby,' not a simple, linear chain as given in the classical model described above. Nevertheless, some kind of strong trophic signal often propagates from fish to algae by way of the zooplankton. Existence of trophic cascades seems to imply a single, coherent trophic level of herbivores, consuming algae and in turn being consumed by fish. There clearly are many interesting questions left to answer in the unraveling of these relationships in aquatic ecosystems.

See also: Algae; Bacteria, Bacterioplankton; Cladocera; Competition and Predation; Copepoda; Cyanobacteria; Egg Banks; Other Zooplankton; Phytoplankton Population Dynamics in Natural Environments; Protists; Rotifera; Trophic Dynamics in Aquatic Ecosystems.

Further Reading

Brett M and Goldman CR (1997) Consumer versus resource control in freshwater pelagic food webs. *Science* 275: 384–386.

Cyr H and Pace ML (1993) Magnitude and patterns of herbivory in aquatic and terrestrial ecosystems. *Nature* 361: 148–150.

Elser JJ, Elser MM, MacKay NA, and Carpenter SR (1988) Zooplankton-mediated transitions between N and P limited algal growth. *Limnology and Oceanography* 33: 1–14.

Hecky RE and Hesslein RH (1995) Contributions of benthic algae to lake food webs as revealed by stable isotope analysis. *Journal of the North American Benthological Society* 14: 631–653.

Lampert W, Fleckner W, Rai H, and Taylor BE (1986) Phytoplankton control by grazing zooplankton: A study on the spring clear water phase. *Limnology and Oceanography* 31: 478–490.

Mazumder A, Taylor WD, McQueen DJ, and Lean DRS (1990) Effects of fish and plankton on lake temperature and mixing depth. *Science* 247: 312–315.

Porter KG (1977) The plant–animal interface in freshwater ecosystems. *American Scientist* 65: 159–170.

Sarnelle O (1999) Zooplankton effects on vertical particulate flux: Testable models and experimental results. *Limnology and Oceangraphy* 44: 357–370.

Shurin JB, Gruner DS, and Hillebrand H (2006) All wet or dried up? Real differences between aquatic and terrestrial food webs. *Proceedings of the Royal Society B* 273: 1–9.

Sommer U, Gliwicz ZM, Lampert W, and Duncan A (1986) The PEG-model of seasonal succession of planktonic events in fresh waters. *Archiv für Hydrobiologie* 106: 433–471.

Sterner RW (1986) Herbivores' direct and indirect effects on algal populations. *Science* 231: 605–607.

Sterner RW (1989) The role of grazers in phytoplankton succession. In: Sommer U (ed.) *Plankton Ecology: Succession in Plankton Communities*, pp. 107–170. Berlin: Springer Verlag.

Microbial Food Webs

J Pernthaler and T Posch, University of Zurich, Kilchberg, Switzerland

Introduction

Microbial food webs are omnipresent in surface freshwater environments ranging from the largest lakes to the smallest puddle (even in the liquid of pitcher plants). The immense growth potential of the unicellular pro- and eukaryotic microorganisms in pelagic, benthic, and littoral habitats is approximately counterbalanced by the action of nano- and microsized protistan predators and by viral lysis. Therefore, the abundances or biomasses of microbial populations in natural waters frequently do not reflect the maximal carrying capacities of aquatic ecosystems as set by resource availability only, but represent a compromise between so-called top-down (mortality) and bottom-up (competition for substrates and nutrients) factors.

The feeding interactions between aquatic microorganisms have attracted the attention of a wider scientific public through the work of ecologists who studied the fate of dissolved organic carbon (DOC) released from primary production, or of both DOC and nutrients originating from zooplankton grazing or allochthonous (terrestrial) sources. The original 'microbial loop' concept suggests that a part of the DOC that is assimilated by heterotrophic prokaryotes is transferred to higher trophic levels by bacterivorous protists, which in turn are consumed by micro- and macrozooplankton. This idea soon gave rise to a heated scientific debate, whether the DOC consumed by bacteria in pelagic systems is indeed significantly transferred to metazooplankton or whether DOC is largely remineralized by respiration, i.e., the microbial loop acts as either a link or a sink for organic carbon. The conflicting views were reconciled to some extent by a concept of microbial food webs that encompasses all auto- and heterotrophic unicellular microorganisms as the nutritional base for higher trophic levels (**Figure 1**). Such a definition allows more adequate appraisal of the complex interactions between different microbial populations, e.g., with respect to mixotrophy (see definition later) or parasitism. Moreover, a wider view of microbial food webs can also address transfer processes of organic carbon to higher trophic levels that are per se not part of the microbial loop, such as protistan predation on autotrophic picoplankton.

This article first gives an introduction to the identity and ecological properties of the major bacterial consumer in fresh waters and discusses their respective roles in a variable and spatially heterogeneous environment. We then outline how the prey selectivity of bacterivorous protists can affect both the phenotypic and genotypic composition of microbial assemblages by favoring particular bacterial species or morphotypes. The last section examines how the simultaneous action of top-down and bottom-up factors in multispecies assemblages may result in the apparent stability of total bacterial cell numbers that is often observed in fresh waters. Discussion will mainly be about the planktonic environment, not because it is more important than other habitats, but because planktonic microbial food webs are relatively well studied compared with, e.g., the littoral or benthic zones.

Protistan Predators

Nanosized flagellated protists (i.e., of a size range of 2–5 μm) are the main bacterivores in many freshwater systems. The taxonomic affiliation of numerically important heterotrophic nanoflagellates is often unclear; members of the genus *Spumella* appear to be widespread in the pelagic zone of lakes. Nanoflagellate abundances may range from a few hundreds per milliliter in oligotrophic mountain lakes to several thousands in eutrophic systems, and they tend to increase with bacterial numbers and productivity. Flagellates are consumed by a variety of zooplankton, including ciliates, rotifers, and crustaceans. Since eutrophic environments can sustain higher concentrations of micro- and metazooplankton, flagellate numbers in more productive systems are hypothesized to be more limited by predation than by the availability of bacterial prey. The magnitude of such top-down control on bacterivorous protists may, however, be substantially influenced by the specific composition of the zooplankton and even fish assemblages, i.e., by trophic interactions cascading through the entire food web.

Flagellate grazers are important agents in recycling limiting nutrients and transferring them to other compartments of the microbial food web. The biomass of prokaryotes contains higher fractions of phosphorus (P) and nitrogen (N) than their eukaryotic predators. Moreover, the largest fraction of the particulate organic carbon ingested by protists is used for respiration, thereby further increasing their relative carbon demand as compared to that for N and P. Heterotrophic

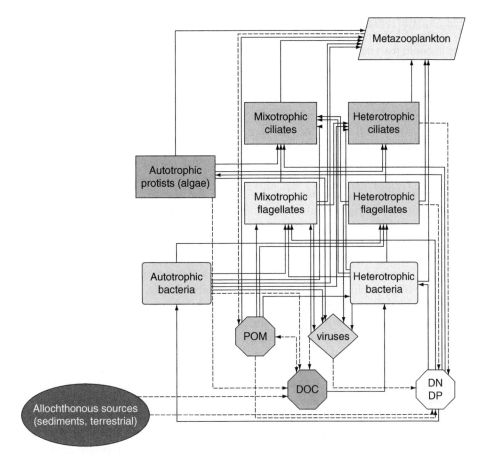

Figure 1 Conceptual depiction of freshwater microbial food webs embedded within bottom-up (resources) and top down (metazooplankton predation) factors. The different shades of grey resemble the differences in connectedness within the food web, i.e., the sum of in- and outgoing connections to other compartments. POM: particulate organic matter; DN, DP: dissolved nitrogen, phosphorus; solid lines represent feeding relationships, slashed lines are element fluxes.

flagellates are thus forced to excrete surplus nutrients into the surrounding water. As a consequence, protistan bacterivory may directly stimulate the growth of both bacteria and phytoplankton. In addition, flagellates represent a source of particulate nutrients for top predators such as micro- and metazooplankton.

The role of ciliated protists in planktonic microbial food webs cannot be generally defined but requires additional taxonomic and ecological differentiation. Some genera (e.g., *Halteria*, *Cyclidium*) may be responsible for most bacterial mortality in eutrophic and humic systems. By contrast, other groups play an important role as consumers of micro- and macroalgae or are predators of nanoflagellates, of other heterotrophic protists, and even of macrozooplankton.

Mixotrophy

Mixotrophic flagellated and ciliated protists are akin to 'carnivorous plants,' acquiring resources and energy autotrophically from photosynthesis as well as heterotrophically from dissolved and particulate

organic matter. These types of protists are a common feature of freshwater microbial food webs. Mixotrophic flagellates are typically bacterivorous, whereas ciliates may consume both pro- and eukaryotes. Many mixotrophic ciliate species actually use chloroplasts acquired from their algal prey for photosynthesis. The particular value of the mixotrophic life style lies in the phagotrophic acquisition of the limiting nutrients (P, N) that are required for growth by autotrophic carbon fixation. However, mixotrophic organisms are often less competitive at optimal growth conditions (i.e., at high nutrient loadings or prey densities) than their purely auto- or heterotrophic competitors. Therefore, mixotrophic protists tend to form largest populations in habitats with (or during periods of) high nutrient limitation, e.g., in Antarctic or high mountain lakes.

Parasitism

More recently, viruses have been recognized as another important element of aquatic microbial food webs.

The total number of virus particles in fresh waters typically exceeds those of bacteria by a factor of 10, and viral lysis may account for a substantial fraction of bacterial and phytoplankton mortality. The ecological consequences of virus-induced mortality probably differ from the effects of protistan grazing, because viral lysis leads to a release of both nutrients and organic carbon from lysed cells. This 'viral shunt' may represent a substantial input (up to 25%) to the DOC and nutrient pools in some aquatic systems.

Viruses are believed to play a particularly important role in eutrophic systems with active bacterial cells, because their reproduction is intimately linked to the metabolic state of their hosts. In addition, viruses are sensitive to deactivation by UV radiation, which is attenuated more rapidly in eutrophic waters. However, viral infection of filament-forming bacteria has also been observed in an oligotrophic highly transparent mountain lake. Viral lysis may moreover be the most important cause of bacterial mortality in the anoxic hypolimnion of lakes. Comparatively few bacterivorous protists are found in anoxic waters, and the typically elevated densities of bacterial host cells in these habitats favor viral reproduction.

Other parasitic relationships between prokaryotic and eukaryotic microbes are less well understood, but may be important elements of aquatic microbial food webs. The interactions between bacterial pathogens of pro- and eukaryotic freshwater phytoplankton species have been investigated in pure culture, but their role in the environment remains obscure. Algicidal bacteria are increasingly regarded as a potential means for biocontrol of harmful algal blooms, e.g., the addition of a *Pseudomonas fluorescens* strain to natural waters was found to terminate blooms of the diatom *Stephanodiscus hantzschii*. Another group of parasitic microorganisms known to infect a variety of different algae species are the chytrid fungi. These microorganisms can control the population dynamics of diatoms (e.g., of *Asterionella*) during spring phytoplankton blooms. The free-living fungal zoospores, in turn, represent food of high nutritional value for cladoceran zooplankton.

Habitat Heterogeneity

Freshwater microbial food webs should not be regarded as an abstract set of biotic interactions only, but must be understood in the context of a heterogeneous physicochemical environment. From a microbial perspective, even an oligotrophic pelagic habitat may exhibit considerable structural variability. For example, phytoplankton cells may be surrounded by a species-specific zone of enhanced DOC concentration (the phycosphere), which can be detected and tracked by motile chemotactic bacteria. On the other hand, colonial diatoms such as *Fragilaria* or *Asterionella* may provide a surface for the attachment of choanoflagellates (e.g., *Salpingoeca*) that are filter-feeding on planktonic bacteria. Senescent algal cells moreover tend to agglutinate because of higher exopolymer secretion and shearing forces, and such slowly sedimenting macroscopic organic aggregates represent hot spots of bacterial abundance and production. It is, therefore, not surprising that heterotrophic flagellates and ciliates actively colonize these 'lake snow' particles in order to profit from the attached bacterial biomass. The elevated bacterial cell densities and activity in such habitats most likely also provide better reproductive conditions for bacteriophages.

Other zones of steep environmental gradients have so far received less attention with respect to microbial food webs, e.g., the benthic boundary layer above sediments. Pronounced layerings of pro- and eukaryotic microbial species are also observed along oxyclines in meromictic lakes. Moreover, there are indications that the microbial food webs in the littoral zone of lakes might differ from the pelagic realms, potentially because of a higher influence of bacteria and protists that originate from biofilms (e.g., on macrophytes).

From Communities to Populations

Current ecological theory assumes that food web structure is not a random phenomenon but governed by a set of general rules. To predict particular aspects of food web topology (e.g., food chain length), detailed knowledge of species richness and connectance (i.e., the fraction of possible predator–prey links that actually occur) would be required. In addition, basic assumptions about the rules governing 'who eats whom' have to be set, such as size differences of predators and prey or the similarity in the prey spectra of phylogenetically closely related organisms.

Unfortunately, theoretical questions, e.g., about the assembly, resilience, or dynamic behavior of aquatic microbial food webs currently cannot be addressed outside highly simplified model microcosms. First, our knowledge of the taxonomic composition of freshwater microbial food webs is rudimentary. An accurate quantitative taxonomic identification of aquatic microbes currently is, by and large, limited to phytoplankton and the ciliated protists. By contrast, it is not possible to assess the community composition of viruses, heterotrophic bacteria, and bacterivorous nanoflagellates at sufficiently high levels of phylogenetic resolution (i.e., species-like units or genotypes). Second, we only possess a vague understanding of the qualitative interactions between the numerically important taxa of aquatic microbes, e.g., of viral host ranges, or the food spectra and preferences of the common species of freshwater bacterivorous flagellates.

As a consequence, research in aquatic microbial food webs has moved from purely community-level analyses (i.e., studying bacteria or flagellates as single conceptual units) to the specific trophic interactions between different genotypic populations of abundant freshwater microbes. This has been facilitated by the advent of a suite of molecular biological techniques aimed at the qualitative and quantitative cultivation-independent identification of microbes in environmental samples, e.g., by analysis of ribosomal RNA genes. Using these approaches, focus can be put on the relationship between those bacterial and eukaryotic species that are numerically important in situ. This approach represents considerable progress compared to laboratory investigations on arbitrary sets of microbial populations that may or may not coexist in natural habitats. In the following sections, we present some aspects of a more population-centered viewpoint to illustrate the heterogeneity and multifariousness of freshwater microbial food webs.

Selective Mortality

Killing the Winner

Sometimes populations of single microbial species escape from control by the food web. For example, bacteria related to the nosocomial pathogen *Stenotrophomonas maltophilia* transiently formed >90% of all cells in a supposedly pristine subtropical lagoon, but virtually disappeared from the pelagic zone within the following 48 h. This observation is one example that highlights the need to better understand the mechanisms that allow for the growth or selective elimination of particular bacterial species or genotypes.

Theoretical analyses suggest that viruses are important for maintaining the diversity of aquatic microbial assemblages. Viral lysis depends on both the density and growth rate of the host and is moreover highly host-specific. Therefore, if a bacterial genotype is particularly successful in outgrowing all its competitors, it should in turn become more sensitive to viral attack leading to a condition referred to as 'killing the winner.' A similar control of specific bacterial populations can also be envisaged for selectively feeding protistan predators.

Particle Uptake Selectivity

Laboratory studies have documented that bacterivorous freshwater flagellate species can feed on a wide range of bacterial species even if they do not coexist with these strains in their natural habitat. However, predation mortality on bacteria in mixed assemblages is neither uniform nor random. For example, bacterial cell size has been recognized as a major selection factor for protistan grazing. Cells within a size range of 1–4 μm are preferably ingested by heterotrophic flagellates and ciliates, whereas the mortality rates of larger and smaller cells are reduced. Depending on their size, bacterial morphotypes are thus either disproportionally sensitive or fully or partially resistant against protistan grazing (**Figure 2**). Moreover, bacteria may become increasingly vulnerable to predation if they are more active or dividing, because higher growth rates are typically associated with larger cell sizes.

To understand this selective mortality, it is important to recognize that even the smallest bacterivorous protists of sizes <5 μm are highly evolved organisms with complex sensory abilities and behavioral adaptation to the predatory life style. The capture, handling, and ingestion of particles by interception-feeding flagellates is a flexible multistep procedure (**Figure 3**). The feeding process allows for a better evaluation of food quality by protists and it provides various possibilities for bacteria to avoid or at least reduce predation. Phagotrophy by filter-feeding as, e.g., found in choanoflagellates and bacterivorous ciliates may also permit substantial selectivity during particle acquisition. For example, particular cell surface properties (hydrophobicity, mucous capsids, or surface-bound chemical deterrents), motility, or cell shape might all interfere with prey handling or ingestion.

Bacterial species consequently differ in their quality as prey, i.e., the grazing of single protistan species on various bacterial strains may yield a range of different growth rates. This might reflect not only the protistan capture success rates for different bacteria, but also their nutritional value (e.g., glycogen content), and it can even be related to specific microbial secondary metabolites (e.g., digestive inhibitors, toxins).

Free-living heterotrophic protists, moreover, follow a so-called optimal foraging strategy. This implies that they flexibly adapt their search behavior and/or their particle uptake selectivity to both the likely success rate and to their physiological condition. Specifically, protists can discriminate against particles or

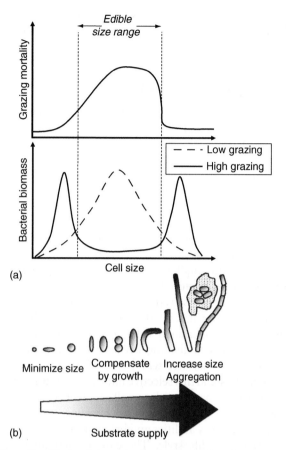

(a)

(b)

Figure 2 Effects of size-selective feeding and nutrients on bacterial phenotypes: (a) The size-selective feeding of heterotrophic flagellates selectively eliminates bacteria of a cell length between approximately 1 and 4 μm. During periods of high grazing, most microbial biomass in freshwater assemblages is thus found in the very smallest and largest size classes. (b) At low ambient levels of nutrients, and DOC grazing-resistant bacteria that are favored are extremely small. By contrast, highly productive waters favor the occurrence of complex or filamentous morphotypes and of microcolonies. At intermediate levels of productivity, bacteria in the grazing-vulnerable size range may still be able to maintain small populations that compensate losses by high growth rates. Combined from various authors.

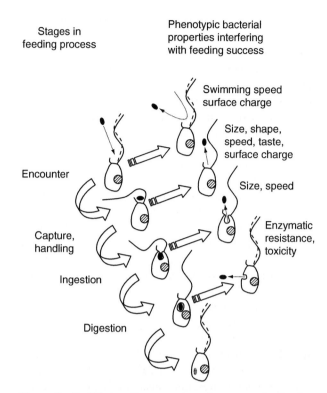

Figure 3 Particle handling procedure of an encounter-feeding flagellate. Bacterial adaptations such as cell shape, motility, or surface properties may influence the success of the predator at difference stages of the feeding process (redrawn with kind permission from Springer Science + Business Media: Jürgens K and Matz C (2002) Predation as a shaping force for the phenotypic and genotypic composition of planktonic bacteria. *Antonie van Leeuwenhoek International Journal of General and Molecular Microbiology* 81: 413–434.

bacteria with lower nutritional value depending on their level of starvation, the overall particle densities, or the relative amount of high-quality prey.

Grazing-Resistant Bacteria

Bacteria with threadlike or branched morphotypes and cell sizes between 10 and >100 μm are typically too large to be ingested by heterotrophic nanoflagellates. Other bacteria achieve the same effect by formation of cell chains or microcolonies that may additionally be embedded in a matrix of exopolymerous organic material. Filamentous bacteria have been found both in very oligotrophic and in highly productive lakes, whereas microaggregate formation appears to be related to elevated ambient substrate levels. Threadlike or colonial morphotypes might even form a dominant component of the microbial assemblages in hypertrophic environments. However, the appearance of these grazing-resistant filaments or cell aggregates is restricted to particular seasons in many freshwater systems (e.g., immediately after the spring phytoplankton bloom in mesotrophic temperate lakes). This reflects the high vulnerability of such morphotypes to grazing by cladoceran metazooplankton, especially by daphnids.

Filamentous bacteria or microaggregates typically do not exceed 1–10% of total microbial cell counts even during bloom situations, but may nevertheless constitute a prominent fraction (more than 50%) of the total microbial biomass. Therefore, such bacteria may represent a substantial fraction of microbially bound organic carbon that bypasses the 'microbial loop' and is directly transferred to higher trophic levels.

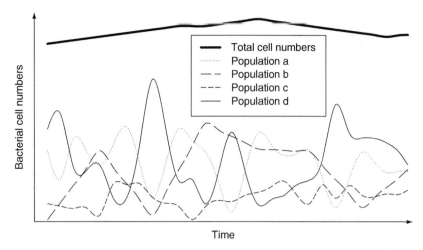

Figure 4 The apparent overall stability of bacterioplankton assemblages may be a consequence of fluctuating individual populations that are affected differently by top-down and bottom-up limitations. In this conceptual depiction, populations a and d are competing for similar resources and are regulated by a 'killing the winner' mechanism. Population b experiences long periods of stability (i.e., successfully compensating mortality by growth), whereas population c is kept below a threshold density by selective predation.

From Phenotypic to Genotypic Selection

It is not likely that protistan predators directly sense the phylogenetic affiliation of their prey; they rather select according to phenotypic criteria (**Figure 4**). However, individual taxonomic groups of aquatic bacteria may substantially differ in their phenotypic properties. For example, some freshwater betaproteobacteria of the genus *Polynucleobacter* and *Actinobacteria* are ultramicrobacteria of a cell length <1 μm. This small cell size is maintained irrespective of growth rates. These microbes consequently suffer less loss by protistan predation than other species. On the other extreme, bacteria from particular phylogenetic lineages related to Saprospiraceae are amongst the most common filamentous bacteria in many freshwater systems. During periods of high protistan bacterivory these bacteria may transiently form planktonic blooms that are terminated by cladoceran grazing.

Interactions Between Mortality Sources

The interplay of viral lysis and protistan predation in shaping the composition of bacterial assemblages is still poorly understood. The presence of grazers can be a stimulating factor for both prokaryotic growth and viral proliferation. On the other hand, some protists even seem to be able to directly consume viral particles (so far this has not been shown outside laboratory cultures). On entirely theoretical grounds, the consumption of virus-infected bacteria by protists is predicted to exert a negative effect on bacterial species richness. This hypothesis has also been recently supported by experimental evidence. Some bacterial populations may even be both resistant against flagellate grazing and immune to viral attack (e.g., filamentous *Flectobacillus* spp.), thus profiting from the combined mortality inflicted on their competitors.

Top-down Versus Bottom-up Control in Mixed Microbial Assemblages

The abundances and biomasses of heterotrophic bacteria in fresh waters are conspicuously more stable than of phyto- (including prokaryotic cyanobacteria) and zooplankton. This phenomenon might be completely unrelated to food webs, as e.g., many researchers doubt that all the particles that are classified as bacteria by direct counting techniques are indeed viable cells. Nevertheless, substantial controversy has centered on whether the apparently tight control of prokaryotic abundances is mediated by top-down or bottom-up mechanisms. However, it might not be entirely appropriate to assess this question at the community level only.

Bacterial Growth Strategies

Prokaryotic assemblages in lakes typically are a mix of different genotypic populations that compete for growth-limiting resources while being affected by unequal mortality rates. As a consequence, individual bacterial species adopt distinct life strategies, i.e., they invest different amounts of energy and resources in minimizing their losses or in maximizing their competitive abilities. Even closely related strains or species may substantially differ in growth-related features or in their sensitivity to predation, e.g., due to subtle differences in cell morphology. Such phenotypic adaptations typically represent trade-offs between increasing growth rates and reducing mortality, but some features might even provide advantages in both areas. For example, bacterial motility allows for the chemotactic tracking of substrate batches or even of motile algal cells in a heterogeneous microenvironment, and extremely high swimming speeds even help to reduce grazing losses. On the other hand, the metabolic costs of swimming might outweigh its benefits, and allometric constraints suggest that useful motility cannot occur in the smallest bacteria. Bacterial species with very small average cell sizes may in turn enjoy the double benefit of high surface-to-volume ratios (and thus of higher substrate diffusion rates) and reduced size-dependent grazing losses.

Currently, knowledge about the growth strategies of the important heterotrophic prokaryotes in the pelagic zone of fresh waters is limited to less than a handful of examples. Some of the smallest and apparently non-motile bacteria in fresh waters, free-living *Actinobacteria* phylogenetically affiliated to the so-called AcI clade, might be classified as typical 'defense specialists.' The direct inspection of the food vacuole content of nanoflagellates shows that AcI *Actinobacteria* are indeed selectively avoided, and their gram-positive cell wall might even provide additional resistance to digestion. On the other hand, these bacteria tend to be rapidly outcompeted by other microbial groups in situations that reward a rapid up-shift in growth rates (e.g., during predator-free incubations with and without substrate addition). By contrast, some betaproteobacteria of the freshwater Beta 1 lineage may rapidly enrich if bacterivores are removed. At the same time, these bacteria suffer the highest loss rates if top-down effects are experimentally magnified (e.g., by exclusion of the microzooplankton that feeds on bacterivorous nanoflagellates). These 'opportunitroph' bacterial groups thus rarely form populations that exceed 5–10% of total microbial counts in freshwater pelagic environments. It is nevertheless conceivable that such small populations of rapidly growing but heavily grazed bacterial species are sometimes responsible for a disproportional fraction of the total organic carbon that is channeled through the microbial food web.

Simultaneous Action of Different Controlling Mechanism

Following the above arguments, it is likely that microbial assemblages as a whole are neither controlled entirely by predation nor by resources: While the concentrations of organic carbon or nutrients in the pelagic zone of a lake might be limiting for one bacterial population, the same bottom-up conditions might still allow for the growth of another one. This implies that bottom-up factors set the lower limits for the growth of bacterial species (e.g., of terrestrial origin) in aquatic environments. However, since top-down control by viruses and protists is density-dependent, bacterial genotypes might nevertheless survive at low cell numbers for considerable time, even if they are not able to form larger populations.

The regulation of the maximal densities of particular bacterial species by top-down and bottom-up processes in mixed microbial assemblages most likely does not result in stable cell numbers as achieved by cultivation of single species in chemostats. The most rapidly growing (least bottom-up controlled) bacterial populations will form short-lived blooms that collapse as soon as selective protistan predation or viral lysis rates exceed their growth potential. By contrast, slowly growing prokaryotic species that are less affected by predation will accumulate more gradually in microbial assemblages, but will be outcompeted in periods of rapidly changing growth conditions, e.g., during phytoplankton blooms. The apparent constancy of microbial numbers in a particular pelagic environment might thus conceal (or even be a consequence of) the contrasting dynamics of more or less stable coexisting bacterial populations that are controlled to a different extent by mortality and growth limitation (**Figure 4**).

Limits of Predictability in Aquatic Microbial Food Webs

Moreover, it appears that top-down and bottom-up factors might not equivalently affect the composition of microbial communities. Over the last decade, the authors have performed extensive studies on experimental freshwater microbial communities fed by organic carbon exudates from a phosphorus-limited *Cryptomonas* sp. The addition of interception-feeding flagellates or filter-feeding ciliates always resulted in a disproportional decline of the most successful bottom-up competitors in this assemblage. By contrast, different bacterial genotypes profited from the presence of the two predators, or even of the same predator in successive experiments. This unbalance of one 'loser' versus various possible 'winners' in this simple microbial food web suggests that the competition between bacterial populations at concomitant top-down control of the most successful bottom-up competitor might lead to various more or less stable community states. Moreover, even the simplest version of an aquatic food web (one predator, two differently selected prey species) can result in aperiodic fluctuations that are highly sensitive to the starting points of the system. Thus, the dynamics of interacting microbial populations may lead to a condition referred to as deterministic chaos, where different outcomes arise depending on initial conditions.

It is thus likely that the population dynamics of individual bacterial species in complex aquatic microbial food webs will never be entirely predictable, but will also feature a stochastic element. A better understanding of the interactions within aquatic microbial food webs at a population level will nevertheless be valuable for the definition of those microbial species that are useful indicators of ecosystem-wide structural changes.

Further Reading

Arndt H, Dietrich D, Auer B, Cleven E, Gräfenhan T, Weitere M, and Mylnikov AP (2000) Functional diversity of heterotrophic flagellates in aquatic ecosystems. In: Leadbeater BSC and Green JC (eds.) *The Flagellates*, pp. 240–268. London: Taylor & Francis.

Boenigk J and Arndt H (2002) Bacterivory by heterotrophic flagellates: Community structure and feeding strategies. *Antonie Van Leeuwenhoek International Journal of General and Molecular Microbiology* 81: 465–480.

Ibelings BW, De Bruin A, Kagami M, Rijkeboer M, Brehm M, and van Donk E (2004) Host parasite interactions between freshwater phytoplankton and chytrid fungi (Chytridiomycota). *Journal of Phycology* 40: 437–453.

Jürgens K and Matz C (2002) Predation as a shaping force for the phenotypic and genotypic composition of planktonic bacteria. *Antonie Van Leeuwenhoek International Journal of General and Molecular Microbiology* 81: 413–434.

Pace ML and Cole JJ (1994) Comparative and experimental approaches to top-down and bottom-up regulation of bacteria. *Microbial Ecology* 28: 181–193.

Pernthaler J (2005) Predation on prokaryotes in the water column and its ecological implications. *Nature Reviews Microbiology* 3: 537–546.

Simek K, Nedoma J, Pernthaler J, Posch T, and Dolan JR (2002) Altering the balance between bacterial production and protistan bacterivory triggers shifts in freshwater bacterial community composition. *Antonie Van Leeuwenhoek International Journal of General and Molecular Microbiology* 81: 453–463.

Simon M, Grossart HP, Schweitzer B, and Ploug H (2002) Microbial ecology of organic aggregates in aquatic ecosystems. *Aquatic Microbial Ecology* 28: 175–211.

Thingstad TF and Lignell R (1997) Theoretical models for the control of bacterial growth rate, abundance, diversity and carbon demand. *Aquatic Microbial Ecology* 13: 19–27.

Weinbauer MG (2004) Ecology of prokaryotic viruses. *FEMS Microbiology Reviews* 28: 127–181.

Trophic Dynamics in Aquatic Ecosystems

U Gaedke, University of Potsdam, Potsdam, Germany

Introduction

An understanding of the role of species populations within an ecosystem is possible only when information on multiple species is systematically joined in a way that reflects mutual interactions. A synthesis of this type produces a view of ecosystem processes that are the by-product of many simultaneous interactions among populations.

Feeding relationships have proven to be an effective means of bridging the gap between populations and ecosystems. Studies of feeding at the ecosystem level may be referred to as 'food-web analysis' or 'trophic dynamics.' Early studies of this type involved diagrammatic portrayal of feeding levels (trophic levels) in an ecosystem (**Figure 1**). Such a diagram begins with photoautotrophs (level 1), which live by using inorganic substances and solar energy to synthesize biomass. Photoautotrophs are consumed by herbivores, which are the first level of consumers. Herbivores in turn are consumed by carnivores of the first level, which are consumed by carnivores of the second level. When the organisms are arranged by trophic level, they can be shown as a vertical stack of boxes that comprise a feeding pyramid (or trophic pyramid), or they can be shown in more detail as a diagram with spatially separated boxes connected by lines indicating the feeding pathways (**Figure 2**). Because there are numerous kinds of organisms at any given trophic level, a feeding diagram takes the form of a web, which gives rise to the term 'food web.' A diagrammatic analysis of a food web may be relatively simple if it focuses only on the dominant members of each level, or it may be much more detailed if it includes the subdominant members as well.

While a qualitative diagram is informative if it is accurate, a quantification of the amount of energy or materials passing through the food web produces more insight (**Figure 2**). Because quantification of a food web involves the linkage of compartments by dynamic pathways representing the flow of materials or energy, the quantitative study of food webs often is conducted through the construction of computer models.

Feeding relationships as portrayed in a food web may be quantified by any of several means, according to the interest of the investigator. For example, the amount of biomass passing from one level to another may be quantified. Because biomass has an energy equivalent per unit mass, quantification of energy flow also is possible, and is especially useful in connecting the flow of energy back to the solar source and to the metabolism of individual organisms or groups of organisms. In addition, it is possible to study food webs quantitatively in terms of the passage of an element or even a class of compounds through the food web. Carbon, a surrogate for biomass, is often traced through food webs, as are phosphorus and nitrogen, the two most critical inorganic nutrients in aquatic ecosystems. Also it would be possible to demonstrate or predict the flow of fatty acids or carbohydrates through food webs. Thus, food-web modeling is applicable to a wide range of interests that involve feeding interactions.

Trophic Levels and Trophic Positions

Modern food-web analysis grew from the concept of the trophic pyramid (**Figure 1**), which was used by the British ecologist Charles Elton (1900–1991) and other founders of modern ecology to summarize information on the roles of organisms in ecosystems. Early portrayals of food webs typically were based on the number of organisms or on biomass. For many environments, such a diagram takes a pyramidal shape. The laws of nature do not require the pyramidal shape, however. For example, a higher trophic level may have more biomass than a lower trophic level adjacent to it.

The concept of the trophic pyramid was advanced significantly toward modern food-web analysis by Raymond Lindeman (1915–1942) who, working with Evelyn Hutchinson (1933–1991), introduced the concept of trophic dynamics. According to Lindeman, the linkages between trophic levels, when portrayed as energy flow, provide a view of the ecosystem that is linked to the solar energy source and to the metabolism of individual organisms. Furthermore, an analysis of trophic relationships based on energy leads to the calculation of energy-transfer efficiencies and other related phenomena that are readily associated with an energy-based analysis.

The early concept of trophic levels contained several important flaws, the solution of which has produced a number of innovations. The first and most obvious flaw is that individual kinds of organisms often cannot be assigned to a single trophic level. For example, an omnivorous organism, such as some types of crayfish, may be able to derive nourishment from both plant and animal matter, and therefore cannot be assigned either to the level of herbivore or to the level of carnivore. It is especially common for high-level carnivores to feed at multiple lower levels, including herbivores and lesser carnivores, at the same time.

The modern view is that species must be assigned fractional trophic levels that reflect their diet. For example, an organism feeding 50% at level 2 and 50% at level 3 would be assigned a level of 2.5. Thus, models can take into account proportionate differences in trophic level.

A second problem arises from organisms that use particulate organic matter (POM, detritus) or dissolved organic matter (DOM; also designated dissolved organic carbon, DOC) as food. Modern food-web analysis views such feeding relationships as a 'detrital food chain,' which is an important complement to the more traditionally recognized feeding relationships, the 'grazer food chain,' based on photoautotrophs, which pass organic matter to herbivores. The detrital food chain, like the grazer food chain, contains multiple levels (**Figure 3**).

Inclusion of the detrital food chain leads to a more realistic view of bacteria in ecosystems. Bacteria are often the direct consumers of detritus (which they dissolve prior to consuming) and dissolved organic matter. Even though they create biomass at the base of the detrital food chain, they cannot do so through use of sunlight, as is the case of production of photoautotrophs, such as algae and aquatic vascular plants at the base of the grazer food chain. Because detritus and dissolved organic matter are abundant in aquatic ecosystems, the realities of the detrital food chain must be incorporated into quantitative models.

The increasingly realistic view of food webs leads away from the traditional pyramid or four-step energy diagram (**Figure 1**) and more toward diagrams that show various groups of organisms occupying noninteger positions in the grazing food chain, and connected to the detrital food chain, which incorporates bacteria as first-level producers of biomass (**Figure 2**).

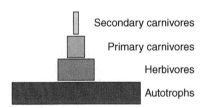

Figure 1 An example of a pyramid of biomass of organisms arranged by trophic level. Such pyramids led to modern trophic diagrams and food-web analysis.

Use of Trophic Guilds to Analyze the Trophic Structure of Food Webs

A trophic guild consists of all organisms within a food web that have similar food sources and predators (e.g., boxes shown in **Figure 2**). In most cases, trophic guilds defined in this way would consist of multiple

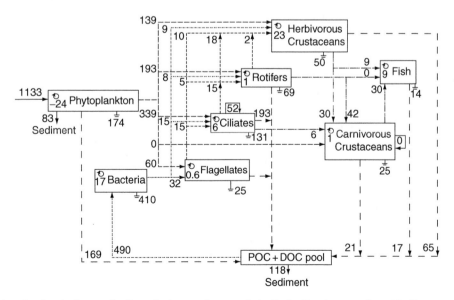

Figure 2 A modern food-web diagram for the pelagic zone (open water) of Lake Constance, adjacent to Germany, Switzerland, and Austria. Fluxes are given as mg C m^{-2} d^{-1}. Source: U. Gaedke.

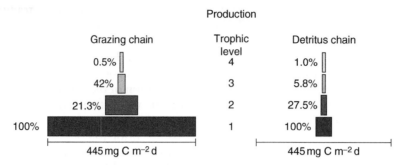

Figure 3 Trophic pyramids for the grazing chain and the detritus chain of Lake Contance, shown in terms of annual production. Source: U. Gaedke.

species, all of which would occupy similar positions in the trophic diagram or food-web model. Thus, application of the concept of guilds simplifies analysis and modeling of food webs.

One complication of the use of guilds is change in feeding habits for a species during its growth and development. For example, a carnivorous fish species may feed on zooplankton when it is small but feed on other fish when it is large. In such cases, it would be justified to treat the younger stages of a certain species as belonging to a different guild than the older individuals of the same species.

Figure 2 shows the structure of a food-web model of the open waters of Lake Constance, which adjoins Germany, Switzerland, and Austria. The model is based on boxes corresponding to guilds of organisms as well as a pool of DOM and POM. Carbon comes into the food web through this DOM and POM pool or through a photosynthetic guild (in this case, phytoplankton) and leaves the food web either as sedimentation of POM to the bottom of the lake or release of CO_2 gas to the water as a by-product of respiration. Circles within each of the boxes indicate net growth. Arrows passing from one box toward another indicate consumption, and arrows passing out of each box indicate grazing, predation, fecal output, or other losses. The numbers indicate fluxes (biomass per unit time for a given unit area of the lake, in this case expressed as mg C m^{-2} d).

A quantitative model such as the one shown in **Figure 2** is either an average for a considerable period of time or snapshot of an instant in time. The flow of mass from one compartment to another can be expected to change substantially from season to season and even week to week in a lake. Thus, modeling that encompasses changes occurring over any substantial period of time (e.g., a year) can require a great deal of information.

Although the progressive change in dynamics of food webs over time is seldom quantified because of the large amount of data that would be required,

averages or snapshot views of food webs can reveal the main fluxes that account for ecosystem metabolism and the efficiency of transfer of mass or energy through the food web.

Trophic Transfer Efficiency

The trophic structure of a food web is affected greatly by loss of energy that occurs between trophic levels. It is not unusual for loss of energy between any two adjacent trophic levels to reach 90% (**Figure 3**) because in **Figure 3** losses are between ca. 70 and 80% throughout. As recognized by Lindeman when he created the concept of trophic dynamics based on energy transfer, the second law of thermodynamics, when applied to trophic transfers, requires that dissipation of energy as heat will be a substantial loss even for the most efficient transfers. For example, the growth efficiency of individual cells seldom exceeds 40%, even when nonmetabolic losses are disregarded. Ecological factors introduce nonmetabolic inefficiencies related to nongrazing mortality, export of organic matter from the system (outflow or sedimentation), and elimination (fecal output).

Nongrazing mortality, which can occur when organisms are exposed to lethal physical or chemical conditions, physiological death related to age, or inadequate nutrition, divert biomass and energy from the transfer process that connects one trophic level to the next in the grazer food chain. Nongrazing mortality passes energy and mass to the detrital food chain. Elimination (fecal output) also is drain on efficiency in that mass or energy passes from the grazer food chain to the detrital food chain. Losses to elimination may be especially high for organisms feeding on organic sources that are difficult to digest. This is particularly the case for herbivores consuming vascular plant tissues, which are rich in complex carbohydrates that are difficult to digest.

All organisms must respire in order to live. Respiration involves the release of chemical energy from organic matter. The released energy is used in part for synthesis of new biomass, and is also used to maintain the integrity of basic metabolic functions that sustain life. Therefore, a considerable amount of organic matter is lost (converted to $CO_2 + H_2O$ and energy) for energy production purposes without passing to a higher trophic level.

The various modes by which energy and mass can be lost are the basis for quantifying several kinds of ecological efficiencies that are used in food-web analysis. The ratio of biomass synthesized (either growth or production of reproductive biomass such as eggs) to biomass ingested as food is symbolized K_1, the gross growth efficiency. K_1 rarely exceeds 0.30–0.35 under natural conditions, and often is considerably lower. The net growth efficiency of an organism (K_2) is defined as the ratio of new biomass produced to the amount of biomass that is assimilated (i.e., passing through the gut wall into the body of the organism). This efficiency has a maximum value of 50% for small organisms but is lower for large organisms (e.g., fish).

Trophic transfer efficiency is the ratio of biomass production at one trophic level to the biomass production of the next lower level. In plankton food webs, trophic transfer efficiencies may be high (0.15–0.30) (**Figure 3**) when compared with webs dominated by a transfer from vascular plants to herbivores. Temporal variability of the trophic-transfer efficiency is high, even within a given ecosystem, because of the instability of the numerous factors that can affect the trophic transfer efficiency. For example, the trophic transfer efficiency is strongly reduced when inedible algae (such as cyanobacteria) dominate or when animals with high respiratory needs (such as vertebrates and especially birds and mammals) prevail.

Food Quality and Quantity

Heterotrophs (consumers, including bacteria) live by consumption of biomass or nonliving organic matter that is derived from biomass. Because the chemical composition of protoplasm in biomass (disregarding skeletal material or support structures) across all heterotrophs falls within a relatively narrow range, heterotrophs that feed on biomass are assimilating approximately the same mixture of elements that they will need in order to synthesize their own biomass (skeletal material and support structure typically pass through the gut, unassimilated). Detritivores also benefit from this carryover of elemental mixtures from one kind of organism to another, although detritus is more likely to show some selective loss of elements that would alter the balance typical of living biomass.

Unlike heterotrophs, photoautotrophs assimilate elements separately from water or, if they are rooted vascular plants, from sediments. For example, carbon is derived from H_2CO_3 and related inorganic carbon forms dissolved in water, and phosphorus is taken up separately as phosphoric acid that is dissolved in water. Because the inorganic substances required to synthesize biomass are taken up separately, large imbalances may develop when some essential components are much more abundant than others. Thus, autotrophs face greater challenges than heterotrophs in assembling the necessary ratios of elements to synthesize biomass, but even heterotrophs can experience imbalances of elements.

The approximate ratios of elements that are characteristic of autotrophic biomass have been extensively studied. Characteristic ratios of carbon to nitrogen and phosphorus are often the greatest focus of analysis because carbon is the feedstock for photosynthesis and phosphorus and nitrogen are the two additional elements that are often in short supply for conversion of photosynthetic products (carbohydrate) to other molecule types that are needed for the synthesis of protoplasm (e.g., amino acids). The importance of C:N:P ratios in aquatic organisms was first brought out by Alfred Redfield (1890–1983), who discovered that healthy oceanic phytoplankton show a characteristic atomic C:N:P ratio of about 106:16:1. Thus, the nutrient status of a phytoplankton community can be judged to some degree from the elemental ratios. For example, a phytoplankton community suffering phosphorus deficiency may show a C:P ratio of 500:1 rather than 106:1, as predicted by the Redfield Ratio for well-nourished phytoplankton. The analysis of elemental ratios for diagnosis of elemental imbalances is termed 'ecological stoichiometry.'

Imbalances in elemental ratios in one trophic level can create imbalances or inefficiencies at the next trophic level. This is particularly true between primary producers and herbivores. For example, phytoplankton suffering phosphorus scarcity may pass biomass with a high C:P ratio to grazers that consume phytoplankton. Because of an imbalance of elements in the food, the grazers must consume extra food in order to obtain the correct balance for the synthesis of their own biomass. Similarly, an especially low C:P ratio (e.g., 50:1) will provide an oversupply of phosphorus (typically this happens when bacteria are consumed), some of which would be released to the environment without generating any biomass.

One strategy that herbivores may employ in improving the elemental balance of food intake is to consume heterotrophs in addition to autotrophs (omnivory, which is feeding at multiple trophic levels). Thus, consumption of a phosphorus-rich food could be offset by consumption of a carbon-rich food, and the combination would provide more efficient use of ingested mass than a single food type.

Trophic Structure of Food Webs in Open Water and Near Shore

In the open water of a lake at some distance from the shore (i.e., in the pelagic zone), the dominant autotrophs are phytoplankton, which live as individual cells or small colonies of cells suspended in the water. Some groups of phytoplankton (e.g., filamentous cyanobacteria) are rejected by most grazers because they are difficult to ingest or difficult to digest. Algae of low palatability may become quite abundant when grazing pressure is high in the pelagic zone. Overall, however, phytoplankton often is composed of a high proportion of edible and digestible species. Vascular plants (macrophytes), which typically grow near the shores of a lake (in the littoral zone), offer a less useful food supply because the digestible component of biomass is embedded in complex carbohydrates that maintain the shape of the plant. For this reason, much of the macrophyte biomass that develops during a growing season in the littoral zone of a lake dies and decomposes rather than being eaten while alive. The detrital food chain benefits from the biomass that was left uneaten by the grazer food chain.

Macrophytes may account for most of the autotroph biomass in a littoral zone, but attached algae colonize macrophytes and all other surfaces that are illuminated in the littoral zone. These attached autotrophs (periphyton) are an important food for herbivores in the littoral zone, even though the macrophytes themselves often are not.

The trophic structure in the pelagic zone often is dominated by four significant trophic levels. The top predator guild, which consists of multiple fish species, is not exposed to substantial predation pressure because of its size and mobility. The top predator guild therefore builds up biomass until it encounters a limitation caused by scarcity of appropriate food. This type of limitation is called a 'bottom up' limitation because the trophic level below is restricting increase in biomass of the trophic level above. While the top predator guild is controlled from the bottom up, its prey are controlled from the top down. 'Top down' control in this situation involves suppression of population biomass of the trophic level that serves as food for the abundant top predator.

In a four-level food web, the third level down from the top consists of herbivores. When top predators are abundant, the intermediate predators (e.g., fish that eat zooplankton) are suppressed, which relieves grazing pressure on herbivores. Therefore, herbivores, which consist mostly of zooplankton, may become abundant because of weak top-down control. This, in turn, will impose a strong grazing pressure on phytoplankton, which can cause a decline in phytoplankton biomass, yielding a higher water transparency. Thus, water clarity may be enhanced by stocking large predatory fish. This technique is an example of biomanipulation.

Physical factors may affect the structure of food webs. For example, the absence of any solid attachment points in the pelagic zone requires that all autotrophs be small. Thus, the food web must be based on grazers capable of harvesting large numbers of small autotrophs. The grazers that can do this work are also small, which creates a niche for small carnivores that eat the small herbivores. This leaves the opportunity for a top carnivore level that feeds on the small carnivores. Thus, predator and prey are related in body size through feeding behavior.

Food webs of the littoral zone in lakes are more similar to those of terrestrial environments or wetlands than to those of pelagic food webs of lakes if their production is dominated by vascular plants. Grazers (herbivores) may then be larger because larger units of food are available. Thus, top carnivores may feed directly on herbivores rather than on intermediate carnivores and body size and trophic level are not as strongly correlated as in pelagic food webs. In this case, the herbivores come under top-down control, and the efficiency of energy transfer from plants to herbivores is reduced because there are not enough herbivores to consume the total plant production. Thus, the length of the food chain in different portions of the ecosystem, or even in different ecosystems, affects the amount of biomass that can be produced by each trophic level.

Transfer Efficiency along the Size Gradient in Pelagic Food Webs

In pelagic food webs, all autotrophs are small and predators exceed the size of their prey. Thus, the flow of matter and energy in pelagic food webs is from small to progressively larger organisms. The entire food web's trophic transfer efficiency along the size gradient of organisms reflects each of the trophic linkages and the number of trophic levels in the web. For example, efficiency of fish production in oligotrophic lakes is lower than the efficiency of fish production in eutrophic lakes when expressed as a proportion of the total amount of photosynthesis.

In unproductive lakes, food is more strongly dispersed, and is gathered by consumers that search for their prey. Consumers that locate their prey by searching usually are closer in size to their prey than consumers that feed without searching (by filtration of water, for example). When consumers are very close in size to their food source, the transfer of energy and materials along the size gradient up to large fish is less efficient because it requires more transfers to reach a given size. Thus, there is a theoretical explanation for the lower transfer efficiency along the size gradient in oligotrophic lakes. This is an example of the use of food-web analysis to explain observations about productivity and efficiency in lakes.

Conclusion

Quantitative studies of food webs greatly magnify the value of information on individual species populations. Analysis of linkages in the food web provide explanations for the efficiency of energy or mass transfer to various points in the food web and the controlling influences that either enhance or suppress production at a given level in the food web. Thus, quantitative food-web studies support the understanding of mechanisms that govern the functioning of aquatic ecosystems.

Glossary

Detrital food chain – Food-web components that begin with particulate organic matter (detritus) or dissolved organic matter, and pass through bacteria to other consumers.

Ecological stoichiometry – Study of element ratios within biomass in relation to element ratios within food or in the surrounding environment.

Food web – A diagram or model that shows the feeding connections between all major groups of organisms in an ecosystem.

Grazer food chain – Components of a food web that begin with photoautotrophs (algae, aquatic vascular plants) and pass through herbivores to carnivores.

Trophic guild – A group of organisms that share common food sources and common predators.

Trophic level – A group of organisms that obtain their food from sources of equal distance from the original source. Autotrophs equal level 1, herbivores equal level 2, primary carnivores equal level 3, and secondary carnivores equal level 4. Species feeding from more than one level may be assigned a fractional trophic level.

Trophic transfer efficiency – The fraction of total production at a given trophic level that is converted to production at the next trophic level.

Further Reading

Begon M, Townsend CR, and Harper JL (2006) *Ecology-From Individuals to Ecosystems*, 4th edn., p. 738. Oxford, UK: Blackwell.

Gaedke U and Straile D (1994) Seasonal changes of trophic transfer efficiencies in a plankton food web derived from biomass size distributions and network analysis. *Ecological Modelling* 77/76: 435–445.

Gaedke U, Straile D, and Pahl-Wostl C (1996) Trophic structure and carbon flow dynamics in the pelagic community of a large lake. In: Polis G and Winemiller K (eds.) *Food Webs: Integration of Patterns and Dynamics*, pp. 60–71, Chapter 5. New York: Chapman and Hall.

Gaedke U, Hochstädter S, and Straile D (2002) Interplay between energy limitation and nutritional deficiency: Empirical data and food web models. *Ecological Monographs* 72: 251–270.

Lampert W and Sommer U (1997) *Limnoecology*, 398 pp. New York: Oxford University Press.

Morin PJ (1999) *Community Ecology*, 424 pp. Oxford, UK: Blackwell.

Sterner RW and Elser JJ (2002) *Ecological Stoichiometry*, 439 pp. Princeton, NJ: Princeton University Press.

Straile D (1997) Gross growth efficiencies of protozoan and metazoan zooplankton and their dependence on food concentration, predator–prey weight ratio, and taxonomic group. *Limnology & Oceanography* 42: 1375–1385.

Harmful Algal Blooms

J M Burkholder, North Carolina State University, Raleigh, NC, USA

Introduction

Harmful algal blooms in inland waters are operationally defined from both ecological and socioeconomic contexts. Ecologically, they produce potent toxins or other bioactive substances that adversely affect beneficial organisms, either causing disease to or death of beneficial aquatic organisms by predation or parasitism, and or causing undesirable changes in habitats or both. From a socioeconomic perspective, they cause undesirable effects to humans, including direct toxicity to humans and other animals (such as wild and cultured fish, waterfowl, livestock, and domestic pets); decreased recreational uses of waterways (via reduced water clarity, fish kills, rotting biomass, reduced waterfront real estate values, and provision of habitat for microbial pathogens, mosquitoes, snails as vectors for schistosomiasis, and other noxious species); increased fouling of pumps, filters, and intake pipes; taste and odor problems in drinking water supplies; increased costs of water treatment and increased costs of managing fresh waters. This review addresses both microalgae of inland waters (cyanobacteria, the haptophyte *Prymnesium parvum* Carter, the diatom *Didymosphenia germinata* (Lyngbye) Schmidt, and others), and macroalgae with emphasis on *Cladophora* and other filamentous green algae as well as the cyanobacterium *Plectonema wollei* Farlow ex Gomont.

Many harmful algae of inland waters are strongly stimulated by increased rates and amounts of nutrient supplies. Cultural eutrophication often leads to high biomass of algal blooms that discolor the water, resulting in oxygen depletion, taste-and-odor problems in potable water supplies, toxin release, fish kills, and food web imbalances. The term 'bloom' is poorly defined, and historically it has referred to accumulations of phytoplankton that discolor the water. In this article, blooms will be considered as planktonic, metaphytic, or benthic algal biomass that is either visible or significantly higher than the average (where known) for the system.

Harmful Phytoplankton

Cyanobacteria

Predominant harbingers of cultural eutrophication Cyanobacteria (formerly known as blue-green algae, Cyanophyta) are abundant in both pelagic and benthic habitats in many inland waters (**Figure 1**). As prokaryotes, they tend to have small cells and rapid growth rates. Planktonic cyanobacteria are widespread in neutral and alkaline habitats, and often they dominate the late summer/early autumn plankton of eutrophic, temperature-stratified lakes and reservoirs as well as slowly flowing lower rivers. Many species can 'fix' or convert dinitrogen gas (N_2) to inorganic nitrogen (N_i, as ammonia), affording a significant competitive advantage over other phytoplankton when N_i becomes limiting in surface waters during summer. In inland fresh, brackish, and salt waters worldwide, they can form major water discoloration from noxious and sometimes toxic blooms with up to several billion cells per milliliter. The blooms commonly deplete dissolved oxygen as a result of high respiration during the night, causing massive fish kills, and may also become toxic to other aquatic organisms.

Nutrient over-enriched conditions stimulate cyanobacteria worldwide. In many enriched north temperate lakes, there is a strong relationship between the total phosphorus (P) concentration at spring overturn and late summer, generally cyanobacteria-dominated algal biomass (**Figures 2–4**). Nitrogen (N) is also important in controlling cyanobacteria blooms. Cyanobacteria generally have a high P optima ('phosphorus-loving') and tend to store ('luxury-consume') P beyond what the cells initially need when excess P becomes available in soluble inorganic form (P_i, as phosphate). They can luxury-consume inorganic N (N_i) as well, and are capable of taking up simple organic nutrient forms.

According to resource ratio competition theory, algal species differ in their competitive abilities for major resources, and the outcome of competition for those resources partly depends on the ratios at which the potentially limiting resources are supplied. Cyanobacteria tend to be competitively superior at low N:P supply ratios (especially <25:1 by atoms), suggesting that low TN:TP ratios favor dominance of some cyanobacterial species, and cyanotoxin production (**Figure 5**). The evidence is conflicting, however: when N_i is limiting, nitrogen-fixing cyanobacteria would have an advantage at low TN:TP ratios. In contrast, some cyanobacteria can dominate when TN:TP ratios are high in surface waters but the P content is insufficient to support growth.

Bloom formation: Buoyancy regulation and other controls In stratified inland lakes and reservoirs

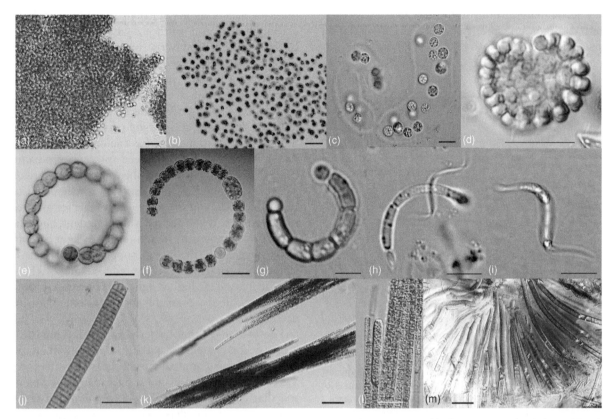

Figure 1 Representative coccoid colonial and filamentous cyanobacteria of inland waters: (a) *Microcystis* sp. (scale bar = 25 μm); (b) *Microcystis aeruginosa* (Kützing) Kützing (scale bar = 15 μm); (c) *Microcystis wesenbergii* (Komárek) Komárek in Kondrateva (scale bar = 10 μm); (d) *Coelosphaerium* sp. (scale bar = 20 μm); (e) *Anabaena* sp. (scale bar = 10 μm); (f) *Anabaena circinalis* (Rabenhorst) Bornet et Flahault (scale bar = 20 μm); (g) *Anabaenopsis circularis* (G.S. West) Woloszynska et Miller (scale bar = 5 μm); (h) *Cylindrospermopsis philippinensis* (Taylor) Komárek (scale bar = 10 μm); (i) *Raphidiopsis curvata* Fritsch et Rich (scale bar = 10 μm); (j) *Oscillatoria* sp. (scale bar = 20 μm); (k, l) *Aphanizomenon flos-aquae*, and (m) *Gloeotrichia* sp. Photos: (a–c, e–j) by the Center for Applied Aquatic Ecology, NC State University; (d, k–m) by J. Oyadomari (http://www.keweenawalgae.mtu.edu), with permission.

during active phytoplankton growth, dissolved N_i and P_i often become depleted in surface waters, whereas deeper waters are more nutrient-rich from decomposition of dead organisms as they settle out of the water column. In deeper waters, however, phytoplankton cannot survive if they sink and remain below the euphotic zone. It would be advantageous for phytoplankton to have a mechanism that enabled them to move to the lower water column to gain access to greater nutrient supplies, then to rise back up into light-replete areas for photosynthesis.

Most bloom formers are gas-vacuolate taxa such as the filamentous forms *Anabaena*, *Anabaenopsis*, *Aphanizomenon*, *Nodularia*, *Cylindrospermopsis*, *Gloeotrichia*, and *Oscillatoria*, and the coccoid colonial forms *Microcystis*, *Gomphosphaeria*, and *Coelosphaerium*. Phytoplankton cells are slightly denser than water, so they tend to sink. Gas vacuoles, found in planktonic cyanobacteria but not in eukaryotic phytoplankton, consist of groups of hollow, rigid, pointed

cylinders (gas vesicles, up to ~10 000 or more per cell) with rigid proteinaceous walls that are permeable to gases such as N_2, but impermeable to water. Their formation is regulated by gene expression, at the physiological level through availability of energy and structural components, and at the environmental level (e.g., induced by low light and surplus N_i; inhibited by N and carbon limitation, and high cell turgor pressure). They are used by cyanobacteria as an efficient mechanism to provide lift. Thus, in a pressure-dependable water column, cyanobacteria can regulate their buoyancy and their vertical position in the water column in response to environmental conditions via the number of gas vesicles in their cells.

Cyanobacteria buoyancy regulation requires a dependable external pressure regime that changes slowly enough to allow cells to adjust their internal pressure. In a pressure-dependable water column, many cyanobacteria photosynthesize under light-replete conditions near the surface during the day.

They accumulate carbohydrates so as to increase their internal turgor pressure; some gas vesicles collapse, causing the cells to sink to deeper, more nutrient-replete waters during late afternoon/evening. There

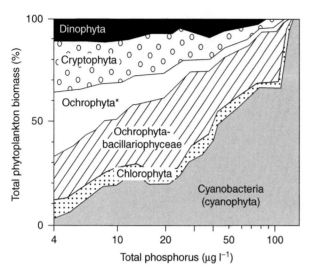

Figure 2 Changes in temperate zone phytoplankton assemblage structure, excluding picoplankton, with changes in water column total phosphorus (TP) in stratified natural lakes. Not shown is a shift that sometimes occurs from cyanobacteria to chlorophyte dominance in highly eutrophic (>500 µg TP per liter) unstratified lakes. Note that the asterisk (*) indicates all ochrophytes ("golden-brown algae", Ochrophyta) except diatoms (Ochrophyta, Class Bacillariophyceae). Modified from Watson *et al.* (1997) in Kalff J (2002) *Limnology*. Prentice-Hall: Upper Saddle River, NJ. Reprinted by permission of Pearson Education, Inc.

the cells access higher nutrient supplies, metabolize carbohydrates, and synthesize additional gas vesicles so that the cells slowly rise up and move toward the nutrient-depleted, light-replete surface waters by daylight. The buoyancy mechanism also interacts with nutrient limitation: when major nutrients limit cell growth, buoyancy tends to decrease because carbohydrates accumulate and turgor pressure increases. N_i limitation leads to reduced gas vacuolation and loss of buoyancy because of reduced protein production needed for gas vesicle assembly. Sustained inorganic carbon (C_i) limitation can also lead to decreased cell buoyancy because less energy is available for synthesis of gas vesicles. When nutrients are abundant, buoyancy is controlled mostly by light intensity. If light is suboptimal for growth, the energy supply will be relatively low, carbohydrate reserves are reduced, and cell buoyancy increases. At higher light, carbohydrate production and turgor pressure increase and buoyancy decreases.

The fact that cyanobacteria have a mechanism to access deeper waters for nutrient supplies without permanently sinking out to light-limited waters gives them a great competitive advantage over many phytoplankton that cannot safely access deeper, nutrient-replete waters. When the dependable pressure gradient is suddenly disrupted, however, cyanobacteria often cannot adjust their buoyancy quickly enough to remain in a favorable environment. Cyanobacterial 'surface scums' occur when conditions change too rapidly from low to high light intensity (corresponding to

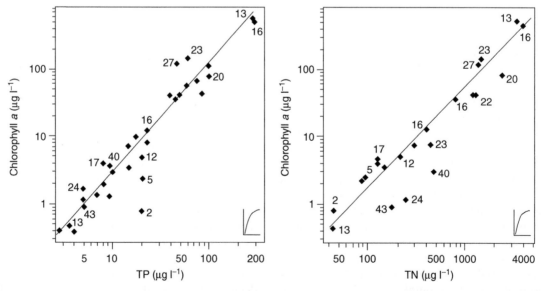

Figure 3 Relationship betwe chl*a* content and (a) TP or (b) TN in surface and near-surface waters of Japanese lakes in May, June, and July. Numbers next to points represent the N:P ratio (by mass). Note that lakes with exceptionally low and exceptionally high N:P ratios are outliers in the TP-chl*a* and TN-chl*a* relationships, respectively. Modified by Sakamoto (1966) in Kalff J (2002) *Limnology*. Prentice-Hall: Upper Saddle River, NJ. Reprinted by permission of Pearson Education, Inc.

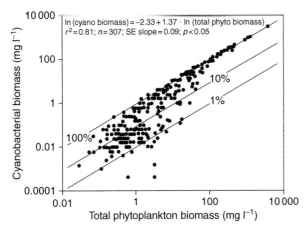

Figure 4 Relationship between cyanobacteria biomass (fresh weight) and total phytoplankton biomass (fresh weight) in 165 lakes of Florida (USA). The solid lines indicate the proportion of phytoplankton biomass comprised by cyanobacteria as 100%, 10%, and 1%. Note that cyanobacteria represent an increasing proportion of the total phytoplankton biomass as total biomass increases. In addition, cyanobacteria can contribute variable portions of the total phytoplankton biomass when the total is less than $\sim$50 mg l^{-1}, but they represent $\sim$100% of the total in highly eutrophic lakes (total phytoplankton biomass more than 100 mg l^{-1}). Modified from Canfield et al. (1989) in Kalff J (2002) *Limnology*, Prentice-Hall, Upper Saddle River, NJ. Reprinted by permission of Pearson Education, Inc.

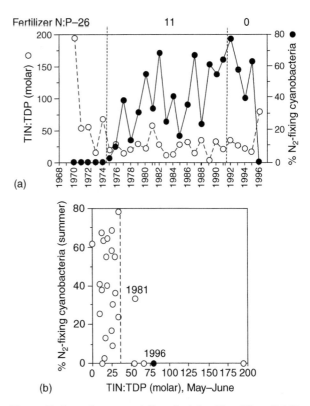

Figure 5 Long-term association of relative N and P availability and dominance of N$_2$-fixing cyanobacteria in experimental lake L227: (a) Dynamics of mean total inorganic nitrogen to total dissolved phosphorus ratio (TIN:TDP) during spring and early summer (May–June), and the summer (June–August) contribution of N$_2$-fixing cyanobacteria to total phytoplankton biomass from 1970–1996. Intervals when different fertilizer N:P ratios were applied to the lake are also indicated. (b) Association of N$_2$-fixer dominance and early season TIN:DIP ratio in Lake L227. The vertical line indicates (except for a datum in 1981) regions with negligible contribution of N$_2$ fixers (TIN: TDP > 38) or with variable but generally high levels (TIN: TDP < 38). Thus, as in previous work (Smith, 1983) low nitrogen to phosphorus ratios favor dominance by blue-green algae in lake phytoplankton (*Science* 221: 669–671), the data suggest the existence of a threshold N:P value above which N$_2$ fixer dominance is suppressed, but below which N$_2$ fixers can dominate. The datum for 1996 is highlighted. Other data values with 0% N$_2$ fixers and high TIN:TDP ratios are from more than 20 years ago during a period when fertilizer was added to the lake at a N:P ratio of 25. Error bars (not discerniable) were $\pm$1 standard error (SE). From Elser JJ, Sterner RW, Galford AE, Chrzanowski TH, Findlay DL, Mills KH, Paterson MJ, Stainton MP, and Schindler DW (2000) Pelagic C:N:P stoichiometry in an eutrophied lake: Responses to a whole-lake food web manipulation. *Ecosystems* 3: 293–307, with kind permission from Springer Science and Business Media.

high turbulence from a storm, then sudden calm) so that cells cannot correct overbuoyancy; when senescent or stressed cells cannot increase their turgor pressure; and/or when photosynthesis becomes limited by carbon dioxide diffusion through calm water. The storm disrupts the dependable pressure gradient and mixes the cells to deeper waters, where they become highly buoyant in the low light and rapidly move to the surface. There, before they can decrease their buoyancy, they may become physiologically damaged by high ultraviolet light and heat. As a thick, senescing surface layer of cyanobacteria accumulates, cells that have lost their buoyancy are blocked from sinking by layers of overly buoyant cells below. Inorganic carbon and light become limiting within the layer, and acidic conditions and anoxia develop. In a surface 'bloom,' many of the cells may be senescent, dying or dead. Surface 'hyperscums' can be several decimeters thick, covered by a crust of photo-oxidized, sometimes-toxic cells, extending over 1–2 hectares for weeks to months.

Although nutrient enrichment and hydrodynamics are clearly important in cyanobacterial blooms, reliable prediction of bloom timing, intensity, and species composition have remained elusive, in part because species responses and even responses of populations (strains) within a given species can vary greatly.

In addition, other factors can be important, or can interact with nutrients and hydrodynamics in controlling cyanobacterial abundance. Biological interactions such as synergisms between cyanobacteria and

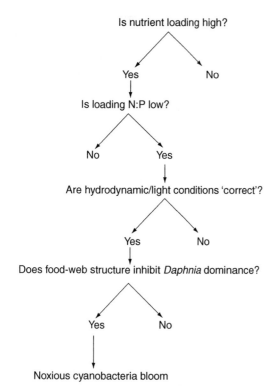

Is nutrient loading high?

Yes — No

Is loading N:P low?

No — Yes

Are hydrodynamic/light conditions 'correct'?

Yes — No

Does food-web structure inhibit *Daphnia* dominance?

Yes — No

Noxious cyanobacteria bloom

Figure 6 Hypothesized four-step pathway to noxious cyanobacteria blooms, with events leading to bloom development viewed as a hierarchical, nested 'decision tree.' The sequence of relative importance places emphasis first on absolute loading rates, second on loading N:P ratios, and third on hydrodynamics. The fourth key step is controls by food web structure and feedbacks from trophic stoichiometry. From Elser JJ (1999) The pathway to noxious cyanobacteria blooms in lakes: The food web as the final turn. *Freshwater Biology* 42: 537–543, with permission from Blackwell Publishing.

certain eubacteria help to create a favorable anoxic microenvironment for nitrogen fixation. Pathogens including various bacteria, fungi (mostly chytrids), and viruses (cyanophages) can affect bloom dynamics and duration. Grazers can be the 'last line of defense' in controlling bloom development (**Figure 6**). Reduced grazing has been related to the large size of cyanobacterial filaments or colonies, high slime production, allelopathic compounds, and altered habitat conditions such as low dissolved oxygen. Filamentous and gelatinous colonies are often selectively avoided by grazers because of mechanical (filter clogging) and chemical inhibition (toxins and other bioactive substances). Nevertheless, if large herbivorous, nonselective zooplankton such as *Daphnia* are abundant at bloom initiation, the grazers can consume small colonies and filaments and control the cyanobacterial populations before the bloom can develop. Moreover, *Daphnia* tends to retain P and release N in its excreta, helping to increase N:P supply ratios in the water

column and to shift the competitive advantage away from noxious cyanobacteria.

Cyanotoxins and their effects The term 'toxic cyanobacteria' refers to species that include some populations capable of producing toxin, that is, *potentially* toxic species. High intraspecific variability in toxin production is well known in cyanobacteria as in eukaryotic 'toxic algae'; populations (called strains) within a species commonly range from benign (with negligible toxin production) to highly toxic. Toxicity often varies among cells collected from the same sample in the same bloom, and even among sub-clones of the same clone.

Toxic cyanobacterial blooms in inland waters cause major economic impacts worldwide, with lethal effects on wildlife, waterfowl, livestock and, rarely, humans. Curiously, cyanotoxins generally are less toxic to aquatic biota, including their natural consumers, than to terrestrial organisms including humans. At least 50 taxa, including benthic as well as planktonic forms, have strains that can produce an array of cyanotoxins, mostly as multiple variants of cyclic peptides, alkaloids, and lipopolysaccharides (**Table 1**). Many strains produce multiple toxins depending upon genetic and environmental controls. Cyanotoxins include lethal neurotoxins or hepatotoxins, and nonlethal cytotoxins with selective bioactivity. They vary (hours to months) in the time required for degradation. Most cyanotoxins appear to be secondary metabolites that may be involved in cellular functions such as N storage or in defense against predators, disease agents, or competitors while promoting the growth of beneficial microbial associations. Cyanotoxins and other bioactive substances from cyanobacteria adversely affect only some types of organisms (e.g., some zooplankton and fish species and life history stages). Organisms at higher trophic levels may be exposed directly to the toxins through ingesting water that contains a toxic bloom, or by consuming prey that have bioaccumulated the toxins, and also may be affected via habitat and food web alteration. There is high variability among fish species in sensitivity to cyanotoxins, but impacts have included damage to liver, kidney, heart, gills, skin, and spleen.

The World Health Organization (WHO) has set 1 μg microcystin-LR per liter in drinking water as a guideline for human health protection. The guideline is based upon one common cyanotoxin, although multiple toxins are often present in affected waters, because supporting animal and human toxicity data are incomplete for most other cyanotoxins. The most common routes of human exposure to cyanotoxins are by direct

Table 1 Some potentially toxic species of cyanobacteria from inland waters, and their toxin(s)[a]

Toxin	Taxon
Anabaena bergii Ostenfeld	Cylindrospermopsin, deoxy-cylindrospermopsin, 7-epi-cylindrospermopsin
Anabaena circinalis (Rabenhorst) Bornet et Flahault	Microcystins, saxitoxins
Anabaena flos-aquae (Brébisson) cylindrospermopsin, Bornet et Flahault	Anatoxin-a, anatoxin-a(S), microcystins, deoxy-cylindrospermopsin, 7-epi-cylindrospermopsin
Anabaena laxa (A. Braun) Bornet et Flahault	Laxatoxin
Anabaena lemmermannii P. Richter	Anatoxin-a, anatoxin-a(S), microcystins
Anabaena planktonica Brunnthaler	Anatoxin-a
Anabaena variabilis (Kützing) Bornet et Flahault	BMMA[a]
Anabaena spp.	Anatoxin-a, microcystins
Anabaenopsis millerii Woronichin	Microcystins
Aphanizomenon flos-aquae (Ralfs) Bornet et Flahault	Aphantoxin, aphantoxinii, saxitoxins
Aphanizomenon ovalisporum Forti	Cylindrospermopsin, deoxy-cylindrospermopsin, 7-epi-cylindrospermopsin
Aphanizomenon sp.	Anatoxin-a
Arthrospira fusiformis (Voronichin) Komárek et Lund	Anatoxin-a, microcystins
Cylindrospermopsis philippinensis (Taylor) Komárek[b]	Cylindrospermopsin, deoxy-cylindrospermopsin, 7-epi-cylindrospermopsin, cytotoxic renal and hepatotoxin, saxitoxins
Cylindrospermopsis raciborskii (Woloszynska) Seenayya et Subba Raju	BMAA, cylindrospermopsin, demethoxy-cylindrospermopsin, cytotoxic renal and hepatotoxin, saxitoxins
Cylindrospermum sp.	Anatoxin-a
Fischerella musicola (Thuret) Gomont	Uncharacterized cytotoxins
Geitlerinema acutissimum (Kufferath) Anagnostidis[c]	Acutiphycin, 20,21-didehydroacutiphycin, homoanatoxins
Gloeotrichia echinulata (J.E. Smith) P. Richter	Microcystins, uncharacterized cytotoxins
Hapalosiphon fontinalis (Agardh) Bornet	Hapalindole A
Hormothamnion enteromorphoides (Grunow) Bornet et Flahault	Hormothamnin A
Microcystis aeruginosa (Kützing) Kützing	Cyanoginosins, microcystins, unidentified volatile sulfur compounds
Microcystis botrys Teiling	Cyanoviridin, microcystins
Microcystis ichthyoblabe Kützing	Microcystins
Microcystis viridis (A. Braun in Rabenhorst) Lemmermann	Cyanoviridin, microcystins
Microcystis wesenbergii (Komárek) Komárek in Kondrateva	Unidentified volatile sulfur compounds
Nodularia spumigena (Mertens) Bornet et Flahault	Nodularin
Nostoc linckia Bornet et Flahault	Uncharacterized cytotoxins
Nostoc rivulare (Kützing) Bornet et Flahault	Microcystins
Nostoc zetterstedtii (Areschoug) Bornet Flahault	Uncharacterized cytotoxins
Oscillatoria limosa (Agardh) Gomont	Microcystins
Oscillatoria sp.	Saxitoxins
Phormidium formosum (Bory ex Gomont) Anagnostidis et Komárek[d]	Homoanatoxin-a
Phormidium nigro-viride (Thwaite ex Gomont) Anagnostidis et Komárek[e]	Asplysiatoxins, debromoaplysiatoxin
Phormidium willei (Gardner) Anagnostidis et Komárek[f,g]	Anatoxin-a, microcystins
Planktothrix. aghardii (Gomont) Anagnostidis et Komárek	Microcystins, BMAA
Planktothrix compressa (Utermöhl) Anagnostidis et Komárek	Microcystins
Planktothrix isothrix (Skuja)[h] Komárek et Komárková	Microcystins
Planktothrix rubescens (DeCandolle et Gomont) Anagnostidis et Komárek	Microcystins
Plectonema wollei Farlow ex Gomont	Cylindrospermopsin, deoxy-cylindrospermopsin, saxitoxin
Phormidium terebriforme (Agardh ex Gomont) Anagnostidis et Komárek[f,i]	Anatoxin-a, microcystins
Pseudanabaena catenata Lauterborn	Unidentified neurotoxin
Raphidiopsis curvata Fritsch et Rich	Cylindrospermopsin, deoxy-cylindrospermopsin, 7-epi-cylindrospermopsin
Schizothrix calcicola Gomont	Asplysiatoxins, debromoaplysiatoxin
Scytonema hofmanni Agardh ex Bornet et Flahaut	Cyanobacterin
Spirulina subsalsa Oersted ex Gomont[f]	Anatoxin-a, microcystins
Synechococcus bigranulatus Skuja[f]	Anatoxin-a, microcystins
Synechococcus nidulans (Pringsheim) Komárek in Bourrelly[j]	Lipopolysaccharide endotoxins

Continued

Table 1   Continued

Toxin	Taxon
Synechococcus sp.	BMAA
Synechocystis sp.	BMAA
Scytonema mirabile (Dillwyn) Bornet	Uncharacterized cytotoxins
Scytonema pseudohofmanni Bharadwaja	Scytophycin
Umezakia natans M. Watanabe	Cylindrospermopsin, deoxy-cylindrospermopsin, 7-epi-cylindrospermopsin
Tolypothrix byssoidea (Agardh) Kirchner	Tubercidin

[a]Modified from Burkholder (2002) – see Further Reading, with additional information for BMAA (β-*N*-methylamino-L-alanine) producers (Cox PA, Banack SA, Murch SJ, Rasmussen U, Tien G, *et al.* (2005) Diverse taxa of cyanobacteria produce β-*N*-methylamino-L-alanine, a neurotoxic amino acid. *Proceedings of the National Academy of Science USA* 102: 5074–5078); *Planktothrix compressa* (from a toxic bloom at Coachman's Trail Reservoir, Raleigh, NC – author's unpublished data); *Plectonema wollei* (Seifert M, McGregor G, Eaglesham G, Wickramasinghe W, and Shaw G (2007) First evidence for the production of cylindrospermopsin and deoxy-cylindrospermopsin by the freshwater benthic cyanobacterium, *Lyngbya wollei* (Farlow ex Gomont) Speziale and Dyck. *Harmful Algae* 6: 73–80); and *Raphidiopsis curvata* (Li R, Carmichael WW, Brittain S, Eaglesham GK, Shaw GR *et al.* (2001) First report of the cyanotoxins cylindrospermopsin and deoxycylindrospermopsin from *Raphidiopsis curvata* (Cyanobacteria). *Journal of Phycology* 37: 1121–1126);
[b]Formerly within *Cylindrospermopsis raciborskii*; separated morphologically (taxonomic reference: Komárková-Legnerová J and Tavera R (1996) Cyano-procaryota (cyanobacteria) in the phytoplankton of the lake Catemaco (Veracruz, Mexico). *Archives für Hydrobiologie (Supplement) Algological Studies* 83: 403–422).
[c]Formerly *Oscillatoria acutissima* (taxonomic reference used for footnotes c–e and g–i: Komárek J and Anagnostidis K (2005) *Cyanoprokaryota. 2. Teil: Oscillatoriales*. Süßwasserflora von Mitteleuropa 19/2. Munich, Germany: Elsevier).
[d]Formerly *Planktothrix formosa, Oscillatoria formosa*.
[e]Formerly *Oscillatoria nigroveridis, Oscillatoria nigroviridis*.
[f]Found in hot springs (Burkholder 2002-see Further Reading).
[g]Formerly *Oscillatoria willei*.
[h]Formerly *Planktothrix mougeotii*.
[i]Formerly *Oscillatoria terebriformis*.
[j]Formerly *Anacystis nidulans* (taxonomic reference: Komárek J and Anagnostidis K (1999) *Cyanoprokaryota. 1. Teil: Chroococcales*. Süßwasserflora von Mitteleuropa 19/1. Heidelberg, Germany: Spektrum Ademischer Verlag).

contact with toxic blooms, ingestion of contaminated water, or aerosol inhalation. Acute health effects, e.g., through water contact sports, have included eye and ear irritation, mouth ulcers, nausea, vomiting, diarrhea, fever, and cold or flu symptoms with fever. A density of 100 000 toxic *Microcystis* cells per milliliter (estimated to correspond to ~50 μg chlorophyll *a* per liter and 100–200 μg microcystins per liter) was recommended by the WHO as a guideline for a moderate health alert in recreational waters. Such guidelines can be misleading, however, since cyanotoxin production depends on the cyanobacterial species, nutrient supplies, and other environmental conditions. Phytoplankton chlorophyll *a* levels, or total cyanobacterial abundance – even cell densities or biomass of known potentially toxic species – often poorly predict the toxicity of blooms (**Figure 7**). In addition, surface and shoreline accumulations or 'scums' can be severe health hazards for humans, wildlife, and domestic animals.

Cyanotoxins are not considered in most countries' guidelines for potable water, and information regarding the extent to which chronic human illness occurs from cyanotoxin exposure is generally lacking. Human gastrointestinal and hepatic illness, and rarely death, linked to cyanobacterial toxin exposure via drinking water supplies has been reported during blooms, often following copper sulfate treatment which causes cyanobacterial cells to lyse and release the toxins.

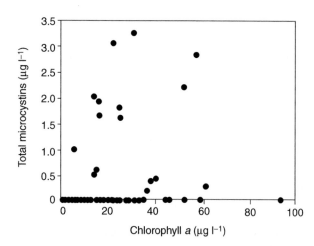

Figure 7   Relationship between microcystins and chlorophyll *a* concentrations in 58 lakes and reservoirs located in New York state, USA ($R^2 = 0.1$, $n = 140$). Chlorophyll *a* would have been a poor predictor of bloom toxicity, at least considering total microcystin cyanotoxins. From Boyer GL (2007) The occurrence of cyanobacterial toxins in New York lakes: Lessons from the MERHAB-Lower Great Lakes program. *Lake and Reservoir Management* 23: 153–60, with permission from the North American Lake Management Society.

Although acute exposure can be minimized by avoiding use of potable water supplies with noticeable blooms, the risk of long term exposure to relatively low concentrations of cyanotoxins is more difficult to avoid because there is no noticeable discoloration, and also because treatment procedures in many water supply plants do not remove these toxins.

Improved assessment of toxic cyanobacterial blooms

Although the factors that regulate expression of toxicity are poorly understood for many cyanobacteria, progress is rapidly advancing through molecular techniques, both in development of species-specific molecular probes that allow rapid detection of potentially toxic species, and in determining the genes that control the synthesis (production) of some cyanotoxins. Microcystin biosynthesis (*mcy*) genes have been completely sequenced from strains of *Microcystis* and *Planktothrix*; molecular probes have been designed for rapid, specific detection of these genes; and microcystin-specific primers are now available to detect cyanobacteria strains that are capable of synthesizing microcystin toxins. Scientists are also beginning to develop molecular techniques to quantify the toxic microcystin-producing strains versus nontoxic strains present in a given water sample.

Toxicity-based monitoring systems targeting at least several common cyanotoxins within a given ecosystem, rather than traditional chlorophyll- or cell count-based monitoring, would significantly advance assessment of the potential risks for aquatic life, drinking water supplies, and recreational activities from specific cyanobacterial blooms. Such programs ideally would involve molecular techniques to detect the presence of genes involved in toxin synthesis as a rapid, routine initial screening method for toxic strains. Although the need is clear, the associated costs generally are still prohibitive. Research is also ongoing to create cyanotoxin microarrays that can be linked to real-time automated monitoring systems. This technology will provide both a powerful early warning system to strengthen protection of potable water supplies and public health, and detailed environmental data to gain understanding about the combination of environmental factors that stimulate cyanotoxin production.

The 'Golden Alga' *Prymnesium parvum*, a Toxic Haptophyte

Ecology The unicellular haptophyte flagellate, *Prymnesium parvum* (**Figure 8**) is found on every continent except Antarctica. It is most often associated with estuarine or marine habitats, but is also common in some inland brackish rivers, reservoirs, and aquaculture facilities. Thus far, its known forms consist of vegetative (asexually reproducing) biflagellate unicells and vegetative cysts that form

under stress. *P. parvum* is obligately photosynthetic, but can also be strongly mixotrophic, ingesting bacteria and various protists. Mixotrophy of bacteria and other prey increases under inorganic nutrient limitation. *P. parvum* can also survive in darkness using dissolved organic carbon sources such as glycerol.

Most *P. parvum* blooms have occurred in nutrient over-enriched, eutrophic waters that are imbalanced in nutrient supplies and supply ratios relative to unenriched habitats. Cell production can be stimulated by P_i enrichment. In contrast, N or P limitation, as well as certain other stressors and suboptimal conditions, significantly enhance toxin production and/or release. *P. parvum* is eurythermal and euryhaline, but it usually develops blooms in quiet waters under suboptimal, lower temperatures and salinities, suggesting that other factors strongly influence this species such as zooplankton grazing and competition with other phytoplankton. Although *P. parvum* generally occurs in alkaline environments, cells can remain active down to pH 5.

Like other 'toxic algae,' *P. parvum* has toxic and nontoxic strains. Genetic controls on toxicity, and the ecological significance of toxin production for this organism (allelopathy toward competitors? grazing defense?), are poorly understood. *P. parvum* blooms cause death of gill-breathing organisms such as fish (all species present), mussels, and larval amphibians, whereas other organisms (e.g., livestock) apparently have not been affected. Fish kills related to *P. parvum* typically occur at high densities of $\geq 50\,000$ to $100\,000$ cells ml^{-1} (less commonly at $\geq 10\,000$ cells ml^{-1}), ranging up to $\sim 800\,000$ cells ml^{-1}. The water is often discolored yellow, gold, or rust, and foams develop where the water is agitated. A common mode of action of *P. parvum* toxins (below) is to destroy the selective permeability of cell membranes and disrupt ion regulation in gills. Affected fish typically bleed from the gills, with reddened epidermis of fins, opercula (gill covers), mouth, and eyes; the fish may also develop a heavy mucus layer. They usually swim slowly, lie on the bottom or congregate near shore; they also sometimes aggregate around a fresh source of water such as springs, or actively leap onto shore. In the western United States, for example, blooms of *Prymnesium parvum* causing the death of millions of fish have been documented in Texas during colder seasons nearly every year since 1985, and evidence suggests that fish kills dating back to the 1960s may have involved this organism. *P. parvum* has also been linked to kills of wild, stocked, and cultured fish in at least nine other southern and western states. In estuaries it has been shown to affect food webs as a phototroph, a microbial predator, and an agent of major fish kills. Its effects on the trophic structure of

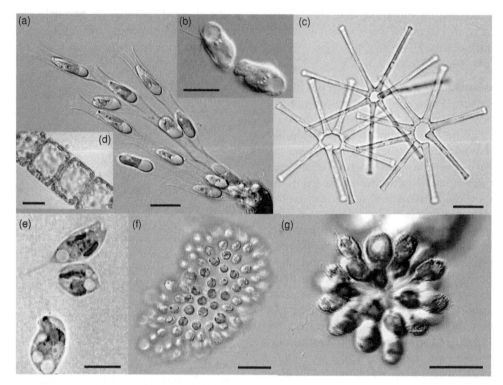

Figure 8 Taste-and-odor producing algae of inland waters, including (a) ochrophyte *Dinobryon cylindricum* Imhoff (scale bar = 10 μm); (b) haptophyte *Prymnesium parvum* Carter, also potentially toxic; (c) *Asterionella formosa* Hassall (scale bar = 50 μm); (d) *Aulacoseira granulata* (Ehrenberg) Simons (scale bar = 20 μm); (e) *Ochromonas danica* Pringsheim (scale bar = 10 μm); (f) *Uroglena* sp. (scale bar = 20 μm); and (g) *Synura petersenii* Korshikov (scale bar = 30 μm). Photos, with permission: (a, f, g) by J. Oyadomari (http://www.keweenawalgae.mtu.edu/...chrysophyceae.htm); (b) by J. Laclaire, courtesy of the Texas Parks and Wildlife Department (http://lutra.tamu.edu/hdlab/Projects/P64.htm); (c) by J.H. Parmentier (http://www.microscopy-uk.org.uk/mag/smallimag/asteri.jpg); (d) by D. Voisin (perso.orange.fr/...melosira%20varians02.html); (e) Photo courtesy of Dr. J.E. Frias, Instituto de Bioquimica Vegetal y Fotosintesis, CSIC-Universidad de Sevilla, Seville, Spain (http://www.ibvf.cartuja.csic.es/Cultivos/microalga/).

inland waters are not well known, but likely are substantial based on available data.

Complex mix of toxins   The toxins of *Prymnesium parvum* have been reported to include lipopolysaccharides (hemolysins), a galactoglycerolipid, polyene polyethers, cyclo amines (fast-acting ichthyotoxins), reactive oxygen species (H_2O_2, O_2^-, OH^-), dimethylsulfoniopropionate (DMSP), and toxic fatty acids with assorted fish-killing, cytotoxic, hemolytic, hepatotoxic, neurotoxic, and/or antimicrobial (allelopathic) activity. Some of the toxins are only partially characterized; moreover, such a complex mixture makes it difficult to assess the influence of each toxic component, and there are conflicting accounts of required conditions (temperature and salinity, role of fish) for toxin production. Detection of toxins other than hemolysins also has been difficult, and some components are highly instable (e.g., the ichthyotoxins and hemolysins are light-sensitive). The toxins remain difficult to quantify as well, despite several decades of research, and routine

methodologies have not been established to isolate and quantify the various toxic fractions. Two glycosidic toxins from *P. parvum* recently were chemically characterized as prymnesin-1 ($C_{107}H_{154}Cl_3NO_{44}$) and prymnesin-2 ($C_{96}H_{136}Cl_3NO_{35}$), and they have similar hemolytic activity. When released from *P. parvum* cells, the toxins form micelles and apparently require activation by cofactors, that can include dissolved potassium, calcium, magnesium, sodium, streptomycin, neomycin, spermine, or other polyamines.

Given this complex situation, what is the 'nutrient stress connection?' The toxins, including a mix of phosphate-containing proteolipids, are thought to be precursors of *P. parvum* structural membrane components, or products of imbalanced cell membrane metabolism. The hemolytic toxins accumulate within cells during exponential growth phase, and are released into the surrounding medium during stationary phase when growth is limited or stressed. High toxin activity occurs under P_i limitation apparently because formation of phospholipids for membrane synthesis is disrupted, leading to leakage of cellular

Table 2 Comparison of some ecological characteristics of five filamentous chlorophytes that can cause harmful effects in inland waters (NF = not found). The first three listed taxa commonly co-occur in temperate habitats; *Pithophora* is a more tropical organism although it can occur in temperate habitats; and *Hydrodictyon* is not usually as problematic as the others. Laboratory studies were performed on unialgal, but not bacteria-free, cultures[a]

Trait or Factor	Cladophora	Mougeotia	Spirogyra	Pithophora	Hydrodictyon
Species	*Cladophora glomerata* (L.) Kützing, *Cladophora fracta* (O.F. Müller ex Vahl) Kützing	*Mougeotia quadrangulata* Hassall, others[b]	*Spirogyra singularis* Nordstedt, *Spirogyra fluviatilis* Hilse, *Spirogyra nitida* (Dillw.) Link, *Spirogyra majuscula* Kützing, others[b]	*Pithophora oedogonia* (Mont.) Wittrock	*Hydrodictyon reticulatum* (L.) Lagerheim
Seasonality	Typically bimodal maxima in late spring and late summer/fall; summer decline	Generally late winter–summer	Mostly spring (cool temperatures)	Maximum in summer–early fall	Summer
Maximum biomass or coverage	100–940 g dr wt m⁻²; filament length 30–50 cm	Filamentous strands up to 90 cm in length	Mats 0.6 m thick washed up on lake shore	163–206 g dry wt m⁻²	Floating mats covering up to 30 hectares; colony length >1 m; 200 g dr wt m⁻²
Growth rate	Maximum 0.77 day⁻¹ (net specific growth); respiration 0.44 day⁻¹; up to 25 cm day⁻¹	Maximum 0.41 day⁻¹	Maximum 0.22 day⁻¹	Maximum 0.25 day⁻¹	Maximum 1.2 day⁻¹ (0.33 g dr wt day⁻¹)
pH	Alkaline conditions (optimum 7–8)	3.0–9.9 (optimum 8); bloom-forming indicator of early acidification (5.0–5.9, sulfuric acid)[c]	≥5–8; can be abundant in acidified lakes (nitric acid)[c]	Alkaline conditions	Alkaline conditions (≥6.8)
Temperature (°C)	Growth at ~9–30 (40); maximum growth at 13–17 (24)	Optimum ~25; range 8–34	Optimum ~25; net photosynthesis mostly negative at 35; mats also tend to lose cohesiveness at high temps.; moderately inhibited at 15	Optimum 25–35; strongly inhibited at 15	Optimum 25; range 5–40
Light (μmol quanta m⁻² s⁻¹)	Optimum 300–600; light compensation point (29) 36–104; ≤35 inhibits photosynthesis	Optimum 200–300; at 10–25 °C, light compensation point 10; can photosynthesize up to 2300 without photoinhibition; at 35 °C, positive net photosynthesis at 40–300.	Optimum ~300 to <1500[b]; net photosynthesis generally negative at 1500; mats tend to lose cohesiveness at high light and in complete darkness	Optimum 500; light compensation point ~20; high tolerance for very low light; growth rates, biomass decrease very slowly in complete darkness (minimal loss of chlorophyll for up to 60 days)	Optimum 100 (12 °C) to 160 (20 °C)

Continued

Table 2 Continued

Trait or Factor	Cladophora	Mougeotia	Spirogyra	Pithophora	Hydrodictyon
Current velocity (cm s⁻¹)	Increased photosynthesis from 0–2.1 (up to 8 for small tufts); decreased at higher levels	NF	~0–30; optimum N:P ratios (by atoms) for S. fluviatilis at 3, 12 and 30 cm s⁻¹ were 50, 58, and 52 (calculated for zero net photosynthesis)	NF	NF
Nutrient regime	Eutrophic to mesotrophic[d] (high P)	Eutrophic to oligotrophic	Eutrophic to mesotrophic[d]	Eutrophic	Eutrophic to mesotrophic
Nutritional ecology	K_s at 20 °C, 250 µg NO$_3^-$ l⁻¹; 8–15, 5–86, or 50–250 µg PO$_4^{-3}$ l⁻¹ (various studies); growth saturated at ~700 µg DIN L⁻¹	NF	K_s at 20 °C, 9–47 µg PO$_4^{-3}$ l⁻¹	K_s at 23 °C, 1230 µg NO$_3^-$ l⁻¹, 100 µg PO$_4^{-3}$ l⁻¹	K_s at 20 °C, 18 NO$_3^-$ l⁻¹, 13 PO$_4^{-3}$ l⁻¹; growth saturated at ~200 µg DIN l⁻¹
Herbivores	Grazed when small; can also be dislodged by some fauna; generally a poor, nonpreferred food source	Apparently poorly grazed	Not grazed (thick mucilaginous covering)	Poorly grazed	Gastropods may graze ~9% of H. reticulatum net production per day

[a]Other chlorophyte taxa that sometimes have been reported as causing adverse effects in inland waters include filamentous green algae Oedogonium, Ulothrix, Stigeoclonium, Zygogonium tunetanum Gauthier-Lievre, Zygnema and Rhizoclonium hieroglyphicum (C.A. Agardh) Kützing; the mat-forming tribophyte (Ochrophyta, Tribophyceae). Vaucheria dichotoma (L.) Agardh; the macroalga, Ulva, found mostly in brackish waters; and the macroalga, Chara. References for this table are as follows:
Cladophora – Lembi et al. (1988), Graham et al. (1996) and Bootsma et al. (2005) and references therein (see Further Reading); Auer MT and RP Canale (1982) Ecological studies and mathematical modeling of Cladophora in Lake Huron: 2. Phosphorus uptake kinetics. Journal of Great Lakes Research 8: 84–92; Auer MT and Canale RP (1982) Ecological studies and mathematical modeling of Cladophora in Lake Huron: 3. The dependence of growth rates on internal phosphorus pool size. Journal of Great Lakes Research, 8: 93–99; Auer MT, Canale RP, Grundler HC and Matsuoka Y (1982) Ecological studies and mathematical modeling of Cladophora in Lake Huron: I. Program description and field monitoring of growth dynamics. Journal of Great Lakes Research 8: 73–83; Canale RP, Auer MT and Graham JM (1982) Ecological studies and mathematical modeling of Cladophora in Lake Huron: 6. Seasonal and spatial variation in growth kinetics. Journal of Great Lakes Research 8: 126–133; Graham JM, Auer MT, Canale RP and Hoffman JP (1982)

Ecological studies and mathematical modeling of *Cladophora* in Lake Huron: 4: Photosynthesis and respiration as functions of light and temperature. *Journal of Great Lakes Research* 8: 100–111; Dodds WK (1991) Factors associated with dominance of the filamentous green alga *Cladophora glomerata*. *Water Research* 25: 1325–1332; Lorenz RC, Monaco ME and Herdendorf CE (1991) Minimum light requirements for substrate colonization by *Cladophora glomerata*. *Journal of Great Lakes Research* 17: 536–542. Ensminger I, Hagen C, and Braune W (2000) Strategies providing success in a variable habitat: II. Ecophysiology of photosynthesis of *Cladophora glomerata*. *Plant, Cell and Environment* 23: 1129–1136; Ensminger I, Forester J, Hagen C, and Braune W (2005) Plasticity and acclimation to light reflected in temporal and spatial changes of small-scale macroalgal distribution in a stream. *Journal of Experimental Botany* 56: 2047–2058; and Higgins SN, Howell ET, Hecky RE, Guildford SJ, and Smith RE (2005) The wall of green: The status of *Cladophora glomerata* on the northern shores of Lake Erie's Eastern Basin, 1995–2002. *Journal of Great Lakes Research* 31: 547–563.

Mougeotia – Lembi et al. (1988) and Graham et al. (1996) (see Further Reading); Mills KH and Schindler DW (1986) Biological indicators of lake acidification. *Water, Air & Soil Pollution* 30: 779–789; Graham JM, Arancibia-Avila P, and Graham LE (1996) Effects of pH and selected metals on growth of the filamentous green alga *Mougeotia* under acidic conditions. *Limnology and Oceanography* 41: 263–270; Turner MA, Howell ET, Robinson GGC, Brewster JF, Sigurdson LJ, and Findlay DL (1995) Growth characteristics of bloom-forming filamentous green algae in the littoral zone of an experimentally acidified lake. *Canadian Journal of Fisheries and Aquatic Science* 52: 2251–2263; Graham MD and Vinebrooke RD (1998) Trade-offs between herbivore resistance and competitiveness in periphyton of acidified lakes. *Canadian Journal of Fisheries and Aquatic Science* 55: 806–814; and Pillsbury RL and Lowe RL (1999) The response of benthic algae to manipulations of light in four acidic lakes in northern Michigan. *Hydrobiologia* 394: 69–81.

Spirogyra – Lembi et al. (1988) and Graham et al. (1996) (see Further Reading); De Vries PJR and Hillebrand H (1986) Growth control of *Tribonema minus* (Wille) and *Spirogyra singularis* Nordstedt by light and temperature. *Acta Botanica Neerlandica* 35: 65–70; Simons J and van Beem AP (1990) *Spirogyra* species and accompanying algae from pools and ditches in The Netherlands. *Aquatic Botany* 37: 247–269; Borchardt MA (1994) Effects of flowing water on nitrogen- and phosphorus-limited photosynthesis and optimum N:P ratios by *Spirogyra fluviatilis* (Charophyceae). *Journal of Phycology* 30: 418–430; Borchardt MA, Hoffmann JP, and Cook PW (1994) Phosphorus uptake kinetics of *Spirogyra fluviatilis* (Charophyceae) in flowing water. *Journal of Phycology* 30: 403–417; Simmons J (1994) Field ecology of freshwater macroalgae in pools and ditches, with special attention to eutrophication. *Aquatic Ecology* 28: 25–33; Graham JM, Lembi CA, Adrian HL, and Spencer DF (1995) Physiological responses to temperature and irradiance in *Spirogyra* (Zygnematales, Charophyceae). *Journal of Phycology* 31: 531–540; Nakanishi M and Sekino T (1996) Recent drastic changes in Lake Biwa bio-communities, with special attention to exploitaton of the littoral zone. *GeoJournal* 40: 63–67; and Berry HA and Lembi CA (2000) Effects of temperature and irradiance on the seasonal variation of a *Spirogyra* (Chlorophyta) population in a Midwestern lake (U.S.A.). *Journal of Phycology* 36: 841–851.

Pithophora oedogonia – Lembi et al. (1988) and Graham et al. (1996) (see Further Reading); Spencer DF and Lembi CA (1981) Factors regulating the spatial distribution of the filamentous alga *Pithophora oedogonia* (Chlorophyta) in an Indiana lake. *Journal of Phycology* 17: 168–173; O'Neal SW and Lembi CA (1985) Productivity of the filamentous alga *Pithophora oedogonia* (Chlorophyta) in Surrey Lake, Indiana. *Journal of Phycology* 21: 562–569; Spencer DF, O'Neal SW, and Lembi CA (1987) A model to describe growth of the filamentous alga *Pithophora oedogonia* (Chlorophyta) in an Indiana lake. *Journal of Aquatic Plant Management* 25: 33–40; and O'Neal SW and Lembi CA (1995) Temperature and irradiance effects on growth of *Pithophora oedogonia* (Chlorophyceae) and *Spirogyra* sp. (Charophyceae). *Journal of Phycology* 31: 720–726.

Hydrodictyon reticulatum – Wells et al. (1999) (see Further Reading).

[b]Specimens used for study rarely have formed the sexual cysts (zygospores) needed for species identifications. The potential for mixes of species used in these studies is especially high for *Spirogyra*, since several studies have shown that *Spirogyra* populations at a single site and time can consist of up to 20 species, sometimes referred to as polyploid complexes.

[c]In experimentally acidified lakes in Canada, *Mougeotia* was dominant in lakes acidified with sulfuric acid, whereas *Spirogyra, Zygogonium, Zygnema* and *Mougeotia* were codominant in lakes acidified with nitric acid.

[d]Flowing-water habitats where these organisms can become abundant are considered here as mesotrophic; although they can be characterized by very low nutrient concentrations, supplies are continually renewed from the flowing water, and boundary layers are minimized.

metabolites including toxins. Nutrient stress (P_i or N_i limitation) has been shown to significantly increase *P. parvum* toxin production/release.

Other Harmful Phytoplankton

Various other phytoplankton species can cause harmful effects on other organisms or habitat quality, but their ecology mostly is poorly known. Chlorophytes such as *Scenedesmus* and *Ankistrodesmus* can overgrow in fresh waters under high nitrate enrichment, and sometimes cause oxygen depletion in ponds and sheltered lake coves. Euglenoids (Euglenophyta) commonly respond to organic enrichment, and carotenoid production of blooms under high light can give water a blood-red appearance, creating aesthetic impacts in recreational areas such as golf courses. Two species of euglenoids found in eutrophic temperate and tropical waters worldwide recently were reported to have strains that can kill fish through production of as-yet uncharacterized toxin(s).

Although dinoflagellates (Dinophyta) are considered the most harmful group of microalgae in estuaries and marine coastal waters, few harmful species are known from inland waters. A strain of the freshwater species *Peridinium aciculiferum* Lemmermann was shown to produce a bioactive substance(s) in both natural habitat and culture that caused cell lysis and death of the co-occurring, benign phytoplankton species *Rhodomonas lacustris* (Cryptophyta). Production of this allelochemical was suggested to be a possible mechanism for outcompeting competitors. On the other hand, in research from the brackish Sea of Galilee spent medium from a culture of *Peridinium gatunense* Nygaard induced settling and massive lysis of the cyanobacterium *Microcystis* sp. It was hypothesized that an uncharacterized bioactive substance from *P. gatunense* was involved; if so, this effect of *P. gatunense* on *Microcystis* would be regarded as a potentially beneficial rather than harmful allelopathic interaction. The freshwater dinoflagellate, *Peridinium polonicum* Woloszynska has also been linked to fish death as an effect of an uncharacterized bioactive substance or toxin.

Some freshwater dinoflagellates are fish parasites. As examples, the ectoparasites *Piscinoodinium* spp. (formerly *Oodinium*; e.g., *Piscinoodinium pillulare* [Schäperclaus] Lom, order Blastiodiniales) cause mortality of various wild and cultured freshwater fish species in temperate and tropical regions throughout the world. In a lake in British Columbia, Canada, the three-spine stickleback (*Gasterosteus aculeatus* L.) has been infected seasonally by the ectoparasitic dinoflagellate, *Haidadinium ichthyophilum* Buckland-Nicks, Reimchen et Garbary. This organism has

photosynthetic, vegetative cysts and is mildly pathogenic in comparison with other known dinoflagellate parasites. Parasitic dinoflagellates are widespread in inland waters, but their effects mostly are poorly known except through work on accidental infestations in cultured fish.

Harmful Benthic and Metaphytic Algae

Cladophora

The chlorophyte *Cladophora* (Chlorophyta, Ulvophyceae; more than 400 species) is widely regarded as the most important harmful filamentous alga of inland waters, and the most abundant alga in alkaline streams throughout the world (**Table 2, Figure 9**). It also thrives in many freshwater and brackish lakes, and among the most famous of its habitats are the Laurentian Great Lakes. As a 'poster child' of the 1960s ecology movement in the United States, *Cladophora glomerata* focused international attention on the west basin of Great Lake Erie where it proliferated in response to phosphorus pollution, then drifted into shore in rotting masses from major seasonal die-offs that were sometimes measured in tonnes of fresh weight. *Cladophora* thrives in P-enriched waters with dependable substrata (e.g., large boulders) for attachment. In some habitats, increased nitrogen enrichment (N_i, urea) can also stimulate increased production. *Cladophora* has a relatively high light optimum for photosynthesis, can rapidly acclimate to low or high light. Like other benthic algae, *Cladophora* is highly patchy in spatial and temporal distribution. Its adverse effects include overgrowing irrigation systems and drainage canals, clogging of water intakes leading to power outages, overgrowth, and displacement of beneficial aquatic plants, reductions in invertebrate densities and fish spawning, reduction in species biodiversity, anoxia from decaying mats resulting in kills of other aquatic life, retention of pathogenic microbes such as *Escherichia coli* Migula and enteroocci fecal bacteria as a potential human health hazard, negative effects on recreational use of affected shoreline and nearshore areas, and depressed property values of adjacent homes.

The Great Lakes recently has become a *Cladophora* story of 'déja vu.' From the 1960s through the early 1980s, massive growth of *C. glomerata* characterized the rocky shorelines of Lakes Erie and Ontario, as well as localized areas of Lakes Michigan and Huron (note: low temperatures of Lake Superior discourage *Cladophora* growth). Research indicated that elevated P_i concentrations were the most important cause of the blooms, and led to multibillion dollar upgrades in wastewater treatment plants and

Figure 9 (a) *Cladophora*, (b) *Cladophora* rotting masses on the shore of Lake Michigan near Cleveland, Wisconsin, USA, October 2003; (c) Late spring *Cladophora* bloom in a stream in Wisconsin, USA; (d, e) *Pithophora oedogonia* (Montagne) Wittrock (scale bar on D, light micrograph of two akinetes = 75 μm); (f) *Spirogyra* sp. (scale bar = 10 μm); (g) *Hydrodictyon reticulatum* (L.) Lagerheim (cells up to 1 cm length × 200 μm width when fully enlarged); (h) *Mougeotia* sp. (scale bar = 40 μm); (i, j) *Plectonema wollei* (scale bars = 30 μm and 60 μm, respectively); (k, l) *P. wollei* bloom. Photos, with permission: (a, h) by J. Oyadomari (http://www.keweenawalgae.mtu.edu); (b) by G.C. Klein, courtesy of the Sheboygan Press; (c) from Sandgren *et al.* in Bootsma *et al.* (2005); (d, e) by Botanic Gardens Trust (http://www.rbgsyd.nsw.gov.au/_data/assets/image)...; (f) by M. Fearn (http://www.southalabama.edu/.../00Horner/micro.jpg); (g) F.J. Garcia Breijo, Universidad Politécnica de Valencia (http://www.euita.upv.es/varios/biologia/images/Figur...); (i-l) Center for Applied Aquatic Ecology, NC State University.

detergent P bans. TP and *Cladophora* markedly decreased in the lower Great Lakes. Ironically, dissolved P_i concentrations during the spring season have increased again in nearshore water throughout Lake Erie since the late 1980s from metabolic wastes and feces of mass invasions of exotic zebra mussels (*Dreissena polymorpha* Pallas) and quagga mussels (*Dreissena bugensis* Andrusov), which now dominate the nearshore benthic environment. Recent surveys have indicated that the mean peak biomass of *Cladophora* is similar to historic values in Lake Erie during the 1960s–1970s, and shorelines along portions of Lakes Ontario, Michigan, and Huron are again being fouled by rotting *Cladophora* growth.

One of the best models to reliably predict harmful alga growth is the Canale and Auer Model (CAM), developed in 1982 for *Cladophora* in Lake Huron (**Figure 10**). Results from the simulations suggested a nonlinear response of *C. glomerata* growth to reductions in P loading, in part because of P luxury consumption. Application of the CAM to the four lower Great Lakes also indicated that P from river

discharges supports localized *Cladophora* production, but that most production is supported by internal P recycling processes. The CAM was later modified as the *Cladophora* Growth Model (CGM) to consider effects of the recent exotic mussel invaders (**Figure 11**), especially increased water clarity and increased P_i concentrations during the spring growing season for *Cladophora*. The CGM was successfully validated on field populations of *Cladophora* in eastern Lake Erie during 2002. This model indicates that *Cladophora* growth has extended to deeper waters post-*Dreissena*, and that *Dreissena*-induced changes in water quality (increased light from removal of phytoplankton, and increased from mussel feces) are responsible for the marked resurgence of *Cladophora*.

The effects of these widespread, re-established blooms on Lake nutrient cycling and food web structure-including the die-off, transport, and decomposition of large amounts of *Cladophora* organic matter to the hypolimnion-remain to be determined. The portion of the blooms that moves shoreward attracts large flocks of gulls that add fecal material

Model I conceptual framework

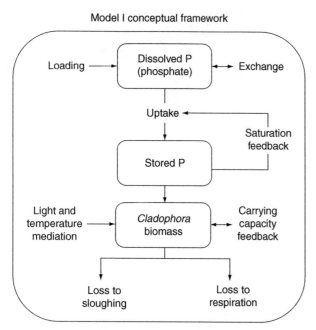

Figure 10 Conceptual framework of the Canule and Auer Model (CAM) to simulate phosphorus (P) dynamics and *Cladophora* biomass in Lake Huron. Experimental data were used to develop the CAM, which simulates the combined influence of temperature, light, and P on *Cladophora* growth and biomass accrual. The model also considered losses from respiration, and from sloughing related to wind direction/speed and current velocity. The Droop equation from classical algal physiology was used to model specific growth as a function of tissue P concentration. The curve from the Droop equation has an inflection point at a tissue P value near 0.1% dry mass. At cellular P concentrations less than that value, the model predicts that specific growth becomes increasingly sensitive to very small shifts in internal P stores. *Cladophora* growth responds strongly to internal P concentrations of 1–2 µg mg^{-1}; within that range, a small increase in cellular P content can support a relatively large increase in growth rate. From Auer (2005) in Bootsma *et al.* (2005), with permission. From Auer M (2005) Modeling *Cladophora* growth: A review of the Auer-Canale framework In: Cladophora *Research and Management in the Great Lakes*, pp. 57–61. Great Lakes Water Institute, Wisconsin Aquatic Technology and Environmental Research, UW Milwaukee.

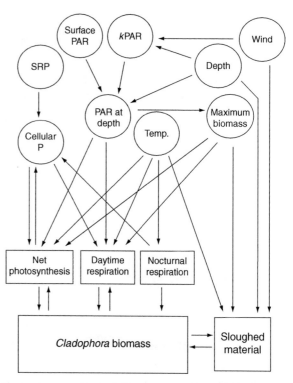

Figure 11 Simplified diagram of the revised *Cladophora* growth model (CGM), used to assess the importance of light, temperature, phosphorus (P), and selfshading on *Cladophora* spatial/temporal variability post-*Dreissena* spp. invasions. The model was developed for the eastern basin of Lake Erie, and extended to other Great Lakes. The CGM also was used to estimate the total P taken up by *Cladophora*, and its significance to nearshore and overall P dynamics in the eastern basin of Lake Erie. Arrows represent mathematical equations linking model parameters (Temp., temperature; SRP, soluble reactive phosphate; PAR, photosynthetically active radiation; kPAR, light extinction coefficient). From Higgins, S.N. (2005) The Contribution of *Dreissena* to the Resurgence of *Cladophora* in eastern Lake Erie. In: Cladophora *Research and Management in the Great Lakes*, pp. 63–71. Proceedings of a Workshop. UW Milwaukee Great Lakes WATER Institute Special Report No. 2005–01, University of Wisconsin Milwaukee.

to the rotting algae. Documented high densities of *Escherichia coli* and other fecal bacteria within the decomposing mats represent a potential human health threat, suggesting that the efficacy of water column *E. coli* as an indicator of fecal contamination from sewage may be compromised in locations where *Cladophora* accumulates along Great Lake shores.

Other Filamentous Chlorophytes

Various other filamentous/thalloid chlorophytes can overgrow inland waters and adversely affect other

aquatic life and local economies. *Cladophora* and *Ulva* (formerly *Enteromorpha*), members of the Ulvo-phyceae, include marine as well as freshwater species. Some *Ulva* spp. can grow in habitats spanning from fresh waters to salt springs and brackish lakes, to the Great Salt Lake, Utah, USA (salinity >50). A recent bloom of *Ulva flexuosa* Wulfen (formerly *Enteromorpha flexuosa*) occurred in Muskegon Lake, Michigan, USA and covered up to 80% of the littoral zone in some areas, mostly as epiphytic overgrowth. The affected lake had low grazing pressure, increased salinity from industrial discharge of chlorinated compounds, and a history of nutrient overenrichment. In brackish Urmia Lake, northwest Iran, *Ulva* has become so abundant that the entire lake was described as having the appearance of a thin vegetable soup.

A third potentially harmful member of the Ulvophyceae, *Pithophora oedogonia*, is restricted to fresh waters. This alga has overgrown shallow lakes, boat channels, and aquaculture facilities, and is resistant to several commonly used algicides. It thrives in shallow littoral areas of eutrophic habitats, and has low affinity for N_i (nitrate) and P_i relative to *C. glomerata*, but generally higher temperature tolerance. This species begins spring growth in benthic habitats, typically in alkaline waters. Oxygen bubbles trapped within the filaments carry them to the water surface where mats develop. Biomass is maximal in late summer/early fall, and photosynthetic rates are positively correlated with temperature and external concentrations of both N and P. Photobleaching by high ultraviolet light affects the upper filaments, while the rest of the mat remains healthy. *P. oedogonia* is also well adapted to low-light conditions and even extended periods of darkness, so it can survive severe selfshading within the mats where ~95% of the available light is absorbed within the first 5 mm of mat thickness. A computer model simulating growth dynamics of *P. oedogonia* was developed for a population from a north temperate lake, based on a modification of the Monod function and N:P resource ratio theory. The model predicted greater biomass reductions of this species from a 50% reduction in nitrate concentrations than from a similar reduction in TP.

The chlorophyte *Hydrodictyon reticulatum* (water net) provides an example of normally beneficial algae that can rapidly take advantage of ideal conditions to form nuisance growth. *H. reticulatum* has formed nuisance blooms in diverse locales such as England, Switzerland, and paddy fields of Asia. Its most problematic occurrence has been in New Zealand from 1988–1994. Where its extensive floating mats caused significant recreational and economic impacts. An isolated strain was eurythermal and capable of maximal photosynthesis at relatively low light levels. It had high affinity for N_i, and low N_i and P_i supported high production in the nutrient-poor waters where it bloomed. Rapid decline of the blooms after 1994 was attributed to high grazing pressure by a contingent of gastropod species that responded over time and were estimated as capable of consuming potentially ~9% of the biomass per day.

Filamentous chlorophytes in the order Zygnematales are common bloom-forming taxa in alkaline and mildly to moderately acidic, eutrophic fresh waters throughout the world. Rapid overgrowth by *Spirogyra* spp. sometimes occurs in lakes, ponds, and slowly flowing streams affected by agricultural runoff or sewage effluent. *Spirogyra* overwinters as zygospores, and forms benthic and floating mats mostly during spring or early summer. It is less tolerant of

low light than *Pithophora oedogonia*, and its mats tend to disintegrate in conditions of high temperature and light, or in darkness. *Spirogyra fluviatilis* can withstand current velocities up to $30 \, cm \, s^{-1}$ if sufficient phosphate is available for uptake to offset the apparent increase in cellular P demand. It can adjust short-term P_i uptake to compensate for the suboptimal conditions imposed by rapid flow.

The zygnematalean *Mougeotia* is common but not usually dominant in freshwater streams, ponds, lakes, and temporary pools worldwide. In contrast, it is a nuisance bloom former and early indicator of acid deposition (pH 5.0–5.9) in fresh waters of Europe and North America. In acidifying waters, *Mougeotia* develops metaphytic blooms as dense, floating mats or amorphous 'clouds' by mid-summer and attains maximal biomass in early fall, sometimes together with zygnematales *Zygogonium* and *Spirogyra*. Potential grazers such as gastropods and crayfish are generally lacking in acidic waters. *Mougeotia* is highly tolerant of the high dissolved concentrations of potentially toxic metals such as aluminum and zinc that generally characterize acidic fresh waters. It can survive low levels of dissolved inorganic carbon, and may augment its carbon supplies by using forms of dissolved organic carbon. This alga can maintain positive net photosynthesis at low light (~0.02% of full sunlight), but can sustain maximal net photosynthesis from 200 to $2300 \, \mu mol \, quanta \, m^{-2} \, s^{-1}$ as well, without evidence of photoinhibition. It can also maintain net photosynthesis at high temperatures (35 °C) if light levels are relatively low (~40–300 $\mu mol \, quanta \, m^{-2} \, s^{-1}$). Thus, in acidifying, clear freshwater lakes, *Mougeotia* can cover most of the littoral zone from 0-2 m depth, and much of the rest of the lake bottom down to ~6 m.

Mat-Forming Cyanobacteria and a Noxious Diatom

Benthic filamentous cyanobacteria commonly form nuisance or potentially toxic growth in inland waters, mostly as the genera *Oscillatoria*, *Plectonema*, *Lyngbya*, and *Phormidium*. Species such as *Phormidium inundatum* Kützing ex Gomont also form and cause problematic growths on cement in swimming pools and hazardous, slippery surfaces on wooden walkways. As an example of this group, *Plectonema wollei* is highlighted here. It occurs in temperate to tropical, alkaline, or mildly acidic fresh waters in southern Europe, Australia, India, and the United States, and its large filaments often dominate the algal flora of lakes and reservoirs in the southeastern United States. *P. wollei* initially grows attached in a benthic habit, later forming dense free-floating mats in lakes and streams. It is strongly stimulated by nutrient overenrichment, and its harmful effects include clogged water intakes, offensive odors, the production of potent toxins, and compromised recreational and potable water use. A massive mat can be ~10 m long, ~1 m wide, and several decimeters thick; other descriptions are of mat thickness up to 0.5 m and biomass as high as 1440 g dr wt m^{-2}. The dense mats block penetration by herbicides such as copper sulfate. Moreover, the filaments are covered by a thick polysaccharide sheath that comprise 50% or more of the total dry mass and can deter grazers. In summer-early fall, *P. wollei* attains maximal biomass at a high optimum temperature for growth. Nevertheless, in warm temperate climates, it has overwintered while maintaining high biomass (~120–440 g dry wt m^{-2}).

P. wollei has an array of other characteristics that enable its proliferation. It is comparably adapted as *Pithophora oedogonia* to low light; net photosynthesis occurs down to 0.3–1% of full sunlight. It is also similar to *P. oedogonia* in that photobleaching of external filaments protects the remainder of the mat from damage by ultraviolet light. It requires low dissolved inorganic carbon concentrations to saturate photosynthesis and growth, enhancing its survival in carbon-limited situations such as the interior of its thick mats. Conditions that enhance photorespiration (high mid-day temperatures, high oxygen concentrations) characterize floating mat environments, but photorespiration is minimal in this organism, probably because it has highly efficient carbon concentrating mechanisms and can use bicarbonate as a carbon source.

Harmful algae of inland waters also include a diatom species. *Didymosphenia geminata* ('didymo') is a large-celled diatom that forms long, gelatinous stalks (**Figure 12**). Historically it was native to nutrient-poor lakes and streams of northern latitudes. Recently it has acted as an invasive species in streams and rivers within its native range, and has also expanded its range to the Southern Hemisphere. For example, it first invaded New Zealand in 2004 and has rapidly expanded to new watersheds despite aggressive control measures. *D. geminata* can remain viable in damp conditions for weeks, and its spread has been related to recreational fishing activities as well as irrigation, water diversions, and waterfowl

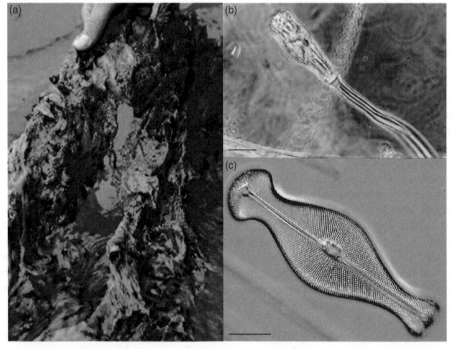

Figure 12 The nuisance diatom, *Didymosphenia geminata* ('didymo,' length 125–140 μm × width 35–45 μm – from Moffat MC (1994) An ultrastructural study of *Didymosphenia geminata* (Bacillariophyceae). *Tranactions of the American Microscopical Society* 113: 59–71). As a native species, *D. geminata* recently has formed massive growths or expanded its range in streams of North America, Europe, and Asia. It also has recently invaded streams of New Zealand as a nonindigenous species. (a) Macroview of a mass of the diatoms and their extracellular, gelatinous stalks; (b) light micrograph of the single-celled organism and its long stalk (scale bar = 40 μm); and (C) light micrograph showing the ornamented pattern in a siliceous frustule (cell wall) from a dead cell (scale bar = 25 μm). Photos: with permission by: (a) MAF Biosecurity, New Zealand (http://www.biosecurity.govt.nz/pests-diseases/plants/didymo-photos-004.jpg); (b) M. Bothwell (web.mala.bc.ca/simmsw/posterbothwell.htm); (c) Allesandro Bertoglio (http://www.microthele.it/micro/immagini/kemp/33.jpg).

hunting. Its present range spans low- and high nutrient, alkaline waters across broad temperature (4–27 °C) and flow regimes (~0 to ~50 cm s^{-1}). In the United States, *D. geminata* has expanded from western states through the Southeast. Its benthic mat-like growths can blanket up to 200 km of length in affected streams for several months each year, with up to 100% coverage several centimeters thick. Such growth can exclude species that formerly occupied the habitat. High densities of *D. geminata* have been related to a decline in beneficial invertebrates such as larval mayflies, stoneflies, and caddisflies in favor of dipterans (chironomids, black flies).

Massive brownish, slimy growths of *D. geminata* frequently have been mistaken for raw sewage. In North America and Europe, blooms commonly occur below impoundments characterized by stable flows and more constant temperatures. *D. geminata* also thrives in canal systems in the western United States that transport water for human consumption, agriculture, and hydroelectric power. In some canal systems, managers already are attempting to regularly remove *D. geminata*.

'Taste-and-Odor' Algae

A wide array of freshwater planktonic and benthic algae, including numerous cyanobacteria and ochrophytes (Ochrophyta: some diatoms, chrysophytes, synurophytes, and tribophytes), but also cryptophytes (*Cryptomonas rostratiformis* Skuja), dinoflagellates (*Peridinium* sp., *Gymnodinium* sp.), and chlorophytes (e.g., *Scenedesmus subspicatus* Chodat, *Chlorella sorokiniana* Shihira et Krauss, *C. homosphaera* Skuja, *Eudorina*, *Chlamydomonas*, *Gonium*, *Volvox*, *Chara*) produce cucumber-like, fishy, rancid, oily, or 'skunk-like' odorous compounds. Many algal volatile organic compounds (AVOCs) have been identified, some of which are also produced by bacteria or fungi. Relatively few AVOCs – certain terpenoids, sulfur compounds, and polyunsaturated fatty acid (PUFA) derivatives – cause most algal-associated taste-and-odor problems, and among these, few AVOCs have been useful as species indicators. High variability in AVOC production occurs both among species and among strains within a given species under similar environmental conditions. As examples, limited strain data are available for AVOC production by strains of cyanobacteria *Anabaena laxa* and *Phormidium calcicola* under high and low levels of light, temperature, P_i, and N_i. Most geosmin was retained inside the cells of *A. laxa* under all conditions except the low P_i treatment, and low cell densities produced high levels of geosmin in that treatment. In contrast, *Phormidium calcicola* released most geosmin to the

surrounding medium, regardless of culture age or environmental conditions. Source waters often contain diverse AVOCs, but resulting odors may show little relationship to the abundance of these components, which vary considerably in potency and can either act synergistically with, or mask, other VOCs. Nutrient-poor systems rarely have detectable odors; rather, PUFAs occur frequently and in higher abundance in eutrophic systems, suggesting that changes in odorous compounds may have potential in assessing source water and basin integrity.

Several biosynthetic pathways are involved in synthesis of AVOCs: Some are synthesized during normal growth (monoterpenes, sesquiterpenes, terpenoids, unique secondary metabolites, fermentation, and bioconversion products). Others are produced when cellular integrity is compromised and lipoxygenases and carotene oxygenases are activated, resulting in extremely potent, low molecular weight cleavage products of fatty acids and carotenoids. Peroxidation reactions induced by light in combination with heavy metals or pesticides can also be involved. These induced reactions follow loss of cellular integrity and often occur during decomposition of cyanobacterial blooms or seasonal succession of algal species, or after lake turnover and stratification and/or changes from oxygenated to anoxic conditions. In addition, for cyanobacteria that contain γ-linolenic acid, copper- or [other] herbicide-mediated peroxidation can lead to liberation of PUFAs. Thus, common chemical treatment of cyanobacteria blooms can sometimes promote production of strong odorous compounds. PUFAs mostly are produced upon cell membrane disruption, usually at the end of an algal population cycle. They can also be produced during growth (e.g., by the planktonic diatom, *Asterionella formosa*), and may function as sexual pheromones or in chemical defense against grazers.

Few mono- and sesquiterpenes have been found in freshwater cyanobacteria and eukaryotic algae, but notable exceptions are cyanobacterial geosmin (*trans*-1,10-dimethyl-*trans*-2-decalol) and 2-methylisoborneol (G-MIB). Most taste-and-odor incidents impart 'earthy, musty or muddy' odors and have been related to G-MIB, which mostly are produced by planktonic and benthic cyanobacteria including some potentially toxic species (**Figure 13**). They can cause taste-and-odor problems in concentrations of only a few nanograms per liter. Production of G-MIB in cyanobacteria has been linked to controls by light quality and quantity. In general, optimum cyanobacterial growth rates coincide with low production of G-MIB. Cyanobacteria synthesize G-MIB throughout growth, and synthesis is related to photosynthesis and pigment synthesis. Algal cells may store and/or release G-MIB, depending on the

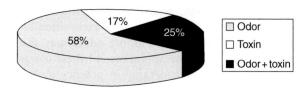

Figure 13 Survey of 45 cyanobacterial species for odor (geosmin, 2-methylisoborneol or β-cyclocitral) and cyanotoxin production (microcystins and/or neurotoxins). From Watson (2004) Aquatic taste and odor: A primary signal of drinking water integrity. *Journal of Toxicology and Environmental Health* 67: 1779–1795, with permission from Taylor and Francis (www. informaworld.com).

species, environment, and growth phase. Cell damage can also increase source-water G-MIB, which is not effectively removed by conventional treatment processes (coagulation-sedimentation-chlorination) and variably removed by activated carbon.

Numerous other odorous substances are produced. As examples, potent alkyl sulfides are excreted by cyanobacteria, such as dimethyl disulfide (many species) and isopropyl sulfides (some strains of *Microcystis*). The major odorous compounds produced by eukaryotic algae are fermentation products that mostly are formed by various flagellates in oxygen-stressed conditions (e.g., ochrophytes *Poteriochromis stipitata* Scherffel, *Synura petersenii*). Sulfur compounds of unique structure, 4-methylthio-1,2-dithiolane and 5-methylthio-1,2,3-trithiane, with pungent 'skunk-like' odors are excreted by *Chara globularis* Thuillier and also have allelopathic properties toward mosquito larvae and certain microalgae. In general, however, environmental influences on AVOC production by species and strains, and the ecological role of most AVOCs, remain to be examined.

Control Measures, Economics, and Prognosis

The physical, chemical, and biological techniques used in attempts to control harmful algae of inland waters unfortunately have not substantially improved over the past several decades. For example, short-term management efforts for cyanobacterial blooms continue to include; use of mechanical mixing or bubbling to disrupt water column pressure gradients, or more rapid flushing; algicides (especially copper sulfate, used for the past several decades), nutrient reductions and increased N:P ratios; and biological manipulations of fish populations to try to increase the abundance of herbivorous zooplankton. Although often economically and politically difficult to achieve, reductions of both N_i and P (N and P co-management) are considered the most effective long-term approach.

Potential strategies suggested for possibly controlling *Prymnesium parvum* mainly have included reductions of nutrient loading, dissolved organic materials, and salinity (the latter, to less than 1). Application of ultrasonic vibrations, barley straw and probiotics (for example, bacterial inoculants) thus far have been unsuccessful in controlling toxic *P. parvum*, whereas ultraviolet light and ozone treatment ($5 \, mg \, l^{-1}$ for 15 min) hold promise for destroying both cells and toxin in aquaculture facilities. Algicides or grass carp (*Ctenopharyngodon idella*) continue to be most frequently used, where feasible, to combat nuisance filamentous algae. Of five biocides tested on the mat-forming diatom *Didymosphenia geminata*, chelated copper may work best if fish health is not compromised. Taste-and-odor producing species generally occur within a mixed algal assemblage that includes many benign organisms (nonproducers). The approach generally taken is to remove their undesirable compounds during water treatment.

Although harmful algae of inland waters cause major ecological and socioeconomic impacts throughout the world, their economic effects are not tracked in any centralized major national or international database, and thus, their quantitative impacts are unknown except for isolated local events. State and federal governmental agencies rarely have conducted surveys to track the affected waters or the related economic costs. Even qualitative information mostly is lacking because the scattered, piecemeal reports and surveys that exist commonly do not distinguish between algae and nuisance aquatic vascular plants. We are left with 'glimpses' of the problems. As examples, *Prymnesium parvum* has caused more than U.S. $20 million in economic impacts in Texas waters. Cyanobacteria blooms are a major problem each year aesthetically, recreationally, and environmentally in many inland waters worldwide, and their potential to cause insidious, chronic effects on human health through potable water supplies are also finally beginning to be examined. Cultured channel catfish (*Ictalurus punctatus* Rafinesque), the largest component of pond-based aquaculture in the United States, have also sustained mass mortality via feeding on planktonic diets or passive assimilation through the gills. Documented losses of fish from toxic algal blooms have occurred in about 5% of the catfish ponds, accounting for about 7% of lost revenue; about 30% of potential catfish revenue is also lost because of off-flavor from taste-and-odor species. It has been estimated that the *D. geminata* will cause NZ $57 to $285 million in New Zealand over the next eight years through its adverse effects on recreation, commercial eel fisheries, water supplies, tourism, and biodiversity.

Inland waters worldwide are increasingly affected by nutrient over-enrichment, not only in economically depressed or developing countries, but also in industrialized nations. It is estimated, for example, that more than 40% of U.S. fresh waters are degraded to the extent that they no longer meet their designated 'fishable/swimmable' uses. Many harmful algae in fresh and brackish inland waters have been shown to be stimulated by increased nutrient loading. Concerted efforts are needed to assess the full range of their environmental and economic impacts, and to substantially reduce the nutrient loading that encourages their growth. Otherwise, increased proliferation and increased impacts can be expected.

Glossary

Allelopathy – Production and release of a chemical substance (allelochemical) by one species that adversely affects other species.

Anoxic – Refers to waters that have no dissolved oxygen.

Aphotic zone – Waters below the depth of sufficient light penetration for net photosynthesis.

Benthic – Growing on the bottom of an aquatic system or attached to various substrata, or beginning growth in that habit.

Bioactive substance – Substance that can affect living organisms.

Boundary layer – Layer of water with reduced velocity (no turbulent flow) that is immediately adjacent to the surface of a solid.

Clone – Group of genetically identical cells, derived from a common ancestral cell by asexual (mitotic) division.

Cultural eutrophication – The accelerated nutrient enrichment of surface waters by human activities.

Cyanotoxin – Toxin produced by cyanobacteria.

Cytotoxins – Toxins that cause damage to various types of cells.

Ectoparasite – Organism that lives externally on another organism (host), from which it obtains food or other resources.

Epilimnion – The upper, warm, wind-mixed or circulating water layer in a thermally stratified lake or reservoir in summer.

Euphotic zone – Waters where light is sufficient for net photosynthesis.

Eurythermal – Capable of tolerating a wide range of temperatures.

Euryhaline – Capable of tolerating a wide range of salinities.

Eutrophic – Referring to waters that contain relatively high levels of nutrients such as inorganic N and P, supporting high biological production.

Fresh waters – Waters with salinity less than one.

Gas vacuole – Aggregate of gas vesicles in cyanobacterial cells.

Gas vesicle – Cylindrical structure in cyanobacterial cells, with protein walls that are permeable to gases but hydrophobic; gas vehicles increase cell buoyancy.

Genome – The complete set of genetic information (DNA, RNA) in an organism; the entire complement of genetic material in the chromosome set of an organism.

Hemolytic – Promoting the destruction of red blood cells; usually used in reference to toxic substances that degrade the bonds between healthy red blood cells and their hemoglobin coat.

Intraspecific – Referring to populations (strains) within a species.

Luxury consumption – Uptake of large quantities of nutrients such as N and P, relative to immediate growth needs.

Macroalga – Macroscopically visible alga.

Mesotrophic – Referring to an aquatic ecosystem of moderate productivity and nutrient enrichment; intermediate between oligotrophic and eutrophic.

Metaphytic – Floating among or over various substrata over the bottom of a lake or reservoir.

Micelle – Colloidal particle consisting of many aggregated small molecules with a layered structure; contains both hydrophobic and hydrophilic molecules.

Microarray – A two-dimensional array of ordered sets of DNA molecules of known sequence, deposited onto defined locations in predetermined spatial order on a solid (usually glass, filter or silicon) surface. Microarray analysis allows simultaneous detection of multiple genes from a small sample, and analysis of expression of those genes.

Mixotroph – Organism that is both photosynthetic and heterotrophic, obtaining its nutrition through photosynthesis as well as consumption of organic carbon produced by other organisms.

Molecular probe – Labeled DNA or RNA used to detect complementary nucleic acid sequences in the species of interest.

Oligotrophic – Referring to waters that have low nutrient supplies and low biological production.

Phagotrophy – Mode of nutrition involving ingestion of particulate food.

Photorespiration – Process by which some of the organic carbon produced during photosynthesis is lost through excretion of a compound called glycolate; increases in some algae under low dissolved organic carbon concentrations, high irradiance, and high dissolved oxygen concentrations.

Phycobilin pigments – Water-soluble blue and red pigments (phycocyanins and phycoerythrin, respectively) that absorb light energy and transfer it to a reaction center of chlorophyll *a* for use in photosynthesis. Both phycocyanins and phycoerythrin occur in cyanobacteria. In addition, a group of eukaryotic algae called cryptophytes (Cryptophyta) also can contain either phycocyanin or phycoerythrin.

Polymictic – Referring to lakes and reservoirs that stratify irregularly and mix frequently throughout an annual cycle.

Salt water – Water that contains a relatively high percentage (salinity greater than 0.5) of salt minerals.

Strain – Population within a species.

Thermal stratification (temperature-stratified) – In the water column of a lake or reservoir, a condition that can develop during the summer wherein the water forms layers that differ in temperature and density. Lighter, warmer water, heated by the sun, forms the top layer (epilimnion); denser, colder water forms the bottom layer (hypolimnion). The temperature within each of these layers is constant. Between these two layers is an intermediate layer (metalimnion) that includes the thermocline, an area where temperature changes most rapidly per unit of increasing depth.

Thalloid – Referring to macroalgae that have a primitive structure called a thallus, which lacks vascular tissue.

Toxin – Poisonous substance produced by certain algae.

Trophic stoichiometry – Branch of science that considers ratios of key elements (mostly carbon, nitrogen, and phosphorus to assess how the characteristics and activities of sources, producers, and consumers influence, and are influenced by the ecosystem where they occur.

Zygospore – Thick-walled resting cyst containing the sexual product (zygote) resulting from gamete fusion.

Further Reading

Bootsma HA, Jensen ET, Young EB, and Berges JA (eds.) (2005) *Cladophora* research and management in the Great Lakes (Special Report No. 2005–01). *Proceedings of a workshop held at Great Lakes WATER Institute (GLWI), University of Wisconsin-Milwaukee, 8 December 2004.* Milwaukee: University of Wisconsin-Milwaukee.

Burkholder JM (2002) Cyanobacteria. In: Bitton G (ed.) *Encyclopedia of Environmental Microbiology*, pp. 952–982. New York: Wiley Publishers.

Chorus I and Bartram J (eds.) (1999) *Toxic Cyanobacteria in Water – A Guide to their Public Health Consequences, Monitoring, and Management.* New York: World Health Organization.

Graham JM, Arancibia-Avila P, and Graham LE (1996) Physiological ecology of a species of the filamentous green alga *Mougeotia* under acidic conditions: Light and temperature effects on photosynthesis and respiration. *Limnology and Oceanography* 41: 253–262.

Higgins SN, Hecky RE, and Guildford SJ (2006) Environmental controls of *Cladophora* growth dynamics in eastern Lake Erie: Application of the *Cladophora* Growth Model (CGM). *Journal of Great Lakes Research* 32: 629–644.

Huisman J, Matthijs HCP, and Visser PM (eds.) (2005) Harmful cyanobacteria. *Aquatic Ecology Series*, vol. 3. Dordrecht, The Netherlands: Springer.

Lembi CA, O'Neal SW, and Spencer DF (1988) Algae as weeds: Economic impact, ecology and management alternatives. In: Lembi CA and Waaland JR (eds.) *Algae and Human Affairs*, pp. 455–481. New York: Cambridge University Press, New York.

Oliver RL and Ganf GG (2000) Freshwater blooms. In: Whitton BA and Potts M (eds.) *The Ecology of Cyanobacteria – Their Diversity in Time and Space*, pp. 149–194. Boston: Kluwer Academic.

Singhurst L and Sager D (eds.) (2003) *Prymnesium parvum Workshop Report for 2003.* Austin: Texas Parks and Wildlife Department.

Spaulding S and Elwell L (2007) *Increase in Nuisance Blooms and Geographic Expansion of the Freshwater Diatom Didymosphenia geminata: Recommendations for Response.* Denver: U.S. Environmental Protection Agency, Region 8.

Watson SB (2004) Aquatic taste and odor: A primary signal of drinking-water integrity. *Journal of Toxicology and Environmental Health Part A* 67(20–22): 1779–1795.

Wehr JD and Sheath RG (2003) *Freshwater Algae of North America.* New York: Academic Press.

Wells RDS, Hall JA, Clayton JS, et al. (1999) The rise and fall of water net (*Hydrodictyon reticulatum*) in New Zealand. *Journal of Aquatic Plant Management* 37: 49–55.

Zurawell RW, Chen H, Burke JM, and Prepas EE (2005) Hepatotoxic cyanobacteria: A review of the biological importance of microcystins in freshwater environments. *Journal of Toxicology and Environmental Health* 8: 1–37.

Subject Index

Notes

Cross-reference terms in italics are general cross-references, or refer to subentry terms within the main entry (the main entry is not repeated to save space). Readers are also advised to refer to the end of each article for additional cross-references - not all of these cross-references have been included in the index cross-references.

The index is arranged in set-out style with a maximum of three levels of heading. Major discussion of a subject is indicated by bold page numbers. Page numbers suffixed by T and F refer to Tables and Figures respectively. vs. indicates a comparison.

This index is in **letter-by-letter** order, whereby hyphens and spaces within index headings are ignored in the alphabetization. For example, acid rain is alphabetized after acidity, not after acid(s). Prefixes and terms in parentheses are excluded from the initial alphabetization.

Where index subentries and sub-subentries pertaining to a subject have the same page number, they have been listed to indicate the comprehensiveness of the text.

Abbreviations

DCAA - dissolved combined amino acids

DFAA - dissolved free amino acids

DIN - dissolved inorganic nitrogen

DOC - dissolved organic carbon

DOM - dissolved organic matter

DON - dissolved organic nitrogen

TDS - total dissolved solids

TSS - total suspended solids

Printed and bound by CPI Group (UK) Ltd, Croydon, CR0 4YY

03/10/2024

01040312-0016